Understanding the Numbers in Depth of Anesthesia Monitors

Ashraf A. Dahaba

Understanding the Numbers in Depth of Anesthesia Monitors

Springer

Ashraf A. Dahaba
Department of Anesthesiology and ICM
Suez Canal University
Ismailia, Egypt

ISBN 978-3-031-88259-3 ISBN 978-3-031-88260-9 (eBook)
https://doi.org/10.1007/978-3-031-88260-9

This Springer imprint is published by the registered company Springer Nature Switzerland AG
The registered company address is: Gewerbestrasse 11, 6330 Cham, Switzerland

To Joline my daughter, Mark Twain said the two most important days in Life are the day you were born and the day you discover why, Joline my daughter you are the reason why.

Prelude

Let me start by saying that anesthesiologists who cannot read their electroencephalography (EEG) is like cardiologists who cannot read their electrocardiography (ECG), and any clinical parameter is only as good as the doctor who interprets the numbers.

Depth Of Anesthesia (DOA) assessment remains a fundamental topic in perioperative medicine. Despite a plethora of clinical signs and available tools, monitoring the hypnotic and antinociceptive components of anesthesia is fundamental good clinical practice. Monitoring the hypnotic component of anesthesia has greatly evolved over the past three decades. Processed Electroencephalography-Depth Of Anesthesia (pEEG-DOA) monitors are available but the algorithms they use have been validated in adult populations within limited anesthetic conditions. Extremes of age can affect the EEG during general anesthesia. The growing elderly population is principally affected by this limitation as they are often at risk of postoperative cognitive adverse events. Whereas the use of multimodal anesthesia utilizes different anesthetic agents that have variable impact on the EEG. Their use can therefore cause the pEEG-DOA indices to display aberrant values. Several monitors now display the EEG spectrogram, which may be part of the solution to these challenges. This three-dimensional (3D) display of the EEG breaks down its signal by wave frequency, intensity, and time. This is a very useful tool in identifying both elderly patients with a "vulnerable brain" phenotype and individual components of multimodal anesthesia.

Over the past decade, several new tools have shown potential in assessing the antinociception component. Many of the available nociception-antinociception monitors focus on assessing the sympathetic response to a noxious stimulus and consequently offer an elegant, but surrogate, measure of nociception. Although several goal-directed approaches coupled with these monitors have shown promise, there remains a need to clarify the impact of guiding antinociception with these tools on patient outcomes.

In many instances we are confronted with bispectral index (BIS) values that do not concur with clinically judged anesthetic state arising from an underlying alteration of the patients' own EEG that is unrelated to anesthesia depth. Numerous clinical conditions that have a direct effect on electroencephalographic cerebral function could also influence the bispectral index value. In very limited conditions we could use the bispectral index off-label outside the scope of the operating room to "grade" other EEG conditions whose EEG changes are somehow like EEG changes with anesthesia.

Using the pEEG-DOA devices in the critical care unit (CCU) setting where there are too much heterogeneity and many disease states in the critically ill patients other than sedatives/hypnotics would strongly influence pEEG-DOA indices that can be misleading. Because most pEEG-DOA sensors are placed on the frontal region; high electromyographic (EMG) activity could create subtle pEEG-DOA artifact signal pollution misinterpreted by the pEEG-DOA algorithms as EEG activity resulting in erroneously elevated pEEG-DOA values.

Ashraf A. Dahaba

Acknowledgment

To the late Professor Werner F. List, you were not just our chairman at Medical University of Graz, Austria, but the captain of the ship and the captain's word was always law. Your knowledge, wisdom, kindness, and charming smile will always be missed.

To the late Professor Shandor Agoston at Groningen University, The Netherlands, my research mentor who taught me how to do research, your research equipment Relaxometer Mechanomyograph to date still appears in many of my publications.

To the late Professor Petr H. Rehak at Medical University of Graz, Austria, our brilliant statistician, may you be in a far better place with only smarter people than me.

Last but not least, to our brilliant Medical English Editor Eugenia Lamont at Medical University of Graz, Austria, who taught us all how to speak medical English.

About the Book

This book represents 30 years of my life. As Don Henley, famous musician of the "Eagles," once said "you have your whole life to write your first album/ book and only few months to write your second." I immensely enjoyed every moment of the journey. Allow me to say that in the darkest periods of my life my research was the only thing that made perfect sense. Starting with an observation, as one of my dear students Robert observed that the bispectral index (BIS) was always low before anesthesia induction in patients with hepatic encephalopathy, he came running to me as now we have an enigmatic research question that had no answer. Next was a very well-designed multicenter study looking at the three main causes of hepatic encephalopathy in three centers in three countries on three continents. Then traveling to each center with my esteemed research colleagues till each research center is standing on its own feet. Research data of each individual recruited patient started pouring into my email, followed by data tabulation and then heated arguments with my esteemed statistician late Prof. Rehak. Finally, the statistical analysis answered the enigmatic research question (usually the answer comes at 5 o'clock in the morning!) and what an adrenaline rush that is only matched with the moment when we all have a high impact publication. We have just added a small piece to a huge research puzzle in a 30-year long journey. Along this long journey just always remember the Arabic culture wisdom of "Travelling Convoy is never bothered by barking dogs"; this you will see a lot!

The Brain

The Brain is wider than the sky – For put them side by side –
The one the other will contain with ease – and you beside
The Brain is deeper than the sea – For hold them blue to blue –
The one the other will absorb – as sponges buckets do
The Brain is just the weight of God – For heft them pound to pound –
And they will differ – If they do as syllable form sound –
Emily Elizabeth Dickinson (December 10, 1830–May 15, 1886) an American poet

Emily Dickinson

Contents

About the Author

Ashraf A. Dahaba is Professor Emeritus of Anesthesiology Suez Canal University, Egypt, Privatdozent (Venia Docendi) Medical University of Graz, Austria; Guest Professor at The Fourth Military Medical University, Xi'an, People's Republic of China; Guest Professor at Zhe Jiang University, Hang Zhou, People's Republic of China; and Consultant of Academic Anaesthesia, General Medical Council, United Kingdom (Appendix 1).

To date Professor Ashraf Dahaba has 63 publications as first author in top impact factor journals with a total of 190 Journal Impact Factor Points (JIF). He has a unique Hirsch Index (H-Index) as first author of 21. Professor Dahaba has more than 240 presentations at International Congresses where he was often the Keynote speaker. Professor Dahaba is currently dedicated to developing Machine-Learning Artificial Intelligence (AI) Autonomous/Augmented Anesthesia that could take the burden off the practicing anesthesiologists from calculating complicated infusion rate adjustments and giving anesthesiologists two free hands.

Depth of Anesthesia (DOA) Monitors: History and Basic Structures

1

1.1 History and Basic Structure

The very first description of electroencephalography (EEG) was actually as far as in 1875 by Richard Canton (1842–1926) [1], an English physician who was working at the time at Liverpool Royal Infirmary, United Kingdom, when he was the first to discover the presence of "currents of varying direction" that were detectable on the skulls of dogs and apes using a sensitive galvanometer and electrodes [1]. He noted fluctuations during sleep and absence of waves after death. He reported his finding in three communications, two of which were in the British Medical Journal (BMJ) and one in the 9th International Congress of Medicine at Washington, DC. He then had no further discoveries as he hid his "work on the brain" which was illegal at the time [1]. In 1929, Hans Berger and colleagues [2], later extended Canton's work to what they called "Elektrenkephalogramm des Menschen" the "electroencephalograph of human beings" what we can consider the first "modern nomenclature" coining the term "Electroencephalogram" [2]. Almost 8 years later in 1937, Frederic A. Gibbs and colleagues [3] noted the effects of certain well-known drugs at the time: barbiturates, scopolamine, morphine, and ether, on EEG [3]. Before the recent introduction in the mid-1990s of the first processed Electroencephalography (pEEG) monitor [4], raw EEG had limited use in the measurement of anesthesia depth, simply because many anesthesiologists did not have the time nor the know-how skills to interpret the complex raw EEG data and make further use of the attained information to titrate the delivery of their anesthetic agents. This opened the door for the efforts to compress and simplify the raw EEG analysis [3]. Nowadays in modern anesthesia practice, numerous monitors are marketed for this purpose; the most popular pioneer Depth of Anesthesia (DOA) monitors/modules include Bispectral Index (BIS™ Medtronic-Covidien, Dublin, Ireland) and M-Entropy™ (General Electric GE Healthcare, Chicago, Illinois, US). However, by far the most extensively studied, particularly in terms of the DOA effect on the of Accidental Awareness under General Anesthesia (AAGA) incidence, is the BIS, but similar principles could largely apply, to all pEEG-DOA based monitors [5].

Electroencephalogram (EEG)-based brain function monitors (BFM) for routine anesthesia has been advocated since 1937 [3], to more effectively titrate anesthetic depth during routine general anesthesia. In principle, such a monitor would better account for inter-individual variability in dose/concentration/effect relationships and help anesthetists /anesthesiologists avoid either insufficient anesthesia and awareness or excessive anesthesia with delayed recovery. Historically, bispectral analysis is a standard high-order statistical analysis first used by oceanographers as a measure of time series to study nonlinearity in ocean waves [6]. This was

A. A. Dahaba, *Understanding the Numbers in Depth of Anesthesia Monitors*,
https://doi.org/10.1007/978-3-031-88260-9_1

further developed in the 1960s by geophysicists/seismologists to study complex seismic waveforms [7]. The BIS was the first EEG-derived monitor, introduced as a monitor of the effects of certain anesthetic/hypnotic agents [4].

Bispectral index monitoring was first commercially introduced in 1994 [8], as a method to evaluate DOA based on pEEG derived from the Barnett and colleagues, 1971 "Bispectum Analysis principle" [9]. The old generation BIS sensors, composed of 3-electrodes to be placed on the forehead, are different than the international standard 10–20 system used in conventional EEG monitoring. After numerous developments, BIS has been established as a widely used, practical monitoring method.

The EEG signal is first acquired via gel electrodes placed on the forehead. Because the acquired EEG signal is weak, the first processing unit is located near the beginning of the connecting cable of the BIS monitor, where it is first digitized, amplified, and subsequently filtered to isolate EEG signals from all other electrobiological potential signals such as ECG waveforms, ocular activity, and mains electric power interference. Scalp electromyography (EMG) is part of the device measurement, that is separately reported [4, 5, 10, 11].

Fast Fourier Transformation (FFT) [12] deconstructs the greatly complex EEG waveform into "individual sine waves" of differing amplitudes and frequencies. The EEG power spectrum is then created by plotting any given frequency signal "power against time" as an EEG graphical display. The "*frequency domain*" can be crudely envisioned as the division of relative proportion of the component waveforms of different EEG wave frequencies. A "*time domain*" can be envisioned as how the pattern of these wave frequencies changes over time. Accordingly, the different pEEG-DOA monitors "proprietary algorithms" are calculated, and the displayed numerical values will then change over time according to the relative proportions of high versus low frequency waveform components within the raw EEG signal. To calculate the third essential component, namely the *burst suppression ratio* (*BSR*), namely EEG activity that does not exceed 5.0 mV and lasts longer than 0.5 s, is identified and the time spent in this state is then calculated as a fraction of the total time. The BSR of the EEG is not only seen during periods of very deep anesthesia, as exposure to severe hypoxia or traumatic brain injuries (TBI) but has also been associated with reduction in cerebral metabolic activity and accordingly can be used to titrate doses necessary to institute barbiturate coma therapy [13, 14]. These three calculated values are then incorporated into the final dimensionless processed Electroencephalography-Depth Of Anesthesia (pEEG-DOA) monitors' displayed numbers that would decline with anesthesia-induced unconsciousness, and is typically between zero cortical electrical silence and 100 normal awake activity. This number is usually averaged/smoothed over the previous 15 s/30 s/60 s data [5].

I will critically review BFM that are based on the pEEG-DOA with regard to their signal quality, signal artefacts, and other limitations in their clinical performance. Several studies have shown that pEEG-DOA monitors have limitations that can lead to anesthesiologists erroneously interpreting the DOA indices. The pEEG-DOA indices could be altered by various non-anesthetic influences ranging from signal-quality and signal-processing artefacts, adverse effects of some non-anesthetic drugs, neuromuscular blocking agents (NMBAs) and temperature, as well as conditions inherent to patients themselves such as cerebral tumors or cerebral ischemia. Anesthesiologists should be aware of limitations of all commercially available pEEG-DOA devices intending to monitor the DOA, that may not reflect the real underlying unconsciousness level [15].

I will also highlight the controversy surrounding the definition of DOA and how to best quantify this phenomenon with currently available existing technologies, the limitation of BFM to exactly describe both the pharmacologically induced EEG changes and other physiological processes that occur during anesthesia, an ongoing debate over the past years despite the rapidly increasing rates of DOA monitors' utilization. Main limitation of these pEEG-DOA indices is the belief that most of the general anesthesia mechanisms are cortical, they are not but are also subcortical [16, 17], and that the EEG only reflects the "cortical activity" of the superficial layers of the cortex, it does not [15].

Accidental Awareness under General Anesthesia (AAGA) could be a very unpleasant experience or even a tragic consequence. The incidence of intra-operative awareness under general anesthesia in 11,785 prospectively examined patients was reported to be 0.18% when muscle relaxants were used [18]. In October 1996, the Food and Drug Administration (FDA) approved the BIS™ monitor (Aspect Medical Systems Newton, MA) for depth of anesthesia monitoring [4]. Although Ira Rampil [10] explained the general principles of the BIS in his Anesthesiology 1998 publication [10], however to date, for obvious commercial reasons, the complete details of the BIS proprietary algorithm's core technology have not been published. Few people have full disclosure to the algorithm used for calculation of the BIS, which would be a requirement for an exhaustive explanation of the "abnormal" BIS variations. That makes the BIS just a "black box" headset. This brings us to the question of whether the pEEG-DOA indices are a true reflection of the EEG signal? Scientifically speaking what we know from all the published studies is not much more than pEEG-DOA indices reflect a "head-related" biosignal that shows a relation to EEG drug effects. Thus, it seems that the only way of better understanding the pEEG-DOA and evaluating their performance is through carefully examining all the numerous published reports appearing in literature of pEEG-DOA indices not indicating the correct depth of anesthesia, the pEEG-DOA indices changes in different settings, or the unexpected and unpredictable changes in the pEEG-DOA indices. I have objectively examined all reports to date of the different situations during which the pEEG-DOA indices did not properly correspond to a clinically judged depth of anesthesia, and I have thoroughly discussed all the different circumstances that might have contributed to this unexpected pEEG-DOA indices change. I evaluated in detail the conditions under which pEEG-DOA indices do not perform as expected, critically analyzed the literature, and evaluated the various studies, objectively looking at the underlying pathophysiology of all the clinical settings. In addition, I objectively discussed all the possible limitations and shortcomings of pEEG-DOA monitoring hoping that this could help a better understanding and use of the pEEG-DOA monitors.

First, to be able to understand the EEG derived parameter, we must understand the fundamental EEG changes with the deepening of anesthesia. Gugino and colleagues [19] documented in full details the topographic quantitative Electroencephalography (qEEG) features in graded steps of increasing sevoflurane or target controlled Infusion (TCI) propofol anesthesia, in combination with TCI remifentanil. The EEG power spectrum is mediated by an anatomically complex and diverse system that involves brainstem, thalamic, hippocampal, and cortical/cortical neuronal pathways that use all known neurotransmitters. Specific portions of this system are involved in the generation of the EEG within each frequency band [19]:

Delta δ (1.5–3.5 Hz)
Theta θ (3.5–7.5 Hz)
Alpha α (7.5–12.5 Hz)
Beta β (12.5–30 Hz)
β_1 (12.75–20 Hz), β_2 (20.25–30 Hz) (Fig. 1.1).

An early step in the algorithm, before calculation of the descriptor of an epoch, is a simple pattern recognition process which then determines the weighing of each descriptor [10]. The proprietary nature of the Bispectral Index algorithm that allows different BIS descriptors to sequentially dominate as the EEG changes its character has never been fully disclosed [10].

This is because typical topographic quantitative sevoflurane, or propofol, EEG features that follow the basic general anesthesia pattern manifest as α wave (7.5–12.5 Hz) predominance when the eyes are closed [19], when there is a posterior α predominance. Frontal power predominance increases with deepening of anesthesia involving α and to a lesser extent the δ and θ power. Light anesthesia is accompanied by decreased posterior α and increased frontal, frontal/central β power.

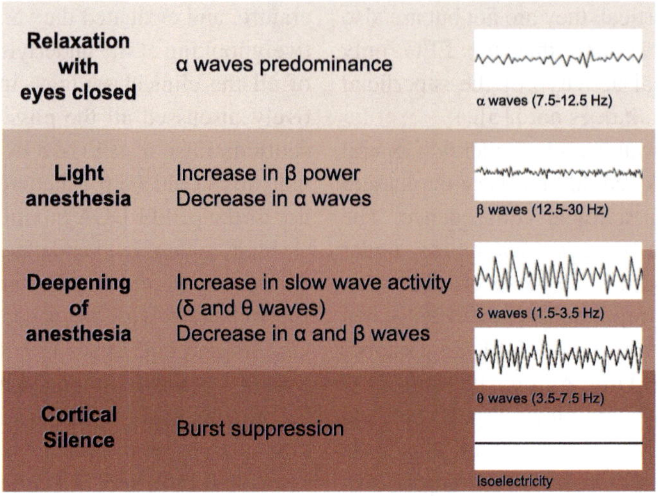

Fig. 1.1 Electroencephalography (EEG) spectral changes with deepening of anesthesia. From α wave (7.5–12.5 Hz) predominance when relaxed with the eyes closed, to an increase in β_1 (12.75–20 Hz) and β_2 (20.25–30 Hz) power with light anesthesia and finally deep anesthesia manifesting as an increase in δ (1.5–3.5 Hz) and θ (3.5–7.5 Hz) slow wave activities. Emergence from anesthesia is associated with the reversal of the EEG spectrum reversal

That proceeds to a decrease in α and increase in β_1 (12.75–20 Hz) and β_2 (20.25–30 Hz) power with light anesthesia. Basic general anesthesia pattern of deep anesthesia manifests as an increase in δ (1.5–3.5 Hz) and θ (3.5–7.5 Hz) slow wave activities, with simultaneous decrease in α and β activities. With loss of consciousness slow wave activity, δ and θ power increase and become more prominent, with a simultaneous decrease in α and β activity in all regions. All these changes are reversed in the same order with the return of consciousness. This represents a decrease in cortical generator activity and a shift towards control by the thalamocortical and hippocampal/septal generators of δ and θ activity [19].

1.2 Background Literature Search

Regarding my search strategy and data analysis plan, my book narrative non-systemic review is not a prospectively registered systematic review and/or Meta-analysis according to Meta-analyses of randomized controlled trials (Quorom Statement) or its update Systematic reviews and meta-analyses (PRISMA) [20]. Because differences between the British English (BrE) and American English (AmE) language terminologies' spelling when using the "diphthongs spelling ae two vowels" or using a single or simple vowel what is known as a "monophthong spelling without the a just the e single vowel" would reveal different results according to the style advocated by different journals, my search strategy was basically, using both spellings, to laboriously examine each individual publication one-by-one using all possible names of currently available commonly used Depth of Anesthesia/Anaesthesia Monitors versus different clinical conditions that might influence the displayed Depth of Anesthesia indices numbers.

Even though I do not consider the PRISMA guidelines [20] as the appropriate search strategy for my book, still I have meticulously taken all steps necessary to avoid bias in my systematic research of my book scientific literature material and scientific data analysis plan. I used all search engines available at Medical University of Graz "Databank Resources" bibliography search engines namely PubMed, Ovid, EMBASE, and CINAHL for my basic literature search on March 24th, 2020, that was last accessed on October 25th, 2020, for my book data collection. However, along the way whenever a new publication appeared that would serve our purpose, this was added to the book manuscript material. I have limited the search strategy to original scientific articles filter and publications in English language filter. This is demonstrated in my PRISMA Visio flow chart diagram to demonstrate my systemic search strategy (Fig. 1.2).

Search words were the "names" of currently available commonly used Depth of Anesthesia/ Anaesthesia Monitors (Bispectral OR SEDline OR Entropy OR Narcotrend OR AEP OR Auditory Evoked potential OR A-Line OR ARX-Index OR Cerebral-State-Monitor OR Patient-State-Index OR Patient-State-Analyzer OR Cerebral-State-Index OR SNAP-II OR qCON

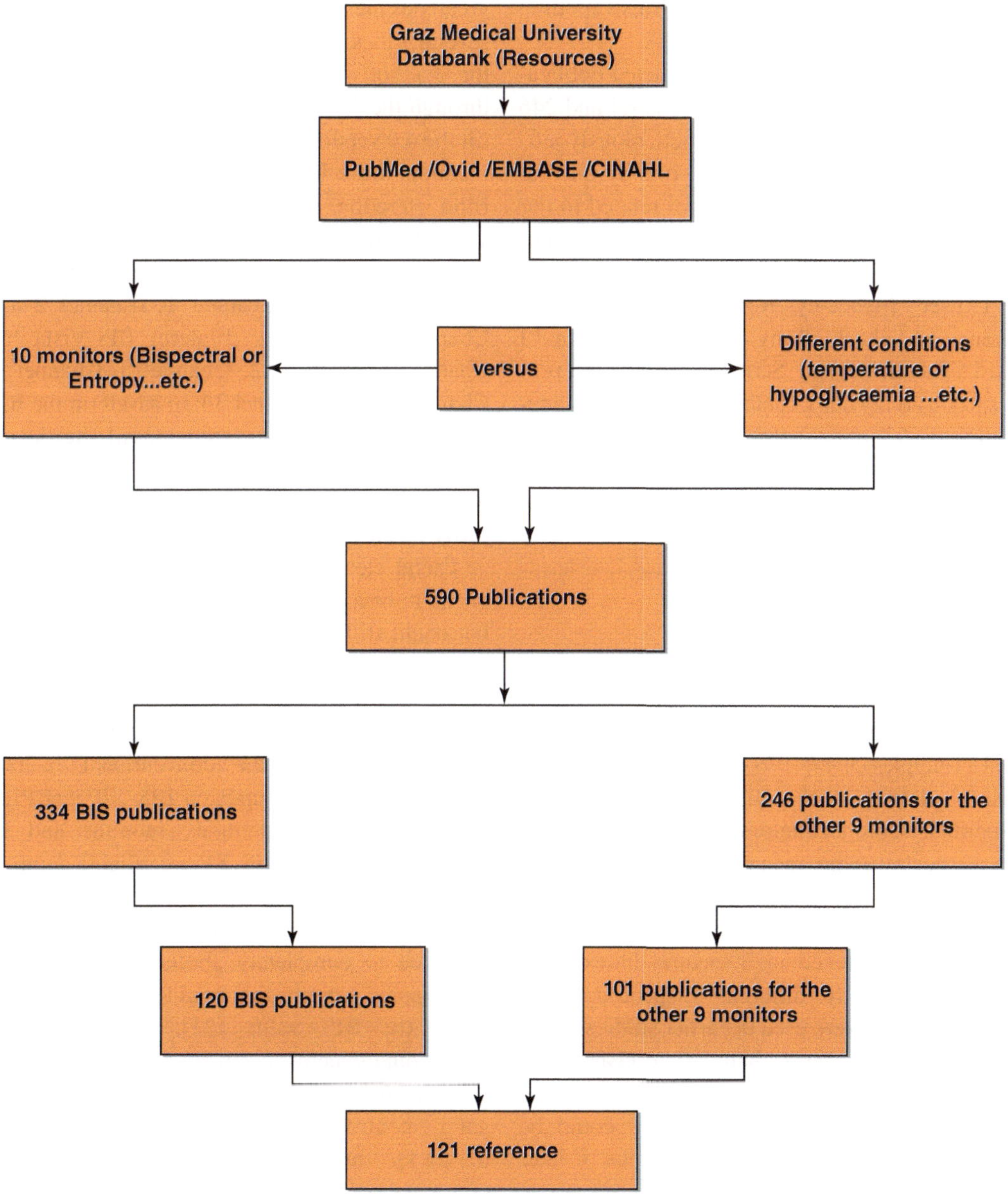

Fig. 1.2 Flow chart of my published literature systematic search, demonstrating my search strategy for my book material and my data analysis prospective plan

OR Index-of-Consciousness OR NeuroSENSE) versus different clinical conditions that would influence the displayed Depth of Anesthesia/Anaesthesia numbers (Nitrous-oxide OR Ketamine OR temperature OR ictal OR seizures OR hypoglycaemia OR hypoglycemia OR hypothermia OR hypovolemia, OR hypovolaemia OR capnography OR Alzheimer OR dementia OR cerebral-ischaemia OR cerebral-ischemia OR neuromuscular-block) using both British spelling (BrE) and American spelling (AmE).

This yielded 590 originally extracted publications divided into 334 mainly bispectral and 246 for the other 9 monitors. After meticulously sifting through all publications one-by-one to exclude publications that were not related to our field, 120 bispectral and 101 for the other 9 monitors publications were ultimately included.

I used Mendeley Reference Manager to arrange and check all my cited references and I used iThenticate and Scribber Plagiarism programs on all my book contents to bring the manuscript to "absolute zero" similarity with any previously published articles, to ensure that every word in my book is authentic.

1.3 Differences Between Different Versions of the Same Monitor

Even though Ira Rampil [10] published the general principles of the BIS, up till now all algorithms' details of the Bispectral Index (BIS) proprietary-core-technology are known only to the Bispectral Index manufacturer. Even though every new release of a Bispectral Index algorithm is undoubtedly considerably created on a great portion of the preceding Bispectral Index algorithms, potentially different Bispectral Index algorithms that were used in earlier studies might be considerably different than the latest Bispectral Index algorithm. Hence one should not assume that older Bispectral Index numbers could be considered as the same values as those of later versions and to carry identical meanings.

Researchers are still confronted with another problem, namely when we discuss the performance of the BIS monitor for example, then the first thing that would immediately come to mind is, which BIS? In addition to the first B-500 prototype, several models of the BIS hardware are still being reported in literature, namely the A-1000, A-1050, A-2000 and the BIS-XP up to the BIS-Vista. Among these models there are several software versions and among these there is still more revision software installed that differs greatly as the BIS software has also been changed several times over the last years. Starting from the 1.1 software version (introduced in 1992) through the 3.1 version (introduced in 1996), up till the 4.0 version (introduced in 2002), big modifications have been made with each BIS algorithm iteration. Bilateral BIS-Vista is a double version of the, at time, latest bilateral BIS using a double version of version 4.0. So, the unilateral BIS versions have Channel 1, Channel 2 and Channel 12, whereas Bilateral BIS-Vista has Channel 1, Channel 2, Channel 3, Channel 4, Channel 12 and Channel 34 all saved on the BIS monitor memory and can be easily retrieved. That is why we must keep in mind that the performance of a certain BIS model might not necessarily be the same as the rest.

Facial electromyography (EMG) is not supposed to be a component of the Bispectral Index but could still greatly influence BIS monitoring. My research team Dahaba and colleagues [21] in Department of Anesthesiology and Intensive Care Medicine, Medical University of Graz, Austria, compared the conventional A-2000 BIS™ ($BIS_{3.4}$) to the updated BIS-XP™ (BIS_{XP}) [21]. The Bispectral index monitor and its updated version the BIS-XP are reliable depth of anesthesia monitors [21]. In our study surgical deep anesthesia and profound analgesia were adequate to completely abolish facial electromyographic activity that could have falsely elevated the BIS value [21]. In 30 patients undergoing general selective surgical procedures expected to last 1–2 h, propofol 3.6 μg mL^{-1} target controlled infusion (TCI) and 0.3 μg kg^{-1} min^{-1} remifentanil infusion to maintain a stable Bispectral Index of around 40 deep surgical anesthesia, mivacurium 0.15 mg kg^{-1} was administered. A conventional Bispectral

Index ($BIS_{3.4}$) 3-electrodes (electrode-1, electrode-2, and electrode-3) sensor and a Bispectral Index BIS-XP (BIS_{XP}) Quatro™ 4-electrodes (electrode-1, electrode-2, electrode-4 and electrode-3) sensor were simultaneously placed one on each side of the forehead (Fig. 1.3). The conventional BIS sensor was horizontally placed on the forehead with electrode 1 in midline, electrode 3 adjacent to the angle of the eye, and then connected to a conventional BIS monitor. On the other side, the XP sensor tangential montage allowed the BIS and XP midline electrodes to be simultaneously placed with the XP fourth electrode just above the eyebrow. The XP sensor was connected to a BIS_{XP} monitor. Channel 12 (CH12), displayed on the $BIS_{3.4}$ and BIS_{XP} monitors, is the combination of the lowest value of Channel 1 (CH1, electrodes 1 + 2 impedance) and Channel 2 (CH2, electrodes 1 + 3 impedance). Both monitors were alternatively placed on the left and right sides in consecutive patients to level out any effect of dominance of one cerebral hemisphere. Bispectral Index and 95% Spectral Edge Frequency (SEF_{95}) recordings were started after verifying absence of EMG activity, signal quality index (SQI) >95%, and electrodes impedance <5 kΩ [21].

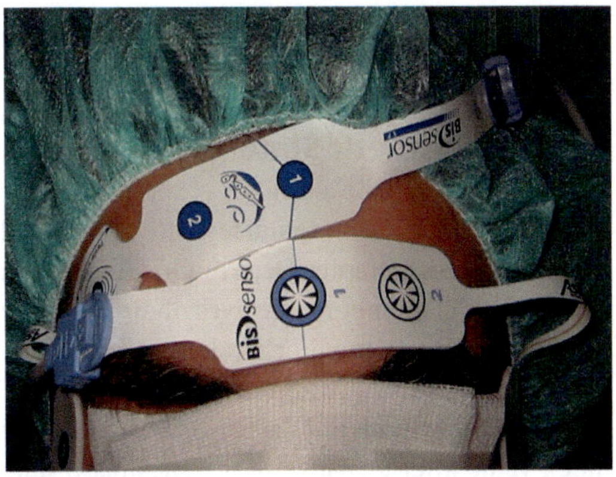

Fig. 1.3 Bispectral Index (BIS) BIS-3.4 conventional Sensor 3-electrodes (1-2-3) and BIS-XP Quatro 4-electrodes (1-2-4-3) sensor simultaneous placements. The BIS-3.4 conventional sensor strip is transversely placed on the forehead with Electrode-1 in the midline, and Electrode-3 adjacent to the angle of the eye. The BIS-XP Quatro sensor strip is inclined on the forehead with Electrode-4 slanting just above the eyebrow, Electrode-1 placed slightly higher on the forehead and Electrode-3 adjacent to the angle of the eye. (Reproduced with permission # 5861390605923 from Dahaba et al. [21])

Data collected from the two monitors were additionally analyzed based on the Bland and Altman format [22]. We have demonstrated a clear bias of a Bispectral Index +7 points, constant differences between the BIS A-2000 version 3.4 and the BS-XP version 4.0 using simultaneously placed Bispectral Index (BIS) conventional and Quatro sensors (Fig. 1.3) with a Bispectral Index difference range Limits of Agreement (LOA) of +22, and − 6 BIS value [21]. At all phases, BIS_{XP} was constantly lower than $BIS_{3.4}$. The $BIS_{3.4}$—BIS_{XP} bias (LOA) was +7 (+22, −6) (Fig. 1.4). Our study clearly indicates that at Bispectral Index 40, remifentanil profound analgesia is effective in preventing EMG activity from falsely elevating the Bispectral Index. However, with the onset of neuromuscular block, the change of muscle tone status can trigger $BIS_{3.4}$ and BIS_{XP} artifact recognition algorithms to manifest as mirror-image transient-artifacts, lasting till patients are fully relaxed. The $BIS_{3.4}$—BIS_{XP} Bias and LOA were still unacceptably wide to allow values given by the two monitors to be used interchangeably for individual patients [21].

With each Bispectral Index hardware and software iteration, attempts have been made to better filter "artifacts." This was very evident in a study by Vuyk and colleagues [23] using a BIS-XP in three volunteers, taking part in another trial, who were responding to oral commands even though the Bispectral Index was 40–45 under low doses of midazolam and TCI propofol of 0.6–1 ng mL^{-1}. The BIS-XP is supposed to undergo better artifact-resistance, especially those from electrocautery usage and to identify and remove the effects of electromyography activity as well as all conditions that usually occur with depth of anesthesia monitoring. However, it seems that the BIS-XP displays lower Bispectral Index values than those displayed by older versions when at the same depth of anesthesia levels in the same subjects. Because EMG activity falsely elevates the BIS, the BIS-XP is supposed to be less vulnerable to EMG-artifacts than previous BIS models. However, it seems that the new BIS-XP

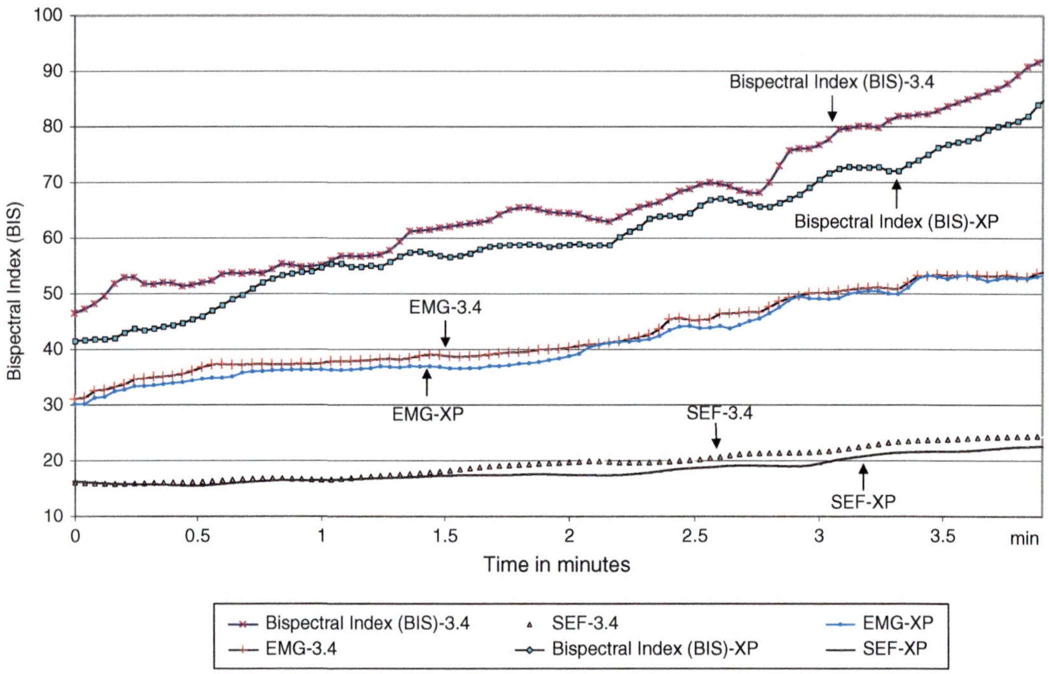

Fig. 1.4 Bispectral Index (BIS) BIS-3.4 and BIS-XP monitors simultaneous recordings in 30 patients at emergence from anesthesia phase starting from termination of anesthesia till full recovery. Electromyography (EMG) in decibels (dB) of BIS-3.4 (EMG-3.4) and BIS-XP (EMG-XP), and 95% Spectral Edge Frequency (SEF_{95}) in Hz of BIS-3.4 (SEF-3.4) and BIS-XP (SEF-XP). (Reproduced with permission # 5861390605923 from Dahaba et al. [21])

model that is less vulnerable to EMG activity artifacts resulted in the displayed lower Bispectral Index values even without the use of neuromuscular blocking agents (NMBAs), as was the case in their volunteers [23].

Renna and colleagues [24] demonstrated that patients showed a normal Bispectral Index value of >95 when EMG was >49 decibels (dB) suggesting that some relationship between the 2 variables exists. They concluded that a biasing effect of the EMG on BIS 3.4 software version exists with the result that patients with high EMG activity >49 dB showed normal awake BIS baseline values. When using the newer version of BIS 4.0, a larger proportion of patients showed a low baseline Bispectral Index value of <93 than with the 3.4 version, A possible explanation for this finding is that the biasing effect of EMG activity is less pronounced in the BIS 4.0 algorithm than it is in BIS 3.4 algorithm, thus "unveiling" any underlying EEG slowing which decreases the calculated baseline BIS [24].

1.4 Time Delays Between Displayed Indices and Actual Anesthetic State

One of the limitations of processed Electroencephalography-Depth Of Anesthesia (pEEG-DOA) monitors involves their lag time delays [25]. Every monitor's calculation and displayed index values require a certain amount of time. This results in a delayed display of a new index value after a change in the anesthetic state of a patient. This is an important area as it relates to the intended pEEG-DOA use and their utility in relation to clinical impact. Time delays are most likely to have a clinical impact on awareness at intubation (most common time for awareness at Cesarean Section) rather than having substantial impact during anesthesia maintenance. All monitors exhibit an inherent delay resulting from the accumulation sum of several "lags" adding up namely time from anesthetic agent administration, time for delivery of an anesthetic agent until exposing the brain to the new effect-site concentration, time for the EEG to change in response to the new effect-site concentration, and time for pEEG-DOA calculation of the index and artefact rejection "smoothing rate" for averaging the previous 15/30/60 s "smoothing time" all added to the calculation time for the index to display. All these lag delays could affect index reliability [25].

A remarkable feature of all pEEG-DOA that are based on the spontaneous or evoked quantitative Electroencephalography (qEEG) is the inevitable time delay pEEG-DOA monitors will take for a "full signal update" required to generate the pEEG-DOA index value on display at a time [26]. This time that could vary between 30 s up to 2.5 min depends upon the pEEG-DOA device. This unavoidable lag/delay phenomenon has led to the introduction of pEEG-DOA monitors various smoothing and moving-window time "retrospective averaging techniques" [15, 26].

Using their perioperatively recorded electroencephalography (EEG) dataset from their pEEG-DOA devices database, Zanner and colleagues [27] found that time delays/lag for all pEEG-DOA indices varied between 24 s up to 122 s before the pEEG-DOA indices new state will be displayed/indicated. They concluded that due to the vastly different time delays for either ascending or descending values, the results of all pharmacodynamic studies will be influenced to varying degrees by that phenomenon, and we can only speculate how this will affect the validity of any anesthetic agent's Pharmacokinetic/pharmacodynamic (PK/PD) model when these PK/PD moles are implemented in effect-site driven target controlled infusion (TCI) devices [27].

To solve this rather very complicated problem, Ionescu and colleagues [28] introduced an online time delay estimation (TDE) calculator for the BIS monitor displayed/indicated Bispectral Index value that will improve the performance of any closed-loop system algorithm for artefact-disturbance rejection, reference tracking, and overall system stability [28].

Three different studies examined time delays to calculate an index representing the time delays for the transition of anesthetic level of five commercially available pEEG-DOA, namely Bispectral Index (BIS), cerebral state index (CSI), and Narcotrend (NCT), [26, 27, 29] State Entropy, and Index of Consciousness (IoC) indices during transitions between "awake," "general anesthesia," and "cortical suppression" at the transition of different states of consciousness [25]. These five monitors have different time delays until the correct index is displayed. For this purpose, they used simulated EEG sequences representing different anesthetic levels and perioperatively recorded real EEG signal sequences indicating stable states "awake," "general anesthesia," and "cortical suppression" played back to the tested monitors and used to evaluate time delays defined as the required time for the monitors to reach the next stable target index. Time delays were measured when switching from one state to another and were defined as the required time span of the monitor to reach the stable target index. Comparable results were obtained using simulated and real EEG sequences. Time delays were not constant, and the five monitors showed different time delays ranging from 18 s to 152 s until the correct index was displayed [25–27, 29].

1.5 Electric Devices Interference

Table 1.1 enumerates and describes all electric devices interferences with depth of anesthesia monitors.

Processed Electroencephalography-Depth Of Anesthesia (pEEG-DOA) indices are derived from a composite of multiple electroencephalographic signal processing techniques. During surgical procedures or in the critical care unit (CCU) setting, a variety of non-EEG signals, generated by the concurrent use of these medical devices, can interfere with electroencephalography (EEG) monitoring. These technical artefacts include other 50 Hz alternating current equipment that is transferred to the patient via either capacitive or inductive current transfer. This kind of electric interference, if not detected, could lead to errors in the pEEG-DOA algorithm calculations resulting in an inaccurate reflection of the

Table 1.1 Electric devices interference with depth of anesthesia monitoring

References	BIS version	*n*	Device	BIS change	Explanation
Gallagher [30]	A-1050	1/1	Atrial pacer	Artifactual BIS ↑	Electric interference
Sloan et al. [31]		1/1	Atrial pacer	Artifactual BIS ↑	Electric interference
Guignard and Chauven [32]	A-1000 (3.12)	1/1	Forced-air-warming-blanket	Artifactual BIS ↑	Air vibrations
Hemmerling et al. [33]	A-2000	5/10	Forced-air-warming-blanket	Artifactual BIS ↑	Air vibrations
Zanner et al. [34]	BIS-XP	1/1	WarmTouch-warming-tent	Artifactual BIS ↑	Air vibrations
Hemmerling et al. [35]	A-2000	1/1	Endoscopic shoulder shaver	Artifactual BIS ↑	Shaver oscillations
Hemmerling et al. [36]	A-2000	2/2	3D positioning system	Artifactual BIS ↑	Electromagnetic field interference
Dahaba et al. [37]	BIS-XP (3.4)	6/6	ESWT	No change in BIS	Energy flux shock waves

BIS Bispectral Index, *n* number of patients per group, *EMG* Electromyography, *3D* 3 dimentional Otorhinolaryngology positioning system, *ESWT* high-energy extracorporeal shockwave lithotripsy therapy

anesthetic hypnotic state in the pEEG-DOA index value. Unfortunately, most of this information has not been studied systematically in large patients' cohorts.

Limitations of pEEG-DOA monitors involve undetected artifacts and sensitivity to a variety of sources of interferences from other electric devices in the operating room (OR). In addition to electrocautery, interferences from several other electric devices could directly affect pEEG-DOA monitoring. Electric devices interference, other than electrocautery, was shown to affect pEEG-DOA indices. During cardiac surgery under fentanyl-isoflurane anesthesia, Gallagher [30] with BIS (A-1050) presented a case of pacer-induced Bispectral Index (BIS) artifact in an 81-year-old patient undergoing a triple-vessel aortocoronary bypass [30]. The patient exhibited pacer-induced erroneous Bispectral Index increase from 50 to 90 each time atrial pacing was started and did not decrease with the deepening of anesthesia but decreased only when pacing was discontinued. During pacing the BIS monitor signal quality bar showed poor signal quality, compared to optimal signal quality before and after pacing, suggesting that electric interference was responsible for the observed artifact [30].

Electric devices interference has been reported to erroneously increase the Bispectral Index value [31]. Spontaneous electromyography (EMG) activity monitoring, via the electromyograph endotracheal tube (EMG-ETT), is used to reduce the recurrent laryngeal nerve (RLN) injury [38] during thyroidectomy surgery. The Xomed EMG-ETT (Medtronic ENT, Jacksonville, FL, USA) consists of an oral endotracheal tube fitted with paired wire electrodes exposed for 30 mm at the glottis level, thus allowing full contact with the vocal cords [39]. With the help of a stimulating probe, surgeons can test whether an anatomic structure is in fact the RLN. While the BIS monitor generally gives reliable information, electrical devices interferences were known to interfere with the accuracy of the Bispectral Index value [40]. Sloan and colleagues [31] report a case of the EMG-ETT erroneously increasing the Bispectral Index values in a 48-year-old ASA-I physical status, 60 kg woman with family history of malignant hyperthermia (MH), undergoing total thyroidectomy for a thyroid nodule. A trigger-free MH general anesthetic technique was chosen with propofol, midazolam, and sufentanil anesthesia induction and tracheal intubation with the Xomed EMG-ETT that was easily done without the aid of neuromuscular blocking agents (NMBAs). Anesthesia was maintained with nitrous oxide (N_2O) in oxygen (O_2), propofol, and sufentanil continuous infusions.

The Bispectral Index value at induction/surgical preparation was 28–40. No NMBAs were used during surgery to allow surgeons to test the integrity of the RLN using the EMG-ETT. The start of the RLN electric stimulation was associated with a sudden increase in the Bispectral Index value to 80, but the heart rate, blood pressure, and end-tidal carbon dioxide (CO_2) all remained constant and normal. Despite hastily increasing propofol and sufentanil infusions, as well as midazolam 4 mg and etomidate 18 mg administrations, the Bispectral Index value remained at 70–80 for 30-min. However, after surgery completion and within seconds from switching off the EMG-ETT stimulator, the Bispectral Index value dropped and remained at 24. It was then realized that the EMG-ETT stimulator had spuriously elevated the Bispectral Index value. Because anesthesia recovery was prolonged, intravenous flumazenil 200 μg and naloxone 80 μg had to be administered. The patient then recovered with no recall of any intraoperative event. In this case, [31] these erroneously high Bispectral Index values, lead to unnecessary anesthetic agents' administration and prolonged anesthesia recovery, requiring the use of benzodiazepine antagonists/opioid antagonists. This case clearly demonstrates that the EMG-ETT stimulation was misinterpreted by the BIS monitor as EEG activity resulting in an erroneously high Bispectral Index value. However, in their case the absence of hemodynamic changes clearly suggested an adequately anesthetized patient, and the immediate decline in the Bispectral Index value as soon as the EMG-ETT stimulation was switched off also suggested that the EMG-ETT stimulation was the cause of the spurious Bispectral Index reading. In case of sudden increases in Bispectral Index values during surgery and anesthesia, anesthesiologists should consider electrical devices artifactual contamination of the Bispectral Index signal, especially when anesthesia clinical judgment would suggest adequate levels of anesthesia [31].

Guignard and Chauven [32] using a BIS A-1000 (version 3.12) monitor reported that Bispectral Index falsely increased from 56 to 80 when a forced-air-warming blanket, placed directly on a patient's forehead was switched on [32]. The Bispectral Index returned to 60 when it was switched off. However, the BIS monitor signal quality index (SQI) bar might not always be helpful, as that is not always the case, as in numerous occasions the subtle signal pollution was not detected by the signal quality bar nor displayed as artifacts, so anesthesiologists should also pay attention to the nature of the artifact in a fashion like the electrocautery artifacts.

Hemmerling and colleagues [33] presented another 5 cases, out of 10 patients, in which BIS A-2000, which has a different sensor electrode than the BIS A-1000, falsely increased from <45 to 70–90 every time a head forced-air-warming blanket placed directly on the BIS electrodes on patients' head, was switched on and decreased to 35–55 when it was switched off. With optimal signal quality in all patients, the raw EEG showed fast moving waves of high amplitudes like those of the awake state. This suggested that the head air-warming blanket's vibrations could have created frequencies, in the BIS electrodes, simulating the EEG waves of light anesthesia or awake state [33]. However, this might not be always true as both forced–air–warming blankets and the new BIS XP A-2000 monitor are both being routinely used by many anesthesiologists all over the world in their daily anesthesia practice with no further reports of disruptions.

Forced-air-warming devices were shown to produce erroneously high Bispectral Index values, when the BIS monitor older versions A-1000 (version 3.12) [32] or the A-2000 [33] were used. Zanner and colleagues [34] report a similar case using the next version BIS-XP monitor, in a 65-year-old male patient with a squamous cell carcinoma scheduled for abdomino-thoracic esophageal resection. After a Thoracic-5 epidural analgesia catheter was placed and 10 mL ropivacaine 7.5 mg mL^{-1} was administered and anesthesia was induced with 2 mg kg^{-1} propofol, 10 μg sufentanil and 0.6 mg kg^{-1} rocuronium were administered for endotracheal intubation. Anesthesia was maintained with 0.7 Minimum Alveolar Concentration (MAC) sevoflurane in 40% O_2 and air. After anesthesia induction, left radial artery and right internal jugular vein catheters were placed. A BIS XP sensor was placed on the patient's forehead according to the manufacturer's instructions and the Bispectral Index tracing was stable at ~40 after surgical incision. For warming, the patient's arms, upper body, and head were surrounded by tent of blankets; WarmTouch air warming system placed above the left arm (Mallinckrodt, Hazelwood, MO, US) [34]. However, as soon as WarmTouch was switched on, within 10 s the Bispectral Index value abruptly increased to ~80 with SQI of 80–100%. With the use of electrocautery, the BIS monitor EMG bar increased slightly. The patient exhibited no clinical signs of inadequate anesthesia. When the WarmTouch air warming device was switched off, the elevated Bispectral Index value declined back to ~40. This phenomenon was repeated every time the WarmTouch device was switched on-off. After extubation and uneventful recovery, the patient did not recall any intraoperative events [34].

Since its introduction in 1994 [4], the Bispectral Index algorithm has been continuously improved to generate a stable processed electroencephalographic signal that is robust and more "resilient" to artefact corruption, to make the Bispectral Index values interpretation much more reliable. Despite the fact that the BIS-XP system has a new electrode-4 that was added to further enhance artefact detection, artifact rejection and reduce the high-frequencies influence on the Bispectral Index algorithm calculations, the authors [34] report a case where the latest version of the BIS-XP system still produced erroneously high Bispectral Index values with the use of a forced-air-warming system, the same problem previously described for older versions of the BIS system [32, 33].

Hemmerling and colleagues [35] presented a case where, despite optimal signal quality, BIS A-2000 suddenly erroneously increased from 40 to 62 with the start of endoscopic shoulder shaver oscillations activity. The Bispectral Index decreased equally abruptly after the shaver device was switched off and the end of the use of the shaver device. They speculated that the shaver oscillations of 1500 rpm at the shoulder joint could have been transmitted via bone contact to the forehead and temporal area bones, causing minimal vibrations of the BIS sensor itself resulting in falsely increased Bispectral Index values [35].

Hemmerling and colleagues [36] reported two cases, under sevoflurane anesthesia in one case and under propofol anesthesia in the second case, in which a 3-dimensional otorhinolaryngology positioning system interfered with BIS A-2000 monitoring [36]. The otorhinolaryngology positioning system consists of a transmitter on a headset that creates a 20 cm electromagnetic field diameter around the patient's head and a handheld receiver. Despite optimal signal quality, when the system was switched on, the Bispectral Index value increased from 40–45 to 60–90. The Bispectral Index value remained high for 2 h although anesthesia increased. The Bispectral Index value immediately decreased to 28–32 when the system was switched off. This led to the assumption that the BIS sensor electrode or the incorporated wires were charged by the electromagnetic field and thus produced falsely increased Bispectral Index value [36]. They proposed that the electromagnetic field could have caused changes in patients' brain electrical activity. Possibly due to the different nature of electromagnetic fields of two different devices, or simply because with the better shielding of the BIS XP A-2000 algorithm iteration (version 3.4) identifying and filtering noise from the algorithm, ESWT electromagnetic interference did not result in disruption of BIS monitoring. As for an electromagnetic field causing changes in brain electrical activity [36], that remains pure speculation and highly unlikely.

In all the above-mentioned reports, except with atrial pacer-induced interference [30], the BIS signal quality indicator displayed "optimal signal quality" in all patients [32, 33, 35, 36]. Whereas in some cases the simultaneously displayed raw EEG showed fast moving waves of high amplitudes resembling those of the awake state [33]. This clearly indicates that the air-warming-blanket vibrations [33], the shoulder shaver oscillations [35], and the electromagnetic field [36] could have all created minimal vibrations or frequencies in the BIS electrodes simulating the EEG waves of light anesthesia or an awake state. All these subtle signal pollutions were not detected by the BIS monitor and hence were not displayed as artifacts, which may lead to inadvertent anesthetic over dosage.

Newer pEEG-DOA monitors are better isolated and new versions of pEEG-DOA algorithms seem to show better artefact rejection against electric interference [37]. In our center we demonstrated that electromagnetic emission of High-Energy Extracorporeal Shockwave Lithotripsy Therapy (ESWT) to the shoulder does not cause disruption of BIS monitoring of propofol or sevoflurane anesthesia. We examined the effect on BIS monitoring of an electromagnetic emitter of an ESWT lithotripter applied to the shoulder [37]. Because Hemmerling and colleagues [36] used both intravenous and inhalational anesthesia in their two case reports [36], we chose to test our hypothesis with both intravenous and inhalational anesthesia. We have clearly demonstrated that electromagnetic high-energy lithotripsy impulses did not disrupt BIS monitoring of either intravenous or inhalational anesthesia.

My research team Dahaba and colleagues [37] in Department of Anesthesiology and Intensive Care Medicine, Medical University of Graz, Austria, in a convenience sample size [37] of six adult patients, undergoing shoulder High-Energy ESWT lithotripsy for symptomatic rotator cuff calcareous tendinopathy under propofol general anesthesia in three patients and sevoflurane-N_2O in another three patients, we placed a Bispectral Index BIS "Quatro" sensor on patients' forehead that was connected to a Bispectral Index BIS-XP monitor (version 3.4). We started Bispectral Index recording when we were insured of more than 95% SQI and less than 5 kΩ electrodes-impedance. The Bispectral Index values and electromyography (EMG) data were continuously collected and stored in the once every 5 s mode, while the smoothing window was set at 30 s [37]. Anesthesia was maintained with propofol 100–150 μg kg^{-1} min^{-1} and remifentanil 0.1–0.3 μg kg^{-1} min^{-1} infusions in cases 1–3 and with sevoflurane 1 MAC with 60% N_2O in oxygen in cases 4–6. All patients underwent electromagnetic induction of high-energy ESWT lithotripsy to the shoulder (Minilith SL1, Storz Medical, Kreuzlingen, Switzerland). The energy flux density in the therapy focus was 0.5 mJ mm^{-2} of a total of 2000–3000 impulses [37].

In all 6 patients [37], the Bispectral Index tracings indicated signal quality index of >95%. With induction of anesthesia, mean ± SD Bispectral Index value declined to 28.4 ± 2.7. Anesthesia was maintained at BIS 36.7 ± 3.1 before the start of ESWT. The Bispectral Index value during ESWT was 37.2 ± 2.7, whereas the Bispectral Index value in the 10 min following the switch of the ESWT off remained at 37.9 ± 3.4 before starting to increase with termination of anesthesia (Fig. 1.5). During the anesthesia maintenance phase, using paired *t*-test, there was no statistically significant difference between mean Bispectral Index values before and during ESWT, or between during and after switching the ESWT off. There was no significant difference in EMG values before, during and after ESWT [37].

The series of reports appearing in literature demonstrating artifactual increases in Bispectral Index values might give the impression that electric devices applied to the shoulder like the shoulder shaver [35], or devices generating an electromagnetic field [36] would all result in erroneous Bispectral Index readings, might not be necessarily true. It seems safe to conclude that the new Bispectral Index algorithm is properly shielded against any shoulder vibrations generated from the electromagnetic shock wave emitter of the lithotripsy impulses causing disruption of BIS monitoring. I would like to emphasize the fact that caution is always recommend in interpretation of Bispectral Index readings for possible artifact signal pollution, while keeping in mind that not all electric devices cause disruptions in BIS monitoring.

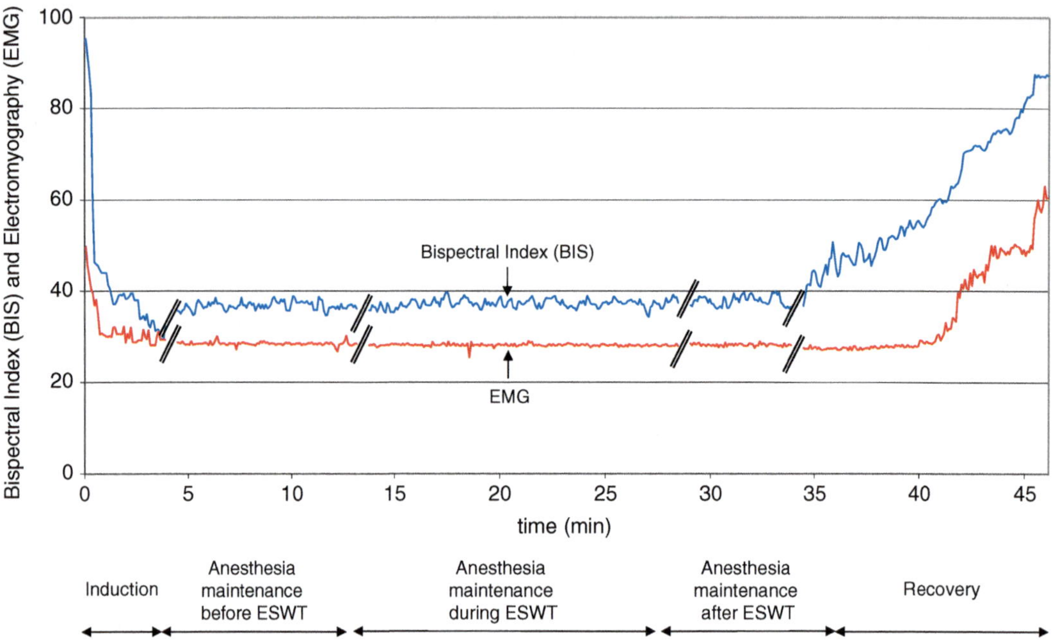

Fig. 1.5 Mean Bispectral index (BIS) and Electromyography (EMG) during High-Energy Extracorporeal Shockwave Lithotripsy Therapy (ESWT) in six patients at anesthesia induction, anesthesia maintenance before ESWT, anesthesia maintenance during ESWT, anesthesia maintenance after ESWT and at emergence from anesthesia. (Reproduced with permission # 5861390528195 from Dahaba et al. [37])

1.6 Operating Room Noise Environment

The performance of the processed Electroencephalography-Depth Of Anesthesia (pEEG-DOA) monitors can be influenced by operating room (OR) surrounding ambient environment and events that occur in the perioperative period. Noise was shown to have a significant effect on Bispectral Index (BIS) monitoring. In elderly patients undergoing knee surgery under spinal-epidural anesthesia and propofol target-controlled-infusion (TCI) sedation, the Bispectral Index value decreased when OR noise was blocked with an even greater effect at lighter sedation level [41, 42]. In a prospective, single-blinded study, Kang and colleagues [42] investigated whether blocking noise or playing music can reduce the Bispectral Index values during propofol sedation in the OR. Sixty-three ASA physical status I–II patients, aged 55–75 year, undergoing total knee replacement, were randomly allocated to 3 groups: noise group who were exposed to ambient OR noise (the electric-saw T_3 80.25 dB and the impact-device T_4 80.98 dB), silence group where patients' ears were tightly plugged to block ambient noise, and music group who were made to listen to music. After combined spinal-epidural anesthesia, TCI propofol 1.2 μg mL^{-1} sedation was started. The Bispectral Index scores were significantly $P = 0.025$ and $P = 0.005$ higher in the noise group (T_3 BIS 77, and T_4 BIS 78) than the Bispectral Index values in the silence group (T_3 BIS 69 and T_4 BIS 68) respectively [42]. However, the Bispectral BIS Index scores were similar in the noise and music groups. Preoperative anxiety level, postoperative pain scores, and comfort level were similar in all groups. The authors concluded that blocking noise is more effective than playing music in reducing the Bispectral Index scores during propofol sedation in a noisy OR environment [42].

Because noise in the OR could interfere with patients' sedation before surgery, Kim and colleagues [41] evaluated the effect of noise before the start of surgery on the Bispectral Index monitoring of propofol sedation in 30 unpremedicated patients on 2 separate occasions in a randomized, crossover study. Under propofol TCI stable baseline Bispectral Index value maintained at either BIS 75 or BIS 80, an external sound source delivered noise at ascending 50-, 80-, 110-, and 120-dB levels while recording the Bispectral Index values at each noise level [41]. In the BIS 75 group, ascending levels of noise did not significantly alter the Bispectral Index value. However, in the BIS 80 group, the Bispectral Index values at 80-, 110-, and 120- dB were significantly higher than the Bispectral Index values at 50-dB. The authors concluded that "noise" increases the Bispectral Index values during propofol sedation in the OR, and noise appears to have a greater effect on the Bispectral Index values at "lighter" levels of propofol sedation [41].

1.7 Commercially Available Depth of Anesthesia Monitors

Currently in daily anesthesia practice there are at least ten commercially available processed Electroencephalography-Depth Of Anesthesia (pEEG-DOA) monitors namely [43], Bispectral Index (BIS), State Entropy (SE), Auditory Evoked potential (AEP), Cerebral State Monitor (CSM) Cerebral State Index (CSI), Narcotrend, SEDLine (formerly Patient State Index PSI), NeuroSENSE, SNAP II, qCON 2000, and Index of Consciousness (IoC); they represent with their frequently updated algorithms the latest developments. Accordingly, more articles have been appearing in literature of pEEG-DOA numbers indicating an incorrect hypnotic state. These monitors utilize pEEG signals to estimate a dimensionless numerical index as a gauge of anesthetic depth [43]. How are these monitors validated depends on the purpose for which those monitors are used. Optimizing the anesthetics dose and to avoid awareness are two quite different purposes. Another challenge to the use of these devices lies in the exclusive proprietary of each algorithm [43].

Table 1.2 enumerates and describes all commercially available depth of anesthesia monitors.

Bispectral Index (BIS) Monitor (Medtronic, Dublin, Ireland)
The Bispectral Index (BIS) monitor utilizes an algorithm based on combining power spectral analysis, bispectral analysis, and burst suppression data. The derivation of the Bispectral Index is achieved through a weighted sum of relevant sub-parameters. The Bispectral Index scale is from 0 to 100. In addition to a single-channel electroencephalography (EEG), it also offers a bilateral sensor for assessment of asymmetry. Density spectral arrays and 95% Spectral Edge Frequencies (SEF_{95}) can be displayed as well as electromyography (EMG) activity and Burst Suppression Ratio (BSR) information [10].

What is the Bispectral Index then? The Bispectral Index algorithm was retrospectively derived from EEG changes with incremental doses of γ-amino butyric acid (GABA)ergic anesthetic agents: isoflurane or propofol while measuring 3 descriptors in a large volunteers' data base. The Bispectral Index is the weighted sum of three sub-parameters descriptors "*Burst suppression,*" "*SynchFastSlow,*" and the "*Relative BetaRatio,*" depending on the depth of anesthesia: The weight of *Relative BetaRatio* is greatest at light sedation. The *Relative BetaRatio*, a frequency–domain feature of the EEG, is calculated as two empirically derived frequency bands spectral power log ($P_{30\text{–}47\,Hz}/P_{11\text{–}20\,Hz}$) which is the ratio of the high-range β to high-α plus low-β activity [10]. The weight of *SynchFastSlow* calculation relates to the degree of EEG activity during general anesthesia. The *SynchFastSlow* is defined as the logarithm of the following ratio: the sum of bispectrum peaks in the 0.5–47 Hz range divided by the sum of the bispectrum peaks in the 40–47 Hz range. The *SynchFastSlow* is a bispectral–domain feature of the Bispectral power wave band log ($B_{0.5\text{–}47\,Hz}/B_{40\text{–}47\,Hz}$). *Burst suppression* has a greater weight in the calculation of Bispectral Index values during general anesthesia when compared to sedation. A time–domain-feature the *Burst Suppression* combines two dif-

Table 1.2 Commercially available depth of anesthesia monitors

Monitor	Manufacturer	Index	Derived from	Range	Characteristics
BIS	Medtronic	BIS	Bispectral analysis	0–100	Weighted sum of 3 subparameters
Entropy	GE Health care	RE/SE	Spectral analysis / EMG	0–100 /0–91	Nonlinear EEG signal complexity
SEDline	Masimo	PSI	Heterogeneity of variance	0–100	Bilateral brain regions coherence
Narcotrend	MonitorTechnik	Narcotrend	Frequency domain	0–100	EEG visual classification of sleep
NeuroSENSE	NeuroWave Sys.	WAVcns	Deterministic approach	1–100	EEG gamma band wavelet analysis
qCON 2000	Quantium Medical	qCON	Spectral analysis	0–99	Artificial neural network /fuzzy logic
AEP/2	Danmeter A/S	AEP /AAI	MLAEP /EEG	0–60 /0–100	Bilateral click stimuli
CSM	Danmeter A/S	CSI	Fuzzy logic inference	0–100	Frequency domain analysis
IoC	Morpheus Medical	IoC	Symbolic dynamics	0–99	Similar to entropy
SNAPII	Stryker	SNAP	Power spectral analysis	0–99	Low: 0–18 /high: 80–420 Hz frequency

GE Health care General Electric Health Care, *EMG* electromyography, *EEG* electroencephalography, *AAI* A-Line Autoregressive Index, *AEP* Auditory Evoked Potential, *MLAEP* Middle Latency Auditory Evoked Potential, *BIS* Bispectral Index, *CSM* Cerebral State Monitor, *CSI* Cerebral State Index, *SE* State Entropy, *RE* Response Entropy, *IoC* Index of consciousness, *PSI* Patient State Index

ferent algorithms together namely an algorithm that measures isoelectrical silence, *the Burst-Suppression-Ratio (BSR)*, and an algorithm that perceives the *Burst Suppression* which is overlaid on wandering low-baseline-voltage the *QUAZI Suppression-Index* [10]. None of these disparate descriptors is particular per se; each has a specific range of anesthetic effects where they perform best. *Relative BetaRatio* being the most influential during light anesthesia [10], *SynchFastSlow* is most influential and predominating during surgical levels of hypnosis anesthesia levels, and *Burst Suppression* detects very deep anesthesia. The Bispectral Index analysis uses a proprietary algorithm that allows the three different descriptors to sequentially dominate as the EEG changes its character with increasing depth of anesthesia [10]. It thus transforms the relative contributions of the nonlinear stages of the anesthetic drug effect on the EEG into an easy-to-use dimensionless number that correlates with the anesthetic drug effect on the EEG depth of anesthesia, ranging from 100 (fully awake) to zero (isoelectric EEG) [10]. The Bispectral Index analysis examines the relationship between sinusoids at two primary frequencies, f_1 and f_2 and a modulation component of the frequency $f_1 + f_2$. The set of these three frequency components is known as a triplet. The bispectrum can then be decomposed to isolate the phase information as bicoherence and the combined magnitude of the members of the triplets as the real triplet product [10]. Interfrequency phase coupling (bicoherence) describes the interaction between the EEG signal components and quantifies the level of synchronization in the EEG signal. Synchronization in neural networks indicates an interaction between different subsystems. Phase synchronization reflects the functional relations of the cortical spatially divergent distinct oscillatory systems [44]. Propofol was shown to systematically change the phase synchronization between the EEG channels in the different frequency bands [44]. In the awake state the frequencies are tightly coupled, and the Bispectral Index is close to a 100 [10]. A word of caution here as Zetterlund and colleagues [45], recently showed that, even in patients with normal cognitive function, the

Bispectral Index might not reflect a specific EEG pattern as there was no statistically significant correlation between the Bispectral Index and EEG at clinical loss of consciousness [45].

Entropy Module (GE Healthcare Technologies, Chicago, Illinois, USA)

By measuring electroencephalography (EEG) and electromyography (EMG) signals [46], the GE (General Electric GE Healthcare Technologies, Chicago, Illinois, USA) Entropy monitor gives information on spectral entropy, response entropy, and state entropy. The Entropy module measures "irregularity" in spontaneous brain and facial muscular activity. The N-Entropy module displays one channel of EEG [47]. The Entropy algorithm uses spectral analysis to produce two main Entropy parameters for assessing Depth of Anesthesia (DOA): the State Entropy (SE) indicates the depth of hypnosis with an index scale from 0 to 91. The Response Entropy (RE) indicates nociception/responsiveness to stimuli derived from the frontal electromyography (fEMG) with an index scale from 0 to 100 [47]. When the difference between SE and RE widens, this indicates inadequate anesthesia. Entropy displays the waveform of SE and RE as well as Burst Suppression ratio (BSR). Entropy uses a single-channel EEG [47].

SEDline Monitor (Masimo, Irvine, California, USA)

The SEDLine monitor [48] uses a symmetrical bilateral array that provides 4-channel data. Bifrontal electrodes measure four channels of raw electroencephalography (EEG) showing separate displays for electromyography (EMG), artifacts such as patient's motion as well as a Burst Suppression Ratio (BSR). Considering coherence between bilateral brain regions and anterior–posterior relationships in the brain, the formerly known as Patient State Index (PSI, Baxter) currently SEDline Monitor (Masimo, Irvine, California, US) algorithm is calculated via a 4-channel EEG incorporating high heterogeneity of variance at different levels of sedation/hypnosis. Burst Suppression data and plausibility analysis are applied for final PSI derivation [48]. SEDline Monitor also displays bilateral four channels of raw EEG waveforms as well as bilateral Density Spectral Arrays (DSA). On the Patient State Index (PSI) scale from 0 to 100, optimal depth is between 25 and 50, in contrast to other monitors with similar scale whose recommended anesthetic depth is between 40 and 60 [48].

Narcotrend Monitor (MonitorTechnik, Bad Bramstedt, Germany)

The Narcotrend Index [49] visually classifies the electroencephalography (EEG) patterns associated with natural sleep stages. The Narcotrend index uses Burst Suppression, time domain analysis, and frequency domain analysis to extract the relevant EEG parameters [49]. These are then classified through plausibility testing into a total of 15 possible substages ranging from A (awake) to F (deep) that include further subdivisions. The recent version provides a Narcotrend Index from 0 to 100 and uses 1- or 2- EEG channels as well as displaying electromyography (EMG) information [49].

The development of the Narcotrend algorithm is based on depictions of EEG changes during human sleep EEG patterns that were first described by Loomis and colleagues [50] in 1937. Loomis and colleagues defined five stages of sleep namely, A–E to distinguish the different human sleep stages [50]. In the late 1990s, Schultz and colleagues [51] developed the Narcotrend, a monitor for anesthetic depth assessment using the *Loomis* classification of the five different hypnotic stages [50]. They added 15 substages for a more precise classification of the EEG. The signals are sampled at 128 samples/sec with 12-bit resolution [49]. These signals are then bandpass filtered to 0.5–45 Hz. The power spectrum is calculated by means of a Fast Fourier Transform (FFT), allowing enormous visual compression of spectral data by stacking and thus simplifying recognition of all time-related changes in the EEG signal [49]. The Narcotrend algorithm continuously verifies impedances and electrode potentials at defined time intervals to ensure high EEG signal quality. Trace segments of 20 s, the so-called epochs, are the units of classification [52] of the EEG parameters that are relevant for the detection of suppression. After filtering all artefacts, the Narcotrend algorithm calculates and updates trace segments that contribute to the discriminant functions. Finally, a smoothing rate of the Narcotrend classification is calculated as a weighted mean depending on the background parameters [49].

The Narcotrend-Compact M, a clinically and scientifically validated processed Electroencephalography-Depth Of Anesthesia (pEEG-DOA) monitor [46, 51, 52], automatically interprets and classifies the EEG during anesthesia. The Narcotrend classification algorithms are based on visually classified EEG epochs. The Narcotrend classification scheme, used for these visual assessments, has its origin in sleep analysis, which was further adapted for EEG during anesthesia. From the awake state to very deep anesthesia, Narcotrend has 15 distinguished stages [A, B (0–2), C (0–2), D (0–2), E (0–2), F (0–1)] [52]. The transformation of these stages into a numerical scale from 100 to 0 is a further Narcotrend monitor refinement for a differentiated presentation of EEG effects. The Narcotrend uses an automatic classification multivariate discriminant function that further incorporates age-related changes of the EEG. The device Narcotrend contains artifacts' identification functions. The Narcotrend typically records EEG from a frontal channel using standard electrocardiography (ECG) electrodes. Anesthesiologists can choose other electrode positions and types [52].

NeuroSENSE Monitor (NeuroWave Systems Inc, Cleveland Heights, OH, USA)

Using a deterministic approach, a method that produces the same output for any given electroencephalography (EEG) interval, the WAVcns index is calculated via wavelet analysis of the EEG signals in the γ frequency band [53]. The NeuroSENSE monitor was intentionally developed for use with closed-loop anesthesia delivery systems. The NeuroSENSE monitor uses bilateral brain monitoring to derive the WAVcns index with a scale of 1–100 [53].

qCON 2000 Monitor (Quantium Medical, Barcelona, Spain)

The qCON index, derived from spectral analysis and Burst Suppression Ratio (BSR) single-channel electroencephalography (EEG), is processed through a fuzzy logic system artificial neural network and has several similarities to the Entropy [54]. The qCON index measures hypnosis, whereas the qNOX index measures nociception. The qCON index and the qNOX index are derived through different frequencies. Both indeces range between 0 and 99. The qNOX reference scale was derived through EEG signals derived from movements in response to noxious stimuli [54].

Auditory Evoked Potential (AEP) Monitor/2 (Danmeter A/S, Odense, Denmark)

The Auditory Evoked Potential (AEP) index, the A-Line Autoregressive Index (AAI) [55], is an index relying on Middle Latency Auditory Evoked Potentials (MLAEPs) and electroencephalography (EEG) signals [55]. After the headphones deliver bilateral click stimuli, the EEG signals are discerned from the background EEG noise and processed for MLAEPs, thus reflecting thalamic and primary auditory cortex neural activity. The AAI is derived mainly from EEG-based spectral parameters when the AEP signals are low in quality. The AEP also displays Burst Suppression Ratio (BSR) and electromyography (EMG) data. The AEP has 2 indices scales: 0–60 and 0–100 [55].

Cerebral State Monitor (Danmeter A/S, Odense, Denmark)

The cerebral state index (CSI) algorithm [56] processes a frequency domain analysis and the Burst Suppression Ratio (BSR) using fuzzy logic methodology for the index inference. The cerebral state monitor (CSM) has a single-channel electroencephalography (EEG) with an index scale of 0 to 100 as well as measures of Burst Suppression percentage and electromyography (EMG) EMG activity [56].

Index of Consciousness Monitor (IoC Morpheus Medical, Barcelona, Spain)

The Index of Consciousness (IoC) [57] is derived from a time domain method via symbolic dynamics that divides the electroencephalography (EEG) signals into partitions and then labels each partition with symbols of 1 and 0. This depends on mathematical determination. Basically, the IoC is conceptually like entropy as IoC can detect nonlinear EEG characteristics thus assessing the level of signal complexity [57]. The algorithm also uses a frequency domain method and Burst Suppression analysis using a fuzzy logic inference system to derive the IoC index. The IoC monitor displays a single channel EEG with an index scale of 0 to 99, as well as Burst Suppression and electromyography (EMG) [57].

SNAP II Monitor (Stryker, Inc, Kalamazoo, MI)

The SNAP index (that snaps into a Personal Data Assistant PDA) [58] is based on the calculations of the low-frequency index of power spectral analysis in the 0–18 and the high-frequency index of power spectral analysis in the 80–420 Hz frequency ranges to derive a single index algorithm, with a shorter lag time that minimizes artifacts, to detect patient awakening using a single-channel EEG and an index scale of 0–99 [58].

1.8 Comparing Different Depth of Anesthesia Monitors

Multiple processed Electroencephalography-Depth Of Anesthesia (pEEG-DOA) monitors were developed to evaluate the Depth of Anesthesia (DOA) during surgical procedures [59]. Whether during surgical procedures or for critically ill patients, evaluation of DOA or depth of sedation is similarly significant to guarantee surgical success and patients' comfort. Hence, pEEG-DOA monitoring was developed to help the clinical evaluation of anesthesia sedation/hypnosis [59]. At present several proprietary commercially available pEEG-DOA devices exist, including the Bispectral Index-VISTA (BIS-VISTA™ Medtronic-Covidien, Dublin, Ireland) [60], approved by Food and Drug Administration (FDA) back in 1996, which uses different wave-frequency bands in a "weighted-algorithm" to form the Bispectral Index [10, 59]. Additional pEEG-DOA devices have been constantly utilized; these include the Patient State Index (PSI, SEDLine®, Masimo, Irvine, California, USA), while the GE S/5 monitor Entropy™ Module (General Electric GE Healthcare, Chicago, Illinois, USA) is derived from different EEG spectral entropy wave-frequency bands that is basically based on anesthetic agents causing either an increase in their regularity, or a decreased EEG entropy [61]. Whereas the PSI implies 4-channels to device an index that leverages the differences in the wave frequency spectrum between the PSI 4-channels [62]. All these devices use different artifact-rejection algorithms to improve their indices [59, 60].

Numerous studies separately assessed the response of many of these pEEG-DOA Bispectral Index, Entropy, and Patient State Index metrics to propofol, volatile anesthetic agents or dexmedetomidine, either individually or in pairwise studies [63–68]. Because almost all these pEEG-DOA sensors are placed on one side of patient's frontal area, for obvious reasons, no studies have compared more than two pEEG-DOA indices when simultaneously placed [59].

The so-called pEEG-DOA monitors [69, 70] claim to measure hypnotic depth from the electroencephalography (EEG) and offer unbiassed guidance for general anesthetic agents' decision-making titration, protecting us from intraoperative Accidental Awareness under General Anesthesia (AAGA) through detection of the EEG wave frequencies accompanying the regaining of consciousness. Accordingly, clinicians should expect reasonable agreement between different DOA monitors if presented with the same EEG signal [70]. The incidence of AAGA is 1:1000 patients [71]; however unfortunately several of these patients thereafter suffer from post-traumatic stress disorder (PTSD) [70, 72]. Almost all commercially available pEEG-DOA monitors displayed indices are from zero 0 (representing very deep anesthesia EEG isoelectricity) to 100 (when the patients become fully awake). All commercially available pEEG-DOA monitor's unique indices are based on the proprietary EEG recognition algorithm, based upon different methods for the detection of muscle activity, burst suppression, and noise elimination from the signal, while using different methodology for statistically validating the indices. All these pEEG-DOA monitors' manufacturers declare that the relevant index range indicates what they consider unwarranted anesthetic depth, the anesthetic depth range of surgical anesthesia, and range of unwarranted light anesthesia [70].

Numerous large randomized controlled trials (RCTs) studied the influence of using these pEEG-DOA indices for general anesthesia stewardship and improve patients' postoperative outcomes including "awareness with explicit recall," cognitive dysfunction delirium, as well as mortality [70]. Anesthesiologists widely agree that using BIS monitoring could effectively reduce, but not eliminate, the occurrence of "awareness with explicit recall" when using total intravenous anesthesia (TIVA) [73], or inhalational-based anesthesia [70, 74]. Regarding the utility of pEEG-DOA indices avoiding possible anesthetic agents overdosage while reducing delirium cognitive dysfunction incidence and death frequency are still debatable, among observational retrospective studies demonstrating an important effect on patients' outcomes [75–77]. We must keep in mind that all these anesthetic maintenance goals using the recommended clinical safety pEEG-DOA indices' ranges were determined by the pEEG-DOA monitors own manufacturers [70].

Published literature extensively debates the utility and clinical bearing of these pEEG-DOA indices; because it was noted that there is limited association with recognized clinical parameters such as loss of consciousness (LOC) or return of consciousness (ROC) [70, 78], and pEEG-DOA indices do not designate awareness (with or without recall) as indicated by using the Isolated Forearm Technique (IFT) [70, 79, 80], more common in the young age patients [81]. Whereas pEEG-DOA indices are affected by advancing age [82, 83], as BIS monitoring fails to indicate deliberately awake state during neuromuscular blockade [84], and that BIS monitoring seems to be insensitive to increases in inhalational anesthetic agents [70, 85, 86]. On the other hand, supporters of pEEG-DOA monitoring advocate that, while these issues draw our attention to the pEEG-DOA limitations, the pEEG-DOA indices are mostly reasonable [87], and that the seemingly "insensitive" Bispectral Index to increasing inhalational anesthetic agents is a phenomenon resulting from "titration paradox" [88], where during Bispectral Index driven inhalational anesthetic agents titration, Bispectral Index driven anesthetic agents titration results in absence of, or at least decline, in correlation between Bispectral index values and inhalational hypnotic drug concentrations, when increasing inhalational agents' concentrations is expected to cause a decline in the Bispectral index values [70]. Irrespective of controversial intellectual views held, commercially available pEEG-DOA monitors are still clinically commonly utilized, and that pEEG-DOA monitors' producers still claim high efficacy in avoiding AAGA [70].

Recently there is a move away from pEEG-DOA monitoring towards focusing on all neurophysiological-mechanisms that are responsible for LOC or ROC, as well as spectral raw EEG wave frequencies that relate with different stages of anesthesia processes [70, 89–91]. A wider knowledge of the EEG patterns [92, 93] would lead us to a personalized anesthesia away from a population-grounded pEEG-DOA monitoring [70, 94]. An enlightening publication [94] presented spectral EEG patterns that are associated with awakening seen during the emergence from isoflurane general anesthesia for cardiac surgery [94]. This was clearly characterized by two crucial features; first, the peak 7–17 Hz α-wave frequency increased by >2 Hz compared with the pre-cardiopulmonary bypass (CPB) peak frequency and second, power in the 3.5–7.5 Hz θ-wave range decreased by >5 Hz compared with pre-CPB power. These distinctive patterns occurred in 5% of their patients with >80% of these distinctive patterns happening during the separation from CPB [94]; clearly indicating that an increasing frontal α-wave frequency represents "thalamocortical-depolarization", thus indicating that the patient is shifting away from surgery-suitable hypnotic state towards "light" hypnotic state, or a "fully conscious reconnected" state [70, 89, 90, 94].

An article by Han and colleagues [59] represents an opportunity for comparing different pEEG-DOA devices [59]. In their study they elegantly show an "in-silico" simultaneous comparison of three of the most commonly used proprietary commercially available pEEG-DOA monitors: SEDLine, BIS, and Entropy, in a cohort of patients undergoing endoscopy with dexmedetomidine [59, 95]. They compared the relative pEEG-DOA indices values after constructing ordinal logistic regression models to predict the measured Richmond Agitation-Sedation Scale (RASS) [96] during sedation for the procedure. All the pEEG-DOA indices showed a significant relationship with the RASS score and increasing differences with deepening of sedation. In particular, the authors found elevated Bispectral Index values during the deepest sedation phases, possibly due to electromyography (EMG) interference, which apparently has far less impact on the PSI and Entropy metrics. The authors concluded that PSI and Entropy are much better suited for assessing deeper levels of sedation with dexmedetomidine [59, 95].

The presence of EMG [95], in the raw EEG trace, is generally felt as an artifact that must be eliminated or at least greatly reduced, because of the risk of a misleading pEEG-DOA indices to increase [40]. Therefore, different pEEG-DOA devices have been continuously developing different filters on their acquired EEG trace, that is usually narrower than the 0.5–70 Hz in case of the of BIS monitor, and concurrently developing different algorithms that would then clean the raw EEG signal and maintain the pEEG-DOA indices much more stable in the non-paralyzed patients [10, 95]. However, this comes with a price to pay, namely to lose the high EMG as an early warning and a close ally when both the EMG and pEEG-DOA values sudden rise can precisely indicate pain or lightening of anesthesia and risk of awakening [97, 98]. Conversely Stewart and colleagues [99] suggested that the BIS-EMG signal could add very useful information in critical care patients when they are subjected to painful stimuli [95, 99].

Interestingly, Hight and colleagues [70] went a step further when they compared five commercially available pEEG-DOA monitors reading the same EEG traces of anesthesia lightning. They found "pEEG-DOA concordance" in just one-third of their study cases. Furthermore, they also found that in 31% of the cases there was at least one monitor, the BIS monitor, that showed excessive anesthesia hypnotic depth when the EEG trace displayed an awakening-like pattern [70]. Their results clearly demonstrated that even though different pEEG-DOA monitors can work well in some conditions or with different anesthetic agents but could be failing in many other situations, concluding that anesthesiologists relying on the pEEG-DOA indices "riding the numbers" alone constitute a high risk for the patients [15, 95].

1.8.1 Comparing PSI, BIS, and Entropy

Dexmedetomidine, α-2 agonist, is widely used in patients under anesthesia and for critically ill patients' sedation, because of its sedative, analgesic, and anxiolytic effects as well as it tolerable respiratory depressive effect [100]. A commonly used tool to assess level of sedation is the Richmond Agitation-Sedation Scale (RASS) [96] that grades the sedative effect in 10 grades of sedation from −5 that is responsive to physical and verbal stimulation, all the way to +4 when the patient is in a violently combative state [96]. The RASS is still a subjective score that requires patients' status assessment that could be not feasible or not desired to make a frequent enough accurate assessment of rapidly changing status of patients' sedation level [59].

Han and colleagues [59] simultaneously assessed the response of three indices BIS, Entropy, and PSI, and their correlation to RASS score [96] dexmedetomidine "procedural sedation." In a prospective observational study, in 51 patients undergoing elective endoscopy under dexmedetomidine sedation, 2% nasal lidocaine gel was applied prior to endoscopy and before dexmedetomidine 1–1.5 μg kg^{-1} bolus over

10-min followed by dexmedetomidine 1–1.5 μg kg^{-1} h^{-1} infusion [59]. Continuous electroencephalography (EEG) was captured using SEDLine® monitor (Masimo, Irvine, California, USA) during the endoscopic procedure and up till the patients recovered. Before dexmedetomidine administration, the authors made a 3-min EEG recording as a baseline as EEG recordings were continuously recorded and electronically stored for subsequent study [59]. As the RASS score [96] was used to measure sedation level during endoscopy, the authors recorded the reference starting RASS score prior to sedation, and RASS score every 5 min throughout endoscopy then every 15 min when patients were recovering from sedation [59].

The authors used five 250 Hz frequency channels to record the EEG, Fp1, Fp2, F7, F8, and AFz as reference. All recorded EEG data were "detrended," 1–50 Hz bandpass filtered and 2-level wave frequency decomposed. They determined the 95% Spectral Edge Frequency (SEF_{95}) corresponding to EEG power 95%, on 30 s EEG segments averaged epochs over all 5 channels [59]. The authors created their playback system that could then retrieve the Bispectral Index, Patient State Index, State-Entropy, and Response-Entropy. The authors evaluated the RASS score that was continuously assessed during endoscopy and recovery. Using logistic ordinal regression, the authors calculated the correlation between the three processed Electroencephalography-Depth Of Anesthesia (pEEG-DOA) indices [59] from the EEG recordings namely the Patient State Index (PSI), left Bispectral Index (BIS-L), right Bispectral Index (BIS-R), State-Entropy (Entropy-S), and Response-Entropy (Entropy-R) [59].

Eagleman and colleagues [101] formulated a procedure to recreate all three previously recorded pEEG-DOA indices [101]. The authors first used the National Instruments (Austin, Texas, USA) PXIe-8821 model to convert raw-EEG into analogue data. Now using the National Instruments CA-1000 model, the authors converted this analogue data to all three monitors input required formats that matches the number and position of the three monitors [101] that is sent to the individual Bispectral Index, Patient State Index, and Entropy monitors, simulating the output of real patients' recordings creating the Patient State Index (PSI), left Bispectral Index (BIS-L), right Bispectral Index (BIS-R), State-Entropy (Entropy-S), and Response-Entropy (Entropy-R) to be further subsequently analyzed. For individual RASS scores, they mimicked Patient State Index (PSI), left Bispectral Index (BIS-L), right Bispectral Index (BIS-R), State-Entropy (Entropy-S), and Response-Entropy (Entropy-R) and SEF_{95} indices' values, at the indicated RASS score time, namely at 30 s before and after, 60 s before and after each RASS score calculation [59, 101].

All pEEG-DOA indices PSI, BIS-R, BIS-L, Entropy-S, Entropy-R well correlated with RASS score, except the SEF_{95} that could not characterize sedation with dexmedetomidine [59]. Entropy 58.3% and PSI 58.3% were highly accurate at predicting deep dexmedetomidine sedation compared with BIS 44.4%. Light sedation RASS scores very well correlated to the three pEEG-DOA monitors' indices [59]. Mimicked patients declined from zero 0 till −5 RASS score over 30 min, with an Entropy decline from 100 till 30, and Bispectral Index decline from 100 till 60, Patient State Index decline from 100 till 30. Approximately at 20 min, the mimicked patient started to experience light sedation, as depicted by an increase in RASS score. All pEEG-DOA monitors' indices, 5 minutes before a rise in the RASS score, were associated with an increase in EEG spectrogram power high-wave frequencies. After 2 hours the mimicked patient suddenly went from RASS score −4 to −1 that was associated with all pEEG-DOA indices rise prior to the RASS score [59].

There was a significant correlation between the decline in RASS score and declining Patient State Index $P < 0.001$, Bispectral Index $P < 0.001$, Entropy-R $P < 0.001$, and Entropy-S $P < 0.001$, but not SEF_{95} $P = 0.42$. The Patient State Index relative likelihood $P < 0.001$ far better performed all the other pEEG-DOA indices [59]. When the authors assessed readily used pEEG-DOA indices compared with the RASS score and with each other [59], the SEF_{95} did not correlate to the RASS score, likely due to artifacts and electromyography (EMG) [59]. On the other hand, pEEG-DOA monitors that employ power artifact detection and filtering all showed well correlation with the RASS score [59].

Though all three pEEG-DOA indices are well correlated to the RASS score, the authors evaluated the three pEEG-DOA indices predictive power. Overall, 37.2% PSI had the best accuracy compared with the other pEEG-DOA indices [59]. Deviating from the RASS score by 1 grade, the 76.8% PSI and 77.5% BIS were almost similar. For prediction of the deep RASS −4/−5 sedation, 58.3% Entropy as well as 58.3% PSI were far better predictors than the 44.4% BIS [59]. When they combined the three pEEG-DOA indices together, this resulted in exact accuracy of 36.7% and for deviating by one RASS score grade this resulted in 74.6% exact accuracy [59].

The pairwise correlation between all pEEG-DOA indices showed that Entropy-R and Entropy-S $r = 0.98$, $P < 0.001$ were well correlated. Similarly the BIS-R and BIS-L $r = 0.92$, $P < 0.001$ were well correlated [59]. The Patient State Index was better correlated to the $P < 0.001$, $r = 0.8$ Entropy-R and $P < 0.001$, $r = 0.82$ Entropy-S than with $P < 0.001$, $r = 0.74$ BIS-R or $P < 0.001$, $r = 0.76$ BIS-L, whereas SEF_{95} did not correlate with any other monitor indices [59].

A very important finding of their study was that for the recommended Bispectral Index 40–60 range for surgical anesthesia, in their study the median pEEG-DOA indices equivalent to a −5 RASS score was BIS-R 73.25 and BIS-L 72.85 [59]. In their mimicked patient −5 RASS score deep surgical anesthesia was equivalent to awake Bispectral Index value most of the time >60, while conversely Entropy indicated the 40–60 range and Patient State Index indicated the 25–50 range both indicating adequate good deep surgical anesthesia [59]. The authors demonstrated that Patient State Index and Entropy both better indicated adequate good deep surgical anesthesia [59].

While previous publications have demonstrated good association between clinical sedation scores such as RASS and the Bispectral Index, as the authors have noted in their study [59], They noticed multiple shortcomings with the Bispectral Index compared with Patient State Index or Entropy because the Bispectral Index was vulnerable to electromyography falsely raising the Bispectral Index in patients who did not receive neuromuscular blocking agents (NMBAs) [59, 102, 103], whereas several publications demonstrated a decline in Bispectral Index when NMBAs were administered to fully awake anesthesiologists volunteering for these kind of studies [59, 84, 104]. On the other hand, Patient State Index was shown to better indicate deeper propofol surgical anesthesia [67], whereas Entropy better indicates overly anesthetized patients' sevoflurane anesthesia that resulted in patients' hypotension when compared with Bispectral Index [105]. Although Bispectral Index and Patient State Index both well indicated patients deep surgical desflurane anesthesia, Patient State Index attained a better receiver operating characteristic area under the curve (ROC-AUC) and was less vulnerable to electric cautery artefacts [59, 64].

In their study, the Patient State Index had a high correlation to Entropy-R $r = 0.82$ and Entropy-S $r = 0.8$. Bispectral Index was moderately correlated $r = 0.7$ to Patient State Index and Entropy. There was a high correlation $r = 0.98$ between Entropy-S and Entropy-R when the mimicked patient was not paralyzed, this is because electromyography would be reflected in the Entropy-R typical high wave frequency and not Entropy-S wave frequency range [59, 60]. An earlier sevoflurane anesthesia study demonstrated that, in the absence of noxious stimuli, the degree of neuromuscular block had no measurable effect on Entropy, and that there was almost no difference between Entropy -S and Entropy-R until noxious stimuli were applied [59, 106]. Entropy's proprietary algorithm may apply in its calculation artefact-rejection filters that might lessen the interference effect of electromyographic activity [59].

1.8.2 Comparing BIS, Entropy-SE, Narcotrend, qCON, and SEDline

The authors [70], in a post hoc analysis of 5 processed Electroencephalography-Depth Of Anesthesia (pEEG-DOA) monitors, Bispectral Index (BIS, Medtronic, Dublin, Ireland), State Entropy (Entropy-SE General Electric, GE Healthcare, Chicago, Illinois, USA), Narcotrend (MonitorTechnik, Bad Bramstedt, Germany), qCON (Quantium Medical, Barcelona, Spain), and SEDline (Masimo, Irvine, California, US), subjected all 5 commercial monitors to a series of similar electroencephalography (EEG) periods of inadvertent light anesthesia of 52 patients who showed EEG emergence patterns from diminished intraoperative anesthesia after surgery [70]. The SEDline was subjected to both hemispheres' frontal EEG. The pEEG-DOA indices were time-aligned to the EEG signal onset using Custom Scripts Matlab (R2017b; MathWorks, Inc., Natick, Massachusetts, USA) program. Erroneous Bispectral Index data were removed. The authors excluded the first 2 minutes from the start of recordings until the pEEG-DOA indices reached a stable state [29, 70].

For monitors' discordance/concordance [70, 107, 108], the authors studied the "clear" EEG emergence spectral patterns "period-of-interest" [70, 94]. The EEG sections from 30-min before until 30-min after the "period of interest," 125 Hz epochs were sent to the 5 pEEG-DOA monitors using a digital to analogue convertor [70, 109]. The authors additionally examined spectral δ-wave EEG "period of interest" clear changes to the spectrum, trying to understand why some pEEG-DOA indices would indicate extremely deep anesthetic state while the EEG spectrum paradoxically indicating light anesthetic state [70]. A word of caution here, because higher EEG sampling rate frequency probably is not included in the estimation of the anesthetic state; the authors used the EEG sampling rate of 125 Hz for their study, that is below the 47-Hz frequency commonly used by most pEEG-DOA monitors for "noise-detection" [70].

The authors investigated whether any of the pEEG-DOA indices remained within, or moved out of, each monitor's maximum or minimal general anesthesia recommended pEEG-DOA index range for ≥2-min during the period of interest, according to EEG spectrum changes seen in a previous study categorized as "below" or "above" surgical anesthesia [70, 94]. The authors rational was that deviations of less than 2 min would be discarded by anesthesiologists as probable noise-in-the-signal, whereas deviations of more than 2 min durations would probably prompt anesthesiologists, attentive to pEEG-DOA indices, to change the anesthetic agent dosage. If surgical anesthesia did not go "below" or go "above," this was considered as staying "inside" surgical anesthesia range [70]. The authors recorded qCON, Narcotrend, and Bispectral Index indices at 1 s intervals, SEDline at 2 s intervals, and Entropy at 5 s intervals. The authors then interpolated Entropy and SEDline indices to the 1 s interval with the help of linear interpolation to make them comparable with the qCON, Narcotrend, and Bispectral Index indices. The authors also interpolated all the numerous Narcotrend index missing 1 s data [70].

Using a 4 s moving window; EEG power spectra were created with a 3 s overlap [70]. The median δ-wave power was in 0.5–4 Hz frequency. The authors then used "linear mixed effects model" of two fixed effects variables. First variable is the time-variable, categorical, before versus during the period-of-interest, and the second variable named "Excessive," also categorical, is whether any given monitor indicated excessive anesthetic depth or not [70]. A random-offset-effect Patient-variable accounts for correlation within patient and allows fixed-factors interactions. The advantage of mixed-effects models is that they account for the within-patient correlations effect. The index-change values denoted as "Δ index" was mean pEEG-DOA indices difference pre and during the period-of-interest for each of the 52 patients [70].

Regarding the 5 pEEG-DOA monitors' concordance and discordance, only 16 /52 (31%) cases exhibited concordance between all 5 pEEG-DOA indices [70] of all 5 monitors agreeing on patients exhibiting deep or light anesthesia. In 13/52 (25%) patients, all 5 pEEG-DOA indices stayed within surgical anesthesia level. In 3/52 (6%) patients, all 5 pEEG-DOA indices increased above surgical anesthesia level. In 16/52 (31%) patients, only 1 pEEG-DOA index increased above surgical anesthesia level that did not agree with the EEG. In 13/52 (25%) patients, 2 pEEG-DOA indices disagreed with the other 3 pEEG-DOA indices. In 4/52 (8%) patients, pEEG-DOA indices showed discordance in depth of anesthesia as some pEEG-DOA indices exhibited light in some patients or excessive deep depth of anesthesia in other patients [70]. Of the 52 patients, 27/52 (52%) had at least one pEEG-DOA monitor warning of a potentially "inadequate hypnosis" index above range and 16/52 cases (31%) had at least one monitor signifying excessive hypnotic depth index below recommended clinical range. In 36/52 (69%) patients, pEEG-DOA indices exhibited disagreement between 1 pEEG-DOA index and the other 4 pEEG-DOA indices [70].

The pEEG-DOA monitors had different incidences of light anesthesia hypnotic state warning, with the qCON warning 22 /52 times, and the SEDline warning 10/52 times. The remaining 3 pEEG-DOA monitors gave warnings 13–16 times [70]. Likewise, pEEG-DOA indicating patients were in excessive deep anesthesia was different among monitors, 12/52 with Narcotrend and 11/53 with Entropy being the most frequent. Converesely the 2/52 with Bispectral Index and the 3/52 with qCON being the least frequent [70].

The mean pEEG-DOA index difference between pre and during the period-of-interest ranged between +40 and −20 index units [70]. Regarding the pEEG-DOA values Bland–Altman change, (each monitor differences between indices versus mean indices change) [22], the one monitor's Bias consistently lower or higher than the index of another second monitor, were as high as 5.8 index units [70]. The 95% Limits of Agreement (LOA) sizeable ranges were often between 40 and 25 units [70].

Many clinical providers still rely on pEEG-DOA index values and the manufacturers' recommended ranges for anesthetic agents' titration decision making [70]. The fact that in their interesting study two-thirds of all cases showed "discordant recommendations" when given identical EEG data, and that one-third of all cases signified excessive anesthetic hypnotic depth when the EEG would suggest a lighter hypnotic state, emphasizes the importance of personalized EEG interpretation as an essential clinical skill in anesthesiology [70]. Their enlightening studies clearly demonstrated that as they installed exactly similar electroencephalographic data in five readily used pEEG-DOA devices as the mimicked patient's anesthesia became lighter, the five devices displayed indices in concordance with their clinical depth of anesthesia suggestions in less than "one-third" of the time and in discordance in the huge majority "two-thirds" of the time. In almost "half of the time," at least one device alerted of possibly awake patients, whereas "one-third" of the time at least one device alerted of possibly excessively deep anesthetic state [70], that were all related with low δ-wave power levels in the EEG [70].

Because of the manufacturers' claims of their pEEG-DOA devices utility in guiding anesthetic agents' titration and in avoiding awareness, we should then expect far more agreements between the five pEEG-DOA devices [70]. However, when there was monitors' discordance in the form of one device indicating good surgical anesthesia level while another pEEG-DOA device indicating awakening state, we cannot conclude, just from the EEG, which pEEG-DOA device was probably "accurate," as in this case the disagreeing pEEG-DOA device could be the "accurate" device while the other pEEG-DOA devices are "inaccurate" [70]. Judging from the EEG patterns, the authors did not make any conclusions regarding memory formation, connectedness or Accidental Awareness under General Anesthesia (AAGA), just suggestions that memory formation, connectedness and AAGA states were all because of EEG patterns indicating recovery from anesthesia [70].

On the other hand, the 16 patients the pEEG-DOA device indicated excessively deep anesthesia, could represent a problem [70]. The authors argue that the pEEG-DOA indices are totally flawed because the pEEG-DOA indices are completely in discordance with the classic emergence from anesthesia recovery EEG spectral patterns [70, 89] namely, increasing α-wave peak frequency, α-wave decline and θ-wave power loss. This is a rather dangerous situation as the devices suggest to anesthesiologists trusting the pEEG-DOA devices, to lighten the anesthesia titration at the exact incorrect time of patients emerging from anesthesia [70]. This might represent a profound impediment for pEEG-DOA devices that are supposed to direct anesthesia delivery and prevent AAGA, and for any automated anesthetic agent's delivery [70].

The authors chose to examine the δ-wave power changes because the "neural-inertia" namely the EEG power remaining the same although the anesthetic agents are declining, occurs very often in that wave frequency [70, 110]. The authors believe that sustaining the δ-wave with a decrease in α-wave when patients are recovering from anesthesia could mislead the pEEG-DOA algorithms with the same EEG pattern that occurs just before Burst-Suppression [70], as the authors mixed-effects model showed a decline in δ-power occurring when one device indicated excessively deep anesthesia [70]. This indicates "neural-inertia" in the δ-wave decline when patients were recovering from anesthesia as one of the possible causes of erroneous pEEG-DOA indices. This hypothesis might help anesthesiologists not to be misled by pEEG-DOA indices and the manufacturers should take this phenomenon into consideration when writing the updated pEEG-DOA monitors algorithms [70].

According to the results of their interesting study, we can conclude that studies of depth of anesthesia levels using different pEEG-DOA devices are not all equivalent because of the discordance among different pEEG-DOA algorithm indices responding to an identical EEG tracing, thus limiting pEEG-DOA devices "comparability" [70]. The authors used physiological electroencephalographic indices as 95% Spectral Edge Frequency (SEF_{95}), Burst Suppression Ratio (BSR), and various wave band power ratios to "inter-study" compare various indices. Thus, it seems that titrating volatile anesthetics to "Minimum lveolar Concentrations (MAC)," instead of relying on pEEG-DOA indices, could be much simpler [70, 111], and could then help explain how anesthetic concentrations and pEEG-DOA indices relate to each other [70].

Because the pEEG-DOA indices were all statistically calculated via population-based algorithms, they might not be accurate for every patient [70]. We can speculate three situations where the pEEG-DOA monitors failed to detect AAGA with recall [70]. The first reason might be a pEEG-DOA device correctly indicating AAGA for a truly occurring AAGA incident, but the attending anesthesiologist may have not noticed that their patient was aware under anesthesia [70, 112], that would be categorized as an "attending-anesthesiologist human-error." Second, a true incident of AAGA clearly displayed electroencephalographically with the pEEG-DOA monitor not detecting this AAGA index, in a so called pEEG-DOA functional-failure [70], resulting in the pEEG-DOA monitor misleading the attending anesthesiologists who completely trust their PEEG-DOA monitor in the so called "riding the numbers" as they do not know how to correctly interpret the numbers based on electroencephalographic patterns. Third, the EEG may not show an AAGA pattern, the so-called electroencephalographic-failure, a situation "epiphenomenal" to AAGA misleading both the pEEG-pEEG device and consequently also misleading an attending anesthesiologist [70, 113]; the authors believe this is the least likely of all three possibilities, an empirical observation. Furthermore, correct electroencephalographic patterns of fully awake patients may be nonfrontal [70, 114], nonspectral [70, 89], or even both, though such situations were not thoroughly studied in a proper research setting [70].

Their study clearly demonstrates that anesthesiologists could and should learn how to read raw EEG [70]. Many resources are readily available to teach anesthesiologists how to interpret raw spectral EEG, specially how to use the spectrogram as a supplement and as a clarification for pEEG-DOA indices [70, 92, 93, 115]. Out of the five studied monitors, three monitors namely, Narcotrend, SEDline, and BIS all clearly simultaneously display, along their pEEG-DOA indices, electroencephalographically color scaled spectrograms to enable anesthesiologists easily interpret the changes in the EEG frequencies and power over time than simply observing the raw EEG [70].

It is unknown whether anesthesiologists diligently consider the manufacturers' instructions, but it is highly probable that a young anesthesiologists would not tolerate a prolonged period of a Bispectral Index value of 65 when they cannot understand an electroencephalographic pattern that clearly says otherwise, and in that case, would immediately deepen anesthesia by increasing anesthetic agents' administration [70]. Most of pEEG-DOA indices are originally derived when patients were under isoflurane anesthesia, and hence the authors current results probably apply to this particular anesthetic agent although all device manufacturers still "falsely claim" that all their devices are "drug-agonists" and are not "differently influenced" by different synesthetic agents' "classes" [70].

The authors demonstrated that when five pEEG-DOA devices were exposed to an exact similar electroencephalographic data taken during surgical procedures, the pEEG-DOA devices agreed only in third of the patients [70]. Furthermore in at least one device, one-third of patients indicated excessive anesthetic hypnotic depth when the spectral electroencephalographic patterns would suggest a lighter hypnotic state [70]. Their study suggests that pEEG-DOA indices should never be considered as "drug-agnostic" quantitatively measured depth of anesthesia, rather these devices have serious limitations thus highlighting the importance of personalized electroencephalographic interpretations should be a diligently acquired clinical skill [70].

The use of a dimensionless index scale of a pEEG DOA monitor to express the "depth of anesthesia" has been questioned by many authors, given the risk of "oversimplification" of the very complicated background neurobiological phenomenon underlying consciousness [95]. This was sharply expressed by Sleigh [69, 95] in his infamous editorial "Perhaps the patient isn't a submarine," when he used the metaphor of the patient under anesthesia state being a "submarine" that goes deep under water should be substituted with the idea of the brain as a household switch board with several switches that we do not need to switch them all -on or -off all at the same time [69, 95].

Introna and colleagues expressed in an editorial [95], that maybe it's about time to accept such a reality without falling in despair [95]. First of all, anesthesiologists should be aware that there is a huge technical difference between different pEEG-DOA monitors analyzing and thus processing the raw EEG and based on such knowledge the anesthesiologists should then make an "educated decision" and critically choose their pEEG-DOA monitor based on their own preferred tradeoff between sensibility and specificity features in each pEEG-DOA; such as favoring the presence of an early warning system of awareness or whether a pEEG-DOA monitor is liable to incorrectly present artifacts [95]. Each commercial pEEG-DOA monitor index is, in fact, based on a unique EEG feature-detection algorithm that use different ways of determining high muscle activity, noise, and burst suppression. Furthermore, each pEEG-DOA monitor employs a different method, displays patented features, for "statistical validation" of their indices [95].

Last, but not least, anesthesiologists should learn how to "read/interpret" the raw EEG and electromyography (EMG) signals [95]. Anesthesiologists are acquainted with the interpretation of electrocardiograms (ECG), that is even mandatory for any board-certified anesthesiologist. I strongly advocate that the same concept should be similarly requested to the interpretation of raw EEG and EMG signals, for anesthesiologist to be able to verify if the calculated pEEG-DOA index is "coherent" with the raw EEG as well as the clinically judged context, i.e., anesthetic agents' dosage, clinically judged parameters [95], along with patients' anthropometric data. For instance, an increase of pEEG-DOA index in the presence of high EMG activity superimposed on the typical δ-wave/α-wave EEG patterns under propofol anesthesia should be interpreted as artifact or increased noxious stimulation and not lightening of anesthesia. That is why nowadays there is a saying that goes like "an anesthesiologist who cannot read his/her EEG is like a cardiologist who cannot read his/her ECG" [95].

Nowadays anesthesiologists have at their disposal many tools with huge potentials but require specific intensive training for the anesthesiologists to be able to understand their wide potentials and limitations [95]. In a metaphor by Introna and colleagues [95], that the discipline of anesthesiology is like "car racing" where hyper technologies cannot take the place of a skilled driver in "delivering the best performance," the same applies to modern anesthesiology. That is why nowadays there is a saying that goes like "any clinical number is only as good as the doctor who interprets the data" [95].

The daunting task now would be to step away from simple comparisons of different pEEG-DOA monitors in different settings, and eventually define all technical features and skills that the anesthesiologists now need for a safe depth of anesthesia monitoring in such a broad spectrum of different clinical conditions, regardless of which specific pEEG-DOA monitor is commercially available in an anesthesia setting [95]. On the other hand, what would also be a great help is a totally transparent approach from the manufacturers of all the pEEG-DOA monitors features regarding their signal filtration and manipulations that would immensely ease the process of pEEG-DOA monitors performance appraisals by the anesthesiologists who are routinely adopting those pEEG-DOA monitors in their daily anesthesia practice [95].

1.9 Bispectral Index (BIS) Occipital Alternative Sensor Montage

The clinical question of Tufegdzic and colleagues' Meta-analysis [116] was to evaluate placing processed electroencephalography-Depth Of Anesthesia (pEEG-DOA) sensors at an alternative sensor position different than the endorsed forehead position and hence their applicability in clinical practice, when the location of conventional sensor electrodes position may not be possible because it interferes with the surgical field [116]. These alternative locations of the sensor electrodes are "off-label" and are not manufacturers' endorsed. There is a strong rationale strongly supporting pEEG-DOA monitoring because the primary site for anesthetic agents is the brain [116–118]. The "level of consciousness/arousal" can be assessed by analyzing the electroencephalography (EEG); however EEG changes occurring with anesthesia should be considered as "epiphenomenal changes" and not "linear changes" [116, 119]. Nevertheless, since 1996, several pEEG-DOA devices are being used in the anesthesia clinical practice, with regular updated algorithms, using near real-time EEG data via dedicated sensors that use their proprietary validated algorithms to reject and remove "artefacts," to process and correlate the raw EEG to indicate patients' Depth of Anesthesia (DOA) in a zero 0 to 100 index value [116]. The pEEG-DOA monitors' algorithms are proprietary, and the details are not published. However, these pEEG-DOA dimensionless scores have numerous restrictions [40, 116]. One major limitation is that the EEG signal acquisition basically relies on the location of the sensor placement [116]. Regarding which sensor placement is best suited for EEG signal acquisition, one must keep in mind that these sensors must sample pure, raw EEG signals, and the anatomical source of any meaningful non-artifactual EEG signal is the cerebral cortex hemispheric convexities that generate currents to be detected by scalp electrodes, and not the gyri or fissures, whereas all the "subcortical structures" do not generate any EEG activity that could be detected/recorded at the scalp [116].

We have always to keep in mind the element of "volume-conduction". The primary EEG signal generator are the pyramidal cells oriented at the cerebral cortex perpendicular to the scalp [116]. Hence the EEG electric signal is "volume-conducted" through cerebral cortical tissue including meninges, cerebrospinal fluid (CSF), bone tissue and finally the skin, to reach the sensor. When the EEG electric signal travels throughout all these tissues, the EEG electric signal is weakened and diffused. The generated EEG electric current amplitude is "directly proportional" to the distribution over the scalp area, that depends on the tissue thickness that lie between the cerebral cortical EEG generators and the electrodes sensors that will record the signal [116]. That is why adjacent sensors could pick up each other's neighboring electric signals. These electric signals could be amplified only when ideally separated by 6 cm^2 from each other; this explains the way EEG 10–20 system sensors are placed separated by maximum possible distance from each. At ideally 6 cm^2 apart, signals obtained are then different because they are originating from different cerebral cortex EEG generators of "synchronized cerebral cortex electric signal" [116, 120]. However, when the scalp sensor's "recording-point" supersedes the diameter of the EEG activity area, this results in under-sampling: "spatial-aliasing" [116, 120]. Basically, all characteristics such as amplitude and frequency are lost when we go further away from the recording electrodes [116, 120]. In the typical "10–20 EEG electrode placement," the sampling-error of around 6% becomes negligible with an extended 10–10 montage. Because "Ohm's law" clearly states that "currents decline when we increase the resistance/distance the current travels," thus electrodes situated away from the EEG source, hence increasing the resistance, will result in a low current reaching the electrodes [116, 120].

Several monitors use quantitative Electroencephalography (qEEG) or processed Electroencephalography (pEEG) to assess the depth of anesthesia. These monitors sensors' frontal montage captures "near field electric potential" from cerebral cortical regions over the fronto-temporal regions (anterior temporal, inferior-frontal, and anterior-frontal regions) [116], capable of capturing anesthesia-produced EEG frontal δ-wave and α-θ wave anteriorization, highlighting the need for electrodes that could be easily placed in an alternative position whereby there is no interference with surgical procedures [116].

The manufacturers' conventional pEEG-DOA frontal electrode sensor strip montage is between eye corner and the hair line [116]. The GE (General Electric, GE Healthcare, Chicago, Illinois, USA) Entropy™ Module EEG signal is a single channel 3-electrodes strip [116]. Reference Electrode-1 montage is on the forehead midline 1-cm from the angle of the nose, additionally acquiring signals from the frontal facial muscles electromyography (EMG). The explorer Electrode-3 fronto-lateral montage (FT 9/10) between the corner of the eye and the hairline is for recording mainly EEG from the cerebral cortex region. The Ground Electrode-2 montage is between Lead 1 and Lead 3 electrodes [116].

The Bispectral Index™ monitor (BIS, Medtronic, Dublin, Ireland) uses a 4-electrodes Quatro sensor strip (1-2-4-3) bipolar-montage corresponding to the 10–20 international system Fz (Fpz or Afz Lead 1); F7 (F9 or FT9 Lead 3) and F3 (AF3 Lead 2) to F7 [116]. Ground Electrode Lead 4 detects and measures EMG from the frontalis muscle.

The SEDLine® (Masimo, Irvine, California, USA) Patient State Index (PSI) is recorded from a 4-channel frontal montage with a 6-electrodes sensor. Ground and Reference Electrodes placed on the forehead positions Fp1, Fp2, F7, and F8. Reference Electrode placed 1-cm above Fpz and Ground electrode at Fpz [116]. In all pEEG-DOA monitors, the Ground Lead prevents the powerline "noise" masking/overwhelming the weak "biopotential EEG signals-of-interest" as the Reference / Ground Lead picks up any unwarranted signal and artefact that could be present. The nearer the Reference/Ground electrode to the Recording Electrode, the greater the "artefact-similarity" thus reducing signal contamination [116].

In the early pioneer study, Glass and colleagues [121] tested several locations for sensor placement: bi-frontal montage (Fp1 and Fp2) with CZ reference electrode for recording Channel 1 and Channel 2, and alternative fronto-central location between the preauricular point and the outer corner of each eye (FPz–At) with Fpz reference electrode for recording Channel 3 and Channel 4 [121]. They [121] reported that BIS A-1000 (version 2.0 and version 3.0) values recorded from the bi-frontal montage provided "similar" values recorded from a fronto-central montage. The authors [121] used the combined data of 4 anesthesia regimes (midazolam, isoflurane, alfentanil, and propofol) in healthy volunteers not undergoing surgical procedures. They demonstrated that Bispectral Index values recorded from different montages provided similar results [121].

For certain neurosurgical procedures, the operating field does not allow frontal BIS sensor electrode placement [122]. Shiraishi and colleagues examined the feasibility of placing the BIS sensor on a different area, such as the occiput, to allow BIS continuous monitoring, in neurosurgical procedures on the frontal area [122]. Twenty-five patients, aged 39–71 year, ASA physical status I, scheduled for unruptured cerebral aneurysms clipping under propofol/fentanyl total intravenous anesthesia (TIVA), were enrolled in the study. To confirm the validity of this occipital alternative approach, using 2 earliest generation BIS® A-1050 (version 3.3) monitors [122], the authors simultaneously placed 2 standard BIS Aspect Medical Systems Part 186–0100 early generation 3-electrodes sensors on the frontal and occipital areas during a neurosurgical procedure and compared Bispectral Index values obtained from the 2 montages [122]. They placed one sensor with Electrode-1 on the center of the forehead 1.5 cm above the angle of the nose, Electrode-2 1.1 cm lateral to Electrode-1, and Electrode-3 on temporal area between corner of the eye and the hairline. They placed a second electrode strip on the occiput with Electrode-1 on the central occipital area 1 cm above the occipital process, Electrode-2 1.1 cm lateral to Electrode-1, and Electrode-3 on the posterior-temporal area [122]. The two electrodes were placed on the same side of the head, either left or right. They meticulously placed the two sensor strips with equal distances between the electrodes, because the distance between each two electrodes determines the amplitude of the recorded signal [122, 123]. They maintained electrode impedance <5000 Ω to ensure high signal quality. Anesthesia was induced with fentanyl 2 mg kg^{-1}. Fentanyl 2 μg kg^{-1} h^{-1} and propofol target-controlled infusion (TCI, STC-525X; Terumo) 6 μg mL^{-1} plasma concentration (C_p) were adjusted to maintain Bispectral Index value between 40 and 60 [122]. They recorded the Bispectral Index value and 95% Spectral Edge Frequency (SEF_{95}) scores from the 2 BIS monitors before anesthesia induction, after induction, after surgical incision, and every 15-min after incision. Using linear regression analysis and Pearson's correlation coefficient with Fisher's r to z conversion, mean ± SD scores of Bispectral Index ($r^2 = 0.9612$; $P = 0.003$) and SEF_{95} ($r^2 = 0.8763$; $P = 0.018$) showed a "strong-overall" correlation between frontal and occipital montages [122]. They concluded that during propofol/fentanyl anesthesia, it may be valid to measure the old generation Bispectral Index values using the old generation BIS 3-electrode sensor placed on the occipital area if necessary, during frontal neurosurgical procedures [122] in cases it is not possible to place it on the frontal area, such as in patients undergoing neurosurgical operations [122].

The BIS electrode strip Quatro might be difficult to apply in certain neurosurgical approaches, yet Bispectral Index monitoring is particularly very useful in neuro-anesthesia for titration of anesthetic agents, because it is looking at the very organ being operated upon [124]. Neurosurgical procedures in the manufacture's recommended area of BIS montage could be a serious hurdle for Bispectral Index montage. My research group Dahaba and colleagues [124] explored the utility of using the BIS-Vista monitor for occipital electrode montage when the surgical procedure requires frontal approach under remifentanil/propofol general anesthesia [124]. In the study conducted by my research team Dahaba and colleagues [124] in Xuan Wu Hospital, Capital Medical University, Beijing, People's Republic of China, two Bispectral Index Quatro electrode strips were simultaneously mounted on the occipital region with Bispectral Index-Vista™ (BIS-Vista™) Quatro sensor Electrode-1 placed at the "Inion" of the Occiput, Electrode-2, Electrode-4, and Electrode-3 placed slanting in the direction of the mastoid-process, and a frontal Bispectral Index-Vista™ (BIS)-Vista™ Quatro sensor placed according to the manufacture guidelines before pinning of the Mayfield head holder fixator (Fig. 1.6) and after pinning of the Mayfield head holder fixator (Fig. 1.7), and connected to 2 BIS-Vista monitors. The Bispectral Index values were recorded at three anesthetic phases: pre-induction, deep surgical anesthesia,

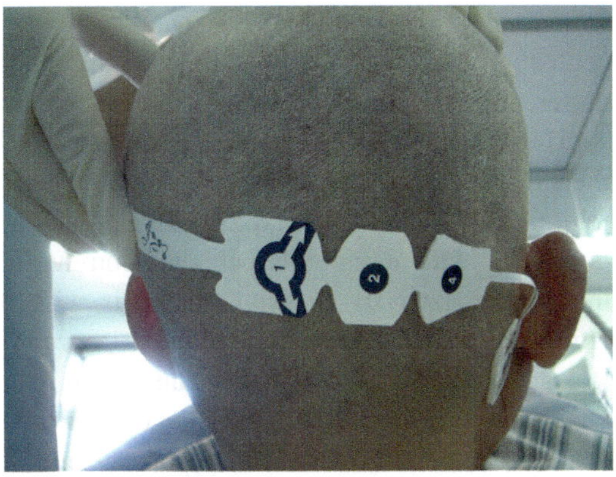

Fig. 1.6 Bispectral Index-Vista™ (BIS-Vista™) monitor Quatro 4-electrodes (1-2-4-3) sensor occipital montage with Electrode-1 placed at the "Inion" of the Occiput, Electrode-2, Electrode-4 and Electrode-3 placed slanting in the direction of the mastoid-process before pinning of the Mayfield head holder fixator

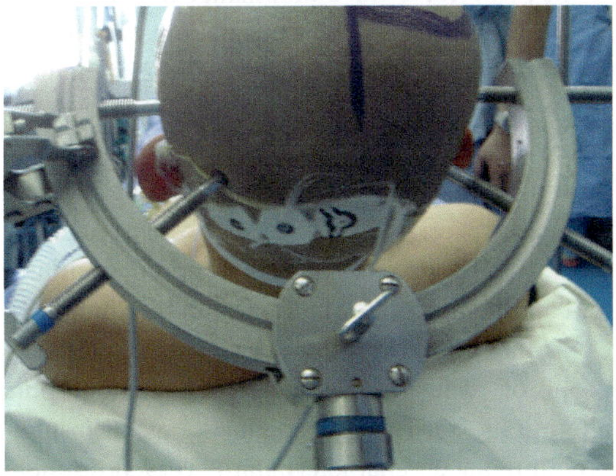

Fig. 1.7 Bispectral Index-Vista™ (BIS-Vista™) monitor Quatro 4-electrodes (1-2-4-3) sensor occipital montage with Electrode-1 placed at the "Inion" of the Occiput, Electrode-2, Electrode-4 and Electrode-3 placed slanting in the direction of the mastoid-process after pinning of the Mayfield head holder fixator

and during emergence from anesthesia [124]. Data collected from the two monitors were additionally evaluated based on the Bland and Altman format [22]. In our study when the simultaneously attained frontal and occipital montage recordings differed by <10 Bispectral Index units, this was an accepted clinical concordance [124, 125].

We demonstrated that BIS sensor occipital montage indicated paradoxically lower Bispectral Index values before induction (84 ± 5) compared with BIS sensor simultaneous frontal montage (93 ± 3) and indicated paradoxically higher Bispectral Index values during maintenance of anesthesia (67 ± 7) when compared with BIS sensor simultaneous frontal montage (57 ± 9) [124]. However, at emergence from anesthesia, the Bispectral Index values recorded from the occipital region (55 ± 9) and frontal region (53 ± 7) were not significantly different. Bland and Altman format showed −9.7 (+1.1, −20.5) Bias and Limits of Agreement (LOA) at preinduction, +9.8 (+22.8, −1.7) at deep surgical anesthesia, and −0.9 bias (+10.9, −12.8) at emergence from anesthesia [124]. Our study showed that the Bispectral Index data recorded from the Frontal and Occipital montages cannot be interchangeably applied and that the difference between the Frontal and Occipital montages is greatly dependent on the depth of anesthesia [124]. Having said that, if we take into consideration that the Bias was seemingly constant and using a Bispectral Index negative −10 units at preinduction, and conversely Bispectral Index positive +10 units at deep surgical anesthesia, and LOA of around ±11 units, (double the generally acceptable <10 BIS units' level of agreement), our research clearly demonstrated that the Bispectral Index "off-label" occipital-montage could be beneficial when we want to follow the trend of remifentanil/propofol anesthesia in patients whose frontal montage is inaccessible [124].

In our study [124], using BIS-Vista™ monitor sensor occipital montage with Electrode 1 sited at the "Inion" and Electrode 2, Electrode 4, and Electrode 3 sited slanting in the direction of the mastoid-process, we clearly showed that the too wide LOA does not allow the Bispectral Index data attained from the Frontal and Occipital montages to be interchangeable also that the differences between the Frontal and Occipital montages are greatly time-domain dependent on the depth of anesthesia. The LOA in our study results were almost double the clinically acceptable difference level of <10 BIS units [116, 124]. We found significant disagreements between the occipital and frontal placement of BIS sensors before anesthesia induction phase and during anesthesia maintenance phase [124]. Because the application to the occipital region is limited by the presence of "hair," this makes the occipital montage only possible in few selected patients who would accept shaving the back of their heads to apply the occipital BIS sensor montage. However, the good news is we would only resort to the alternative occipital montage when the conventional frontal montage is not accessible and that is typically during neurosurgery when patients' heads are already shaved anyway [124].

The electric signal generated from the frontal-montage is employed to calculate a combination of three descriptors through a proprietary algorithm [124]. We demonstrated that the Bispectral Index numeric score generated from an alternate electrode montage differs according to the depth of anesthesia, as we investigated how the data generated from the Frontal and Occipital montages correlate through three phases, anesthesia induction, anesthesia maintenance when there are no noxious stimulation and when neuromuscular blocking agents (NMBAs) are administered with no EMG, and the third phase of emergence from anesthesia [124]. This demonstrated how complex the algorithms calculating these pEEG-DOA indices are and how influential electromyographic activity is [124].

In their leading study, Kochs and colleagues [126] demonstrated that spatial topographic electroencephalographic spectral representation shows that intraoperative EEG electrophysiological changes with varying depths of anesthesia in response to noxious stimuli are not consistent over all cerebral regions [126]. Therefore, we should expect electrodes placed over different regions to detect different EEG patterns, hence generating different pEEG-DOA indices. Surgical noxious surgical stimulation resulted in predominantly frontal increase in δ-wave activity termed the "δ shift" characterized by an α-wave frequency lack of anteriorization, and an upsurge in synchronous frontal δ-wave [126]. Under general anesthesia, these electroencephalographic waves were mostly influential mainly in the frontal area in comparison to the frontal-parietal region and the occipital area [126], as they originate from the "midbrain" and then spread to the frontal cerebral regions. Hence, frontal electroencephalographic waves would represent mainly "afferent-signal-transmissions" [126].

The Bispectral Index platform detects anesthetic agents induced EEG changes at the frontal region [10] and was not devised nor approved to detect EEG alterations at the occipital area. While the standardized 10–20 EEG system is widely comprehended all over various spatial topographical areas, the Bispectral Index platform is still considered a proprietary "black box"; thus we cannot scientifically assess the appropriateness of the utility of Bispectral Index outside the regions the algorithms was devised and approved [10].

In the first stage of our study before induction, my research group Dahaba and colleagues [124] showed that there is a frontal-occipital difference of ≈−10 Bispectral Index units [124] attributed to the posterior α-wave dominance in comparison to the frontal area, when our patients are awake relaxed with their eyes closed [19, 127]. Bianchi and colleagues [128], by means of bispectral and spectral examination, demonstrated that when their patients were wide-awake, comfortable, and their eyes were closed, α-wave were created by two separate oscillators one occipital oscillator and another oscillator in the frontal region [128].

At the second stage of our study [124] at deep propofol anesthesia, electroencephalographic alterations are best defined as "frontal-predominance" characterizing a very prominent constant upsurge of β waves over the frontal region resulting from excitation of the cortical frontal region [19, 124, 129]. That will be ultimately surpassed by noticeable upsurge in δ wave and θ wave low frequency waves traveling to the anterior spatial topographical "anteriorization" patterns [124, 127]. This was previously reported with the early generations BIS devices with general anesthesia [130, 131]. Hall and Lockwood [130] with an A-1000 BIS device (version 2.51) under different anesthesia regimens showed that Bispectral Index values recorded at central-regions are ≈+13 Bispectral Index points more than Bispectral Index values recorded at frontal regions [130]. Pandin and colleagues [131] showed that with gradual increase in target controlled infusion (TCI), BIS A-1000 (version 3.12) displayed consistent central-frontal +12 difference, and parietal-frontal +25 Bispectral Index points difference [131]. That agrees with our study findings of constant occipital-frontal +10 Bispectral Index points difference. The authors reported that frontal-central regions are weakly $r^2 = 0.374$ correlated and that frontal-parietal regions were moderately $r^2 = 0.66$ correlated [131], in agreement with our moderately $r^2 = 0.671$ correlated frontal-occipital regions at anesthesia steady-state [124].

At the second stage of our study of deep steady-state surgical anesthesia [124], there was substantial "loss-of-coherence" with considerable "anterior-posterior-regions-uncoupling" in cerebral hemispheres [127]. This indicates that maintaining adequate surgical anesthesia requires numerous crucial anterior structures, probably superior-frontal, orbital, parahippocampal and anterior-cingulate-gyri [127]. This clearly implies totally separate "anterior" from "posterior" EEG derivation [19]. In our study of a constant occipital-region higher by +10 Bispectral Index points than frontal-region at deep remifentanil/propofol anesthesia, possibly resulting from Bispectral Index *SynchFastSlow* subparameter over the occipital areas detecting, to a lesser extent, something that mostly occurs at anterior structures when α wave and β wave cortical-generators' activity declines with concurrent shifting towards cortical-generators' δ wave and θ wave activity [19].

At the third and last stage of our study of anesthesia recovery [124], the specific EEG alterations that occur with emergence from anesthesia appear as a reverse of all changes that occurred with induction of anesthesia [124]. Recovery from anesthesia is accompanied by general decline in the powers of δ wave and θ waves with a continuous decline in the "frontal-predominance" [19]. Anterior-posterior "coherence" starts to increase because "regional-interactions" resume [127]. The *Relative BetaRatio* subparameter detects the gradual β wave reappearing, and the *SynchFastSlow* descriptor starts to decrease with the gradual decline of δ wave [129]. At emergence from anesthesia until Bispectral Index value of 55 [124], Bispectral Index recorded at the occipital region tends to be higher than Bispectral Index recorded from the frontal region. As the Bispectral Index value increases above 55 the opposite tends to occur with the Bispectral Index value recorded from the frontal region tend to be greater than the Bispectral Index value recorded from the occipital region [124]. The recovery from anesthesia EEG patterns must always be regarded as at the crucial stage of emergence from anesthesia, as the occipital Bispectral Index value will not increase to a value higher than BIS 88 [124].

Our results [124] do not concur with the two previously mentioned studies [121, 122], who obviously considered the first-generation BIS monitors to be "topographically-agonist" [121, 122]. Contrary to our center investigation [124], the two studies did not completely analyze all three phases of anesthesia separately (namely the pre-induction stage, steady-state anesthesia stage, and emergence from anesthesia stage) [121, 122]. The authors did not use Bland and Altman format or scatter plots that would categorize the extent of agreements between Bispectral Index values obtained from the occipital and frontal electrode montage [121, 122]. They offered no rationalization nor justification of the reason the Bispectral Index is "topographic-agonist," even though the explanation of their results [121] is contrary to numerous publications demonstrating distinguished "spatial time-domain topographical EEG variations" [126, 127].

Two case reports [132, 133] used the BIS or Entropy monitors with BIS or Entropy sensors placed on the occipital regions for surgery in 2 patients placed in the ventral position. In the first case, Hemmerling and colleagues [132] reported a case of BIS monitoring in a 52-year-old, ASA physical status II patient undergoing 60-min surgery in the ventral position, for 8% second degree burns skin grafting of the patient's back in the ventral position [132]. The patient came with a completely shaved head; therefore a BIS sensor could be placed on the occipital region simultaneously with a second BIS sensor at the conventional forehead region. After fentanyl 5 μg kg^{-1}

and propofol 2 mg kg^{-1} administrations for anesthesia induction and rocuronium 0.6 mg kg^{-1} administration for endotracheal intubation, initial occipital-frontal BIS readings appeared similar [132]. After 6-min of anesthesia recovery, the patient opened his eyes at Bispectral Index value of 90 and was extubated [132]. The Bispectral Index values from the two regions were compared using Bland and Altman scatter plot [22]. BIS occipital-frontal Bias was +1 with lower and upper LOA of −7 and +9 respectively. Authors concluded that both sites montage could have been used interchangeably because in this patient, occipital montage was easier to apply and was preferable to avoid pressure points on the BIS sensor [132].

Entropy (General Electric GE Healthcare, Chicago, Illinois, USA) generates two indices [47]: the State Entropy (SE), computed over the frequency range 0.8–32 Hz, and the Response Entropy (RE), calculated over the frequency range 0.8–47 Hz. The RE includes additional information of the electromyographic frontalis muscle activity [47]. The SE (range, 0–91) and the RE (range, 0–100) are identical if there is no EMG muscle activity [47].

In the second case report, Sinha and colleagues [133] used the S/5 Entropy Module, in a 56-year-old, ASA physical status I, patient who underwent frontal craniotomy for distal anterior communicating artery aneurysm clipping, where the forehead was inaccessible because of the surgical procedure site, so they placed the Entropy sensors on the occipital area [133]. Anesthesia was induced with propofol 2 mg kg^{-1}, fentanyl 4 µg kg^{-1} and pancuronium for endotracheal intubation and muscle relaxation. Anesthesia was maintained with 1–2% isoflurane, 1:1 Nitrous Oxide (N_2O) in oxygen (O_2), as well as fentanyl 1 µg kg^{-1} h^{-1} and propofol 1–2 mg kg^{-1} h^{-1} infusions [133]. They measured Entropy continuously throughout the neurosurgical procedure and concluded that Entropy values "well corresponded" to clinically judged different stages of anesthesia and surgical stimulation, suggesting that entropy sensors could be applied to the occipital region when the frontal region is inaccessible during surgery [133].

References

1. Canton R. The electric currents of the brain. Br Med J. 1875;2:278.
2. Berger H. Ueber das Elektrenkephalogramm des Menschen. Arch Psychiatr Nervenkr. 1929;87:527.
3. Gibbs FA. Effect on the electro-encephalogram of certain drugs which influence nervous activity. Arch Intern Med. 1937;60:154.
4. Sigl JC, Chamoun NG. An introduction to bispectral analysis for the electroencephalogram. J Clin Monit. 1994;10:392–404.
5. Hajat Z, Ahmad N, Andrzejowski J. The role and limitations of EEG-based depth of anaesthesia monitoring in theatres and intensive care. Anaesthesia. 2017;72(Suppl 1):38–47.
6. Hasselmann K, Munk W, MacDonald GFK. Bispectra of ocean waves. In: Rosenblatt M, editor. Time series analysis. 1st ed. New York: Wiley; 1963. p. 125–39.
7. Haubrich RA. Earth noise, 5 to 500 millicycles per second: 1. Spectral stationarity, normality, and nonlinearity. J Geophys Res. 1965;70:1415–27.
8. Kearse LA, Manberg P, Chamoun N, deBros F, Zaslavsky A. Bispectral analysis of the electroencephalogram correlates with patient movement to skin incision during propofol/nitrous oxide anesthesia. Anesthesiology. 1994;81:1365–70.
9. Barnett TP, Johnson LC, Naitoh P, Hicks N, Nute C. Bispectrum analysis of electroencephalogram signals during waking and sleeping. Science. 1971;172:401–2.
10. Rampil IJ. A primer for EEG signal processing in anesthesia. Anesthesiology. 1998;89:980–1002.
11. Todd MM. EEGs, EEG processing, and the bispectral index. Anesthesiology. 1998;89:815–7.
12. Gallagher TA, Nemeth AJ, Hacein-Bey L. An introduction to the Fourier transform: relationship to MRI. AJR Am J Roentgenol. 2008;190:1396–405.
13. Prins SA, de Hoog M, Blok JH, Tibboel D, Visser GH. Continuous noninvasive monitoring of barbiturate coma in critically ill children using the Bispectral index monitor. Crit Care. 2007;11:R108.
14. Cottenceau V, Petit L, Masson F, Guehl D, Asselineau J, Cochard J-F, et al. The use of bispectral index to monitor barbiturate coma in severely brain-injured patients with refractory intracranial hypertension. Anesth Analg. 2008;107:1676–82.

15. Lobo FA, Schraag S. Limitations of anaesthesia depth monitoring. Curr Opin Anaesthesiol. 2011;24:657–64.
16. Velly LJ, Rey MF, Bruder NJ, Gouvitsos FA, Witjas T, Regis JM, et al. Differential dynamic of action on cortical and subcortical structures of anesthetic agents during induction of anesthesia. Anesthesiology. 2007;107:202–12.
17. Xie G, Deschamps A, Backman SB, Fiset P, Chartrand D, Dagher A, et al. Critical involvement of the thalamus and precuneus during restoration of consciousness with physostigmine in humans during propofol anaesthesia: a positron emission tomography study. Br J Anaesth. 2011;106:548–57.
18. Sandin RH, Enlund G, Samuelsson P, Lennmarken C. Awareness during anaesthesia: a prospective case study. Lancet. 2000;355:707–11.
19. Gugino LD, Chabot RJ, Prichep LS, John ER, Formanek V, Aglio LS. Quantitative EEG changes associated with loss and return of consciousness in healthy adult volunteers anaesthetized with propofol or sevoflurane. Br J Anaesth. 2001;87:421–8.
20. Liberati A, Altman DG, Tetzlaff J, Mulrow C, Gøtzsche PC, Ioannidis JPA, et al. The PRISMA statement for reporting systematic reviews and meta-analyses of studies that evaluate healthcare interventions: explanation and elaboration. BMJ. 2009;339:b2700.
21. Dahaba AA, Mattweber M, Fuchs A, Zenz W, Rehak PH, List WF, et al. The effect of different stages of neuromuscular block on the bispectral index and the bispectral index-XP under remifentanil/propofol anesthesia. Anesth Analg. 2004;99:781–7.
22. Bland JM, Altman DG. Statistical methods for assessing agreement between two methods of clinical measurement. Lancet. 1986;1:307–10.
23. Vuyk J, Lichtenbelt BJ, Vieveen J, Dahan A, Engbers FHM, Burm AGL. Low bispectral index values in awake volunteers receiving a combination of propofol and midazolam. Anesthesiology. 2004;100:179–81.
24. Renna M, Handy J, Shah A. Low baseline Bispectral Index of the electroencephalogram in patients with dementia. Anesth Analg. 2003;96:1380–5.
25. Dahaba AA. Benefits and boundaries of processed electroencephalography (pEEG) monitors when they do not concur with standard anesthetic clinical monitoring: lights and shadows. Minerva Anestesiol. 2020;86:304–16.
26. Pilge S, Zanner R, Schneider G, Blum J, Kreuzer M, Kochs EF. Time delay of index calculation: analysis of cerebral state, bispectral, and Narcotrend indices. Anesthesiology. 2006;104:488–94.
27. Zanner R, Pilge S, Kochs EF, Kreuzer M, Schneider G. Time delay of electroencephalogram index calculation: analysis of cerebral state, bispectral, and Narcotrend indices using perioperatively recorded electroencephalographic signals. Br J Anaesth. 2009;103:394–9.
28. Ionescu CM, Hodrea R, De Keyser R. Variable time-delay estimation for anesthesia control during intensive care. IEEE Trans Biomed Eng. 2011;58:363–9.
29. Kreuzer M, Zanner R, Pilge S, Paprotny S, Kochs EF, Schneider G. Time delay of monitors of the hypnotic component of anesthesia: analysis of state entropy and index of consciousness. Anesth Analg. 2012;115:315–9.
30. Gallagher JD. Pacer-induced artifact in the bispectral index during cardiac surgery. Anesthesiology. 1999;90:636.
31. Sloan PA. Interference of bispectral index monitoring with intraoperative use of the electromyograph endotracheal tube. Can J Anaesth. 2007;54:1028–9.
32. Guignard B, Chauvin M. Bispectral index increases and decreases are not always signs of inadequate anesthesia. Anesthesiology. 2000;92:903.
33. Hemmerling TM, Fortier JD. Falsely increased bispectral index values in a series of patients undergoing cardiac surgery using forced-air-warming therapy of the head. Anesth Analg. 2002;95:322–3. table of contents
34. Zanner R, Schneider G, Kochs EF. Falsely increased bispectral index values caused by the use of a forced-air-warming device. Eur J Anaesthesiol. 2006;23:618–9.
35. Hemmerling TM, Migneault B. Falsely increased bispectral index during endoscopic shoulder surgery attributed to interferences with the endoscopic shaver device. Anesth Analg. 2002;95:1678–9. table of contents
36. Hemmerling TM, Desrosiers AM. Interference of electromagnetic operating systems in otorhinolaryngology surgery with bispectral index monitoring. Anesth Analg. 2003;96:1698–9.
37. Dahaba AA, Bornemann H, Rehak PH, Metzler H. Electromagnetic emission of high-energy extracorporeal shockwave lithotripsy to the shoulder does not cause disruption of bispectral index monitoring of propofol or sevoflurane anesthesia. Anesth Analg. 2006;103:254–6.
38. Farling PA. Thyroid disease. Br J Anaesth. 2000;85:15–28.
39. Eisele DW. Intraoperative electrophysiologic monitoring of the recurrent laryngeal nerve. Laryngoscope. 1996;106:443–9.
40. Dahaba AA. Different conditions that could result in the bispectral index indicating an incorrect hypnotic state. Anesth Analg. 2005;101:765–73.
41. Kim DW, Kil HY, White PF. The effect of noise on the bispectral index during propofol sedation. Anesth Analg. 2001;93:1170–3.
42. Kang JG, Lee JJ, Kim DM, Kim JA, Kim CS, Hahm TS, et al. Blocking noise but not music lowers bispectral index scores during sedation in noisy operating rooms. J Clin Anesth. 2008;20:12–6.
43. Fahy BG, Chau DF. The technology of processed electroencephalogram monitoring devices for

assessment of depth of anesthesia. Anesth Analg. 2018;126:111–7.
44. Koskinen M, Seppänen T, Tuukkanen J, Yli-Hankala A, Jäntti V. Propofol anesthesia induces phase synchronization changes in EEG. Clin Neurophysiol. 2001;112:386–92.
45. Zetterlund E-L, Gréen H, Oscarsson A, Vikingsson S, Vrethem M, Lindholm M-L, et al. Determination of loss of consciousness: a comparison of clinical assessment, bispectral index and electroencephalogram: an observational study. Eur J Anaesthesiol. 2016;33:922–8.
46. Vacas S, McInrue E, Gropper MA, Maze M, Zak R, Lim E, et al. The feasibility and utility of continuous sleep monitoring in critically ill patients using a portable electroencephalography monitor. Anesth Analg. 2016;123:206–12.
47. Viertiö-Oja H, Maja V, Särkelä M, Talja P, Tenkanen N, Tolvanen-Laakso H, et al. Description of the entropy algorithm as applied in the Datex-Ohmeda S/5 entropy module. Acta Anaesthesiol Scand. 2004;48:154–61.
48. Drover D, Ortega HR. Patient state index. Best Pract Res Clin Anaesthesiol. 2006;20:121–8.
49. Kreuer S, Wilhelm W. The Narcotrend monitor. Best Pract Res Clin Anaesthesiol. 2006;20:111–9.
50. Loomis AL, Harvey EN, Hobart GA. Cerebral states during sleep, as studied by human brain potentials. J Exp Psychol. 1937;21:127–44.
51. Schultz B, Kreuer S, Wilhelm W, Grouven U, Schultz A. The Narcotrend monitor. Development and interpretation algorithms. Anaesthesist. 2003;52:1143–8.
52. Panousis P, Heller AR, Burghardt M, Bleyl JU, Koch T. The effects of electromyographic activity on the accuracy of the Narcotrend monitor compared with the Bispectral index during combined anaesthesia. Anaesthesia. 2007;62:868–74.
53. Zikov T, Bibian S, Dumont GA, Huzmezan M, Ries CR. Quantifying cortical activity during general anesthesia using wavelet analysis. IEEE Trans Biomed Eng. 2006;53:617–32.
54. Jensen EW, Valencia JF, López A, Anglada T, Agustí M, Ramos Y, et al. Monitoring hypnotic effect and nociception with two EEG-derived indices, qCON and qNOX, during general anaesthesia. Acta Anaesthesiol Scand. 2014;58:933–41.
55. Weber F, Zimmermann M, Bein T. The impact of acoustic stimulation on the AEP monitor/2 derived composite auditory evoked potential index under awake and anesthetized conditions. Anesth Analg. 2005;101:435–9.
56. Cortínez LI, Delfino AE, Fuentes R, Muñoz HR. Performance of the cerebral state index during increasing levels of propofol anesthesia: a comparison with the bispectral index. Anesth Analg. 2007;104:605–10.
57. Revuelta M, Paniagua P, Campos JM, Fernández JA, Martínez A, Jospin M, et al. Validation of the index of consciousness during sevoflurane and remifentanil anaesthesia: a comparison with the bispectral index and the cerebral state index. Br J Anaesth. 2008;101:653–8.
58. Willmann K, Springman S, Rusy D, Daily E. A preliminary evaluation of a new derived EEG index monitor in anesthetized patients. J Clin Monit Comput. 2002;17:345–50.
59. Han L, Drover DR, Chen MC, Saxena AR, Eagleman SL, Nekhendzy V, et al. Evaluation of patient state index, bispectral index, and entropy during drug induced sleep endoscopy with dexmedetomidine. J Clin Monit Comput. 2023;37:727–34.
60. Musizza B, Ribaric S. Monitoring the depth of anaesthesia. Sensors (Basel). 2010;10:10896–935.
61. Bruhn J, Röpcke H, Hoeft A. Approximate entropy as an electroencephalographic measure of anesthetic drug effect during desflurane anesthesia. Anesthesiology. 2000;92:715–26.
62. Drover DR, Lemmens HJ, Pierce ET, Plourde G, Loyd G, Ornstein E, et al. Patient State Index: titration of delivery and recovery from propofol, alfentanil, and nitrous oxide anesthesia. Anesthesiology. 2002;97:82–9.
63. Balci C, Karabekir HS, Kahraman F, Sivaci RG. Comparison of entropy and bispectral index during propofol and fentanyl sedation in monitored anaesthesia care. J Int Med Res. 2009;37:1336–42.
64. Chen X, Tang J, White PF, Wender RH, Ma H, Sloninsky A, et al. A comparison of patient state index and bispectral index values during the perioperative period. Anesth Analg. 2002;95:1669–74. table of contents
65. Kim D, Ahn JH, Heo G, Jeong JS. Comparison of Bispectral Index and Patient State Index values according to recovery from moderate neuromuscular block under steady-state total intravenous anesthesia. Sci Rep. 2021;11:5908.
66. Ki S, Lee D, Lee W, Cho K, Han Y, Lee J. Verification of the performance of the Bispectral Index as a hypnotic depth indicator during dexmedetomidine sedation. Anesth Pain Med (Seoul). 2022;17:44–51.
67. Soehle M, Kuech M, Grube M, Wirz S, Kreuer S, Hoeft A, et al. Patient state index vs bispectral index as measures of the electroencephalographic effects of propofol. Br J Anaesth. 2010;105:172–8.
68. Absalom AR, Menon DK. BIS and spectral entropy monitoring during sedation with midazolam/remifentanil and dexmedetomidine/remifentanil. Crit Care. 2009;13:137.
69. Sleigh JW. Depth of anesthesia: perhaps the patient isn't a submarine. Anesthesiology. 2011;115:1149–50.
70. Hight D, Kreuzer M, Ugen G, Schuller P, Stüber F, Sleigh J, et al. Five commercial "depth of anaesthesia" monitors provide discordant clinical recommendations in response to identical emergence-like EEG signals. Br J Anaesth. 2023;130:536–45.
71. Mashour GA, Shanks A, Tremper KK, Kheterpal S, Turner CR, Ramachandran SK, et al. Prevention of intraoperative awareness with explicit recall in

an unselected surgical population: a randomized comparative effectiveness trial. Anesthesiology. 2012;117:717–25.
72. Leslie K, Chan MTV, Myles PS, Forbes A, McCulloch TJ. Posttraumatic stress disorder in aware patients from the B-aware trial. Anesth Analg. 2010;110:823–8.
73. Myles PS, Leslie K, McNeil J, Forbes A, Chan MTV. Bispectral index monitoring to prevent awareness during anaesthesia: the B-Aware randomised controlled trial. Lancet. 2004;363:1757–63.
74. Avidan MS, Jacobsohn E, Glick D, Burnside BA, Zhang L, Villafranca A, et al. Prevention of intraoperative awareness in a high-risk surgical population. N Engl J Med. 2011;365:591–600.
75. Fritz BA, Kalarickal PL, Maybrier HR, Muench MR, Dearth D, Chen Y, et al. Intraoperative electroencephalogram suppression predicts postoperative delirium. Anesth Analg. 2016;122:234–42.
76. Wildes TS, Mickle AM, Ben Abdallah A, Maybrier HR, Oberhaus J, Budelier TP, et al. Effect of electroencephalography-guided anesthetic administration on postoperative delirium among older adults undergoing major surgery: The ENGAGES Randomized Clinical Trial. JAMA. 2019;321:473–83.
77. Short TG, Campbell D, Frampton C, Chan MTV, Myles PS, Corcoran TB, et al. Anaesthetic depth and complications after major surgery: an international, randomised controlled trial. Lancet. 2019;394:1907–14.
78. Schneider G, Gelb AW, Schmeller B, Tschakert R, Kochs E. Detection of awareness in surgical patients with EEG-based indices—bispectral index and patient state index. Br J Anaesth. 2003;91:329–35.
79. Schneider G, Wagner K, Reeker W, Hänel F, Werner C, Kochs E. Bispectral Index (BIS) may not predict awareness reaction to intubation in surgical patients. J Neurosurg Anesthesiol. 2002;14:7–11.
80. Sanders RD, Gaskell A, Raz A, Winders J, Stevanovic A, Rossaint R, et al. Incidence of connected consciousness after tracheal intubation: a prospective, international, multicenter cohort study of the isolated forearm technique. Anesthesiology. 2017;126:214–22.
81. Lennertz R, Pryor KO, Raz A, Parker M, Bonhomme V, Schuller P, et al. Connected consciousness after tracheal intubation in young adults: an international multicentre cohort study. Br J Anaesth. 2023;130:e217–24.
82. Dahaba AA, Xiao Z, Zhu X, Dong H, Xiong L, Rehak P, et al. Age progression from vicenarians (20–29 year) to nonagenarians (90–99 year) among a population pharmacokinetic/pharmacodynamic (PopPk-PD) covariate analysis of propofol-bispectral index (BIS) electroencephalography. J Pharmacokinet Pharmacodyn. 2020;47:145–61.
83. Obert DP, Schweizer C, Zinn S, Kratzer S, Hight D, Sleigh J, et al. The influence of age on EEG-based anaesthesia indices. J Clin Anesth. 2021;73:110325.
84. Schuller PJ, Newell S, Strickland PA, Barry JJ. Response of bispectral index to neuromuscular block in awake volunteers. Br J Anaesth. 2015;115(Suppl 1):i95–103.
85. Palanca BJA, Mashour GA, Avidan MS. Processed electroencephalogram in depth of anesthesia monitoring. Curr Opin Anaesthesiol. 2009;22:553–9.
86. Whitlock EL, Villafranca AJ, Lin N, Palanca BJ, Jacobsohn E, Finkel KJ, et al. Relationship between bispectral index values and volatile anesthetic concentrations during the maintenance phase of anesthesia in the B-Unaware trial. Anesthesiology. 2011;115:1209–18.
87. Short TG, Campbell D, Leslie K. Response of bispectral index to neuromuscular block in awake volunteers. Br J Anaesth. 2016;116:725–6.
88. Schnider TW, Minto CF. Beware the drug titration paradox. Comment on Br J Anaesth 2021;127:704–12. Br J Anaesth. 2022;(128):e335–7.
89. Purdon PL, Pierce ET, Mukamel EA, Prerau MJ, Walsh JL, Wong KFK, et al. Electroencephalogram signatures of loss and recovery of consciousness from propofol. Proc Natl Acad Sci USA. 2013;110:E1142–51.
90. Chander D, García PS, MacColl JN, Illing S, Sleigh JW. Electroencephalographic variation during end maintenance and emergence from surgical anesthesia. PLoS One. 2014;9:e106291.
91. Hesse S, Kreuzer M, Hight D, Gaskell A, Devari P, Singh D, et al. Association of electroencephalogram trajectories during emergence from anaesthesia with delirium in the postanaesthesia care unit: an early sign of postoperative complications. Br J Anaesth. 2019;122:622–34.
92. Purdon PL, Sampson A, Pavone KJ, Brown EN. Clinical electroencephalography for anesthesiologists: part I: background and basic signatures. Anesthesiology. 2015;123:937–60.
93. Hight DF, Kaiser HA, Sleigh JW, Avidan MS. Continuing professional development module: an updated introduction to electroencephalogram-based brain monitoring during intended general anesthesia. Can J Anaesth. 2020;67:1858–78.
94. Kaiser HA, Peus M, Luedi MM, Lersch F, Krejci V, Reineke D, et al. Frontal electroencephalogram reveals emergence-like brain activity occurring during transition periods in cardiac surgery. Br J Anaesth. 2020;125:291–7.
95. Introna M, Gemma M, Carozzi C. Improving the benefit of processed EEG monitors: it's not about the car but the driver. J Clin Monit Comput. 2023;37:723–5.
96. Sessler CN, Gosnell MS, Grap MJ, Brophy GM, O'Neal PV, Keane KA, et al. The Richmond agitation-sedation scale: validity and reliability in adult intensive care unit patients. Am J Respir Crit Care Med. 2002;166:1338–44.

97. Carozzi C, Rampil IJ. Bispectral Index for sleep screening: it is time to move on. Minerva Anestesiol. 2011;77:485–7.
98. Dahaba AA, Xue JX, Xu GX, Liu QH, Metzler H. Bilateral Bispectral Index (BIS)-Vista as a measure of physiologic sleep in sleep-deprived anesthesiologists. Minerva Anestesiol. 2011;77:388–93.
99. Stewart JA, Särkelä MOK, Wennervirta J, Vakkuri AP. Novel insights on association and reactivity of Bispectral Index, frontal electromyogram, and autonomic responses in nociception-sedation monitoring of critical care patients. BMC Anesthesiol. 2022;22:353.
100. Naaz S, Ozair E. Dexmedetomidine in current anaesthesia practice – a review. J Clin Diagn Res. 2014;8:GE01–4.
101. Eagleman SL, Drover CM, Li X, MacIver MB, Drover DR. Offline comparison of processed electroencephalogram monitors for anaesthetic-induced electroencephalogram changes in older adults. Br J Anaesth. 2021;126:975–84.
102. Kim D, Ahn JH, Jung H, Choi KY, Jeong JS. Effects of neuromuscular blockade reversal on bispectral index and frontal electromyogram during steady-state desflurane anesthesia: a randomized trial. Sci Rep. 2019;9:10486.
103. Renna M, Wigmore T, Mofeez A, Gillbe C. Biasing effect of the electromyogram on BIS: a controlled study during high-dose fentanyl induction. J Clin Monit Comput. 2002;17:377–81.
104. Messner M, Beese U, Romstöck J, Dinkel M, Tschaikowsky K. The bispectral index declines during neuromuscular block in fully awake persons. Anesth Analg. 2003;97:488–91.
105. Balcı C, Karabekir H, Sivaci R. Determining entropy values equivalent to the bispectral index values during sevoflurane anaesthesia. Arch Med Sci. 2010;6:370–4.
106. Xing Y, Xu D, Xu Y, Chen L, Wang H, Li S. Effects of neuromuscular blockages on entropy monitoring during sevoflurane anesthesia. Med Sci Monit. 2019;25:8610–7.
107. Epstein RH, Maga JM, Mahla ME, Schwenk ES, Bloom MJ. Prevalence of discordant elevations of state entropy and bispectral index in patients at amnestic sevoflurane concentrations: a historical cohort study. Can J Anaesth. 2018;65:512–21.
108. Müller JN, Kreuzer M, García PS, Schneider G, Hautmann H. Monitoring depth of sedation: evaluating the agreement between the Bispectral Index, qCON and the Entropy Module's State Entropy during flexible bronchoscopy. Minerva Anestesiol. 2017;83:563–73.
109. Kreuzer M, Kochs EF, Pilge S, Stockmanns G, Schneider G. Construction of the electroencephalogram player: a device to present electroencephalogram data to electroencephalogram-based anesthesia monitors. Anesth Analg. 2007;104:135–9.
110. Warnaby CE, Sleigh JW, Hight D, Jbabdi S, Tracey I. Investigation of slow-wave activity saturation during surgical anesthesia reveals a signature of neural inertia in humans. Anesthesiology. 2017;127:645–57.
111. Gaskell A, Sleigh J. The quagmire of postoperative delirium: does dose matter? Br J Anaesth. 2021;127:664–6.
112. Berger M, Mark JB, Kreuzer M. Of parachutes, speedometers, and EEG: what evidence do we need to use devices and monitors? Anesth Analg. 2020;130:1274–7.
113. Gaskell AL, Hight DF, Winders J, Tran G, Defresne A, Bonhomme V, et al. Frontal alpha-delta EEG does not preclude volitional response during anaesthesia: prospective cohort study of the isolated forearm technique. Br J Anaesth. 2017;119:664–73.
114. Lee U, Ku S, Noh G, Baek S, Choi B, Mashour GA. Disruption of frontal-parietal communication by ketamine, propofol, and sevoflurane. Anesthesiology. 2013;118:1264–75.
115. Kim MC, Fricchione GL, Brown EN, Akeju O. Role of electroencephalogram oscillations and the spectrogram in monitoring anaesthesia. BJA Educ. 2020;20:166–72.
116. Tufegdzic B, Lobo F, Achi E, Motta S, Carozzi C, Lamperti M. Alternative sensor montage for index based EEG monitoring. A systematic review. J Clin Monit Comput. 2024; https://doi.org/10.1007/s10877-024-01162-1.
117. Lobo FA, Shander A. Modern anesthetic noninvasive monitoring: a deep look into perioperative care. J Cardiothorac Vasc Anesth. 2019;33(Suppl 1):S1–2.
118. Li Y, Bohringer C, Liu H. Double standard: why electrocardiogram is standard care while electroencephalogram is not? Curr Opin Anaesthesiol. 2020;33:626–32.
119. Sepúlveda PO, Vera R, Fernández MS, Lobo FA. Linear thinking does not reflect the newer 21st-century anesthesia concepts. A narrative review. J Clin Monit Comput. 2023;37:1133–44.
120. Buzsáki G, Anastassiou CA, Koch C. The origin of extracellular fields and currents—EEG, ECoG, LFP and spikes. Nat Rev Neurosci. 2012;13:407–20.
121. Glass PS, Bloom M, Kearse L, Rosow C, Sebel P, Manberg P. Bispectral analysis measures sedation and memory effects of propofol, midazolam, isoflurane, and alfentanil in healthy volunteers. Anesthesiology. 1997;86:836–47.
122. Shiraishi T, Uchino H, Sagara T, Ishii N. A comparison of frontal and occipital bispectral index values obtained during neurosurgical procedures. Anesth Analg. 2004;98:1773–5.
123. Muncaster ARG, Sleigh JW, Williams M. BIS monitoring requires proper electrode placement for optimum performance. Anesth Analg. 2003;97:1206.

124. Dahaba AA, Xue JX, Zhao GG, Liu QH, Xu GX, Bornemann H, et al. BIS-vista occipital montage in patients undergoing neurosurgical procedures during propofol-remifentanil anesthesia. Anesthesiology. 2010;112:645–51.
125. Niedhart DJ, Kaiser HA, Jacobsohn E, Hantler CB, Evers AS, Avidan MS. Intrapatient reproducibility of the BISxp monitor. Anesthesiology. 2006;104:242–8.
126. Kochs E, Bischoff P, Pichlmeier U, Schulte am Esch J. Surgical stimulation induces changes in brain electrical activity during isoflurane/nitrous oxide anesthesia. A topographic electroencephalographic analysis. Anesthesiology. 1994;80:1026–34.
127. John ER, Prichep LS, Kox W, Valdés-Sosa P, Bosch-Bayard J, Aubert E, et al. Invariant reversible QEEG effects of anesthetics. Conscious Cogn. 2001;10:165–83.
128. Bianchi AM, Molteni SC, Panzica F, Visani E, Franceschetti S, Cerutti S. Spectral and bispectral analysis of the EEG rhythms in basal conditions and during photic stimulation. Conf Proc IEEE Eng Med Biol Soc. 2004;2006:574–7.
129. Sneyd JR, Samra SK, Davidson B, Kishimoto T, Kadoya C, Domino EF. Electrophysiologic effects of propofol sedation. Anesth Analg. 1994;79:1151–8.
130. Hall JD, Lockwood GG. Bispectral index: comparison of two montages. Br J Anaesth. 1998;80:342–4.
131. Pandin P, Van Cutsem N, Tuna T, D'hollander A. Bispectral index is a topographically dependent variable in patients receiving propofol anaesthesia. Br J Anaesth. 2006;97:676–80.
132. Hemmerling TM, Deschamps S, Michaud G, Trager G. An unusual site for BIS monitoring. Anesth Analg. 2004;99:1264–5.
133. Sinha PK, Suneel PR, Unnikrishnan KP, Smita V, Rathod RC. An alternative site for entropy sensor placement. Anesth Analg. 2006;102:1291.

"Monitors" Terminologies

2

2.1 Are We Just "Chasing" the Numbers?

An editorial by Deflandre and Javillier [1] clarified that a considerable part of the anesthesia literature deals with the prevention and management of anesthesia complications. From the 1980s, anesthesia complications have considerably diminished, largely because of new monitoring techniques of patients under anesthesia vital signs, that are fully trusted by the anesthesiologists performing the anesthesia [1]. However, an undesirable effect of these modern processed Electroencephalography-Depth Of Anesthesia (pEEG-DOA) monitoring techniques is the fact that anesthesiologists are increasingly losing how to skillfully conduct their patients' anesthesia without relying on these equipment, as they become more and more confident in the pEEG-DOA numbers [1]. Conversely, some anesthesiologists are always doubting the pEEG-DOA indices, which might not always be true, and in that case, pEEG-DOA indices do not guarantee safety of their patients but could lead to wrong decisions being taken with various consequences on patients' wellbeing [1].

Nowadays "monitoring" is the crux of doing anesthesia and there is no anesthesia practice without "monitoring" aiming at avoiding exposing patients to unnecessary risks when undergoing modern anesthesia [1]. Monitoring must be accompanied by the anesthesiologists' learning of anesthesia limitations and possible sources of artifactual monitoring signals [1]. Hence, I present in my book all conditions that can influence pEEG-DOA monitoring [2], limiting the pEEG-DOA monitors from performing to their full potential [2]. The purpose of my book is to in depth examine all reports appearing in literature of pEEG-DOA indices that are not concordant with clinically judged Depth of Anesthesia (DOA) hoping that this book could be beneficial to practicing anesthesiologists [1, 2]. I thoroughly reviewed the physiological and pharmacological background factors that could influence pEEG-DOA indices [2]. The book will allow young as well as DOA experienced anesthesiologists to better interpret displayed pEEG-DOA values and all factors that could alter these pEEG-DOA numbers [1, 2].

The purpose of my [2], systematic analysis is to better understand the limits of these monitors via the various published reports of all pEEG-DOA numbers not coinciding with clinically judged standard anesthesia state either because of changes in patients' electroencephalographic cerebral function via factors other than anesthetic agents, underperforming devices or flawed design of these monitors [2]. This narrative will also bring us up to date on pEEG-DOA devices' possible usage outside the operating room. We often see a disparate "clinically judged" hypnotic state that is totally different than the displayed pEEG-DOA numbers. The general concept of just

A. A. Dahaba, *Understanding the Numbers in Depth of Anesthesia Monitors*,
https://doi.org/10.1007/978-3-031-88260-9_2

"riding the numbers" [3] is insufficient when we should properly interpret these pEEG-DOA indices within context and limitations [2]. Nowadays there is a trend towards anesthesiologists' better understanding the displayed real time electroencephalography (EEG) waveforms; a concept that is totally aligned with this book as recognition of specific EEG waveform patterns could offer many sensible explanations [2].

Here I address two areas: different proprietary signal processing [2], and between-drug signal differences. Thus, "depth of anesthesia" monitors cannot be considered a "genuine" depiction of a "uniform-depth-of-anesthesia" or a portrayal of brain functions whatever anesthetics are employed [2, 4]. We should understand that "depth of anesthesia" monitoring is, in fact, a huge "misnomer" [4]. The notion that "depth-of-anesthesia" is only symbolically fitting [4], but not "measurable" [4], as the so-called depth-of-anesthesia monitors do not separately quantify either "depth" or consciousness uniform "anesthesia" [4], based on the very well-known fact that equi-Minimum Alveolar Concentrations (MACs) representing a quantifiable depth of anesthesia resulting from varied inhalational anesthetics produce completely dissimilar Bispectral Index values [4, 5]. I truly believe that a more suitable nomenclature should be "depth-of-anesthetics" and not "depth-of-anesthesia" monitors [2, 5]. I believe that pEEG-DOA monitoring may not be fully understood by practicing clinicians, may be over-marketed, and may not always be used appropriately. The contemporary use of these pEEG-DOA monitors as a simplified EEG monitor that somehow "compressed," into a single number, the complex 10–20 Fast-Fourier Transform (FFT) system recorded from 20 EEG sites, seems to be far too "over simplistic" [2]. The idea of we are "just-riding-the-numbers" appears to be a very bad idea; instead, we ought to properly understand those pEEG-DOA values in their perspective and limits [6]. What is clear, is that this technology should be applied and properly interpreted with the same rigor and background knowledge as any other monitoring device technology [6]. The appropriate terminology to describe pEEG-DOA devices that are used to detect awareness or to guide anesthetic agents' administration is currently lacking [7]. The "depth-of-anesthesia" referring to a global estimate of a "clinically judged hypnotic state" is a tricky concept because "depth of anesthesia" is the result of pharmacological actions of one or more anesthetic agent versus surgical stimulation [7].

To date the complete details of these pEEG-DOA exclusive proprietary algorithms are not fully published [8]. Because of the exclusive proprietaries of pEEG-DOA indices, anesthesiologists only see the displayed pEEG-DOA monitoring indices and to that effect most of these monitors are merely a "black box" head sets and their indices are merely a reflection of a "head-related" biosignal [8], that correlates with changes in the biophase effect-site concentrations of certain hypnotic/sedative drugs that was correlated with summed observations of some endpoint and back calibrated retrospectively derived from EEG variations of gradual increases of γ amino-butyric-acid (GABA)ergic anesthetics: isoflurane or propofol in a large volunteers' data base [8]. That cannot be considered a "true" reflection of a unified "cohesive" uniform "depth-of-anesthesia" or an independent diagnostic tool of the EEG cerebral performance irrespective of the kind of anesthetics used [8]. These pEEG-DOA indices designate ideal anesthetics concentrations helping anesthesiologists prevent inadequate anesthetic state resulting in patients experiencing Accidental Awareness under General Anesthesia (AAGA) or unwarranted deep anesthetic states resulting in unnecessary prolonged emergence from anesthesia and possibly postoperative cognitive dysfunction [8].

When discussing inhalational anesthetics' MAC that does not specify noxious stimulus, age, gender, etc. [2], here MAC represents 50% of the population, and we presume therefore 50% of patients may recall the stimulus, rendering clinical judgment that includes observation of the MAC as an unreliable tool to assess awareness as a comparator for pEEG-DOA indices [2]. Perhaps 1.3 MAC ($MAC_{1.3}$) anesthesia concentration representing 95% of the population would provide greater reliability. Regarding these two values

MAC and Bispectral Index, when MAC represents a clinical value for the level of anesthesia where 50% of patients do not move following a stimulus, when the Bispectral Index is a numerical value related to several pEEG-DOA indices [2], it cannot be assumed that one reflects the other, nor can MAC value represent "depth of anesthesia," rather a guide to the concentration that needs to be delivered to achieve an absence of recall. Several studies reported volatile anesthetic agents yielding different pEEG-DOA values at equipotent MACs due to their uniquely different EEG wavelength patterns [2].

2.2 The Three Terminologies: Brain Function Monitors, Depth of Anesthesia Monitors, and Processed Electroencephalography Monitors

"Brain function monitors (BFM)," "Depth of Anesthesia (DOA)" monitors, and "processed Electroencephalography (pEEG)" monitors are three terminologies often used interchangeably and frequently treated as synonyms when in fact they are not [2]. Brain function monitors have been commercially available since the introduction of the Bispectral Index (BIS) back in 1994 [9]. Most of these monitors utilize pEEG signals to estimate a dimensionless numerical index as a gauge of anesthetic depth in response to certain anesthetics concentrations [2]. Brain Function Monitors is the predominantly broader term that includes both "Depth of Anesthesia" monitors that aim to gauge the effect of sedative/hypnotic medication as well as "nociception/antinociception" monitors [2], sometimes mislabeled as "pain/analgesia" monitors because pain is "the conscious appreciation of nociception" and of course patients under anesthesia are not "conscious"; thus they do not feel "pain" but all their noxious reflexes are indeed intact [2]. Whereas pEEG monitors include those that measure spontaneous pEEG as well as those that measure evoked signals used for another purpose such as spine monitoring for surgical intrusion [2].

The value of the monitors that observe changes in electroencephalography (EEG) is that the anesthesiologists have no other objective mechanism for monitoring the state of awareness when the patient has received enough drugs to prevent the arousal from noxious stimuli. Supported by a Lancet [10] publication, indicating lower incidence of anesthesia awareness with Bispectral Index monitoring, millions of patients are being monitored by processed Electroencephalography-Depth Of Anesthesia (pEEG-DOA) devices calculating easy-to-use indices to give valuable information guiding anesthesia delivery [2, 10]. However, Avidan and colleagues [2, 11] in a New England Journal of Medicine (NEJM) study questioned whether these indices accurately reflect "depth of anesthesia," as awareness is still similarly frequent with the use of these monitoring aids [2]. These pEEG-DOA devices are still not always used appropriately as their underlying scientific principles are not fully understood by practicing anesthesiologists who might not be familiar with technical issues of EEG processing such as Burst Suppression Ratio (BSR) or artefact rejection. Considerable variability in pEEG-DOA devices usage exists among anesthesiologists, in part because of lack of clarity regarding the technical implementation of the EEG monitoring technology [2].

For the last 30-years, pEEG-DOA monitors have become widely popular [3]. The United Kingdom's National Institute for Health and Care Excellence (NICE) in November 2012 issued their official guideline recommendations of how to properly utilize the pEEG-DOA devices, publishing full details of three commonly used pEEG-DOA devices [3, 12]. Though the majority of anesthesiologists are well acquainted with all the fundamental pEEG-DOA technical designs, all the advantages gained from using the pEEG-DOA devices, as well as their limits and boundaries, still most of the anesthetists/anesthesiologists might not know all details that govern the algorithm calculations involved in EEG processing that would additionally contribute to an improved medical reasoning, such as BSR [3], artefacts' rejection, analysis of raw EEG, or displayed electromyography (EMG) muscle activity component contributions/contaminations affecting the precision and correctness of pEEG-DOA devices [3]. Furthermore, practicing anesthesiologists might not grasp how pEEG-DOA devices are precisely related to different aspects of patients' anesthesia. In my opinion, for anesthesiologists to fully understand pEEG-DOA devices' limits as well as all benefits they could gain from using pEEG-DOA devices in their daily anesthesia practice, this would greatly enhance anesthesia delivery to their patients. In my book I will try to disentangle a complicated mix of scientific/clinical facts and conflicting opinions [3].

The use of pEEG-DOA monitors such as the Bispectral Index monitor allows optimizing anesthetic agents dose administration by considering inter-individual variability of dose/concentrations/effect relationships for better anesthesia delivery [3]. These designated pEEG-DOA indices would ideally assist anesthesiologists avert inadequate anesthetic state patients experiencing Accidental Awareness under General Anesthesia (AAGA) or unwarranted deep anesthetic states resulting in unnecessary prolonged emergence from anesthesia [3]. An editorial "from crystal ball towards cognitive anesthesiology" [3, 13] indicated that despite the fact that millions of patients being monitored by "depth of hypnosis/anesthesia" devices calculating EEG indices giving easy-to-use valuable information to guide anesthesia delivery [3], pEEG-DOA devices may not always be used appropriately as we do not fully understand these pEEG-DOA indices because of the propriety nature of these algorithms that is still not completely published and their underlying scientific principles are not fully understood by practicing clinicians who might not be familiar with technical issues of EEG processing [3]. Even when using the pEEG-DOA devices in their intended target population of patients under anesthesia, still we often see a disparate clinical picture from what the pEEG-DOA devices show [3].

2.3 National Institute for Health and Care Excellence (NICE) Training Guidelines

One major problem facing all anesthesiologist dealing with the subject is the fact that there is no recognized standard "comparator" recommended by authorized bodies such as United Kingdom's National Institute for Health and Care Excellence (NICE) against which we can properly precisely evaluate all aspects of the performance of processed Electroencephalography-Depth Of Anesthesia (pEEG-DOA) monitors.

In my book I emphasize and expand on areas of new information, chose and cite selected publications that are highly relevant for explaining the background and interpretation of pEEG-DOA devices' performance [3]. Interestingly the only so-called standard "comparator" to evaluate the performance of pEEG-DOA monitors, that was recommended by NICE that adheres to the current guidelines of the United Kingdom's National Health Service (NHS), was delineated as "standard anesthetic clinical monitoring" or "standard clinically judged hypnotic state" demarcated as "combining routinely used clinical-observations with electronic-monitoring" that is applied in anesthesia conduct for evaluating anesthesia appropriateness [3, 12]. Clinical evaluation of the depth of anesthesia is unreliable, and in line with the NICE Diagnostic Assessment Program manual [3, 12], NICE document considers that awareness is likely to occur less frequently with the use of these diagnostic aids. It also recommends more research in this area of electroencephalography (EEG) analysis [3].

The 2012 NICE guidelines clearly stated that all anesthesiologists using depth of anesthesia monitors should be very "well trained and experienced" for the proper usage of depth of anesthesia monitors [12]. Outcomes of anesthesia largely depends, first and foremost, on how precise what the monitors show, coupled with how anesthesiologists interpret these displayed values in making smart decisions and delivering beneficial and skillful anesthesia [3]. How to combine these two factors of accurate monitoring devices displaying precise depth of anesthesia indices and anesthesiologists that can skillfully utilize these numbers is the name of the game [14]. This is perhaps the anesthesiologists' key professional skill [14]. Published depth of anesthesia studies were never fully implemented or utilized to their full potentials, which is the whole idea of these publications namely to better train and experience anesthesiologists in well utilizing the technology [3, 12]. Even though when these depth of anesthesia studies were conducted, their main objectives were not to train or educate practicing anesthesiologists, still that clearly remains of utmost importance [3, 12].

A study by Lindholm and colleagues [15] demonstrated that despite certified registered nurse anesthetists' (CRNAs) were all well experienced in using the Bispectral index to guide anesthetic agents administration and all CRNAs fully understood what the Bispectral Index values meant, still the anesthetic agents administered by CRNAs were not significantly different when the CRNAs delivered anesthesia guided by Bispectral Index values than anesthetic agents delivered when CRNAs were blinded to Bispectral Index values [3, 15]. I know that the study CRNAs might not be fully autonomous decision makers, so it is not fair to put the blame only on them [3]. The utility of using pEEG-DOA indices as a guide of anesthetic agents' titration is depending on the "efficacy" of this pEEG-DOA device in displaying accurate numbers in addition to the ability of practitioners to take titrating anesthesia decisions based on that monitor displayed data [3, 16]. I have also to add that, although a pEEG-DOA device is "efficacious," i.e., can properly function in achieving a certain required objective such as avoiding AAGA occurrence [3], it is not clear whether this outcome necessarily means that this outcome is attained when this device is integrated in daily anesthesia practice, i.e., "effective," or whether incorporating it into routine anesthesia practice is worthwhile, i.e., "cost-effective" [3, 15].

Avidan and colleagues [11, 17] compared BIS-group patients whose anesthesia was adjusted according to Bispectral Index with End-Tidal Anesthetic Gas (ETAG)-group patients whose anesthesia was adjusted according to volatile anesthetic agents. However, they included in their data analysis all BIS-group patients whose Bispectral Index values were more than BIS 60, in an Intention to Treat (ITT) study design, i.e., effectiveness study, rather than excluding those patients in a Per Protocol (PP), i.e., efficacy study design [3, 11, 17], that could have produced a different outcome. Interestingly in their comparisons of BIS-guided versus ETAG-guided anesthesia administration [3], Avidan and colleagues [3, 11, 17] did not reproduce the results of previous studies that reported a lower incidence of Accidental Awareness under General Anesthesia (AAGA) when Bispectral Index was maintained between 40 and 60 [3, 10, 18], as they did not exclude BIS-group patients whose Bispectral Index was more than 60 representing 55.0% of patients, possibly because of CRNAs' difficulty in adhering to the prescribed protocol or reluctance on the CRNAs part to follow the protocol [3, 11]. Intention to Treat studies [11, 17] should not determine cost-effectiveness as ITT is a mere indication of how good practitioners are in keeping Bispectral Index values below 60 [3]. The Bispectral Index device "efficacy" could be just fine, but the CRNAs trained in "how to operate" the Bispectral Index device [11, 17] might not be adequately trained in "competently" maintaining the Bispectral Index values between BIS 40 and BIS 60. In many situations the Bispectral Index device might correctly recognize and display either an over-anesthetized or under-anesthetized state, but the anesthesiologist who is trained in just "how to operate" the Bispectral Index device might not competently titrate his anesthetic agents' administration to sustain the Bispectral Index between BIS 40 and BIS 60 [3].

References

1. Deflandre EP, Javillier BX. Anesthesiologists and their monitoring: benefits, limits and confusing factors. Minerva Anestesiol. 2020;86:250–1.
2. Dahaba AA. Benefits and boundaries of processed electroencephalography (pEEG) monitors when they do not concur with standard anesthetic clinical monitoring: lights and shadows. Minerva Anestesiol. 2020;86:304–16.
3. Dahaba AA. Brain function monitors: are we just "chasing" the numbers? Anesth Analg. 2019;128:1042–4.
4. Shafer A. Metaphor and anesthesia. Anesthesiology. 1995;83:1331–42.
5. Sleigh JW. Depth of anesthesia: perhaps the patient isn't a submarine. Anesthesiology. 2011;115:1149–50.
6. Andrzejowski JC, Wiles MD. Was NAP5 "NICE" enough; where next for depth of anaesthesia monitors? Anaesthesia. 2015;70:514–8.
7. Servin FS, Billard V. IV. Surrogate measures, do they really describe anaesthetic state? Br J Anaesth. 2014;112:787–90.
8. Vuyk J, Lichtenbelt BJ, Vieveen J, Dahan A, Engbers FHM, Burm AGL. Low bispectral index values in awake volunteers receiving a combination of propofol and midazolam. Anesthesiology. 2004;100:179–81.
9. Sigl JC, Chamoun NG. An introduction to bispectral analysis for the electroencephalogram. J Clin Monit. 1994;10:392–404.
10. Myles PS, Leslie K, McNeil J, Forbes A, Chan MTV. Bispectral index monitoring to prevent awareness during anaesthesia: the B-aware randomised controlled trial. Lancet. 2004;363:1757–63.
11. Avidan MS, Zhang L, Burnside BA, Finkel KJ, Searleman AC, Selvidge JA, et al. Anesthesia awareness and the bispectral index. N Engl J Med. 2008;358:1097–108.
12. Smith D, Andrzejowski J, Smith A. Certainty and uncertainty: NICE guidance on "depth of anaesthesia" monitoring. Anaesthesia. 2013;68:1000–5.
13. Jäntti V. From crystal ball towards cognitive anaesthesiology. Acta Anaesthesiol Scand. 2005;49:273–6.
14. Smith AF, Mort M, Goodwin D, Pope C. Making monitoring "work": human-machine interaction and patient safety in anaesthesia. Anaesthesia. 2003;58:1070–8.
15. Lindholm M-L, Brudin L, Sandin RH. Bispectral index monitoring: appreciated but does not affect drug dosing and hypnotic levels. Acta Anaesthesiol Scand. 2008;52:88–94.
16. Escallier KE, Nadelson MR, Zhou D, Avidan MS. Monitoring the brain: processed electroencephalogram and peri-operative outcomes. Anaesthesia. 2014;69:899–910.
17. Avidan MS, Jacobsohn E, Glick D, Burnside BA, Zhang L, Villafranca A, et al. Prevention of intraoperative awareness in a high-risk surgical population. N Engl J Med. 2011;365:591–600.
18. Sebel PS, Bowdle TA, Ghoneim MM, Rampil IJ, Padilla RE, Gan TJ, et al. The incidence of awareness during anesthesia: a multicenter United States study. Anesth Analg. 2004;99:833–9.

3 "Anesthesia" Definitions

3.1 Wakefulness Versus Awareness

The definition and assessment of "consciousness" is still by far one of the most challenging areas of modern contemporary neuroscience. Empirically speaking, consciousness is separated into its two components namely "wakefulness," i.e., "arousal," referring to the "level of consciousness." The other component of consciousness is "awareness" referring to "the contents of consciousness" [1]. Wakefulness is a state in which we "purposefully open our eyes with motor arousal," whereas awareness is the "ability to have some kind of experience" [2]. Wakefulness is relatively easy to measure by confirming that the resting state electroencephalography (EEG) recordings exhibit a typical EEG pattern for the normal waking state. Measuring awareness is far more difficult. Indeed, "objective functional biomarkers" of awareness are still lacking and clinically the ability to follow verbal or behavior commands is just a proxy measure for awareness. However, consistent "following commands" is not always possible in severely brain injured subjects, whose subjective nature undoubtedly contributes to the very well documented high rate of misdiagnosis in those patients' population [3–6].

Proper assessment of Depth of Anesthesia (DOA) ensures optimal conditions for patients' safety and patients' comfort during surgical procedures. However, the DOA definition is still "very vague" and may be simply defined as the interaction of different types of anesthetic agents' effects that induce "unconsciousness" or the complex interaction between several brain neural networks including the autonomous nervous system, hemodynamic system, and motor system all resulting in different states of responsiveness and arousal [7].

To date we have no single cohesive neurophysiological definition of anesthesia end point. We are simply monitoring the associated unconsciousness, unawareness, amnesia, or lack of response to noxious stimulation. A unified standard of a precisely defined end point of anesthesia that corresponds to all anesthetic agents does not exist. Although there is a positive correlation, EEG alone at this time seems incapable of providing all necessary information needed to reliably assess various anesthetic end points. The currently available processed Electroencephalography-Depth Of Anesthesia (pEEG-DOA) devices can only infer some of these end points through EEG and other surrogate parameters such as electromyography (EMG).

What are we exactly quantifying by the "recommended dimensionless Bispectral Index (BIS) 40-60"? In fact, the widely advocated BIS range of 40–60 is a post hoc value, merely suggested by the early studies as a recommendation to prevent Accidental Awareness under General Anesthesia (AAGA) [8, 9], and was not based upon Bispectral Index electroencephalographic analysis [10].

A. A. Dahaba, *Understanding the Numbers in Depth of Anesthesia Monitors*,
https://doi.org/10.1007/978-3-031-88260-9_3

Since the introduction of the Bispectral Index in 1994 [11], we are living in an overcrowded market of fierce competition from at least ten commercially available pEEG-DOA monitors namely, Bispectral Index, State Entropy (SE), Auditory Evoked potential (AEP), Cerebral State Monitor (CSM) Cerebral State Index (CSI), Narcotrend, SEDLine (formerly Patient State Index PSI), NeuroSENSE, SNAP II, qCON 2000, and Index of Consciousness (IoC). To the extent that Massimo brought the discarded Baxter's monitor PSI and revived it under a new name SEDLine.

Let me start by the Gordian Knot that is supposed to be impossible to untie, as Alexander the Great entered the city and saw the Gordian Knot tied to a cart in the market; Alexander the Great solved the problem by slicing the Gordian Knot with his sword (Fig. 3.1). How can we "quantify" the unquantifiable Anesthesia or Analgesia? Anesthesia = Ane +sthesia (no sensation = unconsciousness), Analgesia = An +algesia (No pain) these are absolute negatives = lack of. You can never "quantify" the absolute negative like "I have no money" that is an absolute negative impossible to quantify, so it is impossible to quantify unquantifiable absolute negatives anesthesia or analgesia.

Above Bispectral Index 60, you are trying to quantify the failure of anesthesia or analgesia, as now the patient has neither adequate anesthesia nor analgesia with Bispectral Index >60 when the patients are conscious or feeling pain, so it is not depth of anesthesia but depth of consciousness (= GCS Glaxo Coma Scale) and not depth of analgesia but extent of pain (=VAS visual analogue scale), i.e., measuring failure of anesthesia or measuring failure of analgesia.

On the other hand, if the Bispectral Index set-point is BIS-50 (40–60), which is the drug efficacy level, then what is Bispectral Index between 40 and zero? This is the overdose area of the Burst Suppression (BS) and the Iso-electricity but not the real phenomenon of "unconscious."

Once you surpass your set point (BIS-50), you cannot quantify the overdose, i.e., if I am totally "unconscious" at Bispectral Index 40 or am totally "pain-free," there is nothing beyond that to "quantify" as below BIS-40 I cannot be "unconscious-er"? Or in case of pain, I cannot be "pain-free-er"? So, BIS between 40 and zero is now quantifying something totally different than

Fig. 3.1 Gordian Knot and Alexander the Great. The Gordian Knot is supposed to be impossible to untie. Alexander the Great solved the problem by slicing the Gordian Knot with his sword when he first saw the Gordian knot tied to a cart at the market

"unconsciousness," i.e., the "Bispectral Index between 100-0" is not measuring the same phenomenon from 100 to zero. Rather measuring 3 totally different phenomena namely, 100-60 measuring failure of unconsciousness, 60-40 measuring unconsciousness, and 40-zero measuring overdosing of anesthetic agents. Although processed electroencephalography (pEEG) monitoring devices are calibrated to clinical endpoints like loss of eyelash reflex or response to command, we do not know what the numbers below these thresholds mean in terms of cerebral processing as these numbers were not based on an anesthetic /hypnotic state that goes beyond the recommended Bispectral Index 40 level [12].

What are we monitoring, when we monitor the brain for "anesthesia"? Any answer would probably depend on what exactly do we mean by "anesthesia." That was how an excellent editorial by Pandit [13] questioned few years ago [13]. Pandit editorial [13] stated that before trying to find very specific precise objectives that can quantify brain hypnotic "state of anesthesia" we have first to define what exactly is anesthesia. There is growing consensus that "intra-operative awareness is a wide spectrum of brain states" [13].

What are we exactly monitoring? Are we monitoring (un)consciousness as Pandit's editorial title suggests [13], AAGA, nociceptive arousal, amnesia with explicit-recall, memory loss, prevention of patients moving when under anesthesia in response to surgical stimulation, hindering all the sympathetic reactions to surgical stimuli, Triple-Low effect [14], post-operative cognitive dysfunctions (POCD) [15], prediction of future anesthetic events, or enhanced recovery after anesthesia (ERAS) [15, 16]? Processed electroencephalography monitors currently seem incapable of providing all the information necessary to reliably and separately assess every anesthetic endpoint. This was and still is a heated debate because of the cost-effectiveness involved, as highlighted by the recent fifth National Audit Project (NAP5) recommendations [16].

Are we using pEEG- DOA monitors to help us predict patients' outcomes such as POCD or mortality outcome? An extensive Cochrane-database search revealed that using the Bispectral Index considerably reduced the incidence of "postoperative-delirium" outcome in hospitalized non-Critical Care Unit (CCU) patients [17]. Other trials revealed that Bispectral Index guided anesthesia influenced several patients' outcome-parameters. Infamous Sessler and colleagues [14] "Triple-Low" trial demonstrated that combining a Low-Bispectral Index, Low-Blood Pressure, and Low-Minimum Alveolar Concentration (MAC) correlates to increased death rate [14, 18], indicating that patients' who subsequently perished were more "sensitive" to their anesthetic agents. This was followed by a larger trial "The Balanced-Study" aiming at determining the effects of light-anesthesia versus deep-anesthesia on postoperative one-year survival [7, 19].

If Bispectral Index is specifically quantifying something "quantifiable," then what is the Bispectral Index unit? Like blood pressure (BP) unit is mmHg or temperature unit is degrees Celsius °C or Fahrenheit °F? Why do the Bispectral Index units never correspond to Entropy units or other pEEG-DOA monitors units and why equi-MAC, an acceptable comparable depth of anesthesia, of different inhalational anesthetic agents never yield the same Bispectral Index values?

Now imagine you are driving home, and the coordinates of your home (Point A) and your work (Point B) are on your car's Global Positioning System (GPS). The very first prototype GPS would simply show you the route from Point A to Point B. That was more than 20 years ago; now all fancy GPS Navigators on your fancy Dashboard Screens will offer you the following GPS choices:

1. The shortest land distance between Point A and Point B.
2. Fastest route from Point A to Point B.
3. The best paved way between Point A and Point B.
4. The best scenery view route between Point A and Point B.
5. Avoid the most crowded way, the heaviest traffic between Point A and Point B.

6. What is the most "economic gas consumption" way between Point A and Point B that might be going downhill and not climbing uphill.
7. The "Toll Free" route between Point A and Point B.

Nociception monitors are now spreading like wildfire. What exactly are they monitoring for a patient under anesthesia and cannot appreciate "pain"? Under anesthesia of course there is no "comparator like VAS? Most of these monitors' sensors detect phenomena like changes in galvanic skin resistance or plethysmography that could be "at best" just measuring an event that simply "happens" when "nociception" happens! So, with these monitors, you are simply lacking "direct causality" to pain or noxious stimulation; keeping in mind that once the patient is under anesthesia, everything simultaneously changes at the same time, which of course does not mean direct causality to noxious stimuli rather correlation at best. What do we call the Bispectral Index? The bottom line is the Bispectral Index is a 30-year-old pEEG-DOA algorithm that correlates /corresponds to the increasing doses of two anesthetic agents: isoflurane or propofol when there are no EEG confounding factors.

3.2 Definition of "Depth-of-Anesthesia"

The famous editorial by Sleigh [20], entitled "Perhaps the patient isn't a submarine," starts by saying that the anesthesiology discipline is packed with different "metaphorical descriptions" [21]; however no metaphor is as unescapably alluring as the Depth of Anesthesia (DOA) concept. Everyday anesthetists regularly titrate their anesthetics' dosing aiming at putting their patients at the optimal "DOA" suitable for the surgical procedure they are undergoing. The Sleigh [20] metaphor describes the patient as an analogous of going "under anesthesia" being like a "submarine" that will "dive under water." In that case when we adjust our anesthesia administrations, is just like changing submarine's weight-tanks, to be able to hold patients different "DOAs" equivalent metaphor as how deep submarines are submerged under water [20].

The article by Whitlock and colleagues [22] sets out rigorous criteria for evaluation of the performance of processed Electroencephalography-Depth Of Anesthesia (pEEG-DOA) monitors [23]. Essentially, they advocate that pEEG-DOA monitors must possess two essential features: a sharp pEEG-DOA cut-off point that distinguishes between responsiveness and unresponsiveness, e.g., above /below Bispectral Index 60, and at the same time they should possess another feature of a consistent "gradient curve" representing the relation between anesthetic agents' concentrations and the corresponding pEEG-DOA range at surgical anesthesia, e.g., Bispectral Index between 40 and 60. Unfortunately, the presently available pEEG-DOA monitors seem not to meet these two essential criteria as they do not tend to produce curves that exhibit the expected consistent gradient curves representing the relationship between the increasing anesthetic agents' concentrations such as volatile anesthetics and the supposed gradually decreasing pEEG-DOA-indices [20, 24, 25]. Previous small sample size studies [20, 24, 25] concentrated on basically assessing the first feature of the cut-off point separating the unresponsive status from the awake status, while very few trials have exclusively examined the second feature of how reliable the pEEG-DOA indices track anesthetic /hypnotic agents' concentrations during the surgical phase of anesthesia, e.g., Bispectral Index between 40 and 60 when the patient is already unresponsive [20, 26]. Although there was a clear statistically significant relationship between anesthetic agents' concentrations and Bispectral Index values in the vast majority of patients, still in $\approx 10\%$ of cases the Bispectral Index rises paradoxically with deepening of anesthesia [22, 26]. A clarification of that odd phenomenon would be that all the current pEEG-DOA indices' algorithms are originally intended to fulfill the first criteria of the cut-off point separating the unresponsive state from the awake state, with little regard to the surgical phase of anesthesia, e.g., Bispectral Index between 40 and 60 Dose Response Curves second criteria. Nonetheless, this dysfunction might not be due to the pEEG-DOA devices or the algorithms processing their signals, instead the dysfunction lies with how we as anesthesiologists interpret the entire anesthetic state [20, 26]. This might indicate that, over all these specific anesthetic agents' Bispectral Index 40-60 limits, the cerebral function does not change when the anesthetics' administration is increased, deepening the hypnotic/anesthetic state as the Bispectral Index does not change much between 40 and 60. In other words, deepening of anesthesia via administering higher doses of anesthetics was not the cause of Bispectral Index value decline simply, as Sleigh [20] explained in his own words: "One cannot additionally Switch Off what is at present already Switched Off" [20].

Shafer and Stanski [20, 27] proposed the most insightful multi-dimensional definition of DOA as an alternative to the conventional no-response possibility when a certain stimulus is inflicted. Using the "possibility" when they defined DOA, changes what is "fundamentally" a two-way function of either "yes" or "no" of "consciousness" or "unconsciousness," into a totally different metaphor of a "household-switchboard" and completely does away with the idea of "depth-under-water." In other words, various anesthetics agents perform an act of closing a series of semi-binary switches of several neurological sub-systems. These general anesthesia neurological sub-systems need to be separately switched on-off. Anesthesiologists should not ask themselves the murky question of "anesthesia depth" rather they should be asking a succession of precise questions of all the specific neurologic subsystems most likely to be suppressed when anesthesia is successfully delivered. Each question should be individually answered. Clearly in the current form it is very unlikely that the pEEG-DOA monitors could simultaneously answer all these questions at the same time; hence researchers must attempt looking for a separate individual index for each question:

"Patients' consciousness system"
"Patients' anti-noxious noxious arousal response"
"Patients' memory system"
"Patients' autonomic response"
"Patients' somatic response"
"Patients' post-surgery hyperalgesic response"
"Patients' immune-inflammatory response".

Shafer and Stanski [20, 27] stated that defining "anesthesia depth" requires assessment of three elements: strength of the surgical stimulus, success of response suppression, and effect site concentrations of the drugs that suppress the responses to surgical stimulation. Unfortunately, neither the "stimulus" nor the "response" could be precisely measured to capture the essence of DOA in a scientific or clinical meaningful manner [20, 27]. Hence the "art of clinical anesthesia" requires calibration of the interaction of the stimulus with various responses such as hemodynamic, verbal, and movement responses, while constantly titrating/adjusting the administered anesthetic agents' doses to reduce the probability of these responses. Shafer and Stanski [20, 27] state that the anesthetic state has two components: "hypnosis effect" created via hypnotic / anesthetic agents, volatile anesthetic agents or propofol together with "analgesic effect," Nitrous Oxide (N_2O) or opioids. They demonstrated that very intense degrees of "sedation/hypnosis" without effective "anti-noxious effect" can't obtund response to profound painful stimulation, whereas intense levels of "anti-noxious effect" can't produce "loss of consciousness." There is a certain DOA "sweet spot" where the effects of hypnotic agents and analgesic opioid interact to suppresse hemodynamic response to noxious stimulation and at the same time guarantees unconsciousness [20, 27]. That is why our knowledge of the precise attributes of each anesthetic agent/opioid used in clinical anesthesiology practice must be incorporated into our DOA understanding. Here, the careful combination of anesthetic agent /opioids by skillful anesthesiologist can create a very efficient clinical anesthesia with high safety margin [20, 27].

Anesthesia "Sine Qua Non," "without which not" is "unconsciousness." However, core difficulties lie in the fact that "unconsciousness" cannot be directly measured or quantified. What we can measure are patient's rise in hemodynamic heart rate (HR) or blood pressure (BP) parameters, in responses to noxious surgical stimuli, or patients' response as we call their names, all suggesting conscious perceptions and whether patients retain memory recall of surgical events, conversations, or pain? In 1987, Prys-Roberts [28] broadly defined anesthesia as "non-responsiveness" determined by three factors: the intensity of the noxious stimulation, the strength of the responses, in addition to the type of anesthetic agents' effect site concentrations administered to obtund the responses to noxious stimulation. Consciousness could be assessed, though not in a direct fashion, via analyzing processed Electroencephalography (pEEG) using spectral-EEG based techniques, such as the Bispectral Index that cannot quantify unresponsiveness or unconsciousness. Intense scientific validations of these pEEG-DOA indices could predict whether there is a "possibility of responding" when under propofol or inhaled anesthetics, the two anesthetic agents that were originally used to calibrate the pEEG-DOA indices. The main differences between no-response because of anesthesia and no-response because of natural physiological sleep lie basically in the surgical noxious stimulation that could pierce /infiltrate / breach natural physiological sleep no-response and result in brain's "arousal."

3.3 Hunting of the Snark. Sweetspot of Propofol-Remifentanil Interaction

In Sleigh and Sanders editorial [29], entitled "The Hunting of the Snark", the "Snark" here is a "Chimera" mythological creature composed of two different animal parts half snake and half shark, first coined by Lewis Carroll in his 1876 poem by the same title "The Hunting of the Snark" (Fig. 3.2). In their editorial, Sleigh and Sanders [29] use the Snark to describe combining two different drugs remifentanil opioid with propofol hypnotic to create the perfect "combined drug" that can entirely obtund responses to noxious surgical stimulation while the patient is utterly unconscious [29]. When anesthetic medications fail in suppressing surgical stimulation, this will provoke an arousal reaction manifesting as wakefulness, increased skeletal muscular activity, somatic muscle response like bronchospasm, and autonomic system responses sweating, hypertension, and tachycardia. Patients' responses to surgical stimulation, most of the time is precise, complex and very well defined [29]. However, the question remains of what is the optimal dose combination of analgesics and hypnotics? There is a huge variation in practice among clinical anesthesiologists. Should we give high doses of hypnotics and just a small dose of opioids in "Hypnotic-based anesthesia," or high doses of opioids and small dose of hypnotics in "Opioid-based anesthesia" like in cardiac anesthesia as our opioid ampule preparations appear "Homeopathic" compared to the cardiac anesthesia much larger opioid ampule preparations. The main thing is to determine the exact optimal, quantitatively measured, methods to intricate between the anesthetic agents and the opioids [29].

Modern clinical anesthesiology, that is composed of four main components, Antinociception /Hypnosis /Relaxation /Areflexia, is like four blind-folded people each touching one foot of an elephant? What would each of the four blind-folded people say? (Fig. 3.3). The four answers would depend on what each one of the four individuals considers which of the four anesthesia components Antinociception /Hypnosis / Relaxation /Areflexia comes first and would answer accordingly. A most interesting Sahinovic and colleagues [30] trial investigated electroencephalography (EEG) and skeletal muscle movement responses as a result of the balance between nociception-antinociception and hypnosis in patients under general anesthesia. Sahinovic and colleagues [30] studied four clinically relevant remifentanil concentrations at three degrees of propofol hypnotic /anesthetic states namely, "sedation," "surgical anesthesia," and "profound anesthesia," to evaluate the opioids-hypnotics interactions [29, 30]. On one end, at the level of just light propofol "sedation," all remifentanil doses could not block the EEG measure and skeletal muscle movement responses in all patients.

Fig. 3.2 The Hunting of the Snark. Lewis Carroll's 1876 poem. The Snark here is a Chimera mythological creature composed of two different animal parts, half snake and half shark, first coined by Lewis Carroll in his 1876 poem with the same title "The Hunting of the Snark". Here the Snark is used to describe combining two different drugs remifentanil opioid with propofol hypnotic to create the Snark representing the perfect drug that can completely obtund all responses to noxious surgical stimulation while the patient is perfectly unconscious under general anesthesia

On the other end at "profound anesthesia," very high doses of propofol alone could not completely block all noxious surgical stimulation responses while very high doses of remifentanil could not ensure full deep hypnosis /anesthesia. Interestingly, a "Sweet Spot" existed at mid-grade anesthetic level created via maximum synergistic interaction between remifentanil and propofol where a minimal of 2 ng mL^{-1} remifentanil concentrations were sufficient in obtunding skeletal movements response and EEG response, but what are the basis of these neuropharmacological mechanisms causing these hypnotics-opioids interactions [29]? Sahinovic and colleagues [30] model could explain the synergy between propofol and opioids as opioids significantly shifted propofol dose-response-curve to the left and at the same time "steepened" the curve [31]. Opioids could block the nociceptive inputs reaching the cerebral cortical structures; thus, patients could be maintained at the "unresponsiveness state" using much less propofol concentrations than without remifentanil opioid. Apparently, this confirmed the view that propofol is acting "top-down cortical-subcortical" to explicitly obliterate cerebral response [32], while opioids act predominantly "bottom-up brain stem-cortical" [33, 34]. In clinical concentrations, remifentanil alone is incapable of completely blocking response reactions to noxious stimulation [31]. Despite not being an "analgesic," propofol has an antinociceptive effect, as it interacts with brainstem's opioid systems [35, 36] possibly via breaking the hippocampal-antinociception connectivity [37], whereas remifentanil antinociception positive modulation is in the form of ventral tegmental activation of dopaminergic neurons [38], or by breaking the con-

Fig. 3.3 Four blind-folded people each touching one foot of an elephant. Modern clinical anesthesiology, composed of four main components: Antinociception, Hypnosis, Relaxation and Areflexia, is like four blind-folded people each touching one foot of an elephant. What would each say?

nectivity between the brainstem center and anterior cingulate cortex [29, 39, 40]. Sahinovic and colleagues [30] study demonstrated that, to some extent, when increasing propofol concentrations more than what is necessary for patients to lose verbal command responses, propofol is capable of greatly amplifying remifentanil antinociceptive effect. We can say that, not only remifentanil is increasing the potency of propofol hypnotic effect, but propofol is also increasing the potency of remifentanil antinociceptive effect. Sanders and colleagues, [41] in their editorial "unresponsiveness ≠unconsciousness" suggested that we could achieve intense antinoxious level meaning being capable of totally ignoring the surgery noxious stimulation without losing conscious like that in Isolated Forearm Technique (IFT) studies [29].

Kern and colleagues [42] characterized the pharmacodynamic synergy of remifentanil and propofol drug interactions over a full range of clinically relevant concentrations using a "response-surface study paradigm" of surrogate noxious stimulation in 12 female and 12 male volunteers. In each subject, they used remifentanil and propofol Target-Controlled-Infusions (TCI) to create three different increasing staircase drug concentration profiles: one profile using a single agent and two profiles using different combinations of remifentanil and propofol [42]. They used four different surrogate measures to assess the drug effects at each staircase-profile plateau namely laryngoscopy response, Observer's Assessment of Alertness/Sedation (OAA/S) Scale, electrical tetany, and tibial pressure algometry. In all subjects they created "response surfaces" mapping of remifentanil and propofol interactions to those four surrogate effect measures. Using an interaction parameter, they assessed whether remifentanil and propofol behaved synergistically to blunt responses to noxious stimulation. In their study [42], the response-surfaces demonstrated substantial synergy between propofol and remifentanil in blunting responses to noxious stimulations. Using the Greco interaction model, the "Interaction Index," a measure of synergy between propofol and remifentanil, was highly significant ($P < 0.001$) 33.2

for response to laryngoscopy, 14.7 for response to electrical tetany, 8.2 for response to algometry, and 5.1 for response to OAA/S sedation, when the 4 surrogate stimuli were mapped for clinically relevant propofol and remifentanil combinations [42]. They concluded that their "response surface models" demonstrated enormous synergy between propofol and remifentanil as their surface morphologic features clearly revealed the relative contributions of analgesia and sedation to response blunting [42].

A far better analogy describing the cerebral mental status of general anesthesia is to use the analogy with which we describe the cardiovascular system that consists of numerous variables that are separately monitored by skilled well experienced physicians using different endpoints. This intellectualization of general anesthesia being an amalgamated mixture of various pharmacologic influences implies that our "concept of conscious" should be approached just like we approach all the other comprehensive multifold physiologic systems. For instance, we handle the hemodynamic cardiac system as an amalgamation of numerous fundamental parameters like venous return, cardiac filling, blood pressure, heart rate, and cardiac output each requiring its own monitoring technique and each measured using its own individual parameter [13], that requires individual monitoring [20]; the same may be true for "unconsciousness."

A cerebral monitoring device that depicts the "conscious state" could be best described as a device that can generally make an educated estimate of the possibility /probability of being "conscious." Such a device is certainly not telling us all the different factors that contribute to the state of "consciousness" or which of these factors, at any given time, are missing and which are still preserved. Because there is an array of factors contributing to the brain states under anesthesia, then these "numerical ranges" are possibly very good at delineating some of these elements but very poor at demarcating others, i.e., a certain "index range" could accurately quantify a certain parameter and is totally incapable of accurately demarcating all the other. We do not expect one size (range) to "fit all" or a single Bispectral Index "assigned range" to accurately demarcate all the other endpoint parameters. The good enough Bispectral Index 40–60 for Accidental Awareness under General Anesthesia (AAGA) might not be good enough for suppression of movement responses as Professor Werner F. List (my former Boss) always said. Those different "parameters" measuring diverse endpoints are then characterizing separate cerebral functions; just like a system with multiple "light switches" that need not be switched on/off together all at the same time, you might choose to switch on some and switch off some other [20]. The results of the fifth National Audit Project (NAP5) study on intraoperative awareness [43] clearly suggest that, instead of being a singular "anesthetic state," there is in fact a spectrum of experiences associated with "not being properly anesthetized," and a third of which involve distress or pain [13]. Shafer and Stanski [27] have brilliantly argued that anesthetic depth can be regarded as separately attaining several different endpoints relevant to anesthesia. Sleigh [20] has made the analogy of anesthesia when he portrayed the process of anesthesia as the process of the switch-off progression of a series of separate light-switches, each has its own distinct purpose and performs a certain function, for example nociception -antinociception, recall, or autonomic response. All of this makes our approach to unconsciousness should closely resemble our approach to other physiological systems such as the cardiovascular system [20].

The results of the analysis by Mathews and colleagues [44] of a variable that could accurately measure nociception /antinociception balance consists of the weighed contributions of three variability metrics namely Standard Deviation of the Bispectral Index (sBIS), Standard Deviation of Electromyography (sEMG), and the Bispectral Index (BIS), measuring the electroencephalogram and electromyogram variability thus quantitatively delivering an index combining variabilities of both the Bispectral Index and Electromyography (EMG) in the so-called Composite Variability Index (CVI) a single metric that measures variability in the signal and is scaled from 100 to 0 zero [44].

Two of the CVI descriptors the sEMG and sBIS [44], via measuring their Standard Deviations could quantify the variability of electromyography and the Bispectral Index respectively, as the Standard Deviations of the two metrics calculated over the preceding 3-min provides the optimal discriminative ability between "somatic occurring events" and "no-events" [44].

3.4 Accidental Awareness Under General Anesthesia (AAGA) Without Recall

Tunstall [45], Isolated Forearm Technique (IFT) first introduced in 1977 [45], showed that patients undergoing Cesarean Section could be "Awake" during halothane-Nitrous Oxide (N_2O) general anesthesia as indicated by purposefully squeezing the anesthetist's hand. This was achieved in 39 out of 56 subjects, yet none of the patients had an explicit or implicit recall of any of the intraoperative events [45]. Using the "Tunstall ME Isolated Forearm Technique" [45, 46], that inflates a sphygmomanometer cuff to block the effect of Neuromuscular Blocking Agents (NMBAs) from reaching patients' arm, we find that patients who were under deep general anesthesia could smoothly respond by purposefully squeezing the anesthetists' hand when the Bispectral Index (BIS) values at induction, laryngoscopy, and intubation were 37, 45, and 46 respectively [45, 47], yet none of the patients had an explicit or implicit recall of any of the intraoperative events [47].

Ekman and colleagues [48], in an extended study in 4945 patients maintained within a Bispectral Index value of 40–60, reported only 2 cases (0.04%) of awareness when the Bispectral Index value was >60. This was significantly lower than the previously reported [49] 0.18% incidence of awareness in the non-Bispectral Index monitored patients, demonstrating that Bispectral Index monitoring reduces the incidence of awareness [49].

Using the semi-direct questioning "Brice interview methodology" [50], for actively seeking Accidental Awareness under General Anesthesia (AAGA) cases, yielded an incidence of ≈1 in every 600 general anesthesia [8, 9, 49–52]. The fifth National Audit Project (NAP5) [53] reported a far lower incidence of ≈1 in every 19,000 general anesthesia [53]. Avidan and Sleigh [54] editorial attributed the disparity between results among different studies to the different methodologies used, as the NAP5 was reliant on population sampling of all general anesthesia survivors from the year 1946 who did not report intra-operative recall when asked by the study researchers [54]. Interestingly, the NAP5 stated that just 2.8% of general anesthesia in the United Kingdom involved Depth of Anesthesia (DOA) monitoring [7, 53].

Because the incidence of AAGA without recall is far more common than the incidence of AAGA with-recall, the best method to systematically examine such a "phenomenon" is to use the IFT first described by Tunstall [45] in 1977, where the arm becomes isolated from the circulating administered NMBAs' effect via applying a sphygmomanometer-cuff on the arm before NMBA administration and the subject is then asked to wiggle their fingers on command to detect AAGA [45]. The IFT is appropriately considered the Gold-Standard for measuring the wakeful state under general anesthesia [45], still the NAP5 [53] reported that just 0.03% of United Kingdom anesthetics utilized the IFT [16, 45]. Lately, the Association of Anaesthetists of Great Britain and Ireland (AAGBI) [55, 56] acknowledged that IFT might be useful when the responses are carefully interpretated [45, 55].

Russell [56] explored whether the Bispectral Index monitoring is capable of detecting AAGA in comparison with the IFT [56], in 22 patients who underwent gynecological surgical procedures under propofol/remifentanil Target-Controlled-Infusion (TCI) and atracurium neuromuscular block. Total intravenous anesthesia (TIVA) was adjusted to maintain a reportedly "difficult to maintain" Bispectral Index at 55–60. The Bispectral Index was >60 during 47 /80 IFT hand movements seen in 16 /22 women, giving the Bispectral Index monitor 59% awareness sensitivity [56]. Postoperatively, no study patient experienced "awareness-with-recall" during their surgical procedure. The study demonstrated that a Bispectral Index closely maintained at a tight predetermined 55–60 range could not be a reliable monitoring for patients being awake when applied in that fashion, and Bispectral Index maintained around BIS 60 could easily result in patients being awake but, in that case, "without recall" [7, 56].

Several researchers investigated AAGA without-recall using the IFT [45, 46]. Schneider and colleagues [57] investigated whether a Bispectral Index maintained in the range between 50 and 60 could prevent arousal resulting from tracheal-intubation in 20 patients under general anesthesia induced with alfentanil 10 μg kg^{-1}, and followed by alfentanil 2 μg kg^{-1} min^{-1} together with propofol infusions titrated for maintaining Bispectral Index at the range of 50 and 60 for 5-min. Using Tunstall's IFT, subjects were assessed every 1-min pre-endotracheal-intubation and for 3-min after intubation [57]. Eight /20 (40%) patients showed an awareness reaction, squeezing the investigator's hand in response to command. None of the study patients recalled these responses postoperatively. There was no significant difference in Bispectral Index values between patients with and without awareness reaction [57]. They concluded that the Bispectral Index values between 50 and 60 before endotracheal intubation is not sufficient in preventing wakefulness reactions during endotracheal intubation under alfentanil propofol general anesthesia. They concluded that because the Bispectral Index was not designed or validated to distinguish between subjects who exhibit wakefulness reactions, its monitoring for awareness significance must be questioned [7, 57].

On the other hand, three volunteers, under propofol-midazolam anesthesia [58], were reported to remain responsive to verbal command at a Bispectral Index value of as low as 40 [58]. Similarly, Russell and colleagues [59] demonstrated in 12 women, using IFT, that subjects responded to command during their surgical procedures with no noticeable change in their vital signs. The 12 subjects showed a total of 92 responses during their surgical procedures [59]. They concluded that that a Bispectral Index value of 40–60 is inadequate in preventing wakefulness and to ensure that awareness or explicit recall did not occur [49].

Zand and colleagues [47] compared Bispectral Index values with IFT responses in 61 women undergoing Cesarian Sections under standardized anesthetic techniques. Forty-one percent, 46%, and 23% of parturient subjects exhibited a positive IFT response even though the Bispectral Index mean values at laryngoscopy, endotracheal intubation, and surgical incision were 45, 48, and 46 respectively [47]. They concluded that the Bispectral Index was not able to consistently distinguish the IFT-responder subjects from non-IFT-responder subjects at these three phases. As the subjects were questioned after 12–24 h following the surgical procedure, no women recalled any intraoperative incidences [7, 47].

Because Bispectral Index monitoring seems to be not effective in tracking rapid changes during anesthesia induction, and a poor detector of wakefulness at this dynamic stage, but a good detector of wakefulness in the steady state of anesthesia maintenance, one can always claim that because "recall" was not reported in any IFT trial, this means that the Bispectral Index was effective in performing the expected function it was designed and validated for. However, because IFT positive subjects did not spontaneously move indicating their wakefulness [60], this means that the cognitive function of IFT positively responding to command could not be considered a true "awareness state" like a fully awake subject that was not given any anesthetics /sedatives /hypnotics. This contradiction could signify a newly devised unique cognitive status coined as "Dysanesthesia" [61, 62], that the Bispectral Index monitor was never constructed or validated to identify [7].

3.5 Accidental Awareness Under General Anesthesia (AAGA) with Implicit or Explicit Recall

The first case by Mychaskiw and colleagues [63] of Bispectral Index (BIS) monitored "explicit recall" of intra-operative events under sevoflurane-Nitrous Oxide (N_2O) anesthesia took place during the sternal splitting for a cardiopulmonary bypass (CPB) cardiac surgery despite a Bispectral Index value of 47 [63]. However, Ira Rampil [64] pointed out that reporting a Bispectral Index value of 47 by an anesthesia resident handwritten anesthesia record is not a reliable way of retrieving data and pinpointing time points since the Bispectral Index value is a continuously changing variable that is updated every 2 s with great moment-to-moment variability [64]. The time of sternal splitting is also the time of frequent application of electrocautery affecting the Bispectral Index signal quality [64]. The Bispectral Index values at the time of the supposed explicit awareness episode were, in fact, Bispectral Index 63 when the sevoflurane concentrations could have been just in the range of 0.8% [64].

Large multicenter trials investigated the use of Bispectral Index monitoring for preventing Accidental Awareness under General Anesthesia (AAGA). The B-Aware trial [9] examined 2463 patients who were at high risk of AAGA under general anesthesia and neuromuscular block [9]. In postoperative interviews, among 13 confirmed cases of AAGA, only 2/1225 patients were in the Bispectral Index monitored group compared with 11/1238 in the not Bispectral Index monitored Standard of Care (SOC) group. Thus, Bispectral Index guided general anesthesia resulted in 82% AAGA incidence reduction in "high-risk" patients [9].

Ekman and colleagues [48], in the Swedish SAFE-2 prospective trial, interviewed for AAGA, 4945 surgical patients maintained at Bispectral Index 40–60 and requiring Neuromuscular Blocking Agents (NMBAs), on 3 separate occasions [48]. Two /4945 Bispectral Index monitored patients (0.04%) had awareness with explicit intraoperative recall during intubation when their Bispectral Index values were > 60 for 4-min and >10 min. However, periods with high Bispectral Index >60 for >4-min were also seen in other patients with no awareness with explicit recall [48]. Their results [48] indicated that patients in the Bispectral Index monitored group had significantly lower incidence of AAGA than the historical 17/11785 (0.18%) similar cases in a previous study with 77% statistically significant reduction ($P < 0.038$) when no Bispectral Index monitoring was used in the Sandin and colleagues SAFE-1 previous study [49]. They concluded that the use of Bispectral Index monitoring during general anesthesia requiring endotracheal intubation and NMBAs was associated with a significantly lower incidence of awareness, compared with a historical control population [48, 49].

Avidan and colleagues [52], in their New England Journal of Medicine 2008 groundbreaking B-Unaware [52] publication, compared a Bispectral Index maintained between 40 and 60 in the BIS-group, with End-Tidal Anesthetic Gas (ETAG) maintained between 0.7 and 1.3 Minimum Alveolar Concentration (MAC) in the ETAG-group, in 2000 patients considered at high AAGA risk [52]. In their postoperative interviews, 2/1000 definite cases of AAGA with explicit recall were found in each of the two groups. The authors were the first to conclude that the use of Bispectral Index monitoring during general anesthesia requiring endotracheal intubation and NMBAs did not significantly decrease the incidence of awareness when compared with the use of ETAG concentration at 0.7–1.3 MAC guided anesthesia [52]. The study at the time of publication provoked many criticism due to the fact that the 2 patients in the BIS-group were both inadvertently allowed to be above their strict Bispectral Index 60 cut-off target and that the Certified Registered Nurses of Anesthesia (CRNAs), who might have not been adequately trained in maintaining the Bispectral Index between 40 and 60, are the ones who administered the study's general anesthesia [52]. The authors subsequently followed their groundbreaking study with a bigger 2011 three centers multicenter trial, the BAG-RECALL trial [65, 66], where they fully addressed previously identified weaknesses of their first B-Unaware 2008 study as they increased their sample-size 3 folds in 3 centers. Still outcome of the new multicenter study [65, 66] revealed no statistically significant differences in AAGA incidence between the ETAG-group 0.7–1.3 MAC guided and the

BIS-group Bispectral Index 40–60 guided group. This further emphasized their earlier deduction of ETAG 0.7–1.3 MAC guided was as efficient as Bispectral Index 40–60 guided in preventing AAGA [65, 66]. Accordingly, an alarm set to sound for anesthesiologists administering inhalational general anesthesia when MAC declined below 0.7 MAC appears to be a sensible procedure in avoiding AAGA with explicit or implicit recalls. However, the matter is still not so “simple” as one would think because 1 MAC is the ETAG concentration that prevents skeletal muscle movement, in response to noxious surgical stimuli, in 50% of patients; then the real aim is the 1.3 MAC which is the ETAG concentration that prevents skeletal muscle movement in response to noxious surgical stimuli in 95% of the patients. Obviously, the aim is to prevent skeletal muscle movement in response to noxious surgical stimuli in 95% and not 50% of the patients. The 1.3 MAC is conventionally achieved with 0.7 MAC inhalational gas +0.5 MAC (with an additional analgesic effect) of Nitrous Oxide (N_2O) 1:1 in Oxygen (O_2), which could then be easily and routinely substituted by “0.5 MAC-equivalent” opioid analgesics in adequately high doses. That is why there is still a debate of whether this “0.5 MAC-equivalent” should be substituted with adequately high opiate doses, or whether patients could hemodynamically tolerate that [7].

A literature search systematic review meta-analysis examined all available 36 Randomized Controlled Trials (RCT) that compared Bispectral Index guided and Standard of Care guided general anesthesia [67]. The meta-analysis systematic review outcome revealed a Bispectral index guided anesthesia significantly decreased AAGA incidence in high risk AAGA subjects, when compared with the AAGA incidence in high risk AAGA subjects in trials that utilized just ETAG guided anesthesia [67]. Their meta-analysis systematic review revealed that the use of Bispectral Index guided anesthesia also required significantly fewer intravenous or volatile anesthetic agents, while resulting in significantly shortening the emergence time from general anesthesia [7, 67].

The United Kingdom National Institute for Health and Care Excellence (NICE) in their 2012 guidelines [12] recommend that the Bispectral index, E-Entropy, or Narcotrend should be considered in any patient who is at high risk of over- or under-dosing of anesthesia, and specifically recommended Depth of Anesthesia (DOA) monitoring in patients receiving Total Intravenous Anesthesia (TIVA) [12]. The fifth National Audit Project (NAP5) recommendations were broadly in agreement and recommended routine DOA use for TIVA when NMBAs were used [53].

References

1. Plum F, Posner JB. The diagnosis of stupor and coma. Contemp Neurol Ser. 1972;10:1–286.
2. Working Party of the Royal College of Physicians. The vegetative state: guidance on diagnosis and management. Clin Med (Lond). 2003;3:249–54.
3. Andrews K, Murphy L, Munday R, Littlewood C. Misdiagnosis of the vegetative state: retrospective study in a rehabilitation unit. BMJ. 1996;313:13–6.
4. Childs NL, Mercer WN, Childs HW. Accuracy of diagnosis of persistent vegetative state. Neurology. 1993;43:1465–7.
5. Schnakers C, Vanhaudenhuyse A, Giacino J, Ventura M, Boly M, Majerus S, et al. Diagnostic accuracy of the vegetative and minimally conscious state: clinical consensus versus standardized neurobehavioral assessment. BMC Neurol. 2009;9:35.
6. Fernández-Espejo D, Owen AM. Detecting awareness after severe brain injury. Nat Rev Neurosci. 2013;14:801–9.
7. Hajat Z, Ahmad N, Andrzejowski J. The role and limitations of EEG-based depth of anaesthesia monitoring in theatres and intensive care. Anaesthesia. 2017;72(Suppl 1):38–47.
8. Sebel PS, Bowdle TA, Ghoneim MM, Rampil IJ, Padilla RE, Gan TJ, et al. The incidence of awareness during anesthesia: a multicenter United States study. Anesth Analg. 2004;99:833–9.
9. Myles PS, Leslie K, McNeil J, Forbes A, Chan MTV. Bispectral index monitoring to prevent awareness during anaesthesia: the B-aware randomised controlled trial. Lancet. 2004;363:1757–63.
10. Rampil IJ. A primer for EEG signal processing in anesthesia. Anesthesiology. 1998;89:980–1002.
11. Sigl JC, Chamoun NG. An introduction to bispectral analysis for the electroencephalogram. J Clin Monit. 1994;10:392–404.
12. Smith D, Andrzejowski J, Smith A. Certainty and uncertainty: NICE guidance on “depth of anaesthesia” monitoring. Anaesthesia. 2013;68:1000–5.

13. Pandit JJ. Monitoring (un)consciousness: the implications of a new definition of "anaesthesia". Anaesthesia. 2014;69:801–7.
14. Sessler DI, Sigl JC, Kelley SD, Chamoun NG, Manberg PJ, Saager L, et al. Hospital stay and mortality are increased in patients having a "triple low" of low blood pressure, low bispectral index, and low minimum alveolar concentration of volatile anesthesia. Anesthesiology. 2012;116:1195–203.
15. Chan MTV, Cheng BCP, Lee TMC, Gin T, CODA Trial Group. BIS-guided anesthesia decreases postoperative delirium and cognitive decline. J Neurosurg Anesthesiol. 2013;25:33–42.
16. Cook TM, Andrade J, Bogod DG, Hitchman JM, Jonker WR, Lucas N, et al. The 5th National Audit Project (NAP5) on accidental awareness during general anaesthesia: patient experiences, human factors, sedation, consent and medicolegal issues. Anaesthesia. 2014;69:1102–16.
17. Siddiqi N, Harrison JK, Clegg A, Teale EA, Young J, Taylor J, et al. Interventions for preventing delirium in hospitalised non-ICU patients. Cochrane Database Syst Rev. 2016;3:CD005563.
18. Leslie K, Myles PS, Forbes A, Chan MTV. The effect of bispectral index monitoring on long-term survival in the B-aware trial. Anesth Analg. 2010;110:816–22.
19. Short TG, Leslie K, Chan MTV, Campbell D, Frampton C, Myles P. Rationale and design of the balanced anesthesia study: a prospective randomized clinical trial of two levels of anesthetic depth on patient outcome after major surgery. Anesth Analg. 2015;121:357–65.
20. Sleigh JW. Depth of anesthesia: perhaps the patient isn't a submarine. Anesthesiology. 2011;115:1149–50.
21. Shafer A. Metaphor and anesthesia. Anesthesiology. 1995;83:1331–42.
22. Whitlock EL, Villafranca AJ, Lin N, Palanca BJ, Jacobsohn E, Finkel KJ, et al. Relationship between bispectral index values and volatile anesthetic concentrations during the maintenance phase of anesthesia in the B-unaware trial. Anesthesiology. 2011;115:1209–18.
23. Palanca BJA, Mashour GA, Avidan MS. Processed electroencephalogram in depth of anesthesia monitoring. Curr Opin Anaesthesiol. 2009;22:553–9.
24. Bruhn J, Kreuer S, Bischoff P, Kessler P, Schmidt GN, Grzesiak A, et al. Bispectral index and A-line AAI index as guidance for desflurane-remifentanil anaesthesia compared with a standard practice group: a multicentre study. Br J Anaesth. 2005;94:63–9.
25. McKay IDH, Voss LJ, Sleigh JW, Barnard JP, Johannsen EK. Pharmacokinetic-pharmacodynamic modeling the hypnotic effect of sevoflurane using the spectral entropy of the electroencephalogram. Anesth Analg. 2006;102:91–7.
26. Kreuer S, Bruhn J, Larsen R, Bauer C, Wilhelm W. Comparison of BIS and AAI as measures of anaesthetic drug effect during desflurane-remifentanil anaesthesia. Acta Anaesthesiol Scand. 2004;48:1168–73.
27. Shafer SL, Stanski DR. Defining depth of anesthesia. Handb Exp Pharmacol. 2008:409–23.
28. Prys-Roberts C. Anaesthesia: a practical or impractical construct? Br J Anaesth. 1987;59:1341–5.
29. Sleigh JW, Sanders RD. Intraoperative analgesic titration: the hunting of the Snark. Anesth Analg. 2014;119:234–6.
30. Sahinovic MM, Eleveld DJ, Kalmar AF, Heeremans EH, De Smet T, Seshagiri CV, et al. Accuracy of the composite variability index as a measure of the balance between nociception and antinociception during anesthesia. Anesth Analg. 2014;119:288–301.
31. Bouillon TW, Bruhn J, Radulescu L, Andresen C, Shafer TJ, Cohane C, et al. Pharmacodynamic interaction between propofol and remifentanil regarding hypnosis, tolerance of laryngoscopy, bispectral index, and electroencephalographic approximate entropy. Anesthesiology. 2004;100:1353–72.
32. Monti MM, Lutkenhoff ES, Rubinov M, Boveroux P, Vanhaudenhuyse A, Gosseries O, et al. Dynamic change of global and local information processing in propofol-induced loss and recovery of consciousness. PLoS Comput Biol. 2013;9:e1003271.
33. Wagner KJ, Sprenger T, Kochs EF, Tölle TR, Valet M, Willoch F. Imaging human cerebral pain modulation by dose-dependent opioid analgesia: a positron emission tomography activation study using remifentanil. Anesthesiology. 2007;106:548–56.
34. Wanigasekera V, Lee MC, Rogers R, Kong Y, Leknes S, Andersson J, et al. Baseline reward circuitry activity and trait reward responsiveness predict expression of opioid analgesia in healthy subjects. Proc Natl Acad Sci USA. 2012;109:17705–10.
35. Lee J-J, Hahm E-T, Min B-I, Han S-H, Cho J-J, Cho Y-W. Roles of protein kinase A and C in the opioid potentiation of the GABAA response in rat periaqueductal gray neuron. Neuropharmacology. 2003;44:573–83.
36. Xiao C, Zhou C, Atlas G, Delphin E, Ye JH. Labetalol facilitates GABAergic transmission to rat periaqueductal gray neurons via antagonizing beta1-adrenergic receptors–a possible mechanism underlying labetalol-induced analgesia. Brain Res. 2008;1198:34–43.
37. Liu X, Pillay S, Li R, Vizuete JA, Pechman KR, Schmainda KM, et al. Multiphasic modification of intrinsic functional connectivity of the rat brain during increasing levels of propofol. NeuroImage. 2013;83:581–92.
38. Li K-Y, Xiao C, Xiong M, Delphin E, Ye J-H. Nanomolar propofol stimulates glutamate transmission to dopamine neurons: a possible mechanism of abuse potential? J Pharmacol Exp Ther. 2008;325:165–74.
39. Långsjö JW, Alkire MT, Kaskinoro K, Hayama H, Maksimow A, Kaisti KK, et al. Returning from oblivion: imaging the neural core of consciousness. J Neurosci. 2012;32:4935–43.
40. Gili T, Saxena N, Diukova A, Murphy K, Hall JE, Wise RG. The thalamus and brainstem act as key hubs in alterations of human brain network connectivity

induced by mild propofol sedation. J Neurosci. 2013;33:4024–31.
41. Sanders RD, Tononi G, Laureys S, Sleigh JW. Unresponsiveness ≠ unconsciousness. Anesthesiology. 2012;116:946–59.
42. Kern SE, Xie G, White JL, Egan TD. A response surface analysis of propofol-remifentanil pharmacodynamic interaction in volunteers. Anesthesiology. 2004;100:1373–81.
43. Pandit JJ, Cook TM, Jonker WR, O'Sullivan E. 5th National Audit Project (NAP5) of the Royal College of Anaesthetists and the Association of Anaesthetists of Great Britain and Ireland. A national survey of anaesthetists (NAP5 baseline) to estimate an annual incidence of accidental awareness during general anaesthesia in the UK. Anaesthesia. 2013;68:343–53.
44. Mathews DM, Clark L, Johansen J, Matute E, Seshagiri CV. Increases in electroencephalogram and electromyogram variability are associated with an increased incidence of intraoperative somatic response. Anesth Analg. 2012;114:759–70.
45. Tunstall ME. Detecting wakefulness during general anaesthesia for caesarean section. Br Med J. 1977;1:1321.
46. Tunstall ME. The reduction of amnesic wakefulness during caesarean section. Anaesthesia. 1979;34:316–9.
47. Zand F, Hadavi SMR, Chohedri A, Sabetian P. Survey on the adequacy of depth of anaesthesia with bispectral index and isolated forearm technique in elective Caesarean section under general anaesthesia with sevoflurane. Br J Anaesth. 2014;112:871–8.
48. Ekman A, Lindholm M-L, Lennmarken C, Sandin R. Reduction in the incidence of awareness using BIS monitoring. Acta Anaesthesiol Scand. 2004;48:20–6.
49. Sandin RH, Enlund G, Samuelsson P, Lennmarken C. Awareness during anaesthesia: a prospective case study. Lancet. 2000;355:707–11.
50. Brice DD, Hetherington RR, Utting JE. A simple study of awareness and dreaming during anaesthesia. Br J Anaesth. 1970;42:535–42.
51. Wennervirta J, Ranta SO-V, Hynynen M. Awareness and recall in outpatient anesthesia. Anesth Analg. 2002;95:72–7. table of contents
52. Avidan MS, Zhang L, Burnside BA, Finkel KJ, Searleman AC, Selvidge JA, et al. Anesthesia awareness and the bispectral index. N Engl J Med. 2008;358:1097–108.
53. Pandit JJ, Andrade J, Bogod DG, Hitchman JM, Jonker WR, Lucas N, et al. The 5th National Audit Project (NAP5) on accidental awareness during general anaesthesia: summary of main findings and risk factors. Anaesthesia. 2014;69:1089–101.
54. Avidan MS, Sleigh JW. Beware the Boojum: the NAP5 audit of accidental awareness during intended general anaesthesia. Anaesthesia. 2014;69:1065–8.
55. Checketts MR. AAGBI recommendations for standards of monitoring during anaesthesia and recovery 2015. Anaesthesia. 2016;71:470–1.
56. Russell IF. The ability of bispectral index to detect intra-operative wakefulness during total intravenous anaesthesia compared with the isolated forearm technique. Anaesthesia. 2013;68:502–11.
57. Schneider G, Wagner K, Reeker W, Hänel F, Werner C, Kochs E. Bispectral index (BIS) may not predict awareness reaction to intubation in surgical patients. J Neurosurg Anesthesiol. 2002;14:7–11.
58. Vuyk J, Lichtenbelt BJ, Vieveen J, Dahan A, Engbers FHM, Burm AGL. Low bispectral index values in awake volunteers receiving a combination of propofol and midazolam. Anesthesiology. 2004;100:179–81.
59. Russell IF. The Narcotrend "depth of anaesthesia" monitor cannot reliably detect consciousness during general anaesthesia: an investigation using the isolated forearm technique. Br J Anaesth. 2006;96:346–52.
60. Pandit JJ, Russell IF, Wang M. Interpretations of responses using the isolated forearm technique in general anaesthesia: a debate. Br J Anaesth. 2015;115(Suppl 1):i32–45.
61. Pandit JJ. Acceptably aware during general anaesthesia: 'dysanaesthesia' – the uncoupling of perception from sensory inputs. Conscious Cogn. 2014;27:194–212.
62. Pandit JJ. Isolated forearm — or isolated brain? Interpreting responses during anaesthesia — or "dysanaesthesia". Anaesthesia. 2013;68:995–1000.
63. Mychaskiw G, Horowitz M, Sachdev V, Heath BJ. Explicit intraoperative recall at a Bispectral index of 47. Anesth Analg. 2001;92:808–9.
64. Rampil I. False negative BIS? Maybe, maybe not! Anesth Analg. 2001;93:798–9.
65. Avidan MS, Jacobsohn E, Glick D, Burnside BA, Zhang L, Villafranca A, et al. Prevention of intraoperative awareness in a high-risk surgical population. N Engl J Med. 2011;365:591–600.
66. Avidan MS, Palanca BJ, Glick D, Jacobsohn E, Villafranca A, O'Connor M, et al. Protocol for the BAG-RECALL clinical trial: a prospective, multicenter, randomized, controlled trial to determine whether a bispectral index-guided protocol is superior to an anesthesia gas-guided protocol in reducing intraoperative awareness with explicit recall in high risk surgical patients. BMC Anesthesiol. 2009;9:8.
67. Punjasawadwong Y, Phongchiewboon A, Bunchungmongkol N. Bispectral index for improving anaesthetic delivery and postoperative recovery. Cochrane Database Syst Rev. 2014;2014:CD003843.

4 Paradoxical Inhalational Anesthetics, Nitrous Oxide, Ketamine, and Propofol

There are large differences between volatile anesthetic agents and other drug types such as ketamine or Nitrous Oxide (N_2O) [1]. This is where the processed Electroencephalography-Depth Of Anesthesia (pEEG-DOA) monitors start to diverge from clinical signs of anesthesia. Cortical electroencephalography (EEG) is mainly the target of γ-amino butyric acid (GABA) receptors agonists that modulate sedation and consciousness. Ketamine and N_2O, two N-methyl-D-aspartate (NMDA) receptor inhibitors, induce loss of consciousness without depressing the cortical EEG. Ketamine induces loss of consciousness through a completely different mechanism of "dissociative anesthesia" other than GABA receptors [1].

4.1 Equipotent /Paradoxical Inhalational Anesthetics

Fahlenkamp and colleagues [2] assessed the Bispectral Index (BIS) monitoring of Depth of Anesthesia (DOA) in patients randomized to receive either sevoflurane or xenon general anesthesia for noncardiac surgical procedures [2]. Forty elderly subjects aged between 65 and 75 years of age underwent general anesthesia with 1.6 ± 0.1% sevoflurane or 53.2 ± 0.8% xenon in 30% Oxygen (O_2) and remifentanil infusion titrated to clinical signs by an attending anesthesiologist blinded to the Bispectral Index values that were recorded every 1-min during induction, every 5-min during anesthesia maintenance, and every 20 s during recovery [2]. Emergence from anesthesia was when subjects were capable of eye opening, responding to verbal orders, extubation, and becoming orientated with their surroundings. At the phases of anesthesia induction and maintaining of anesthesia, Bispectral Index value of 40 in subjects under sevoflurane anesthesia was not significantly different than Bispectral Index values in subjects under xenon general anesthesia. Time to full recovery from general anesthesia till becoming oriented to the surroundings was statistically shorter in the xenon group than in the sevoflurane group [2]. At the phase of recovery from xenon general anesthesia, the Bispectral Index value was statistically less than Bispectral Index values with recovery from sevoflurane anesthesia. The authors concluded that during sevoflurane or xenon anesthesia, Bispectral Index values show sufficient "concordance" with clinical signs of DOA in the elderly. With clinically guided xenon general anesthesia, the Bispectral Index was at the lowest recommended range of around 40 [2].

Because the Bispectral Index was empirically derived from electroencephalography (EEG) data of subjects under commonly used conventional anesthetic agents isoflurane and propofol, these EEG algorithm data probably do not apply to rare or unusual anesthetics like xenon [3]. Goto and colleagues [3] compared the effect of isoflurane or

A. A. Dahaba, *Understanding the Numbers in Depth of Anesthesia Monitors*,
https://doi.org/10.1007/978-3-031-88260-9_4

xenon anesthesia on Bispectral Index A-1050 (version 3.22) monitoring in 9 subjects under 0.8 Minimal Alveolar Concentration (MAC) 0.92% isoflurane in air general anesthesia or 11 subjects under 0.8 MAC 56% xenon in air general anesthesia [3]. At 0.8 MAC, median (range) Bispectral Index values with isoflurane or xenon were 40 (36–53) and 36 (30–61) respectively [3]. At the end of surgery, MAC concentrations were decreased in 0.1 MAC decrement steps (isoflurane 0.12% or xenon 7% decrement steps) for a period of 15 minutes up till subjects' eye opening or hand squeezing when asked by the anesthesiologists. With decreasing isoflurane concentrations, the Bispectral Index value progressively increased, until median Bispectral Index value of 96 (range 90–98) as the subjects emerged from anesthesia. Conversely, 4 /11 subjects in the xenon group could respond when asked at median Bispectral Index value of 45, <50 with (range 41–49) [3]. Whereas the other 7 /11 subjects emerged from xenon anesthesia when their Bispectral Index >80 with median Bispectral Index value of 96 (range 82–98), four of these 7 patients at xenon of just 7% 0.1 MAC. The authors concluded that decreasing the Bispectral Index value below 50 does not ensure deep surgical anesthesia in patients receiving xenon anesthesia [3].

Detsch and colleagues [4] reported that the Bispectral Index A-1000 (version 3.12) and simultaneously recorded real time EEG were shown to paradoxically increase in 27 /70 patients from Bispectral Index value of 35 to 46 when isoflurane concentration was increased from 0.8% to 1.6% [4]. When the isoflurane was restored back to 0.8%, the Bispectral Index recovered back to baseline values. Because the Bispectral Index paradoxical increase occurred in the relatively elder study patients, the authors attributed the paradoxical Bispectral Index increase to EEG patterns of elderly patients under isoflurane anesthesia [4]. However, this could not have been the case because elderly patients, 70–85 years of age, show the opposite effect of decrease in the Central Nervous System (CNS) sensitivity to isofluran, manifesting as EEG with a greater proportion of total time in electric silence, as well as greater number of isoelectric periods under isoflurane anesthesia [4, 5]. An increase in isoflurane concentration results in the rise of α-wave and β-wave "paradoxical EEG arousal reaction" [4, 6].

A simpler explanation was presented by Clark and colleagues [4, 6], in which they reported that in some patients an increase in the isoflurane concentration from 0.79% to 1.26% would result in a rise in α-wave (8–12 Hz) and a rise in β-wave (15–35 Hz), what they called "isoflurane paradoxical EEG arousal reaction" [6]. This increase in the spectral or bispectral power in the higher frequency bands >30 Hz *Relative BetaRatio* and >40 Hz *SynchFastSlow descriptors* "tricked" the Bispectral Index into a higher value [6]. In a later study by Kochs and colleagues [7], with isoflurane-Nitrous Oxide (N_2O) anesthesia, the start of surgery under end-tidal isoflurane 0.6% resulted in a paradoxical "δ-wave shift" phenomenon. However, doubling the isoflurane anesthesia from 0.6% in 23 patients to 1.2% in 23 patients attenuated the "δ-wave shift" [7].

Another paradoxical phenomenon was described by Bimar and colleagues [8], who detected the paradoxical "δ-wave shift" phenomenon that was first described under halothane anesthesia, in which paroxysmal EEG δ-wave activity, interpreted as signs of deepening anesthesia, was noted in the frontal area of surgical patients when the concentration of halothane anesthetic was low relative to the degree of noxious surgical stimulation [8]. This intraoperative "δ-wave shift" has been interpreted as an electrophysiological reverse or "paradoxical arousal reaction" [8]. This phenomenon is entitled "paradoxical" because arousal following surgical stimulation should result in high EEG frequencies, but instead it results in frequencies indicating deeper anesthesia. This EEG change occurs predominantly in the frontal area [8].

Bispectral Index values are not the same for equipotent-MAC of different inhalational anesthetics; for instance, Ryu and colleagues [9] demonstrated that one-MAC desflurane Bispectral Index value of 36, in 44 patients was shown to be significantly lower than equipotent one-MAC sevoflurane Bispectral Index value of 41, in 45 patients [9].

Davidson and colleagues [10], using Bispectral Index A-2000 (version 3.21) monitor, demonstrated different Bispectral Index values for equipotent-MAC isoflurane and halothane anesthesia in 34 children [10]. In a crossover study, they administered one-MAC isoflurane followed by one-MAC halothane in 17 children, and halothane followed by isoflurane in another 17 children [10]. They demonstrated the exact effect at the second crossover leg of the study of one-MAC Bispectral Index value of 57 being statistically significantly more for halothane than the Bispectral Index value of 33 with isoflurane. While at awakening, there was no difference between Bispectral Index value of 81 for halothane and Bispectral Index value of 83 for isoflurane [10].

Similarly, Edwards and colleagues [11], using the Bispectral Index BIS-XP (version 4.0) monitor, demonstrated the same effect with halothane and sevoflurane [11]. They showed that at 1.3 Minimum Alveolar Concentration ($MAC_{1.3}$) adjusted for patients' age [11, 12], the Bispectral Index value of 57, in 8 patients is statistically significantly more when using halothane compared with the Bispectral Index value of 32–33, in 8 patients when using sevoflurane general anesthesia [11]. Unexpectedly during emergence, Bispectral Index value of 87 for halothane at eye opening was the same Bispectral Index value of 84 for sevoflurane [11].

Each individual anesthetic agent exhibits a unique characteristic EEG "footprint" signature. The EEG "footprint" signature of halothane is totally different than that of sevoflurane [13]. Constant and colleagues [13] found that halothane inhalational anesthetic in 13 patients, in the conventional anesthetic doses, produces slow-waves electroencephalographic fast-rhythm spindles [13], while sevoflurane in 11 patients was shown to produce sharp slow-waves on the electroencephalogram tracings associated with far fewer fast-rhythm waves [13]. The Bispectral Index values are not identically similar for different inhalational anesthetic agents "equipotent doses." The Bispectral Index algorithm was written mainly using volunteers under isoflurane or propofol anesthesia and not under halothane anesthesia [13]. Because the algorithm of the Bispectral Index was not devised nor validated for halothane, desflurane, or sevoflurane inhalational anesthetic agents, hence the Bispectral Index algorithm is not an accurate representation of the status of halothane, desflurane, or sevoflurane general anesthesia and adjusting the inhalational anesthetic agents' concentrations accordingly could result in unintended halothane, desflurane, or sevoflurane over- or under-dosage [13].

Table 4.1 enumerates and describes all Equipotent /Paradoxical inhalational anesthetics effects on depth of anesthesia monitors.

Table 4.1 Equipotent /paradoxical inhalational anesthetics

Reference	Monitor	*n*	Agent	Change	Explanation
Fahlenkamp et al. [2]		20 /40	Xenon or Sevoflurane	Xenon ↓ BIS at recovery	Xenon ↓ awake at BIS 40
Goto et al. [3]	BIS A-1050 (3.22)	11 + 9	Xenon or Isoflurane	Xenon ↓ BIS at recovery	Xenon ↓ awake at BIS 45
Detsch et al. [4]	BIS A-1000 (3.12)	27 /70	Isoflurane ↑ 2X MAC	Paradoxical ↑ BIS	α, β waves ↑
Kochs et al. [7]	10–20 EEG	23 /46	Isoflurane 0.6% or 1.2%	↑ attenuated δ-wave	Attenuated δ-wave shift
Ryu et al. [9]		44 + 45	1-MAC Desflurane /Sevoflurane	BIS 36 < BIS 45	Different EEG
Davidson et al. [10]	BIS A-1000 (3.21)	17 /34	1-MAC Halothane/Isoflurane	BIS 57 > BIS 33	Different EEG
Edwards et al. [11]	BIS-XP (4.0)	8 /16	1.3-MAC Halothane /Sevoflurane	BIS 57 > BIS 33	Different EEG
Constant et al. [13]	10–20 EEG	13	Halothane atypical EEG	Deep anesthesia	Slow waves + fast spindles
Constant et al. [13]	10–20 EEG	11	Sevoflurane typical EEG	Deep anesthesia	α, β waves ↓, δ, θ waves ↑

BIS Bispectral Index, *n* number of patients per group, *EEG* elecrtroencephalography, *MAC* Minimum Alveolar Concentration

4.2 Paradoxical N-Methyl-D-Aspartate (NMDA) Nitrous Oxide (N_2O)

Each individual anesthetic agent, such as sevoflurane or propofol, has their own characteristic electroencephalographic "footprint" profile for calculating processed Electroencephalography-Depth Of Anesthesia (pEEG-DOA) algorithms, that is different than other anesthetic agents: xenon, nitrous oxide (N_2O), or ketamine with unique electroencephalography (EEG) patterns than that commonly seen with other conventional general anesthetic agents.

Hirota and colleagues [14] demonstrated in 7 patients, using the Bispectral Index A-1050 monitor, that escalating doses of N_2O from 20% to 70% did not significantly change the Bispectral Index or 95% Spectral Edge Frequency (SEF_{95}) [14]. When N_2O was used as the sole anesthetic agent it induced unconsciousness without modifying the Bispectral Index value [15, 16], Cerebral State Index™ (CSI) [17], State Entropy, or Response Entropy values in 20 patients randomly allocated to propofol or N_2O used as a sole agent [18].

Barr and colleagues [15], using the Bispectral Index A-1000, demonstrated that 70% N_2O as a sole agent, in 10 healthy volunteers, resulted in loss of consciousness with no change in the Bispectral Index value, whereas escalating N_2O concentrations up to 66% N_2O as an adjunct to midazolam fentanyl intravenous cardiac anesthesia in 10 patients resulted in no change in Bispectral Index value [15]. Coste and colleagues [16], using Bispectral Index A-1000 (version 3.12), in 15 patients, showed that there was no difference in Bispectral Index value with the addition of 60–70% N_2O to propofol-remifentanil anesthesia compared with a control group [16].

Similarly, Anderson and colleagues [17] demonstrated that with loss of consciousness (LOC), the Cerebral State Index progressively decreased with propofol anesthesia induction but not with N_2O induction in 20 patients randomly allocated to propofol or N_2O as the sole agent [17]. Anderson and colleagues [18] further reported that with loss of consciousness, State Entropy and Response Entropy progressively decreased with propofol but not with N_2O in 20 patients randomly allocated to propofol or N_2O as a sole agent [18]. This could explain why numerous trials failed to reveal that when N_2O is used as a single agent or in combinations with other intravenous anesthetic agents results in subjects losing conscious without any visible alteration in the Bispectral Index values [14–16], using Bispectral Index A-1050 [14], or A-1000 monitors [15, 16], Cerebral State Index [17], State Entropy (SE), or Response Entropy (RE) [18].

The first report of Bispectral Index paradoxical changes appeared shortly after the introduction of the Bispectral Index. This was a report by Rampil and colleagues [19], in which the Bispectral Index A-1000 (version 3.22) simultaneously recorded with real time EEG exhibited an obvious rise in δ-wave and θ-wave activity with N_2O discontinuation [19]. They demonstrated, in 13 volunteers, that Bispectral Index transitionally, in a paradoxical fashion, declined from a mean awake value of 95 to 81 upon N_2O sudden discontinuation. Maximum suppression of Bispectral Index value occurred on average at 6 min after the discontinuation of N_2O and returned to awake values 10–15 min after discontinuation [19]. Withdrawal of N_2O was characterized by an increase in the δ-wave and θ-wave activity and a decrease in the high EEG frequency range activity [19].

Nitrous Oxide unique electroencephalographic phenomenon of "withdrawal suppression" was first described, back in 1961, by Henrie and colleagues [20], when 7–10 min following the abrupt termination of N_2O administration, they reported that N_2O withdrawal was associated with what they described as the electroencephalographic phenomenon of paroxysmal "overswinging" low-frequency δ-wave and θ-wave (4–7 Hz) frequencies that appeared as diffuse electroencephalographic eruptions very much representative of deep general anesthesia EEG patterns [20].

Similarly, Puri and colleagues [21], using the Bispectral Index A-1000, demonstrated that the Bispectral Index paradoxically declined from >90 in 2 /300 patients to <30 and < 50, 2–5 min following N_2O discontinuation [21]. This paradoxical change in the Bispectral Index value was attributed to an alteration in the *Relative BetaRatio descriptor* that typically indicates light levels of anesthesia [21].

Because N_2O analgesic effect is mainly through activating the descending inhibitory noradrenergic pathway in the brainstem and spinal cord [22], N_2O has mainly sub-cortical and spinal anti-nociceptive action through spinal-cord dorsal-horn neurons releasing noradrenaline, endogenous encephalin, β-endorphins, or dynorphin all stimulating the opiate-receptors located at the brainstem's periaqueductal gray matter regions producing a strong anti-nociceptive effect [22]. However, N_2O secondary mild cortical effect nor N_2O primarily stimulating spinal cord and brainstem inhibitory descending noradrenergic pathway effects [22] are not detectable by the Bispectral Index algorithms [16].

Nitrous oxide is reported to be associated with "controversial effects" on pEEG-DOA indices. Soto and colleagues [23] demonstrated in 60 subjects that supplementing N_2O 65% to sevoflurane 1.3 Minimum Alveolar Concentration ($MAC_{1.3}$) general anesthesia did not result in any statistically significant change in Bispectral Index value, nor Patient State Index (PSI), but resulted in State Entropy statistically significant decrease from 31.1 ± 7.3 to 18.9 ± 3.7 [23]. However, they demonstrated in 20 subjects that Bispectral Index and State Entropy did increase when N_2O was discontinued [23].

Ozcan and colleagues [24] showed that, N_2O effect on pEEG-DOA monitoring is exclusively dependent on the anesthetic agent used for general anesthesia [24], i.e., the effect of N_2O on two pEEG-DOA indices is different if volatile anesthesia is used instead of propofol intravenous anesthesia [24]. The addition of N_2O 60% in 14 patients did not alter Bispectral Index or State Entropy during propofol general anesthesia, but conversely both Bispectral Index and State Entropy indices exhibited a statistically significant but clinically small additive effect of reducing Bispectral Index (36 ± 7 to 32 ± 7, $P < 0.001$) and State Entropy (37 ± 10 to 31 ± 11, $P = 0.001$) when sevoflurane was used for general anesthesia in 14 patients [24]. Nitrous Oxide monitoring with Bispectral Index or State Entropy might result in an unintentionally deep level of anesthesia [24]. The authors attributed their findings to either the so-called Nitrous Oxide second-gas effect" phenomenon and /or to the well-known powerful N_2O antinociception effect rather than a primary extremely weak N_2O anesthetic hypnotic effect. The authors concluded that neither pEEG-DOA monitors Bispectral Index nor State Entropy can reliably detect the anesthetic hypnotic effect of N_2O when added to propofol or sevoflurane anesthesia [24], because N_2O is often used in combination with a volatile agent N_2O: Oxygen (O_2) 1:1 where N_2O 0.5 Minimum Alveolar Concentration (MAC) is used to constitute the commonly used 1.3 MAC combination with volatile inhalational anesthetic agents aiming at increasing the depth of anesthesia while concurrently benefiting from the N_2O powerful antinociceptive effect. Thus, anesthesiologists must be aware of the combined effect of this N_2O /volatile inhalational anesthetic agents' interaction on Bispectral Index, Patient State Index, or State Entropy pEEG-DOA monitoring [24].

Table 4.2 enumerates and describes all paradoxical Nitrous Oxide (N_2O) effects on depth of anesthesia monitors.

Table 4.2 Paradoxical nitrous oxide (N_2O)

Reference	Monitor	*n*	Agent	Index change	Explanation
Hirota et al. [14]	BIS A-1050	7 /7	N_2O 70% ↑	No change	Lack of cortical effect
Barr et al. [15]	BIS A-1000	10 /20	N_2O 70% ↑	No change with LOC	Lack of cortical effect
Barr et al. [15]	BIS A-1000	10 /20	N_2O 66% ↑ + midazolam fentanyl	No change	Lack of cortical effect
Costa et al. [16]	BIS A-1000 (3.12)	15 /30	N_2O 70% ↑ + propofol remifentanil	No change	Lack of cortical effect
Anderson et al. [17]	CSI	10 /20	N_2O ↑	No change with LOC	Lack of cortical effect
Anderson et al. [18]	SE and RE	10 /20	N_2O ↑	No change with LOC	Lack of cortical effect
Rampil et al. [19]	BIS A-1000 (3.22), EEG	13	N_2O termination	Paradoxical BIS ↓	δ, θ waves ↑
Puri et al. [21]	BIS A-1000	2/300	N_2O termination	Paradoxical BIS ↓	δ, θ waves ↑
Soto et al. [23]	BIS, SEDLine PSI	60	N_2O 65% +1.3-MAC Sevoflurane	BIS, PSI no change	No cortical effect
Soto et al. [23]	SE	60	N_2O 65% +1.3-MAC Sevoflurane	Paradoxical SE ↓	Different cortical effect
Ozcan et al. [24]	BIS, SE	14 /28	N_2O 60% + propofol	BIS, SE no change	N_2O second gas effect
Ozcan et al. [24]	BIS, SE	14 /28	N_2O 60% + Sevoflurane	BIS ↓, SE ↓	N_2O second gas effect

BIS Bispectral Index, *n* number of patients per group, *N_2O* Nitrous Oxide, *LOC* Loss of Consciousness, *SE* State Entropy, *RE* Response Entropy, *CSI* Cerebral State Index, *PSI* Patient State Index, *MAC* Minimum Alveolar Concentration, *EEG* Electroencephalography

4.3 Paradoxical N-Methyl-D-Aspartate (NMDA) Ketamine

Following midazolam administration [25], ketamine modulated both the Bispectral Index (BIS) and 95% Spectral Edge Frequency (SEF_{95}) as it caused a rise in fast β-wave (21–30 Hz) activities that was associated with a decline in the δ-wave activities [25]. Ketamine does not exhibit the conventional electroencephalographic general anesthesia patterns of β-wave activity decline with an increase in δ-wave power, instead a paradoxical rise in the β-wave activity that is associated with decline in the *δ*-wave activity [25, 26]. Vereecke and colleagues [26], using Bispectral Index A-2000 (version 3.4) in 15 patients, demonstrated the exact same effect of Bispectral Index value significantly increasing, following ketamine 0.4 mg kg^{-1} bolus followed by ketamine 1 mg kg^{-1} h^{-1} continuous infusion, from Bispectral Index baseline value of 57 ± 17 at 1 min of ketamine administration to Bispectral Index value of 60 ± 15 at 3–8 min from ketamine administration that was followed by a subsequent decrease for the rest of the study period, whereas the A-Line Autoregressive Index (AAI) remained unchanged and showed no significant increase or decrease following ketamine administration [26].

Ketamine may also change the electroencephalography (EEG) power spectrum by characteristically paradoxically causing a rise in θ-wave activity that will greatly influence the way the Bispectral Index algorithm is calculated [27]. Suzuki and colleagues [27] assessed the effect of a series of low doses ketamine 0.5 mg kg^{-1}, 0.25 mg kg^{-1}, and 0.1 mg kg^{-1} administrations on

Bispectral Index levels of sedation as well as on other EEG variables α-wave, β-wave, θ-wave, δ-wave frequencies, and percentile frequencies $f_{95\%}$ and $f_{50\%}$, when patients were instructed to relax and keep their eyes closed [27]. There was a strong correlation between dose-dependent ketamine administration and Observer's Assessment of Alertness/Sedation (OAA/S) Scale level of sedation but no correlation between ketamine dose administration and Bispectral Index value remaining at around 97 [27]. However, there was a significant correlation between relative θ-wave power ($r = -0.57$) significantly increasing and the relative δ-wave power ($r = 0.43$) significantly decreasing with increasing doses of ketamine administration, clearly demonstrating that the dose dependent ketamine administration level of sedation does not correlate with the Bispectral Index value [27].

Ketamine is an anesthetic agent with unique dissociative characteristic properties that distinguishes ketamine from almost all other anesthetic agents. Although the rational for deepening of anesthesia would suggest that the concurrent use of ketamine as part of a multimodal anesthetic regimen would deepen anesthesia and thus lower Bispectral Index values, however the opposite is true [26, 28–30]. Interestingly, Hans and colleagues [30] demonstrated in 22 women undergoing gynecological procedures that with ketamine administration under sevoflurane anesthesia a significant increase in Bispectral Index value from 35 to 46, Response Entropy (RE) from 33 to 52, and State Entropy (SE) from 32 to 50 without alteration in the RE–SE gradient [30]. This increase is paradoxical because it is associated with a deepening of anesthesia level [30]. That unique EEG-pattern would be exhibited by the processed Electroencephalography-Depth Of Anesthesia (pEEG-DOA) monitors in the form of ketamine administration paradoxically increasing the State Entropy, Response Entropy, but not the A-Line® Auditory Evoked Potential when in fact ketamine has inevitably created a "deeper" anesthetic state [26, 31].

Vereecke and colleagues [31] reported in 20 patients in 2 groups of 10 patients who received a ketamine 0.4 mg kg^{-1} bolus followed by a ketamine 1 mg kg^{-1} h^{-1} continuous infusion and another group of 10 patients who received rocuronium 0.9 mg kg^{-1} and ketamine 0.4 mg kg^{-1} bolus followed by ketamine 1 mg kg^{-1} h^{-1} continuous infusion, that ketamine administration provoked no response from the A-Line® Auditory Evoked Potential Autoregressive Index nor the Bispectral Index as they did not change significantly, but State Entropy and Response Entropy indices both significantly increased following ketamine infusion [31].

Hirota and colleagues [14] demonstrated in 7 patients that the Bispectral Index A-1050 and SEF_{95} paradoxically increased with ketamine administration from Bispectral Index value of 44 to 59 and SEF_{95} from 16 to 20 Hz [14].

Whereas Chaaben and colleagues [32, 33] reported an increase in the Bispectral Index value in two patients induced by an anti-hyperalgesic dose of ketamine [32], despite deepening the level of anesthesia with propofol or sevoflurane anesthetics [32]. Conversely Faraoni and colleagues [34] reported no Bispectral Index A-2000 (version 3.12) increase in 30 patients randomly allocated to receive a placebo saline infusion or ketamine administered as an infusion with no bolus dose; as they showed that a 5-min ketamine 0.2 mg kg^{-1} infusion with no bolus dose, during stable propofol /remifentanil anesthesia without any noxious surgical stimulation [34], was not associated with any increase in the Bispectral Index value over the following 15-min, strongly suggesting that ketamine duration and mode of administration plays a key role in the effect of ketamine on Bispectral Index [33, 34].

Table 4.3 enumerates and describes all paradoxical ketamine effects on depth of anesthesia monitors.

Table 4.3 Paradoxical ketamine

Reference	Monitor	*n*	Agent	Index change	Explanation
Vereecke et al. [26]	BIS A-2000 (3.4)	15	Ketamine infusion + bolus	Paradoxical BIS ↑	β wave ↑, δ waves ↓
Vereecke et al. [26]	AEP	15	Ketamine infusion + bolus	AEP no change	β wave ↑, δ waves ↓
Suzuki et al. [27]	BIS	24	Ketamine ↑ series	BIS 97 no change	θ wave ↑, δ waves ↓
Hans et al. [30]	BIS, SE, RE	22	Ketamine + Sevoflurane	Paradoxical BIS ↑, SE ↑, RE ↑	Dissociative anesthesia
Vereecke et al. [31]	SE, RE	20	Ketamine infusion + bolus	SE ↑, RE ↑	Dissociative anesthesia
Vereecke et al. [31]	BIS, AEP	20	Ketamine infusion + bolus	BIS, AEP no change	Dissociative anesthesia
Hirota et al. [14]	BIS A-1050	7	Ketamine	Paradoxical BIS ↑	β wave ↑, δ waves ↓
Chaaben et al. [32]	BIS	2	Ketamine	Paradoxical BIS ↑	Anti-hyperalgesic effect
Faraoni et al. [34]	BIS A-2000 (3.12)	15 /30	Ketamine infusion no bolus	BIS no change	EEG slowly accommodating

BIS Bispectral Index, *n* number of patients per group, *EEG* elecrtroencephalography, *SE* State Entropy, *RE* Response Entropy, *AEP* Auditory Evoked Potential

4.4 Paradoxical Propofol

Bruhn and colleagues [35], using the Bispectral Index a A-1000 (version 3.22) and a simultaneously recorded raw electroencephalography (EEG) signal, demonstrated in two volunteers that with doubling and then tripling propofol Target Controlled Infusion (TCI) from 3 μg ml^{-1} to 6 μg ml^{-1} and then to 9 μg ml^{-1}, Bispectral Index value of 35 remained unchanged in one volunteer. Similarly, in the second volunteer Bispectral Index value of 40 remained unchanged with doubling and then tripling propofol TCI from 2 μg ml^{-1} to 4 μg ml^{-1} and then to 6 μg ml^{-1}, although the raw EEG indicated Burst Suppression (BS) in both cases [35]. The Bispectral Index gave a value that failed to indicate propofol induced *Burst Suppression* as the Bispectral Index Burst Suppression Ratio (BSR) failed to indicate additional Central Nervous System (CNS) suppression; thus, the Bispectral Index value did not change. The authors attributed this phenomenon to the "range of uncertainty" of Bispectral Index 35–40 that lies between the end of *Relative BetaRatio descriptor* and the onset of *Burst Suppression descriptor* at exactly the so-called quantification of the phase coupling of the Bispectral Index being less sensitive to drug effect at this level [35]. Thus, at the onset of burst suppression with Bispectral Index values between 35 and 40, an increase in drug pharmacodynamic effect may not be reflected by the Bispectral Index [35].

References

1. Servin FS, Billard V. IV. Surrogate measures, do they really describe anaesthetic state? Br J Anaesth. 2014;112:787–90.
2. Fahlenkamp AV, Krebber F, Rex S, Grottke O, Fries M, Rossaint R, et al. Bispectral index monitoring during balanced xenon or sevoflurane anaesthesia in elderly patients. Eur J Anaesthesiol. 2010;27:906–11.
3. Goto T, Nakata Y, Saito H, Ishiguro Y, Niimi Y, Suwa K, et al. Bispectral analysis of the electroencephalogram does not predict responsiveness to verbal command in patients emerging from xenon anaesthesia. Br J Anaesth. 2000;85:359–63.
4. Detsch O, Schneider G, Kochs E, Hapfelmeier G, Werner C. Increasing isoflurane concentration may cause paradoxical increases in the EEG bispectral index in surgical patients. Br J Anaesth. 2000;84:33–7.
5. Schwartz AE, Tuttle RH, Poppers PJ. Electroencephalographic burst suppression in elderly and young patients anesthetized with isoflurane. Anesth Analg. 1989;68:9–12.
6. Clark DL, Hosick EC, Adam N, Castro AD, Rosner BS, Neigh JL. Neural effects of isoflurane (forane) in man. Anesthesiology. 1973;39:261–70.

7. Kochs E, Bischoff P, Pichlmeier U, Schulte am Esch J. Surgical stimulation induces changes in brain electrical activity during isoflurane/nitrous oxide anesthesia. A topographic electroencephalographic analysis. Anesthesiology. 1994;80:1026–34.
8. Bimar J, Bellville JW. Arousal reactions during anesthesia in man. Anesthesiology. 1977;47:449–54.
9. Ryu K-H, Song K, Lim T-Y, Choi W-J, Kim Y-H, Kim H-S. Does Equi-minimum alveolar concentration value ensure equivalent analgesic or hypnotic potency?: a comparison between Desflurane and Sevoflurane. Anesthesiology. 2018;128:1092–8.
10. Davidson AJ, Czarnecki C. The Bispectral index in children: comparing isoflurane and halothane. Br J Anaesth. 2004;92:14–7.
11. Edwards JJ, Soto RG, Thrush DM, Bedford RF. Bispectral index scale is higher for halothane than sevoflurane during intraoperative anesthesia. Anesthesiology. 2003;99:1453–5.
12. Eger EI. Age, minimum alveolar anesthetic concentration, and minimum alveolar anesthetic concentration-awake. Anesth Analg. 2001;93:947–53.
13. Constant I, Dubois MC, Piat V, Moutard ML, McCue M, Murat I. Changes in electroencephalogram and autonomic cardiovascular activity during induction of anesthesia with sevoflurane compared with halothane in children. Anesthesiology. 1999;91:1604–15.
14. Hirota K, Kubota T, Ishihara H, Matsuki A. The effects of nitrous oxide and ketamine on the bispectral index and 95% spectral edge frequency during propofol-fentanyl anaesthesia. Eur J Anaesthesiol. 1999;16:779–83.
15. Barr G, Jakobsson JG, Owall A, Anderson RE. Nitrous oxide does not alter bispectral index: study with nitrous oxide as sole agent and as an adjunct to i.v. anaesthesia. Br J Anaesth. 1999;82:827–30.
16. Coste C, Guignard B, Menigaux C, Chauvin M. Nitrous oxide prevents movement during orotracheal intubation without affecting BIS value. Anesth Analg. 2000;91:130–5.
17. Anderson RE, Barr G, Jakobsson JG. Cerebral state index during anaesthetic induction: a comparative study with propofol or nitrous oxide. Acta Anaesthesiol Scand. 2005;49:750–3.
18. Anderson RE, Jakobsson JG. Entropy of EEG during anaesthetic induction: a comparative study with propofol or nitrous oxide as sole agent. Br J Anaesth. 2004;92:167–70.
19. Rampil IJ, Kim JS, Lenhardt R, Negishi C, Sessler DI. Bispectral EEG index during nitrous oxide administration. Anesthesiology. 1998;89:671–7.
20. Henrie JR, Parkhouse J, Bickford RG. Alteration of human consciousness by nitrous oxide as assessed electro-encephalography and psychological tests. Anesthesiology. 1961;22:247–59.
21. Puri GD. Paradoxical changes in bispectral index during nitrous oxide administration. Br J Anaesth. 2001;86:141–2.
22. Zhang C, Davies MF, Guo TZ, Maze M. The analgesic action of nitrous oxide is dependent on the release of norepinephrine in the dorsal horn of the spinal cord. Anesthesiology. 1999;91:1401–7.
23. Soto RG, Smith RA, Zaccaria AL, Miguel RV. The effect of addition of nitrous oxide to a sevoflurane anesthetic on BIS, PSI, and entropy. J Clin Monit Comput. 2006;20:145–50.
24. Ozcan MS, Ozcan MD, Khan QS, Thompson DM, Chetty PK. Does nitrous oxide affect bispectral index and state entropy when added to a propofol versus sevoflurane anesthetic? J Neurosurg Anesthesiol. 2010;22:309–15.
25. Hering W, Geisslinger G, Kamp HD, Dinkel M, Tschaikowsky K, Rügheimer E, et al. Changes in the EEG power spectrum after midazolam anaesthesia combined with racemic or S- (+) ketamine. Acta Anaesthesiol Scand. 1994;38:719–23.
26. Vereecke HEM, Struys MMRF, Mortier EP. A comparison of bispectral index and ARX-derived auditory evoked potential index in measuring the clinical interaction between ketamine and propofol anaesthesia. Anaesthesia. 2003;58:957–61.
27. Suzuki M, Edmonds HL, Tsueda K, Malkani AL, Roberts CS. Effect of ketamine on bispectral index and levels of sedation. J Clin Monit Comput. 1998;14:373.
28. Sakai T, Singh H, Mi WD, Kudo T, Matsuki A. The effect of ketamine on clinical endpoints of hypnosis and EEG variables during propofol infusion. Acta Anaesthesiol Scand. 1999;43:212–6.
29. Sandler NA. Additional clinical observations utilizing bispectral analysis. Anesth Prog. 2000;47:84–6.
30. Hans P, Dewandre P-Y, Brichant JF, Bonhomme V. Comparative effects of ketamine on Bispectral index and spectral entropy of the electroencephalogram under sevoflurane anaesthesia. Br J Anaesth. 2005;94:336–40.
31. Vereecke HEM, Vanluchene AL, Mortier EP, Everaert K, Struys MMRF. The effects of ketamine and rocuronium on the A-line auditory evoked potential index, Bispectral index, and spectral entropy monitor during steady state propofol and remifentanil anesthesia. Anesthesiology. 2006;105:1122–34.
32. Chaaben K, Marret E, Lamonerie L, Lembert N, Bonnet F. Increase in bispectral index induced by antihyperalgesic dose of ketamine. Ann Fr Anesth Reanim. 2004;23:513–6.
33. Lobo FA, Schraag S. Limitations of anaesthesia depth monitoring. Curr Opin Anaesthesiol. 2011;24:657–64.
34. Faraoni D, Salengros J-C, Engelman E, Ickx B, Barvais L. Ketamine has no effect on bispectral index during stable propofol-remifentanil anaesthesia. Br J Anaesth. 2009;102:336–9.
35. Bruhn J, Bouillon TW, Shafer SL. Onset of propofol-induced burst suppression may be correctly detected as deepening of anaesthesia by approximate entropy but not by bispectral index. Br J Anaesth. 2001;87:505–7.

5 Clinical Conditions That Could Alter Processed Electroencephalography Depth of Anesthesia Monitoring

We are often confronted with processed Electroencephalography-Depth Of Anesthesia (pEEG-DOA) monitors' readings that do not concur with "clinically judged hypnotic state" arising from an underlying pathophysiology alteration of the patients' own electroencephalography (EEG) cerebral function or those due to shortcomings in the performance and design of the pEEG-DOA monitors. Numerous clinical conditions that have a direct effect on EEG cerebral function could thus directly influence pEEG-DOA indices. In my book I thoroughly reviewed all publications of pEEG-DOA indices that did not concur with medically assessed anesthetic status. However, I must stress the fact that for obvious reasons, the clinical and physiological alterations that occur with dangerous life-threatening medical conditions like cardiac stand still because of hypovolemia or hypoglycemia could not be, for obvious reasons, meticulously assessed in Randomized Controlled Trials (RCT) conclusive reasons for the associated pEEG-DOA indices' alterations. This renders all possible clarifications that I will offer in such extreme medical conditions as assumptions founded on an established concrete scientific basis.

My well-researched narrative of possible benefits and problems facing pEEG-DOA indices' interpretations of EEG in the presence of factors that could change pEEG-DOA indices for reasons other than depth of anesthesia allows us to investigate the various medical presentations that could directly affect brain conditions, thus influencing the reliability of the pEEG-DOA indices used as a guide to the depth of anesthesia. My main objective is to critically analyze these published reports of disparate depth of anesthesia numbers in a manner that could be beneficially utilized in daily anesthesia and critical care practices.

One challenge with pEEG-DOA monitors assessment is that there is no conventional comparator technology for pEEG-DOA indices assessment accurate enough for assessing anesthetic depth. The standard comparator chosen by United Kingdom National Institute for Health and Care Excellence (NICE) was "standard clinical monitoring" [1]. Because pEEG-DOA indices are EEG derived parameters, thus any condition that can "modify" the raw EEG signal during general anesthesia is likely to reduce the ability of these pEEG-DOA monitors to truly reflect the anesthetic state.

When using pEEG-DOA monitors one should keep in mind that, in addition to the hypnotic anesthetics drug effects, some clinical conditions and /or medications could influence the EEG signal either by directly altering the EEG frequencies or through a change in the anesthetic drugs pharmacokinetics, which the pEEG-DOA monitors might not accurately detect and display. Furthermore, some electric devices, under certain conditions, could also result in unnoticed artifactual signals' "contaminations," which would not

A. A. Dahaba, *Understanding the Numbers in Depth of Anesthesia Monitors*,
https://doi.org/10.1007/978-3-031-88260-9_5

necessarily appear in the form of artifact in the signal and could falsely influence pEEG-DOA indices.

During acute pathological cerebral conditions, changes in pEEG-DOA indices would be highly beneficial to a trained well-informed clinician as it alerts them to something, they would not otherwise be aware of, was happening. Because the pEEG-DOA indices are EEG derived parameters; it goes without saying that conditions that can "alter or modulate" the raw EEG signal during general anesthesia would affect the reliability of the processed Electroencephalography (pEEG) indicators of anesthesia, thus not truly reflecting the primary purpose to which they are designed, validated, and licensed and would render these monitors incapable of truly reflecting the hypnotic anesthetic state. Such EEG changes would subsequently cause pEEG-DOA indices to basically reflect unrelated EEG changes of patients' own pathophysiologic perturbations, thus rendering these monitors not truly reflecting changing anesthetic states [2].

On the other hand during anesthesia, the pEEG-DOA monitors could also act as a reliable alert of various medical situations that have a direct effect on cerebral function, not so much as hypoglycemia that is relatively rare during anesthesia maintenance or acute hypovolemia which is usually associated with hemodynamic signs, but perhaps more covert conditions such as cerebral ischemia, air embolism, or hypercapnia in spontaneously breathing patients, as specialized monitors specifically designed and validated for such purposes are still lacking.

5.1 Hypoglycemia

5.1.1 Hypoglycemia Electroencephalographic Changes

Bjorgaas and colleagues [3], in controlled hypoglycemic conditions, demonstrated that hypoglycemia of 72 mg dl^{-1} (4 mmol L^{-1}) caused a small increase in the very low-frequency δ and θ waves. This proceeds to a widespread increase in the very low-frequency δ and θ waves at 54 mg dl^{-1} (3 mmol L^{-1}) blood glucose level [3]. At a median blood glucose concentration of 36 mg dl^{-1} (2.0 mmol L^{-1}), α activity decreased abruptly in the electroencephalography (EEG) tracings, concomitant with an increase in θ activity, reflecting neuronal dysfunction in the cortex [3].

Similarly, Tribl and colleagues [4] showed that a further decline to 34.2 mg dl^{-1} (1.9 mmol L^{-1}) deeper hypoglycemia resulted in an increase in the very low-frequency δ and θ EEG activities associated with a significant decrease in α waves activity, a pattern very much like that induced by general anesthesia [4]. When the blood glucose concentration was further lowered, changes were observed in the EEG indicating that deeper brain structures were affected [4]. A normal electroencephalogram was re-established at a blood glucose concentration of 36 mg dl^{-1} (2.0 mmol L^{-1}) [5].

5.1.2 Bispectral Index (BIS) and Hypoglycemia

As expected, these raw electroencephalographic alterations were also shown in 2 different publications with the Bispectral Index A-2000 model by Wu and colleagues [6], as well as the Bispectral Index BIS-XP model by Vivien and colleagues [7], in which two hypoglycemic coma patients at 34.7 mg dl^{-1} (1.93 mmol L^{-1}) [6] and 20.7 mg dl^{-1} (1.15 mmol L^{-1}) [7] blood glucose levels were reported to manifest Bispectral Index values of as low as 45. The Bispectral Index values rapidly increased to 80 along with the increase of blood glucose levels and the return of consciousnesses [6, 7].

Thus, blood glucose levels should be taken into consideration as a contributing factor at interpretation of the Bispectral Index values. Furthermore, functional brain imaging using the ^{18}fluorodeoxyglucose technique showed that the Bispectral Index A-1000 (version 3.2) monitor as an electroencephalographic (EEG) *descriptor* correlated linearly with the magnitude of cerebral brain glucose metabolic rate reduction caused by propofol and isoflurane anesthesia [8].

This was recently confirmed in another two case reports. The first report was by Alkire and colleagues [9], where in a 74-year-old woman, Bispectral Index A-2000 XP value was in the range of 38–46 when her hyperglycemia was 264.6 mg dl^{-1} (14.7 mmol L^{-1}); immediately actrapid 6 IU (purified human neutral insulin) was given for her elevated blood sugar [9]. Following insulin administration, and over the next 8–12 min, a rapid decrease in Bispectral Index value to 12–16 was observed as her blood glucose measurement showed severe hypoglycemia of 41.4 mg dl^{-1} (2.3 mmol L^{-1}) [9]. Glucose 35 mL 50% over 1 min was administered. A rapid increase in Bispectral Index value of 35–39 over 4–8 min was noted as glucose measurement revealed a blood sugar of 129.6 mg dl^{-1} (7.2 mmol L^{-1}) [9].

Whereas in a 67-year-old man undergoing coronary artery bypass grafting second report by Mahaldar and colleagues [10], the Bispectral Index A-2000 XP value was 40–46 when his blood glucose was 102.6 mg dl^{-1} (5.7 mmol L^{-1}) and rapidly declined to 22–26 with severe hypoglycemia 39.6 mg dl^{-1} (2.2 mmol L^{-1}) blood glucose overcorrection [10]. Blood glucose restoration to 129.6 mg dl^{-1} (7.2 mmol L^{-1}) resulted in a rapid increase in Bispectral Index value of 45–49 over 3–5 min [10].

Table 5.1 enumerates and describes all hypoglycemia situations where cortical electroencephalography is modified from anesthesiologists' perspective.

5.2 Hypovolemia

The Index of Consciousness (IoC) and Bispectral Index [11] were interchangeably lower during normotension and hypotensive Cardiopulmonary Bypass (CPB) in patients undergoing Coronary Artery Bypass Graft (CABG) [11]. Interestingly, Chakravarthy and colleagues [11] assessed the interchangeability of the Bispectral Index and Index of Consciousness techniques during normotension, hypotension and during pulseless state CPB in 15 patients undergoing CABG surgery [11]. The Bispectral Index and IoC values were interchangeable. The interchangeability was better appreciated during normotension and hypotension but not during non-pulsatile state of CABG [11].

In another case, Honan and colleagues [12] reported that Bispectral Index value decreased from a steady value of 35 to a Bispectral Index value of 20 prodroma, that took place 10-min well preceding the hypotension from Systole / Diastole blood pressure of 120/70 down to 65/30 indicating deadly massive hemorrhage where the aorta was grafted [12]. A possible explanation could be the Bispectral Index prompt "prodromal" recognition of the alterations in propofol's pharmacokinetic changes of reduced Volume of distribution (V_d) resulting in increased propofol concentrations [13], or a change in propofol's pharmacodynamics end-organ sensitivity arising from a profound decease in cerebral perfusion [13].

Table 5.1 Hypoglycemia

References	BIS Monitor	*n*	Change in blood glucose level	BIS change	Explanation
Wu et al. [6]	BIS A-2000	1	Hypoglycemia 1.93 mmol L^{-1}	BIS 45 ↓	δ, θ waves ↑, α waves ↓
Vivien et al. [7]	BIS-XP	1	Hypoglycemia 1.15 mmol L^{-1}	BIS 45 ↓	δ, θ waves ↑, α waves ↓
Alkire et al. [9]	BIS A-2000 XP	1	Hypoglycemia 2.3 mmol L^{-1}	BIS 12–16 ↓	δ, θ waves ↑, α waves ↓
Mahaldar et al. [10]	BIS A-2000 XP	1	Hypoglycemia 2.2 mmol L^{-1}	BIS 22–26 ↓	δ, θ waves ↑, α waves ↓

n number of patients per group, *BIS* Bispectral Index

Table 5.2 Hypovolemia

References	Monitor	*n*	Change	Index change	Explanation
Chakravarthy et al. [11]	BIS, IoC	15	CPB	BIS ↓ IoC ↓	CPB pulseless state
Honan et al. [12]	BIS	1	Massive hemorrhage	Prodromal BIS 20 ↓	Change in propofol PK or PD

n number of patients per group, *BIS* Bispectral Index, *IoC* Index of Consciousness, *CPB* Cardiopulmonary Bypass, *PK* Pharmacokinetics, *PD* Pharmacodynamics

Table 5.2 enumerates and describes all hypovolemia situations where cortical electroencephalography is modified from anesthesiologists' perspective.

5.3 Deep Hypothermia and Isoelectric Zero Display

My research team Dahaba and colleagues in West China Hospital, Sichuan University, Cheng Du, Sichuan, People's Republic of China, often demonstrated that the Bispectral index (BIS) value would suddenly decrease to Zero "0" Isoelectricity with 100% Burst Suppression Ratio (BSR) at 17 °C with the institution of Deep Hypothermic Circulatory Arrest (DHCA) as this was often captured in the somehow blurry images taken by my research team Dahaba and colleagues as Bispectral Index value suddenly decreased to Zero Isoelectricity with 100% BSR when temperature reached 17 °C with DHCA in the first case (Fig. 5.1) and in the second case (Fig. 5.2).

Meybohm and colleagues [14] studied State Entropy (SE), Response Entropy (RE), and Bispectral Index BIS-XP in 25 patients undergoing Coronary Artery Bypass Grafting (CABG). They demonstrated that patients' processed Electroencephalography-Depth Of Anesthesia (pEEG-DOA) indices were significantly lower during hypothermic (31–34 °C) Cardiopulmonary Bypass (CPB) than normothermic (34–37.5 °C) CPB [14]. Entropy variables' SE and RE exhibited more pronounced decline than the Bispectral Index [14]. The Bispectral Index value of 33, SE value of 20 and RE value of 22 indices signifi-

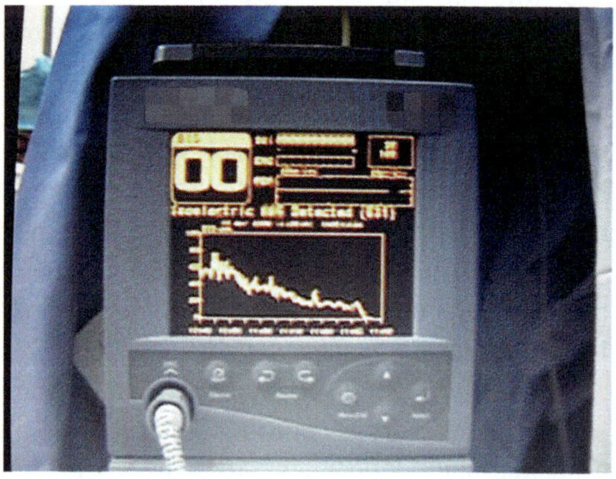

Fig. 5.1 Captured in the somehow blurry image of the first case of Bispectral Index (BIS) value sudden decrease to Zero Isoelectricity with 100% Burt Suppression Ratio (BSR) at 17 °C Hypothermia Deep Hypothermic Circulatory Arrest (DHCA). (©2025 Medtronic. All rights reserved. Used with the permission of Medtronic)

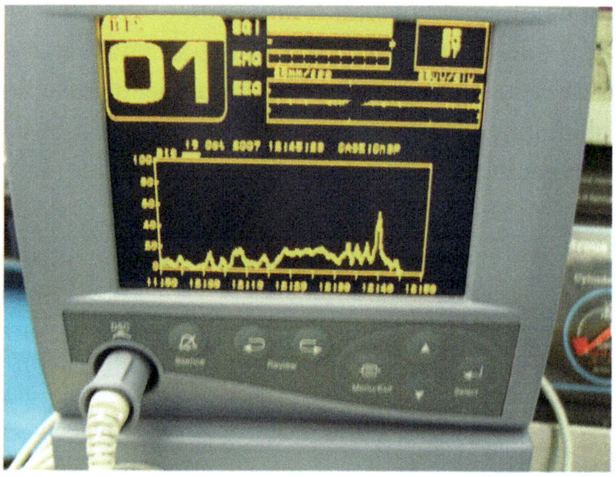

Fig. 5.2 Captured in the somehow blurry image of the second case of Bispectral Index (BIS) value sudden decrease to Zero Isoelectricity with 100% Burt Suppression Ratio (BSR) at 17 °C Hypothermia Deep Hypothermic Circulatory Arrest (DHCA)

cantly decreased during the hypothermic phase, but the decrease was more significantly pronounced for the Entropy variables compared with the Bispectral Index value [14]. Thus, there seems to be some evidence that the pEEG-DOA monitors might inconsistently perform under Hypothermic Cardiopulmonary Bypass (HCPB) at the phase when the patients are rewarmed [14]. Hypothermia's alterations in electroencephalography (EEG) cerebral patterns appear as slow cortical rhythms, that is totally different than the anesthetic agents induced γ-Amino Butyric Acid (GABA) receptors EEG suppression patterns. Obviously, the Bispectral Index algorithm does not detect or reflect the resulting hypothermic "EEG de-synchronization" patterns, that seem to be detected by the Response Entropy [15–17].

Similarly, the Approximate Entropy of the electroencephalogram changed in a sigmoidal fashion during 24.7 °C cooling and 28 °C rewarming [18]. The temperature corresponding to 5% Entropy was 18.7 °C. The temperature corresponding to 95% Entropy was 31.3 °C during cooling and 38.2 °C during rewarming [18].

Honan and colleagues [19] demonstrated that moderate Hypothermic Cardiopulmonary Bypass (HCPB 28–30 °C) in 14 patients was associated with significantly lower mean Bispectral Index value of 18 when compared with mean Bispectral Index value of 40 in 16 patients undergoing mild HCPB (32–34 °C) under at a steady state of inspired enflurane [19].

Similarly, Hayashida and colleagues [20] demonstrated in 20 patients undergoing thoracic aortic surgery with Deep Hypothermic Circulatory Arrest (DHCA) under sevoflurane fentanyl anesthesia, that the Bispectral Index A-1050 (version 3.3), connected to 4 disposable silver /silver chloride electrodes (reference Fpz and ground Fp1 10-20 international system), value slowly decreased in a biphasic manner till 26°C then suddenly dropped to Zero "0" at 17 °C as Burst Suppression Ratio (BSR) rose rapidly to 100% in inverse proportion to the Bispectral Index value with the institution of DHCA [20]. With rewarming Bispectral Index value slowly rose and Burst Suppression Ratio declined to reach precooling levels as time to recovery corresponded to duration of DHCA [20].

Mychaskiw and colleagues [21] reported a case of "falsely elevated" Bispectral Index A-2000 in a 2-month-old patient undergoing Ventricular Septal Defect (VSD) repair [21]. At DHCA of 18 °C, the Bispectral Index value decreased from 40 to Zero "0" indicating isoelectricity [21]. After 15 min from the start of profound DHCA Bispectral Index value suddenly started to increase to 98–100. Immediately CPB was resumed and after 10 min of re-warming Bispectral Index value returned to 40 [21]. This Bispectral Index "erroneous readings" was probably an artefactual product of the algorithm's misinterpretation of electromyographic (EMG) activity from electrical equipment, or the toweling ice wrapping surrounding the patient's head increasing the electrode impedance [21]. A possible explanation would be the fact that in circumstances of extreme EEG suppression, as in DHCA, it is possible that some radio frequency interference noise from electrical equipment result in the Bispectral Index algorithm's calculation being misled to consider as an electroencephalographic activity and thus allocating a very high Bispectral Index value. When EEG activity resumed with re-warming the Bispectral Index monitor then re-interpreted the data appropriately, discounting the noise from the algorithm.

Burst Suppression is considered of great value in indicating adequate brain protection during DHCA [22]. In that regard, detection of the Burst Suppression Ratio (BSR) by the Bispectral Index monitor adds great value when used to monitor anesthesia involving DHCA [22]. Vretzakis and colleagues [23] present a report of a 76-year-old man with a history of under control chronic hypertension, in which a decline of Bispectral Index (BIS-XP) value to Burst Suppression occurred with the initiation of DHCA support, during emergency surgical repair of an Ascending Aorta Aneurism Stanford type-A dissection with pericardial fluid collection [23]. Before anesthesia induction, the patient was calm and cooperative with blood pressure 120/70 mm Hg after 5 mg atenolol and continuous infusion of nitroglycerin β-blockade. Transesophageal Echocardiography (TEE) revealed 2+ aortic regurgitation and a well performing heart [23]. A Bispectral Index BIS-XP monitor, connected to a BIS Quatro sensor placed on his forehead, showed a Bispectral Index value of 95 [23]. Anesthesia was induced via etomidate 16 mg, pancuronium 10 mg, fentanyl 200 μg, and midazolam 3 mg, for tracheal intubation, and the lungs were ventilated to achieve $ETCO_2 \simeq 32$ mm Hg slight hypocapnia [23]. A Pulmonary Artery Catheter (PAC) was inserted in the left jugular vein. Anesthesia was maintained with propofol $\simeq 5$ mg kg^{-1} h^{-1}, remifentanil $\simeq 20$ μg kg^{-1} h^{-1} and pancuronium neuromuscular block [23]. Surgery proceeded with median sternotomy after placement of left femoral arterial cannula for the heart-lung machine [23]. The Bispectral Index value was maintained around 40 and bladder temperature T = 34.9 °C until the initiation of the hypothermic extracorporeal circulation [23]. After midazolam 10 mg, hydrocortisone 1 g, and slight "head-down" position, the patient's head was covered with ice and the anesthetic agents' infusion pumps were turned off [23]. Over the next 25 min, hypothermic perfusion of the brain resulted in gradual decline in the Bispectral Index value till the Bispectral Index value reached and stayed at Zero "0," with patient's pupils fully dilated for the remaining 85 min of hypothermic perfusion at T = 18.5 °C [23]. When T = 29.4 °C, the Bispectral Index showed more realistic values of 0–20. Brain reperfusion resulted in an abrupt increase in the Bispectral Index value to >80 and remained high >60 for the next 60 min, despite the restart of propofol /remifentanil infusion pumps as his pupils gradually returned to miosis [23]. Using moderate inotropic support, the patient was weaned off the CPB. In the following 1 h 45 min until the patient was transferred to the Critical Care Unit (CCU), at T = 34.9 °C, his Bispectral Index value gradually rose from 10 to 45. The postoperative course was uneventful, and the patient fully recovered without any neurologic damage, and he was discharged 12 days later [23].

The Bispectral Index tracings during rewarming in this case [23] show many similarities to the case previously reported by Mychaskiw and colleagues [21], in which realistic Bispectral Index values replaced falsely elevated Bispectral Index values after 15-min of rewarming with the resumption of normal EEG activity allowing the Bispectral Index algorithm to reinterpret EEG appropriately, discounting artefact frequency activity from the algorithm [21]. The authors believe that anesthesiologist using Bispectral Index monitoring should have at least basic EEG knowledge to be able to recognize high-frequency EEG artifacts. When these high-frequency EEG artifacts appear on the raw EEG tracings, anesthesiologist can then ignore the false Bispectral Index readings. Simultaneous Bispectral Index monitoring together with real-time EEG waveform examination may aid the diagnosis of this kind of Bispectral Index artifacts, thus avoiding any unnecessary interventions [23].

Nevertheless, this effect was not reportedly consistent among different trials as Doi and colleagues [24], using a Bispectral Index BIS A-1000 (version 3.12), Spectral Edge Frequency (SEF), and Auditory Evoked Potential index (AEPIndex), examined the effect of hypothermia (<32 °C) during CPB under anesthesia with alfentanil propofol Target Controlled Infusion (TCI) in 12 patients undergoing uneventful cardiac anesthesia [24]. Other than the Bispectral Index value somewhat declining with lower temperatures, all three variables were not affected by the transition to hypothermic CPB. A probable explanation of the inconsistent correlation between Bispectral Index values and measured temperature's changes might be that the number of patients included in this study was inadequately minor thus falling short to reveal such constant correlation between the Bispectral Index values and temperature's changes.

Hypothermic CPB significantly increases the propofol plasma concentrations and decreases propofol clearance [25]. In addition, metabolism of propofol may be decreased, because of hepatic blood flow decline. Furthermore, hemodilution and heparin administration could increase the protein unbound fraction of propofol [25]. Hypothermic conditions between 25 °C and 27 °C were found to significantly raise propofol's serum levels via a decline in the activity of the liver microsomal enzymes resulting in a decline in propofol's biotransformation activities thus ultimately resulting in a decrease in Bispectral Index values [25].

Similarly, Schmidlin and colleagues [26], using a Bispectral Index A-1000, demonstrated in 14 patients under fentanyl propofol anesthesia, that mean Bispectral Index of 41 was statistically significantly lower with Hypothermic CPB compared with mean Bispectral Index of 49 with normal temperature CPB [26]. This clearly shows that CPB per se did not affect the Bispectral Index value, only hypothermia significantly decreased the Bispectral Index value. The significantly lower Bispectral Index values with hypothermia were a result of brain hypothermia or that of increased serum propofol concentrations, resulting from propofol decreased biotransformation with hypothermia, since the hepatic microsomal enzyme activity was shown to decline in a temperature dependent fashion [25].

Mathew and colleagues [27], using Bispectral Index A-1050 (version 3.3) monitor in a large study of 100 patients undergoing Hypothermic CPB under fentanyl isoflurane general anesthesia, reported a significant overall correlation between temperature and the Bispectral Index values, with Bispectral Index values steadily declining with hypothermia. The Bispectral Index value decline was calculated to be a decline of 1.12 Bispectral Index points for every 1 °C Celsius degree decline in core temperature [27]. Hypothermic effects appear as a steady regular decline in inhalational volatile anesthesia doses required. As a matter of fact, the hypothermic effect itself could function as full anesthetic effect when temperature declines down to 20° degrees Celsius body temperature, then patients will require no further anesthetic agents administration [28]. Multiple mechanisms may be responsible for this effect. This could be attributed to a rise in solubility of inhalational volatile anesthetic agents in lipid-membranes when temperatures decline, that would result in higher concentration of volatile inhalational anesthetics at the cellular level, resulting in the Bispectral Index values decline [28].

Table 5.3 enumerates and describes all hypothermia situations where cortical electroencephalography is modified from anesthesiologists' perspective.

Table 5.3 Hypothermia

References	Monitor	*n*	Change	Index change	Explanation
Meybohm et al. [14]	BIS-XP, SE, RE	25	Hypothermic CPB 31–34 °C	BIS 33, SE 20, RE 22	↓ BIS decline <SE, RE decline
Honan et al. [19]	BIS	14-16	CPB 28–30 °C/32–34 °C	BIS 18 /BIS 40	↓ BIS decline with ↓ °C
Hayashida et al. [20]	BIS A-1050 (3.3)	20	DHCPB 17 °C	BIS ↓ to Zero 0	↓ BIS decline with ↓ °C
Mychaskiw et al. [21]	A-2000	1	DHCPB 18 °C	BIS ↓ to Zero 0	↓ BIS decline with ↓ °C
Vretzakis et al. [23]	BIS-XP	1	DHCPB 18.5 °C	BIS ↓ to Zero 0	↓ BIS decline with ↓ °C
Doi et al. [24]	BIS A-1000 (3.12), AEP	12	Hypothermic CPB <32 °C	BIS, AEP no change	Small sample size
Schmidlin et al. [26]	BIS A-1000	14-14	Hypothermic / Normothermic CPB	BIS 41 < BIS 49	Hypothermic Propofol enhancement
Mathew et al. [27]	BIS A-1050 (3.3)	100	1.12 BIS units /°C ↓	↓ BIS decline with ↓ °C	Isoflurane enhancement

n number of patients per group, *SE* State Entropy, *RE* Response Entropy, *BIS* Bispectral Index, *CPB* Cardiopulmonary Bypass, *AEP* Auditory evoked potential, *DHCPB* Deep Hypothermic Cardiopulmonary Bypass

5.4 Hypercapnia

Thompson [29, 30] in 1912 labelled the country doctor who discovered Inhalational Anesthesia Henry Hill Hickman (1800–1830), as "A Forgotten Pioneer of Anesthesia" [29, 30] because he described the first experiments of Carbon Dioxide (CO_2) possessing anesthetic potency [29, 30]. Nowadays, CO_2 is still used in laboratory animal science to perform short lasting anesthesia. Electroencephalography (EEG) was shown to change, in a fashion comparable to anesthetic agents, because of elevated levels of CO_2 [31]. The influence of CO_2 on the Bispectral Index is not yet satisfactorily investigated. Although a single randomized controlled study could not find an influence of altered CO_2 on Bispectral Index values during stable anesthesia [32], in contrast numerous case reports showed a significant effect of hypocapnia on Bispectral Index monitoring during induction, maintenance, and recovery from anesthesia. However, hypercapnia should always be considered as a possible factor that could influence Bispectral Index monitoring [31, 33, 34].

In a study in swine, Martoft and colleagues [31] reported that hypercapnia was shown to be associated with EEG changes. Spectral Edge Frequency (SEF) significantly declined, and episodes of Burst Suppression (BS) occurred after induction of CO_2 narcosis [31]. Acute hypercapnia with Partial Pressure of Carbon Dioxide (P_aCO_2) above 10.7 kPa was reported to potentially initiate narcosis [31]. Although hypercapnia is one of these conditions associated with significant EEG changes [31], however in humans, there is no clearly defined threshold [35].

Narcosis from hypercapnia has been previously described by Takahashi [36] and Smyth [37] in adults with an End-tidal Carbon Dioxide ($EtCO_2$) concentration of 70 mmHg (9.3 kPa) [36, 37]. Takahashi [36] reported a case of hypercapnia caused by a small dose of 0.25 mg midazolam in a 44-year-old patient with myotonic dystrophy undergoing hysterectomy under lignocaine epidural anesthesia, when at the end of surgery, she became unarousable. Her $EtCO_2$ rose to 70 mmHg (9.3 kPa) [36]. Flumazenil 0.2 mg administration resulted, after 5 min, in a decrease of her $EtCO_2$ to 45 mmHg (6 kPa) as she woke up [36]. Smyth [37] similarly reported another case in which morphine in combination with intravenous diazepam delivered via Patient Controlled Analgesia (PCA), resulted in complete apnea and CO_2 narcosis in a patient admitted for cervical traction. A patent airway and high flow oxygen face mask ventilation successfully maintained her oxygenation despite complete apnea [37].

Jain and colleagues [38] reported a sudden decrease in Bispectral Index value during emergence from general anesthesia in a one-year-old female patient undergoing an elective ventriculoperitoneal shunt insertion for meningomyelocoele excision and repair under general anesthesia [38]. At the end of surgery, when her anesthesia was discontinued, her inadequate initial respiratory efforts resulted in a rise of her $EtCO_2$ concentration from 37 mmHg (4.9 kPa) to 52 mmHg (6.9 kPa), that was associated with a sudden Bispectral Index decrease from 54 to 5. As her ventilation was immediately assisted; her EtCO2 decreased to 32 mmHg (4.3 kPa). Within 2 min, the Bispectral Index value increased to 48 and with further awakening increased to 90 [38].

This phenomenon was similarly described by Yoshida and colleagues [33], who reported a case of Bispectral Index suddenly declining to as low as 4 with P_aCO_2 reaching 137 mmHg (18.3 kPa) in a patient under stable anesthesia suffering from capnopertineum [33]. Burst Suppression Ratio (BSR) increased to 90. Bispectral Index returned to original levels with discontinuation of the laparoscopic procedure and resumption of normal ventilation [33].

Another case of Bispectral Index monitoring of hypercapnic coma was reported by Lee and colleagues [34], after accidental administration of a Neuromuscular Blocking Agent (NMBA) to an awake patient. With the patient unable to breath and difficult mask ventilation, P_aCO_2 increased to 130 mmHg (17.3 kPa), when Bispectral Index value dropped to 35 [34]. However, the sudden decrease of Bispectral Index value could also be attributed to the effect of the NMBA on the Bispectral Index [39].

However, this does not seem to be always the case, rather only in extreme conditions. An extended randomized controlled prospective study by Turnbull and colleagues [32], evaluated the effect of varying $EtCO_2$ levels on Bispectral Index BIS-XP (version 3.12) monitoring in 24 patients. Under propofol anesthesia, Bispectral Index was recorded while patients were subjected, in a randomized manner, to hypocapnia 22.5–30 mm Hg (3–4 kPa), normocapnia 33.8–45 mm Hg (4.5–6 kPa), or hypercapnia 67.5–90 mm Hg (9–12 kPa) for 5 min. Turnbull and colleagues [32] reported that hypercapnia did not alter Bispectral Index monitoring as they failed to demonstrate an effect of different $EtCO_2$ levels on Bispectral Index values during deep propofol anesthesia. Obviously, in their study design, propofol potent deep anesthesia strongly influenced Bispectral Index monitoring with an average Bispectral Index value of 38, that would have probably masked any possible effects of hypercarbia. In addition, a study limitation was they used $EtCO_2$ and not P_aCO_2, which could be less sensitive to hypercapnia [32].

Not only "Hypercapnia" might not influence Bispectral Index monitoring [32] rather a decline in Bispectral Index is not always the case with "Hypercapnia" as another study demonstrated the exact opposite of a decline in Bispectral Index with the opposite "Hypocapnia" and not conventionally with "Hypercapnia" as I previously mentioned [33, 34, 36–38]. To assess features of hyperventilation, Hayashi and colleagues [40] examined quadratic phase coupling features using bicoherence analysis. Twenty-two patients were anaesthetized using sevoflurane 1.5%, remifentanil anesthesia or epidural anesthesia [40]. After a stable normocapnic period, hypocapnia was induced by hyperventilation, and the raw EEG signals were recorded. Bispectral analysis (bicoherence) and power spectrum analysis were performed before and after hypocapnia [40]. Hayashi and colleagues [40] showed an opposite effect of a significant decline of the Bispectral Index BIS A-2000 (version 4.0) in hypocapnic hyperventilated patients under anesthesia. Hyperventilation, with the resulting hypocapnia, reduces cerebral blood flow and causes slowing of the EEG activity, although neuronal oscillating properties of the thalamocortical network during hyperventilation was not demonstrated [40]. Mean ± SD peak bicoherence in the δ-frequency significantly increased from normocapnia 35.6 ± 10.9% to hypocapnia 43.8 ± 10.9%. Whereas mean peak bicoherence in the α-frequency significantly decreased from normocapnia 42.8 ± 14.4% to hypocapnia 37.5 ± 12.3%. Normalized power in the δ-θ frequencies on the power spectrum significantly increased from 60.2 ± 13.1% to 72.5 ± 12.7% [40]. Normalized power in the α-frequencies on the power spectrum decreased from normocapnia 27.8 ± 10.6% to hypocapnia 20.3 ± 9.6%. Mean ± SD Bispectral Index values significantly declined from 45.9 ± 7.0 to 40.1 ± 5.6 and Spectral Edge Frequency (SEF) from 15.0 ± 2.3 to 14.0 ± 2.5 Hz. No significant differences in these values were observed between the two types of anesthesia. They concluded that hypocapnia enlarged bicoherence growth in the δ-θ frequency range, suggesting the contribution of subcortical oscillating mechanisms in regulating EEG during hyperventilation [40].

Table 5.4 enumerates and describes all hypercapnia situations where cortical electroencephalography is modified from anesthesiologists' perspective.

Table 5.4 Hypercapnia

References	BIS Monitor	*n*	Change	BIS change	Explanation
Jain et al. [38]	BIS	1	Hypercapnia 6.9 kPa	CO_2 ↑ BIS 5 ↓	CO_2 Narcosis ↑
Yoshida et al. [33]	BIS	1	Hypercapnia 18.3 kPa	CO_2 ↑ BIS 4 ↓	CO_2 Narcosis ↑
Lee et al. [34]	BIS	1	Hypercapnia 17.3 kPa	CO_2 ↑ BIS 35 ↓	CO_2 Narcosis ↑
Turnbull et al. [32]	BIS-XP (3.12)	24	Hypercapnia 9-12 kPa	CO_2 ↑ BIS 39 no change	propofol anesthesia
Hayashi et al. [40]	BIS A-2000 (4.0)	22	Hypocapnia BIS 46	CO_2 ↓ BIS 40 ↓	δ-θ waves ↑, α waves ↓

n number of patients per group, *BIS* Bispectral Index, CO_2 Carbon Dioxide

5.5 Cerebral Ischemia

5.5.1 Aortic or Carotid Artery Cross-clamping

Kakinohana and colleagues [41] reported a case of a rapid Bispectral Index decline from a stable Bispectral Index value of 60 to 0 during propofol 6 μg kg^{-1} h^{-1} ketamine fentanyl anesthesia 15 min after the cross-clamping of the descending thoracic aorta for elective thoracic Descending Aortic Aneurysm repair followed by the start of a normothermic Cardiopulmonary Bypass (CPB) [41]. Mean arterial pressure (MAP) was 72 mmHg. Propofol was immediately stopped, and 15 min later, Bispectral Index value increased to 30. Propofol was started at the same original rate and 25 min later the Bispectral Index re-declined to <10. This was repeated several times [41]. Radial artery, proximal to the cross clamp, propofol levels were >7 μg ml^{-1}, 4–5 times higher than those at the femoral artery. Cerebral hypoperfusion or cerebral ischemia was ruled out since blood pressure and cardiac output were within normal and patient had a full uneventful recovery [41]. A simple explanation for the Bispectral Index decline would be the fact that blood flow from the area proximal to the cross clamp, including hepatic blood flow the main route of propofol excretion [42], is stopped, leading to a reduction in propofol Volume of distribution (Vd) and Clearance (Cl). This would result in a significant increase in propofol concentrations proximal to the cross-clamp that includes the brain [42].

Numerous reports demonstrated that lack of blood to the brain results in Electroencephalography (EEG) changes consistent with a decrease in the Bispectral Index value. Billard [43], in an editorial, reported that the Bispectral Index monitor could be successfully used to assess the brain function or detect brain injury that would otherwise go undetected [43]. A precipitous Bispectral Index decline has been suggested as a clear indicator of a decline in Cerebral Blood Flow (CBF) following aortic or carotid cross-clamping [44]. Mérat and colleagues [44] has shown that the Bispectral Index A-2000 (version 2.1) monitor was successfully used to detect cerebral ischemia and brain injuries. Under stable propofol sufentanil Target Controlled Infusion (TCI), carotid artery cross clamping, in a 58-year-old patient undergoing Carotid Endarterectomy (CEA), caused the Bispectral Index value to decline from 40–60 to below 10 that did not subsequently increase [44]. In another case of a 62-year-old patient undergoing an Aortic Aneurysm resection, the Bispectral Index decreased from 40–50 to 8 with aortic cross clamping, as the Cardiac Index and central venous Oxygen Saturation (SO_2) decreased [44]. The Bispectral Index returned to normal values when the low Cardiac Output was pharmacologically corrected. Patient's recovery with no neurological injury revealed a Bispectral Index value of 90. They attributed the Bispectral Index declines to severe cerebral ischemia [44].

Patients undergoing Carotid Endarterectomy (CEA) for extracranial internal carotid artery stenosis are at high risk of cerebral ischemia or hypoperfusion. According to Ackerstaff and colleagues [45], the European and American committees surgical criterion to determine whether a shunt is to be placed or not consists of a decline in Transcranial Doppler (TCD) ultrasonography Middle Cerebral Artery Blood Flow Velocity (MCBFV) to less than 30–40% of intraoperative pre-clamp level [45]. Similarly, Halsey and colleagues [46], in an 11 centers multicenter study with 1495 CEA patients, categorized Severe Ischemia to be MCBFV of 0–15%, Mild Ischemia to be MCBFV of 16–40%, and Absent Ischemia to be MCBFV of less than 40% of pre-clamp value [46]. Obviously, false positives, i.e., shunting when not necessary, could perhaps be of less concern, but even one false negative, i.e., not shunting, when necessary, is unacceptable.

Estruch-Pérez and colleagues [47] demonstrated that the Bispectral Index A-2000 BIS-XP monitor could be useful in diagnosing intraoperative cerebral ischemia. In patients undergoing awake CEA under regional anesthesia, they conducted a study to investigate the Bispectral Index changes in 35 consecutive patients with neurological deficits compared with 35 patients without neurological deficits. They compared patients' neurological deficits with the Bispectral Index values measured at baseline, after sedation, at start of surgery, at preclamping, at 3-min clamping test, during shunt insertion, at de-clamping, at 15-min after de-clamping, and after end of surgery [47]. A decrease in the Bispectral Index value of >10 was considered as the cut-off for the classification of patients with neurological deficits logistic regression model. Thirteen/70 patients (18.6%) developed clinical cerebral ischemia, though the Bispectral Index values decreased in just 10 /13 of these patients (76.9%) [47]. The mean Bispectral values of 85 ± 13 in patients with neurological deficits were significantly lower than the mean Bispectral Index values of 93 ± 6 in patients without neurological deficits ($P < 0.05$). The odds ratio of 8.5 (95% confidence interval 2.1–35.1) decline in the Bispectral Index value was associated with neurological deficits. They concluded that there is a significant relationship between the Bispectral Index decline and neurological deficits during carotid endarterectomy in awake patients [47].

Absalom and colleagues [48], recently noted that Body Mass Index (BMI) and Lean Body Mass (LBM) could be confounding factors for our Target Controlled Infusion (TCI) Diprifusor's pharmacokinetic model [48]. In the study conducted by my research team Dahaba and colleagues in Xuan Wu Hospital, Capital Medical University, Beijing, People's Republic of China [49], we decided to tighten the Ackerstaff's criteria [45], from 60% to just 40% MCBFV decline compared with pre-clamp values, and we only included patients with BMI 20–26 kg m^{-2}. We assessed the discriminative power of the Bilateral Bispectral Index BIS-Vista monitor (version 1.4 which is the bilateral version of the single version 4.1) using the Bilateral Bispectral Index sensor for detecting a 40% decline in MCBFV with carotid cross-clamping [48]. Twenty patients with extracranial Internal Carotid Artery (ICA) stenosis, undergoing Carotid Endarterectomy (CEA) with or without carotid artery enlargement and shunt insertion under remifentanil /propofol anesthesia, were recruited in our Bispectral Index BIS-Vista Doppler simultaneous monitoring study [49]. We used a two channel Pioneer TC 8080 Transcranial Doppler with bilateral 2 MHz frequency probes mounted on a Welder Probe Fixation head mount (Viasys Healthcare, Wisconsin, USA) (Fig. 5.3) to bilaterally monitor MCBFV [49]. My research team Dahaba and colleagues also recorded Gosling Pulsatility Index (PI) [50], a blood-flow waveform index that quantifies the oscillations of the waveform as an indicator of vessel's flexibility [49]. After identification of the optimal Middle Cerebral Artery (MCA) depth and insonation angles, all TCD variables were digitally recorded and stored on a laptop computer. Before surgery, using a separate handheld probe, we recorded the luminal diameter and blood velocity of the stenosed ICA [49]. In one patient, satisfactory bilateral TCD measurements during surgery were not possible. We excluded this patient from the study and this patient was replaced by another study patient. Bispectral Index BIS-Vista data using a bilateral Bispectral Index BIS-Vista sensor shown in

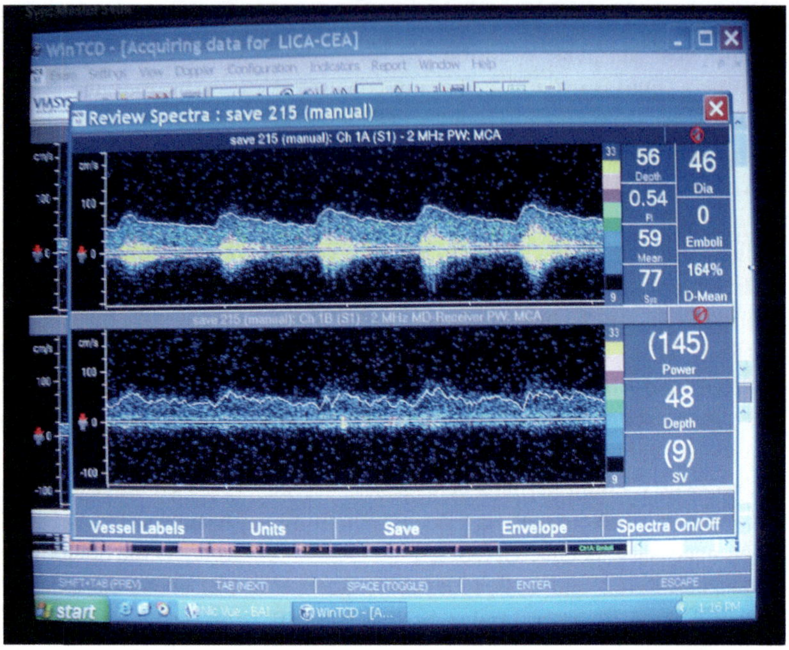

Fig. 5.3 Pioneer TC 8080 Transcranial Doppler. Two channel Pioneer TC 8080 Transcranial Doppler that uses bilateral 2 MHz frequency probes mounted on a Welder Probe Fixation head mount

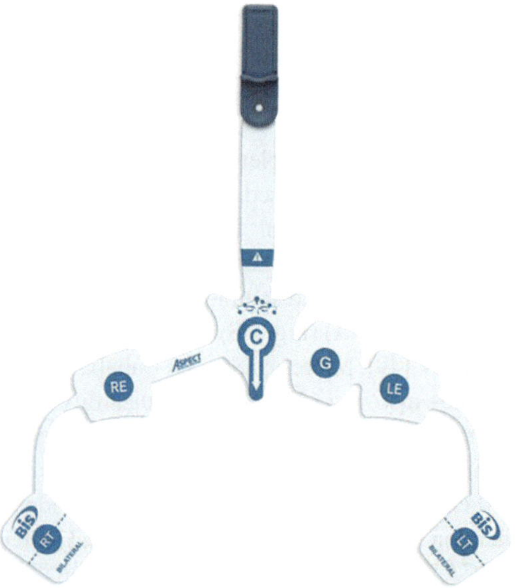

Fig. 5.4 Bilateral Bispectral Index BIS-Vista Sensor

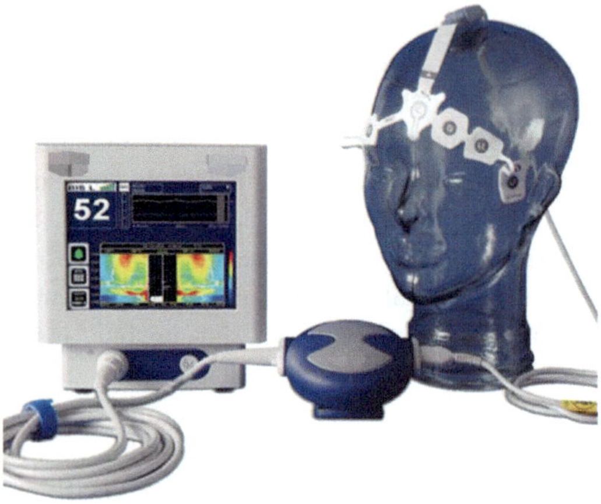

Fig. 5.5 Bilaterally mounted Bispectral Index BIS-Vista Sensor. (©2025 Medtronic. All rights reserved. Used with the permission of Medtronic)

(Fig. 5.4) that was bilaterally mounted as shown in (Fig. 5.5), MCBFV and PI from the TCD monitors were continuously recorded [49].

In our study, coefficient of determination revealed good correlation ($r^2 = 0.763$) between ipsilateral Bispectral Index BIS-Vista and MCBFV following Carotid cross-clamping [49].

The Bispectral Index BIS-Vista exhibited high discriminative power Sensitivity versus 1-Specificity Area Under Receiver Operating Characteristic (ROC) Curve of 0.85 (Confidence Intervals 0.455–0.966) in detecting an ipsilateral 40% Middle Cerebral Artery Blood Flow Velocity (MCBFV) decline [49] (Fig. 5.6). Amazingly, Two-way Analysis of Variance (ANOVA, Location x Time) showed that the Bispectral Index BIS-Vista exhibited a "Global Decline,"

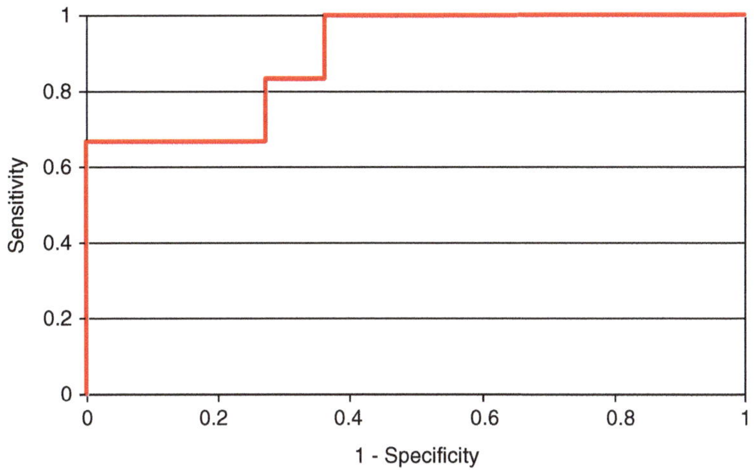

Fig. 5.6 Sensitivity versus 1-Specificity Receiver Operating Characteristic of Bispectral Index BIS-Vista versus Middle Cerebral Artery Blood Flow Velocity (MCABFV) decline. (Reproduced with permission # 5885780888974 from Dahaba AA, et al. The utility of using the bispectral index-Vista for detecting cross-clamping decline in cerebral blood flow velocity. Neurosurgery. 2010;67 3 Suppl Operative:ons102–7; discussion ons107)

i.e., both sides ipsi- and contra-lateral Bispectral Index BIS-Vista declined when only one Carotid Artery on either side was clamped, as there was no significant interhemispheric difference in mean Bispectral Index BIS-Vista values over time [49] (Fig. 5.7). We demonstrated a never reported before amazing new phenomenon of a bilateral Bispectral Index BIS-Vista both sides exhibiting a bilateral global decline when only one Carotid Artery was clamped, i.e., both sides ipsi- and contra-lateral BIS-Vista Bispectral Index values declined with just a unilateral decline in TCD ultrasonography of MCBFV when just one Carotid Artery on either side was clamped for Carotid Endarterectomy [49]. Our results clearly indicate, for the first time, that the Bispectral Index does not distinguish cerebral changes confined to a certain area, rather reveals a global decline in overall level of the cortical activity. Although we demonstrated good correlation and high discriminative power of the Bispectral Index BIS-Vista value in depicting a MCBFV decline that could serve as an indicator of a decline in cerebral activity, still the Bispectral Index BIS-Vista cannot be considered a reliable indicator of cerebral ischemia or cerebral hypoperfusion as the Bispectral Index does not distinguish cerebral changes confined to a certain area, rather indicates a global decline in overall level of cerebral activity [49]. Our study clearly indicates that The Bispectral Index monitor, albeit exhibiting high specificity and high sensitivity in detecting cerebral ischemia or cerebral hypoperfusion, still the Bispectral Index bilateral global decline renders Bispectral Index monitoring incapable of replacing TCD monitoring in determining whether to place a shunt or not [49].

Table 5.5 enumerates and describes all aortic or carotid artery cross-clamping situations where cortical electroencephalography is modified from anesthesiologists' perspective.

5.5.2 Cardiac Arrest and Cardiopulmonary Resuscitation (CPR)

England [51] and Azim et al. [52] reported two cases in which the sudden loss of blood pressure in a hypovolemic cardiac arrest [51, 52] was reported to produce a decline in the Bispectral Index value to a full Zero "0" isoelectricity that was associated with bioelectricity electroencephalographic silence.

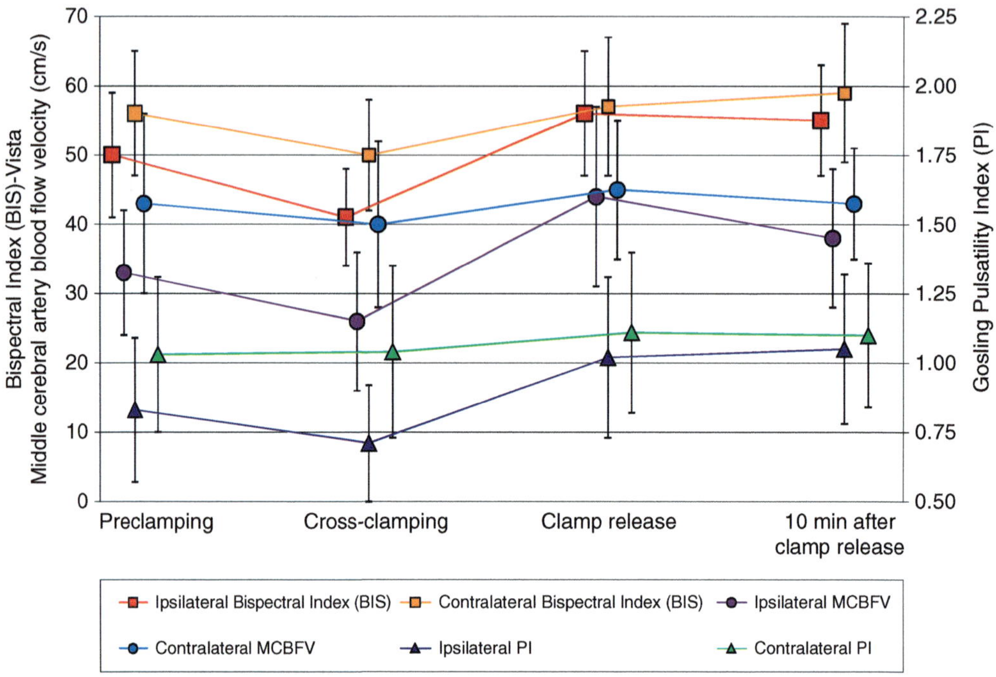

Fig. 5.7 New Phenomenon of Global Decline both sides ipsi- and contra-lateral Bispectral Index BIS-Vista declined when only one Carotid Artery on either side was clamped. (Reproduced with permission # 5885780888974 from Dahaba AA, et al. The utility of using the bispectral index-Vista for detecting cross-clamping decline in cerebral blood flow velocity. Neurosurgery. 2010;67 3 Suppl Operative:ons102–7; discussion ons107)

Table 5.5 Cerebral ischemia aortic or carotid artery cross-clamping

References	BIS Monitor	*n*	Change	BIS change	Explanation
Kakinohana et al. [41]	BIS	1	Descending aortic cross clamp	BIS = 0 ↓	↑ Propofol concentrations
Merat et al. [44]	BIS A-2000 (2.1)	1/2	Carotid cross clamp	BIS < 10 ↓	↓ Cerebral ischemia
Merat et al. [44]	BIS A-2000 (2.1)	1/2	Aortic cross clamp	BIS 8 ↓	↓ Cerebral ischemia
Estruch-Pérez et al. [47]	BIS-XP A-2000	35/70	Awake CEA	BIS 85 < BIS 93	CEA neurological deficit
Dahaba et al. [49]	Bilateral BIS Vista	20	Carotid cross clamp CEA	Bilateral BIS ↓ TCD ↓	Global BIS ↓ Unilateral TCD ↓

n number of patients per group, *BIS* Bispectral Index, *CEA* Carotid Endarterectomy, *TCD* Transcranial Doppler

England [51] report a case of 22-year-old woman undergoing Tricuspid Valve Replacement (TVR) with Bispectral Index monitoring. Transient hypovolemic cardiac arrest resulted when the sternotomy oscillating saw resulted in right atrium laceration, with rapid loss of blood volume in 2 min [51]. Cardiac filling could not be maintained as patient's Mean Arterial Pressure (MAP) dropped to 20 mm Hg with a decline in the Bispectral Index value to a full Zero "0" that was associated with isoelectricity high Burst Suppression Ratio (BSR). With immediate femoral cannulation and Cardiopulmonary Bypass (CPB), the damage to the atrium was repaired and Blood Pressure (BP) was restored with femoral arterial volume replenishment [51]. After 2 min Bispectral Index value returned to premorbid level restoring neuronal function [51]. With reinstitution of Cerebral Blood Flow (CBF), the Bispectral Index increased, and Burst Suppression declined [51]. This suggested that the patient was not going to suffer from a permanent brain injury, which was the case [51].

Azim and colleagues [52] report a second case of a 62-year-old male patient who underwent an uneventful Coronary Artery Bypass Grafting (CABG). Thirty minutes after admission to Critical Care Unit (CCU), his BP progressively dropped with tachycardia and rapidly accumulating chest drainage blood, accompanied by a parallel decline in the Bispectral Index values, culminating in a pulseless isoelectrical activity and a plunge of the Bispectral Index value to Zero "0" [52]. Cardiopulmonary Resuscitation (CPR) was promptly commenced with sternotomy internal cardiac compressions. There was no cardiac tamponade, and the Bispectral Index value picked up to a high of 42 with still no palpable carotid pulse [52]. After 6 min, circulation was restored with MAP of 35–50 mm Hg as the Bispectral Index also gradually increased to 60 after 15 min of resuscitation. Hemostasis was secured in a bleeding point found in one of the grafts. He was then taken off cardiopulmonary bypass with Intra-Aortic Balloon Pump (IABP) support [52].

When normal Blood Pressure (BP) was reinstated via fluid volumes infusions, the Bispectral Index values rose back to pre-morbid values [51, 52]. The main and foremost dispute between Bispectral Index displayed values and raw electroencephalography tracings in the first patient could be attributed to the fact that the decline then rise in the Bispectral Index value lag was delayed after hypotension by around two minutes as a result of the 60 s lagging of the Smoothing Rate (SR) that the Bispectral Index uses to estimate the Bispectral Index value in addition to the delay between the restoration of the BP and CBF autoregulation [51]. Nevertheless, this did not appear with the other patient because of the slow hypotension that took around 20 min would have ameliorated the delay between the hypotension and the apparent displayed decline in the Bispectral Index value [52].

Mérat and colleagues [44] reported a case of cardiac arrest due to electrolyte imbalance in a 62-year-old patient undergoing an Aortic Aneurysm resection, where institution of Bispectral Index A-2000 (version 2.1) monitoring while performing CPR revealed a Bispectral Index decline from 40–50 to a Bispectral Index value of 8 that was associated with a sharp decline in Cardiac Index and central venous Oxygen Saturation (SO_2) indicating severe cerebral ischemia [44]. Following successful CPR, the Bispectral Index returned to normal ordinary value as cerebral perfusion returned to normal when the low Cardiac Output (CO) was pharmacologically corrected. Bispectral Index value of 90 indicated patient's recovery with no neurological injury. They attributed the Bispectral Index decline to severe cerebral ischemia [44].

Szekely and colleagues [53] report a case of a 67-year-old man who underwent left thoracotomy for pulmonary metastasis resection under general anesthesia, combined with thoracic epidural. The immediate postoperative period was uneventful. Six hours postoperatively, suddenly the patient became comatose within few seconds and his BP was unmeasurable [53]. External chest compression, intubation, and ventilation were initiated. Around and through the chest tubes, extensive bleeding was noted. Because there was no jugular venous pulse, internal cardiac compression was performed, and the surgeon clamped the left pulmonary artery with his fingers [53]. While performing CPR to the hypovolemic cardiac arrest, Bispectral Index A-2000 monitor was placed, that revealed a Bispectral Index value of 80 and a with the administration of anesthesia, the Bispectral Index declined to 40 [53]. Bleeding was stopped, and hemodynamic stabilization was achieved 15 min later. After the end of surgery [53], the patient started to recover and 1 h later the Bispectral Index recovered to a value of 90. The patient recovered with no neurological injury [53].

The cardiac arrest case reported by England [51] demonstrated a noticeable lag delay in the Bispectral Index decline immediately following cardiac arrest with another noticeable lag delay in the Bispectral Index to return to normal value after restoration of hemodynamic stability. Another case report by Losasso and colleagues [54] also demonstrated a similar real-time raw electroencephalography (EEG) lag delay in a patient undergoing Carotid Endarterectomy (CEA), who suffered from a cardiac arrest asystole because of Carotid Artery surgical manipulation [54]. Approximately 20 s from asystole onset, the raw EEG showed generalized suppression isoelectric patterns because of complete global ischemia [54]. With the institution of manual chest compression for 15–20 s, Low-Voltage High-Frequency EEG activity started to appear, which signaled the return of cerebral perfusion oxygenation [54]. With successful resuscitation, which lasted for approximately 2-min, the EEG tracings returned to appear as pre-asystole EEG [54]. The rapid return of the EEG signal was because the "well oxygenated blood" was essentially "standing by" in the arterial circulation waiting for the cardiac compressions to push in circulation to restore cerebral blood flow and cerebral oxygenation. Restoration of normal heart contractions was accompanied with restoration of normal pre-asystole EEG [54].

Even though processed Electroencephalography-Depth Of Anesthesia (pEEG-DOA) monitors are not meant or validated to detect "cerebral injury," numerous studies and reports, demonstrating their utility in recognizing cerebral injury, appeared in literature [55]. Goodman and colleagues, [55] described two patients in whom the Bispectral Index abruptly decreased to Zero "0" isoelectricity following cardiac arrest during cardiothoracic surgery. Despite successful Cardiopulmonary Resuscitation (CPR), the Bispectral Index value remained low for the remainder of surgery. Postoperatively, both patients were found to suffer from severe brain damage [55]. The first case was a 63-year-old man with a history of 3-vessel Coronary Artery Bypass Grafting (CABG), 50 pack-year smoking, hypertension, bronchial asthma, and diabetes medical history. The patient with 60% ischemia of his myocardium and 24% Ejection Fraction (EF) was scheduled for a redo CABG. Anesthesia was induced after central catheter, Pulmonary Artery Catheter (PAC), and arterial catheter placements [55]. A Bispectral Index monitor was placed, and a Transesophageal Echocardiography (TEE) probe showed severe global hypokinesis. Shortly after sternotomy, the patient developed severe systemic hypotension, third-degree Atrioventricular (AV) block and pulmonary hypertension that progressed to cardiac arrest [55]. Direct cardiac massage resuscitation was started, epinephrine and atropine were administered, with isoproterenol and dopamine infusions. Transvenous pacemaker and Intra-Aortic Balloon Pump (IABP) were placed. The surgery was stopped, and the chest was closed. Before the cardiac arrest, Bispectral Index values maintained at 40–50 dropped to Zero "0" isoelectricity [55]. After a successful 10-min resuscitation, the Bispectral Index value remained between 10 and 20 despite normothermia and anesthesia discontinuation. Postoperatively, the patient suffered from severe neurological damage as Computerized Tomography (CT) scan revealed left frontal cerebral infarct and multiple basal ganglia infarcts consistent with sustained cerebral hypoperfusion [55].

The second case reported by Goodman and colleagues [55] was a 57-year-old man with a history of uncontrolled hypertension, who underwent an emergency left thoracotomy for repair of Descending Aortic Aneurysmal dissection with contained rupture [55]. A Bispectral Index monitor and TEE probe were placed. Anesthesia was induced with midazolam, etomidate, sufentanil, succinylcholine, and maintained with isoflurane, sufentanil and vecuronium infusions. One-lung ventilation was facilitated and left atrial-femoral CPB was initiated [55]. After aortic-cross-clamp removal, the patient developed hypovolemic cardiac arrest. Direct cardiac massage resuscitation was started with fluid volume replenish and epinephrine administration. After cardiac arrest lasting for 6 minutes, a sustainable cardiac rhythm and BP were restored. However, the Bispectral Index, previously maintained between 40 and 60, abruptly declined to Zero "0" isoelectricity during the cardiac arrest and remained low after successful resuscitation, despite normothermia and inhalational anesthetics discontinuation [55]. Postoperatively, the patient remained unresponsive, and a CT scan revealed a hypoperfusion left hemisphere cerebral infarct. After a 2-month hospital stay, the patient was transferred with severe neurological damage to a long-term care facility [55].

The Bispectral Index has also been successfully used to assist medical resuscitation teams via assessing the resuscitative maneuvers adequacy during cardiac arrest's CPR [52, 53, 56]. Numerous reports suggest that Bispectral Index monitoring during cardiac arrest's CPR may be very helpful, not only in ensuring the resuscitation adequacy, but also assessing the effectiveness of changing techniques such as switching from external chest compressions to a more aggressive open cardiac massage [56]. A Bispectral Index >30 during CPR may be ideal and could also be reassuring of an adequate cerebral perfusion [56]. This is exactly what Goodman and colleagues [55] did not encounter with their two case reports when neither patient reached that level of Bispectral Index of >30 after their cardiac arrest unsuccessful resuscitation. In their two case reports, the Bispectral Index value did not recover to pre-arrest Bispectral Index values, and both patients developed severe permanent brain damage [55]. A low Bispectral Index value is an indicative sign of cerebral hypoperfusion anoxia [43, 51, 57, 58]. On the other hand, a high Bispectral Index value of >30 during CPR [56], reflects the presence of cerebral activity and should encourage the resuscitation team to continue with their efforts. This finding is reported in the other case reports where the Bispectral Index value returned to pre-arrest levels, and the patients did not suffer from permanent neurological damage because of cerebral hypoperfusion anoxia [43, 51, 57, 58]. Although the Bispectral Index monitor lacks the sensitivity and specificity of the traditional raw EEG in successfully detecting cerebral ischemia events, it may be another useful tool in predicting cerebral neurological recovery after intraoperative cardiac arrest.

However, a low Bispectral Index value is more difficult to interpret; as Myles and Cairo, [59] using Bispectral Index (version 3.4) monitor in another case of cardiac arrest resulting from pulmonary embolism; in which following 60 min of CPR patient's pupils were fixed and dilated while Bispectral Index was 0–5 with Burst Suppression Ratio (BSR) 100% indicating anoxic global cerebral injury. They reported an "unexplained" Bispectral Index level rising from Zero "0" to 38 after anoxic brain injury. They concluded that the pattern was consistent with Bispectral Index artifact and was not indicative of presence of cerebral activity of successful resuscitation [59]. Movement Bispectral Index artifacts that occur during CPR external chest compressions is always a concern and has a major impact on interpretation of Bispectral Index tracings during CPR, as indicated in an interesting study by Fatovich and colleagues [60], who placed a Bispectral Index A-2000 BIS-XP on 21 Out-of-Hospital Cardiac Arrest patients and one on cadaver for comparison during Chest compressions CPR [60].

Bispectral index values sharp decline following cardiac arrest, because of cerebral global ischemia, was also shown in observational studies [61]. Burjek and colleagues [61], in a cardiovascular Critical Care Unit (CCU) observational study, prospectively tested the premise of a Bispectral Index value decline in patients who received therapeutic hypothermia with fewer hypnotic agents doses following cardiac arrest, could predict poor neurological outcome. In 146 consecutive cardiac arrest patients treated with therapeutic hypothermia, 84/141 subjects (60%) who exhibited Bispectral index values of <31 ($P < 0.001$), and received 17% fewer sedative agents requirements ($P < 0.001$), were discharged from the CCU with Cerebral Performance Category (CPC) scores of 3, 4, or 5 representing poor neurological outcomes; compared with 57/141 (40%) subjects discharged with CPC scores of 1, 2 representing good neurological outcomes [61]. To early predict a good neurological outcome was superior at 7 h from the start of CCU stay. Each 10-point Bispectral index score decline is accompanied with 59% rise in poor-outcome odds (Confidence Interval 32–76%; $P < 0.001$) [61]. They concluded that Bispectral Index scores and sedative requirements early during hypothermic therapy for cardiac arrest episodes would better identify subjects that would have a high probability of recovery from cerebral hypoxia following cardiac arrest [61, 62].

Similarly, Stammet and colleagues [63], in anther observational prospective trial in comatose subjects who received hypothermic therapy following a cardiac arrest, utilized mean Bispectral Index BIS-XP values measured over the first 24 h after CCU admission to predict neurologic outcome [63]. They demonstrated that mean Bispectral Index values were statistically significantly higher in subjects with favorable outcome than subjects with unfavorable outcome [63]. Blinded neurological outcome assessment, using CPC scores, was undertaken six months following the cardiac arrest. In their study, 46/96 subjects (48%) exhibited CPC scores of 1–2 good neurological outcome at 6-month of CCU stay, compared with 50/96 patients (52%) with a CPC scores of 3–5 poor neurological outcome [63]. Over 24 h, mean ± SD Bispectral Index recordings revealed that the Bispectral Index value of 38 ± 9 over 24 h was statistically significantly ($P < 0.001$) more in subjects with favorable recovery than subjects with unfavorable recovery Bispectral Index value of 17 ± 12. The Bispectral Index recording at 30 min intervals revealed that to optimally predict neurological outcome at 12.5 h, Bispectral Index cut-off value was 23, with a Receiver Operating Characteristic (ROC) Area Under the Curve (AUC) of 0.89, specificity 89% and sensitivity 86% [63].

Seder and colleagues [64], in a retrospectively designed assessment of subjects who underwent CPR following cardiac arrest, postulated what they called a "Neuroradiological Triage" based on the initial post-cardiac arrest using the Bispectral Index (BIS A-2000 and Vista) value to predict patients' post-resuscitation outcome [64]. In their 171 patients cohort, they devised a "Bispectral Index identification (BISi)" parameter, defined as the mean Bispectral Index plateau value recorded during 10-min following the administration of the first dose of Neuromuscular Blocking Agent (NMBA) [64]. Those patients with a BISi < 10 points suffered 91% overall mortality, of which 82% of patients will die because of neurologic etiology, while BISi > 20 points resulted in just 36% overall mortality, of which 12% of patients will die because of neurologic etiology [64].

Table 5.6 enumerates and describes all Cardiopulmonary Resuscitation (CPR) situations where cortical electroencephalography is modified from anesthesiologists' perspective.

5.5.3 Increased Intracranial Pressure (ICP)

Changes in electroencephalography (EEG) patterns correlate well with changes in Cerebral Perfusion Pressure (CPP) and hence processed Electroencephalography-Depth Of Anesthesia (pEEG-DOA) monitors such as Entropy may also correlate with CPP. To highlight the potential utilization of Entropy in such situations for indicating Cerebral Perfusion in patients with increased

Table 5.6 Cerebral ischemia cardiac arrest and cardiopulmonary resuscitation (CPR)

References	BIS Monitor	*n*	Change	BIS change	Explanation
England et al. [51]	BIS	1	Hypovolemic cardiac arrest	CPR BIS = 0 ↓	Isoelectricity
Azim et al. [52]	BIS	1	Hypovolemic cardiac arrest	CPR BIS = 0 ↓	Isoelectricity
Merat et al. [44]	BIS A-2000 (2.1)	1/2	Electrolyte imbalance cardiac arrest	CPR BIS 8 ↓	Cerebral ischemia
Szekely et al. [53]	BIS A-2000	1	Hypovolemic cardiac arrest	CPR BIS 80	Effective cardiac massage
Goodman et al. [55]	BIS	1/2	CABG cardiac arrest	CPR BIS = 0 ↓	Isoelectricity
Goodman et al. [55]	BIS	1/2	Aortic Aneurism rupture cardiac arrest	CPR BIS = 0 ↓	Isoelectricity
Myles and Cairo [59]	BIS (3.4)	1	Pulmonary Embolism cardiac arrest	CPR BIS = 0 BIS 38	Movement Artifacts

n number of patients per group, *BIS* Bispectral Index, *CABG* Coronary Artery Bypass Graft, *CPR* Cardiopulmonary Resuscitation

Intracranial Pressure (ICP), Khan and colleagues [65] report two cases of decreased Cerebral Perfusion because of elevated ICP. The first case is of a patient undergoing emergency neurosurgical procedure, when development of a large intracranial hematoma following temporal lobe tumor resection resulted in general anesthesia titrated to low Response Entropy (RE) /State Entropy (SE) RE/SE values of 37/33 rapidly increased to 94/83 with cranial decompression and evacuation of the hematoma, reflecting an increase in CPP [65]. The second case is of an acute hydrocephalus secondary to subarachnoid hemorrhage with low Entropy RE/SE values of 56/50, rapidly increased to 80/72 with cerebral decompression and insertion of external ventricular drain [65]. Maintaining systemic blood pressure to preserve the CPP is the anesthetic goal when managing patients with raised ICP. The pEEG-DOA monitors may hold valuable information on guiding anesthetic management in patients with decreased CPP because of increased ICP, for a better neurological outcome [65].

5.5.4 Balloon Occlusion/Coiling

An interesting case report of a 56-year-old man undergoing coiling of an accidentally discovered unruptured approximately 1.2 cm wide necked intracranial left Internal Carotid Artery (ICA) aneurysm. Under sevoflurane 1.2–1.4%, and Target Controlled Infusion (TCI) remifentanil plasma concentration (C_e) 4 ηg ml^{-1} Bispectral Index Vista was maintained at 40–55 range and 30 s smoothing interval [66]. Coiling was via a right femoral artery vascular access guide catheter in the extra-cranial segment of the left ICA, and a balloon catheter navigated into the intracranial circulation. The 4 mm inflatable /deflateable balloon was placed across the aneurysm neck and during inflation there was occlusion of the aneurysm neck and the proximal segment of the left Middle Cerebral Artery (MCA). The balloon was intermittently inflated across the aneurysm neck to prevent coil prolapse during the 11 platinum microcoils placements into the aneurysm. Each inflation lasted for 30–120 s, and each time the balloon was inflated, after an initial 30–40 s delay, the Bispectral Index value declined to 25 over a period of 60 s. Upon balloon deflation, the Bispectral Index value returned rapidly to its pre-inflation value. There was no electromyography (EMG) interference nor any hemodynamic variations that could have influenced these changes. The patient's sudden, reversible, and repeated decreases in intraoperative Bispectral Index values after a 30 s lag that took another minute to reach nadir during a balloon occlusion /coiling of the intracerebral MCA aneurysm could be attributed to repeated periods of cerebral hypoperfusion [66]. Each time the balloon occluded the cerebral flow to MCA, this resulted in sudden abrupt reversible decreases in the Bispectral Index values that was not associated with changes in anesthetic technique, EMG, or hemodynamics [66].

5.5.5 Syncope Bradycardia and Tet-Spells

Win and colleagues [67] reported 2 interesting cases of Bispectral Index A-2000 decline with Syncope before sedation in 2 patients with needle phobia undergoing dental procedures under propofol sedation and local anesthesia. The first patient was a 27-year-old male with suspected needle phobia, an awake Heart Rate (HR) 96 beats min^{-1}, Blood Pressure (BP) 135 /67 mm Hg and awake Bispectral Index score of 95 that declined to 42 when he lost consciousness and developed bradycardia 35 beats min^{-1} and hypotension 78 /30 mm Hg from a vasovagal reflex [67]. The authors immediately raised his legs, administered i.v. atropine 0.5 mg, 100% oxygen, and infused i.v. fluids. After recovery, the Bispectral Index score returned to normal. The Bispectral Index decline was attributed to cerebral hypoperfusion [67]. In the second case of 29-year-old male, before sedation, his HR was 80 beats min^{-1}, BP was 130 /70 mm Hg, and his Bispectral Index score was 98. He developed bradycardia of 68 beats min^{-1}, as the Bispectral Index score decreased to 35 when the fainting episode occurred, although the BP was maintained at 130 /80 mmHg. The patient was treated with i.v. atropine 0.5 mg, raising his legs and infusing i.v. dextrose /saline. After recovery, the patient's consciousness level, BP, HR, and Bispectral Index score returned to normal. They attributed the Syncopal attacks to an increase in slow-wave electroencephalography (EEG) activity δ-waves during syncope and before their syncopal episodes [68].

Win and colleagues [67] reported another interesting case of a patient who lost consciousness with decreasing Bispectral Index A-2000 score from 98 to 35 while his Heart Rate (HR) and Blood Pressure (BP) were maintained at 70 beats min^{-1} and 130 /80 mm Hg [67]. Shortly thereafter, a seizure of the entire body occurred and continued for few minutes, during which the Bispectral Index value increased to 98 because of the increased electromyography (EMG) activity. In that patient the Bispectral Index decline cannot be attributed to cerebral hypoperfusion [67]. A possible explanation of the original decline would be that shown by Mercader and colleagues [68], in which they demonstrated an increase in the EEG slow wave activity specially the δ waves, that lateralized to the left side of the brain, before and during a syncopal episode [68].

Mercader and colleagues [68] conducted a study in 18 near syncope or syncope patients whom they placed in head up tilt table test, with EEG monitoring, to determine the part the cerebral cortex plays in neurally mediated syncope [68]. They showed that, before syncope, EEG activity lateralizes to the left hemisphere of the brain [68]. They showed that this left hemisphere lateralization precedes syncope and is associated with loss of consciousness, sweating, hypotension, and bradycardia, suggesting a cerebral cortical role in the neurally mediated syncope [68].

Bleeker and colleagues [69] reported another interesting case of pre-induction for cardiac surgery, Bispectral Index A-2000 BIS-XP values in an alert and communicative 76-year-old patient who, without provocation, suddenly complained of dizziness, stopped breathing, became unresponsive, his eyes rolled upwards and deviated laterally as he suffered from brainstem Transient Ischemic Attacks (TIA) [69]. The Bispectral Index registered the TIA when the Bispectral Index declined to 60 before induction of anesthesia. Lack of input from the brainstem to the frontal cortex resulted in reduced cortical electrical activity that was clearly registered by the Bispectral Index. The Bispectral Index increased back to 85 when he spontaneously regained consciousness [69]. They attributed the Bispectral Index decline to a decrease in the frontal cortical activities because of TIA brainstem ischemia resulting in diffuse low voltage EEG activities and bilateral slowing of the EEG while preserving the posterior α waves rhythms [69].

Hashimoto and colleagues [70] reported a case of a 75-year-old female patient undergoing right vitrectomy under propofol, fentanyl Total Intravenous Anesthesia (TIVA). Twenty minutes after the start of surgery, with a stable BP 175 /90 mmHg and HR 60 beats min^{-1}, she started to develop bradycardia 40 beats min^{-1}, but the Bispectral Index value stayed stable at 70. She then developed sudden severe bradycardia 33 beats min^{-1}, without hypotension, that did not well respond to atropine administration [70]. This sudden severe bradycardia event was accompanied by a marked decrease in the Bispectral Index value from 70 to 40 that was associated with an increase in the Burst Suppression Ratio (BSR) to 33. Following the initiation of an isoproterenol infusion, the HR promptly increased to normal values along with an increase in the Bispectral Index to 70 [70]. After surgery the patient recovered from TIVA without any neurological consequences. The authors speculated that severe bradycardia during anesthesia causing cerebral hypoperfusion affected patient's cerebral function. They concluded that Bispectral Index monitoring in such situations may be a useful tool for the detection of "bradycardia related" cerebral hypoperfusion [70].

Chauhan and colleagues [71] conducted a study in pediatric patients with congenital heart disease undergoing corrective heart surgeries [71]. While using the Bispectral Index BIS-XP A-2000 monitoring on children with cyanotic

Table 5.7 Cerebral ischemia syncope bradycardia and Tet-spells

References	BIS Monitor	*n*	Change	BIS change	Explanation
Win et al. [67]	BIS A-2000	2 +1	Needle Phobia Syncope HR ↓ BP ↓	BIS 35-42 ↓	δ waves ↑ CBF ↓
Bleeker et al. [69]	BIS-XP A-2000	1	TIA	BIS 60 ↓	Posterior α waves
Hashimoto et al. [70]	BIS	1	HR 33 ↓ normal BP	BIS 40 ↓ BSR 33 ↑	Cerebral Hypoperfusion
Chauhan et al. [71]	BIS-XP A-2000	All	Fallot Tetralogy Tet Spells	BIS 25-15 ↓	Hypercyanotic Tet Spells

n number of patients per group, *HR* Heart Rate, *BP* Blood Pressure, *BIS* Bispectral Index, *BSR* Burst Suppression Ratio, *CBF* cerebral blood flow, *TIA* Transient Ischemic Attacks, *Tet Spells* Fallot Tetralogy cyanotic Spells

heart disease, mainly Fallot Tetralogy patients, authors found that Bispectral Index values very consistently warned of the occurrence of hypercyanotic "Tet Spells" totally not related to Depth of Anesthesia (DOA) [71]. Whenever any child with Fallot Tetralogy manifested a Tet Spell, the Bispectral Index value suddenly and dramatically always decreased, in all children, from the pre-Tet Spell Bispectral Index values of 50–60 to very low Bispectral Index values of 25–15 that would persist at these very low Bispectral Index levels until the Tet Spell was successfully resolved with an α-receptor stimulant such as phenylephrine [71]. This sudden decrease in Bispectral Index during a Tet Spell first started to occur in the preoperative holding area after the patients were premedicated and when an Oxygen (O_2) supplementation nasal cannula was placed on patients' face with Bispectral Index and O_2 saturation pulse oximetry monitoring [71]. Likewise, the sudden dramatic decrease in the Bispectral Index during a Tet Spell continued to occur under anesthesia when surgery was in progress, before the institution of Cardiopulmonary Bypass (CPB). While the medical staff in the preoperative holding area would question the oximetry low O_2 saturation value by rechecking O_2 flow or repositioning the pulse oximeter probe, there was no doubt when the Bispectral Index suddenly and dramatically decreased with the onset of the Tet Spells under anesthesia [71]. Interestingly there was always a lag-period in the Bispectral Index recovery to the pre-Tet Spell levels that was only reached after the O_2 saturation returned to pre-Tet Spell O_2 saturation oximetry values. The authors concluded that if Bispectral Index is being monitored in a child with Fallot Tetralogy, a sudden decrease in Bispectral Index in the absence of sedative or anesthetic agents could be attributed to a Tet Spell requiring an immediate intervention [71].

Table 5.7 enumerates and describes all Syncope Bradycardia and Tet-Spells situations where cortical electroencephalography is modified from anesthesiologists' perspective.

5.5.6 Stroke Cerebral Infarction

Welsby and colleagues [72] reported that the Bispectral Index was the earliest indicator of acute perioperative stroke cerebral infarction during the removal of an Abiomed BVS 5000 Left Ventricular Assist Device (LVAD) in a 69-year-old man undergoing emergency Coronary Artery Bypass Grafting (CABG) after acute myocardial infarction complicated by post-infarction angina and cardiogenic shock [72]. The Bispectral index precipitously decreased from its previously steady state of 50–60 to <10 and remained <15 until the end. Despite heparinization, a cerebral embolus was suspected, and a Computerized Tomography (CT) of the brain confirmed presence of an extensive cerebral infarction [72]. The patient was managed conservatively and quickly succumbed to complications consistent with intracranial hypertension [72]. Up to 3% of cardiac surgical patients suffer serious, typically embolic, neurological complications, with an associated 20% mortality rate. The opportunity to make an earlier diagnosis may help improve outcome in this group as novel therapeutic options become available [73].

5.5.7 Air Embolism

Villacorta and colleagues [74] reported a precipitous Bispectral Index decline that was suggested as a clear indicator of a decline in Cerebral Blood Flow (CBF) following showers of air emboli in a 46-year-old woman undergoing her third redo Aortic and Mitral Valve Replacements. Propofol, sufentanil anesthesia was titrated to maintain the Bispectral Index between 25 and 45 [74]. On release of the Aortic cross-clamp the Bispectral Index value dropped to Zero "0" and remained between 0 and 25, with Burst Suppression Ratio (BSR) >95% until the end of operation. Showers of air emboli resulted in a sudden, severe, unexplained, and sustained fall in the Bispectral Index. Postoperatively, the patient's Glasgo Coma Score was 5, without sedation with decerebrate movements. A Computerized Tomography (CT) scan was consistent with multiple ischemic lesions presumed to be due to showers of air emboli as the patient succumbed 2 days later [74].

Kin and colleagues [75] presented a case report of a 38-year-old man with Marfan syndrome who underwent an Aortic Valve Replacement under Bispectral Index A-1050 monitoring of sevoflurane general anesthesia maintained at mid-50s Bispectral Index value. Aortic cross-clamp release resulted in sudden [75], severe, and sustained Bispectral Index decline from mid 50s to 30 that lasted for more than an hour because of multiple ischemic lesions presumed to be due to large amount of air in the Aortic Arch and descending Aorta as revealed by Transesophageal Echocardiography (TEE). The patient suffered from delayed wakening and temporary hemiparesis but recovered without any neurological deficit [75].

5.5.8 Dislodged Subclavian Arterial Cannula

Ellenberger and colleagues [76] present an interesting case report of a 62-year-old female with an Ascending Aortic Aneurysm (AAA) and stable Aortic Valve Regurgitation scheduled for elective valve replacement and aortic prosthetic graft insertion under morphine, propofol general anesthesia monitored by Bispectral Index BIS-XP (version 4.0) and maintained between 40 and 60 [76]. Normothermic Cardiopulmonary Bypass (CPB) was instituted with cannula inserted in the right subclavian artery and the right atrium. After Aortic cross-clamping, Bispectral Index value abruptly decreased from 40 to Zero "0" along with a steep increase in the Burst Suppression Ratio (BSR) to 100%, that was not resolved [76]. Transcranial color and pulsed Doppler failed to demonstrate blood flow in the Middle Cerebral Artery (MCA) and Anterior Cerebral Artery (ACA) [76]. The subclavian arterial cannula was dislodged. Postoperatively, no clinical sign of awakening was noticed and Magnetic Resonance Imaging (MRI) documented severe and diffuse cortical lesions consistent with anoxic encephalopathy and Persistent Vegetative State (PVS) that was confirmed by a flat electroencephalography (EEG) line [76]. This was confirmed after life support withdrawal and postmortem autopsy confirming global ischemic brain injury [76].

5.5.9 Cerebral Hemorrhage

Uhrig and colleagues [77] present a case of a 52-year-old man, American Society of Anesthesiologists (ASA) classification physical status I, suffering from severe headache. Two days after the initial headache, he was conscious with World Federation of Neurosurgical Societies Subarachnoid Hemorrhage Score (WFNS SAH) grade 1. Angio-Computerized Tomography (CT) scan showed a spontaneous subarachnoid hemorrhage of the anterior interhemispheric fissure, basal cisterns, and sylvian fissures, Fisher scale 3 [77]. He underwent intracranial basilar neuroendovacular coil embolization of the Anterior Communicating Artery (ACA) under propofol 3 μg ml^{-1}, remifentanil 2 ηg ml^{-1} Target Controlled Infusion (TCI) [77]. During embolization, Bispectral Index values maintained at 60-40 suddenly dropped to Zero "0" along with a steep increase in the Burst Suppression Ratio (BSR) to >90, because the aneurysm ruptured during the insertion of the first coil, as clearly shown by the contrast medium leakage through the aneurysm wall [77]. Two hours later, the Bispectral Index stabilized at 45–50, as the coiling session was continued with successful placement of a second coil to secure the aneurism. Patient fully recovered from anesthesia with Glasgow Coma Scale (GCS) 14, without sensory or motor deficits, and he regained full consciousness GCS 15, 1 day later [77]. While undergoing a coil embolization of ruptured intracranial aneurysm, a sudden transient Bispectral Index decline to Zero "0" suggests that the change in the Bispectral Index values could provide useful information about the cerebral hemodynamic during aneurysm treatment and might provide indications of a serious cerebral event.

Prabhakar and colleagues [78] reported a similar case in a 55-year-old male. Digital subtraction angiography revealed Interpendendicular Cistern hemorrhage from a 1.5 cm neck basilar top aneurism [78]. He underwent coil embolization under fentanyl, isoflurane anesthesia maintained to keep the Bispectral Index between 40 and 60. Before coiling the left Posterior Cerebral Artery (PCA) could be stented, while the right PCA could not be stented when the Bispectral Index suddenly decreased to Zero "0" as the contrast medium injection showed aneurism rupture [78]. Immediate coiling of the aneurism was successfully done, and the Bispectral Index started to rise to 0–25. A Computerized Tomography (CT) scan showed a ruptured aneurism acute subarachnoid hemorrhage and hematoma in the right occipital region. Two hours later the electrocardiography (ECG) showed electromechanical dissociation, and the patient did not survive [78]. Interpreting Bispectral Index value of Zero "0," corresponding to an isolelectric electroencephalography (EEG) to indicate a ruptured intracranial aneurysm, after ruling out technical issues such as leads disconnection, can be difficult [78].

5.5.10 Cervical Hematoma

Umegaki and colleagues [79] reported a case of a 77-year-old man presenting with a ruptured Abdominal Aortic Aneurysm (AAA), and severe hemorrhage that was further complicated by an attempted central venous catheter mis-puncture. His Blood Pressure (BP) was 40/20 mm Hg that increased with fluid volume resuscitation and an insertion of an Aortic Occlusion Balloon (AOB) [79]. The patient underwent aneurysmectomy with a Y-Graft placement under ketamine 1 mg kg^{-1} h^{-1} anesthesia and Bispectral Index A-1050 maintained at 55, with Burst Suppression Ratio (BSR) at 0. First severe Bispectral Index reduction with elevation of BSR was observed following severe hypotension after deflation of the AOB. The re-inflation of the AOB and rapid blood transfusions improved his hemodynamics, Bispectral Index and BSR values [79]. However, as a second severe Bispectral Index reduction to Zero "0" and BSR to 100% elevation occurred, surgeons discovered a heavy massive subcutaneous cervical hematoma swelling around the previously mis-punctured right Carotid Artery extending throughout the whole neck and depressing the bilateral jugular veins, with his tongue enlargement and projection, but without hypotension as his systolic BP was 80–100 mm Hg [79]. Immediate cervical relief incision improved the Bispectral Index and BSR. This case suggests that Bispectral Index monitoring may be a simple and convenient indicator for cerebral ischemia detection [79].

5.5.11 Cotton Ribbon Tightened Around Patient's Neck

Lauwick and colleagues [80] present a very interesting case of a 63-year-old patient undergoing laparoscopic prostatectomy under desflurane anesthesia. With the gradual positioning of the patient in Steep Trendelenburg position head lowering, Bispectral Index A-2000 started gradually declining from 38 to 16 and finally to 6 and Burst Suppression Ratio (BSR) indicating isoelectricity when a cotton ribbon used to secure an endotracheal tube, instead of tape because of facial hair, slid and accidentally tightened around patient's neck [80]. The ribbon was promptly cut and immediately the Bispectral Index returned to the previous level of 38 [80].

5.5.12 Hemodynamic Changes with Patient Positioning

Several factors were shown to alter processed Electroencephalography-Depth Of Anesthesia (pEEG-DOA) indices that are not related to Depth of Anesthesia (DOA). Patient position was found to be one of the factors that could influence the pEEG-DOA indices [81]. Changing the subject's position to Heads Down position from Neutral Supine position increased Bispectral Index score, the Bispectral Index score decreased when the patient's head was again repositioned upwards [81]. Kaki and colleagues [81] conducted a study to assess the impact of changing patients' position on Bispectral Index A-2000 (version 3.21) in 40 patients undergoing minor surgeries under general anesthesia. Patients were first kept in the Neutral Supine position for 15-min, with Bispectral Index score, Mean Arterial Pressure (MAP), and Heart Rate (HR) continuously recorded [81]. Patients were then inclined to a 30° Head Down position for 15-min, back to the Neutral Position for 15-min, and lastly inclined to a 30° Head Up position for 15-min. There was a significant increase in median Bispectral Index score of 47 in the Head Down position compared with median Bispectral Index value of 40 in the Neutral Supine position ($P < 0.05$) [81], whereas the Head Up position significantly decreased median Bispectral Index score to 39 compared with Neutral Supine position median Bispectral Index score of 41 ($P < 0.05$). The authors concluded that changing patients' position significantly changes the Bispectral Index score, which might affect interpretation of pEEG-DOA indices data [81]. The authors ascribed these changes in the Bispectral Index to be caused by the increase and decrease in cerebral perfusion with MAP changing during Head Down and Head Up positions, respectively [15, 81].

Many situations were shown to alter pEEG-DOA readings, independent of the DOA, because of disturbances in Cerebral Perfusion on top of the anesthetic agents' effect. Numerous publications showed a DOA index decline with cerebral perfusion impairment [82]. Soto and colleagues [82] demonstrated that the Bispectral Index score was the first warning sign of a serious neurological event taking place in a 65-year-old patient with Intramedullary Cervical Neoplastic Protrusion, undergoing biopsy taking and Cervical Spine Laminectomy [82]. The patient had a history of hypertension that was under control with anti-hypertensives. After fentanyl 5 $\mu g\ kg^{-1}$ and propofol 2 $mg\ kg^{-1}$ anesthesia induction, anesthesia was maintained with propofol and remifentanil continuous infusions. A Central Venous Catheter (CVC) was situated in the right Internal Jugular Vein (IJV) and an arterial cannula was placed in the left Radial Artery for invasive Blood Pressure (BP) monitoring with transducer placed at the level of the External Auditory Meatus. Bispectral Index score readings and hemodynamic parameters were continuously recorded [82]. After applying a tight band around the legs and infusion of Ringer Lactate fluid load, the subject was very slowly placed in the sitting position while maintaining normal BP and HR. However, before surgery, when the surgeons were fixing the subject's head in an upward position looking for an appropriate head position, as they slightly flexed her cervical column, abruptly the Bispectral Index score of 47 suddenly dropped down to Zero "0" with Burst Suppression Ratio (BSR) increasing to to 97%, Heart Rate declined from 87 min^{-1} to 38 min^{-1} and BP of 122/68 decreased to 55/33 [82]. Administering atropine 1 mg and phenylephrine 0.15 mg resulted in a dangerous 210/100 hypertension and 38 beats min^{-1} increasing to 47 beats min^{-1}. As the brainstem was compressed by surgeon's head manipulation in the sitting position, the patient was immediately placed back in the supine position that resulted in a prompt stabilization of all her hemodynamic parameters, and the Bispectral Index score slowly increased as the patient woke up without any neurologic damage [82]. The

Bispectral Index can represent an early warning system to highlight cortical brain incidents when unexpected, serious, extensive and ongoing change in the Bispectral Index score takes place, particularly if associated with significant hemodynamic changes [82]. Authors presumed that severe cerebral hypoperfusion could have caused the Bispectral Index to decline to Zero "0." With a notably wide adoption of the Bispectral Index monitoring in daily anesthesia practice, this could help detecting critical conditions that could potentially lead to neurological injuries, motivating swift interventions to avoid permanent neurologic damage from taking place [15, 82].

5.5.13 Twisting of the Heart

Mourisse and colleagues [83] report a case of a 61-year-old man undergoing Coronary Artery Bypass Graft (CABG) of the left Marginal Coronary Artery under propofol Target Controlled Infusion (TCI) effect-site concentration (C_e) 2.0 $\mu g\ mL^{-1}$, titrated to maintain Bispectral Index A-2000 (version 1.08) between 40 and 50 [83]. The heart was difficult to expose because of fatty tissue and adhesions. During off-pump, to gain access to the anastomosis site, the heart was twisted to the right, when suddenly the Bispectral Index dropped from 40 to 11, as the heart was twisted for Coronary Artery grafting [83]. The surgeon was unable to return the heart to its position before the grafting was finished. The Bispectral Index returned to 40 when the heart was returned to position and the circulation restored [83].

Hemmerling and colleagues [84] report a similar case of Bispectral Index decline in a 58-yearr-old man undergoing off-pump Coronary Artery Bypass Graft (OPCAB), under sevoflurane anesthesia, as the heart was being repositioned compressing the carotid arteries [84], representing a case of possible reversible global cerebral hypoperfusion during distal grafting of the left Circumflex Coronary Artery [84]. The Bispectral Index A-2000 (version 2.10) suddenly decreased from 45–50 to Zero "0" during distal grafting for a period of 3 min during OPCAB. There was immediate return to Bispectral Index values of 40–50 after the heart was repositioned to the normal configuration [84]. Neurologic evaluation after tracheal extubation in the operating room was normal and the patient did not suffer any neurologic sequelae. Postoperative recovery was uneventful, and the patient was discharged 5 days after surgery [84]. Cerebral hypoperfusion is a possible complication of heart repositioning during OPCAB; thus Bispectral Index monitoring in OPCAB could be an early indicator of cerebral hypoperfusion [84].

5.5.14 Cerebral Hemoglobin Saturation (SrO_2)

Hayashida and colleagues [85, 86], in two separate publications, demonstrated a decrease in cerebral blood flow measured by cerebral hemoglobin Oxygen Saturation (SrO_2) Near-Infrared Spectroscopy (NIRS) that coincided with an acute decrease in Bispectral Index A-1050 and slowing of the raw electroencephalography (EEG) waveforms in 65 children undergoing cardiac surgery in one study [85], and in 10 children undergoing cardiac surgery in another study [86]. They noted 72 episodes in 38 children in one study [85], and 14 episodes in 5 children in the other study [86], of simultaneous decreases in SrO_2 and acute decrease in Bispectral Index during acute hypotension in children [85, 86], suggesting a reduction in cerebral blood flow, associated with acute slowing of the raw EEG waveforms indicating cerebral hypoperfusion [85, 86]. Cerebral hypoperfusion caused by hypotension may occur frequently during pediatric cardiac surgery. The bispectral index may indicate changes in cerebral activity when the cerebral circulation is affected by acute hypotension [85, 86].

5.6 Neurological Disorders

Not all processed Electroencephalography-Depth Of Anesthesia (pEEG-DOA) monitors have pediatric /neonatal software or sensors small enough for infant/neonatal heads. Depth of Anesthesia (DOA) monitors seem to be particularly sensitive to pediatric and elderly populations extremes of age which brings in question the reliability of DOA monitors in pediatric age, especially in neonates, and in the elderly since these ages of life, for pharmacokinetic and pharmacodynamics reasons, are particularly at risk of incurring unwanted side effects of drugs commonly used in anesthesia.

Table 5.8 enumerates and describes all neurological disorders where cortical electroencephalography is modified from anesthesiologists' perspective.

5.6.1 Cerebral Palsy and Mental Retardation

Choudhry and colleagues [87], using Bispectral Index A-2000 (version 3.21) monitor, compared 20 Cerebral Palsy Mentally Retarded (CPMR) children with 21 normal children under sevoflurane 8% induction and sevoflurane–Nitrous Oxide (N_2O) anesthesia maintenance [87]. At midazolam premedication sedation, at 1% sevoflurane concentrations, and after emergence from anesthesia, the mean Bispectral Index values were statistically significantly lower in CPMR

Table 5.8 Neurological disorders

References	Monitor	*n*	Change	Index change	Explanation
Choudhry et al. [87]	A-2000 (3.21)	20/21	Cerebral palsy Mental Retardation	BIS ↓	abnormal mental function
Valkenburg et al. [89]	BIS	18/35	Intellectually Disabeled Children	BIS ↓	abnormal mental function
Ponnudurai et al. [90]	BIS	20 x4	Mild, Moderate, Severe and Profound MR	BIS no change	No effect of MR on BIS
Faraoni et al. [91]	BIS, SE, RE	1	Intracranial Mass	BIS 67, SE 17, RE 16	Artifact rejection allgorithms
Ferreira et al. [92]	BIS-XP A-2000	13/26	Intracranial Mass	BIS ↑ at LOC	Abnormal mental function
Fudickar et al. [93]	BIS, AEP	1	Ipsi- contra- intracranial Mass Bias	BIS >AEP Bias	AEP algorithm
Dahaba et al. [95]	BIS-Vista	15-15	Pediatric hydrocephalus	BIS ↓	Contur, ICP ↑, CBF ↓
Dahaba et al. [96]	BIS-Vista	15-15	Adult hydrocephalus	BIS ↓	ICP ↑, CBF ↓
Urfalioglu et al. [98]	BIS	1	CIPA syndrome movements	BIS 19-58	No pain feeling hyperesthesia
Kim et al. [100]	BIS	1	Rett syndrome cerebral effect	delayed recovery at BIS 60	Abnormal mental function
Zabala et al. [102]	BIS	1	Scimitar syndrome cerebral effect	BIS lag rewarming	Abnormal mental function
Renna et al. [103]	BIS A-2000 (3.4, 4.0)	36/36	Alzheimer dementia	89 awake BIS↓	β wave ↓, δ-θ waves ↑
Erdogan et al. [109]	BIS	47/45	Dementia	awake BIS ↓	No BIS change at maintenance
Saeki et al. [110]	BIS	1	Huntington Disease cerebral effect	↓ BIS 55 awake	Abnormal mental function

n number of patients per group, *BIS* Bispectral Index, *RE* Response Entropy, *SE* State Entropy, *AEP* Auditory evoked potential, *ICP* intracranial pressure, *MR* Mental Retardation, *CBF* cerebral blood flow, *CIPA* Congenital Insensitivity to Pain with Anhidrosis, *LOC* Loss Of Consciousness children (92 ± 8, 49 ± 13 and 91 ± 12) compared with normal children (97 ± 3, 58 ± 7 and 96 ± 2), while there was no significant difference at 8% sevoflurane induction or at 3% sevoflurane steady state maintenance between CPMR children (23 ± 9 and 30 ± 10) compared with normal children (28 ± 8 and 33 ± 8) [87]. They concluded that, in children with CPMR, Bispectral Index values exhibit a pattern of change like that observed in normal children. However, absolute Bispectral Index values obtained in such children are significantly lower than those in normal children while awake and at light sevoflurane concentrations [87]. It would be, as expected due to the damage of the cerebral palsy to the developing brains, the repeated epileptic fits and the antiepileptic medications, all strongly influencing the Bispectral Index monitoring [87]. The fact that the Bispectral Index values reach a plateau when high sevoflurane concentrations are administered [88] could explain the leveling out of the statistical differences between the CPMR and normal children at high sevoflurane concentrations [88].

Valkenburg and colleagues [89] found significantly lower Bispectral Index values, recorded throughout the whole diagnostic gastroduodenoscopy under isoflurane general anesthesia in 18 intellectually disabled children compared with 35 control children aged 2–13 years of age [89]. Median (interquartile range) Bispectral Index values in the intellectually disabled children were significantly lower than those for normal intellectual function children in the awake state 72 (4–77) versus 97 (84–98), $P < 0.001$, during stable intraoperative anesthesia 34 (21–45) versus 43 (33–52), $P = 0.02$, and during return of consciousness 59 (36–68) versus 73 (64–78), $P = 0.009$ [89]. The Receiver Operating Characteristic (ROC) curves showed that the Bispectral Index could successfully discriminate consciousness equally in intellectually disabled children compared with normal children. However, the optimal Bispectral Index discriminative cutoff-value between unconsciousness and consciousness was 28 Bispectral Index points less in the intellectually disabled children [89].

Interestingly Ponnudurai and colleagues [90] demonstrated in a prospective clinical study that the mental retardation severity degree according to their intellectual Intelligence Quotient (IQ) score = Mental age/Chronological age × 100, ranging from IQ 55–69 Mild mental retardation, IQ 40–54 Moderate mental retardation, IQ 25–39 Severe mental retardation, to IQ < 25 Profound mental retardation, does not statistically significantly affect the Bispectral Index score during general anesthesia [90]. Eighty American Society of Anesthesiologists (ASA) classification physical status I, II and III patients with varying degrees of mental retardation, according to the criteria of the American Psychiatric Association (APA), were grouped into Mild, Moderate, Severe, or Profound degrees of mental retardation by an independent registered research nurse [90]. All patients undergoing dental surgical rehabilitation were given a standard sevoflurane in oxygen general anesthesia. A research assistant, blinded to the study group assignment, continuously recorded the Bispectral Index scores and compared the Bispectral Index scores at awake, anesthesia induction, catheter placement, endotracheal intubation, start and end of surgery, awakening to verbal commands, and endotracheal extubation [90]. The authors found no significant difference in Bispectral Index scores among the study groups at any time point [90]. However, the endotracheal extubation slope of emergence from anesthesia showed significantly longer emergence time in the severe mentally retarded group. The authors concluded that mental retardation severity degree does not affect Bispectral Index values during general anesthesia [90].

5.6.2 Intracranial Mass Lesions

Brain mass lesions have been associated with processed Electroencephalography-Depth Of Anesthesia monitoring (pEEG-DOA) monitors' artefacts. Faraoni and colleagues [91], reported pEEG-DOA artefactual readings resulting from two Fronto-Occipital metastasis on the right side of the brain in a patient undergoing emergency laparoscopy under Target Controlled Infusion (TCI) propofol remifentanil anesthesia simultaneously monitored via Entropy on the metastasis right ipsilateral side of the brain and Bispectral Index on the left contralateral side of the brain [91]. During anesthesia maintenance Response Entropy of 16 and State Entropy of 17 differed widely from the simultaneous Bispectral Index value of 67, as the difference between the 2 monitors of 50 points before the start of surgery decreased to a difference of just 20 points during maintenance of anesthesia [91]. The wide differences between the Bispectral and Entropy indices were attributed to the differences between the Entropy and Bispectral Index on the metastasis ipsilateral Entropy and contralateral Bispectral Index artefact rejection algorithms' calculations that could result in erroneous anesthesia management [15, 91].

Ferreira and colleagues [92] conducted a prospective study, comparing the Bispectral Index A-2000 BIS-XP monitoring of propofol predicted effect-site concentrations (C_e) at Loss of Consciousness (LOC) as indicated by failing to respond to name calling or gentle tapping on the forehead during propofol TCI anesthesia induction in 13 American Society of Anesthesiologists (ASA) classification physical status I and II patients undergoing surgery for small brain tumor removal compared with 13 ASA I and II patients undergoing nontumor spinal surgeries [92]. The Bispectral Index sensor, placed at the contralateral side of the head of the small brain tumor, and propofol effect-site C_e were analyzed at the beginning of the propofol infusion and every 30 s during induction. They concluded that, in the period from 150 s to 210 s from start of anesthesia induction, at the same propofol effect-site C_e concentrations, subjects who suffered from small cerebral tumors exhibited statistically significantly higher Bispectral Index values compared with patients without brain tumors undergoing spine surgery [92]. There was no significant difference in time to LOC or Bispectral Index value at LOC between patients with brain tumors (3.9 ± 0.7 min, 59 ± 22) compared with patients without brain tumors (3.6 ± 0.8 min, 49 ± 11) respectively [92]. They concluded that, at the same propofol effect-site C_e concentrations during anesthesia induction, subjects who suffered from small cerebral tumors exhibited higher Bispectral Index values but similar time to LOC and similar Bispectral Index values at LOC [15, 92].

Fudickar and colleagues [93] studied the impact of intracranial mass lesions on simultaneously monitored Bispectral Index module of the S/5 GE Monitor (General Electric GE Healthcare, Chicago, Illinois, USA) and Mid-Latency Auditory Evoked Potentials (AEP) module of the S/5 GE Monitor. Bispectral index and Mid-Latency AEP were simultaneously recorded bilaterally in 20 patients undergoing unilateral supratentorial lesion removal under general anesthesia [93]. Authors tested the hypothesis that Mid-Latency AEP (Pa wave and Nb wave) bilaterally recorded in patients with unilateral brain mass lesions result in less side of the lesion, ipsi—contra-lateral bias than Bispectral index ipsi—contra-lateral bias. Authors considered a "clinically relevant bias" to be Bispectral index or Mid-Latency AEP Indices ipsi- and contralateral differences of >10% [93]. Using Bland and Altman [94] format Bias and Limits of agreement (LOA), they found that the number of patients (18/20) with discordant Bispectral index ipsi—contra-lateral Bias (LOA) was far more significantly higher than the number of patients (Pa wave 1 /20 patient, Nb wave 3/20 patients) with discordant Mid-Latency AEP ipsi—contralateral Bias (LOA). Mean ± SD Bispectral Index ipsi—contra-lateral Bias (LOA) of -18.8 (±25.7) was significantly higher ($P < 0.0001$) than mean Mid-Latency AEP Pa wave Bias (LOA) of +1.5 (±3.4) msec or AEP Nb wave Bias (LOA) of -1.5 ± 6.1 ms [93]. They concluded that Mid-Latency AEP exhibited far less intracranial lesion side ipsi—contra-lateral Bias (LOA) than Bispectral Index and therefore seem to be far superior to Bispectral Index in monitoring patients with intracranial lesions [15, 93].

5.6.3 Hydrocephalus

Enlarged skull circumference in pediatric patients with hydrocephalus might alter Bispectral index values [95], because of distorted skull contour distancing the Bispectral index (BIS) sensors from a shifted cerebral mass, low electroencephalography (EEG) voltage conduction, or because of the associated hydrocephalus cerebral insult. Hydrocephalus could also affect Bispectral index values because of delayed mental development resulting from sustained elevated Intracranial Pressure (ICP) or Cerebrospinal Fluid (CSF) accumulation. My research team Dahaba and colleagues conducted a study [95], in the Department of Anesthesiology, Critical Care and Pain Medicine, The Second Affiliated Hospital and Yuying Children's Hospital of Wenzhou Medical University, Wen Zhou, Zhejiang, People's Republic of China, where we tested the hypothesis that Bispectral index BIS-Vista values are altered in pediatric patients with hydrocephaly [95]. In a prospective, observational study, we compared 15 pediatric patients less than 3 years of age with gross hydrocephaly, undergoing ventriculo-peritoneal shunt placement under propofol Target Controlled Infusion (TCI) anesthesia, with 15 matched control pediatric patients of the same age. In pediatric patients with hydrocephalus, mean ± SD 10-min before induction, during stable propofol 4 μg ml^{-1} TCI anesthesia maintenance and at emergence from anesthesia, ipsilateral Bispectral index values (76 ± 12, 39 ± 13, 64 ± 10) and contralateral (77 ± 10, 40 ± 12, 64 ± 9) Bispectral index values were significantly lower than control ipsilateral (93 ± 4, 62 ± 7, 72 ± 3) and contralateral (93 ± 1, 63 ± 5, 72 ± 3) Bispectral index values [95]. We depicted Bispectral index monitoring trends exhibiting a Bispectral index pattern of change that is consistently significantly lower in pediatric patients with hydrocephalus compared with normal development matched pediatric patients, indicating a consistently deeper hypnotic state than in normal subjects (Fig. 5.8) [95]. This should be taken into consideration when

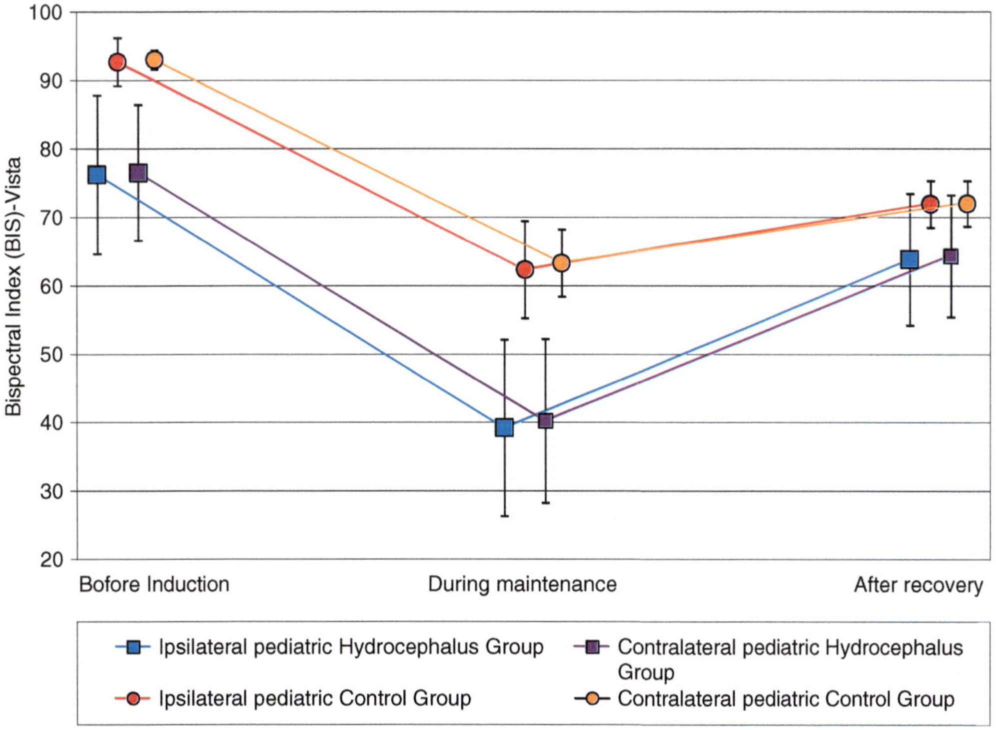

Fig. 5.8 Mean Ipsi– and contra–lateral Bispectral Index (BIS)-Vista values (SD) at 3 time points of 10-min before anesthesia induction, during 1 h of maintenance and at 20-min emergence from anesthesia in Pediatric patients with Hydrocephalus and in matched control patients. (Reproduced with permission # 5885800546470 from Dahaba AA, et al. Bispectral index monitoring of propofol anesthesia in pediatric patients with hydrocephalus. A prospective observational study. Paediatr Anaesth. 2014;24:1190–2)

adjusting anesthesia levels based upon Bispectral index values in pediatric patients with hydrocephalus [95].

Titrating hypnotic agents for patients who suffer from cerebral insults is a challenging task [96]. To date there is no real gold standard to precisely quantify EEG response in a fashion that could be utilized for adult patients with post-cerebral hemorrhage hydrocephaly [96]. In practice, when we care for patients with neuronal injury and impaired consciousness, we must administer "as per usual" analgesics for attenuation of sympathetic responses to noxious stimuli, while we usually must administer the hypnotic agents carefully and often more "sparingly" due to lack of objective monitoring [96]. However, despite anecdotal evidence, this effect is till now not precisely quantified as we have previously described altered Bispectral index values in pediatric patients with hydrocephaly [95]. This raises the question of what is then the hypnotic agent dose in patients with post-cerebral hemorrhage hydrocephalus that would ensure the desired unconsciousness [96].

In the second leg of our study [95, 96], my research team Dahaba and colleagues [96], in the Department of Anesthesiology, Critical Care and Pain Medicine, The Second Affiliated Hospital and Yuying Children's Hospital of Wenzhou Medical University, Wen Zhou, Zhejiang, People's Republic of China, compared propofol-BIS EEG Pharmacokinetic/Pharmacodynamic (PK/PD) model in 15 adult post-cerebral-hemorrhage hydrocephalus trauma patients, undergoing ventriculo-peritoneal shunt placement, with 15 adult controls patients matched for gender and approximate age undergoing non-neurosurgical procedures [96]. In our study we examined whether the Bispectral index BIS-Vista erroneously identifies an actual deep hypnotic state in adult patients who suffer from cerebral insults. We set propofol TCI estimated Plasma Concentrations (C_p) to gradually reach 4 μg ml^{-1} over 4-min [96]. To precisely quantify adult trauma patients' post-cerebral hemorrhage mental dysfunction, we used Bispectral index and propofol C_p data points to create individual Inhibitory Monophasic Mathematical Model (IMMM) of each patient that incorporates an independent hysteresis "lag" function [96]. Because our "C_p plasma concentrations" is a continuous variable, we based our PK-PD continuous data model analysis on the "Mass Action Law" of the asymmetry between the Bispectral index ipsi- *Versus* contra -lateral sides of patients with hydrocephalus *Versus* control patients as the 2 covariates [96]. The Effect (E) is The Bispectral index value; E_0 represents the Bispectral index value before propofol administration; E_{max} is the eventual nadir Bispectral index value maximal depression. In our model C_{50} represents propofol concentration at 50% of maximum Bispectral index effect, and γ is Hill coefficient "sig" Sigmoidicity factor that influences the slope and steepness of the concentration–response curve and describes Sigmoidicity of the relation [97].

In our adult post-cerebral hemorrhage patients' E_{max} Bispectral Index (BIS) versus propofol estimated plasma concentrations (C_p) in (μg ml^{-1}) ±95% Confidence Intervals PK/PD model in adult patients with and without hydrocephalus C_p-BIS curve [96], C_{50} (propofol concentration associated with inhibitory 50% Bispectral index response) cut off point was significantly shifted to the left by 39% (Fig. 5.9) [96], whereas mean ± SD Bispectral index values in adult patients with hydrocephaly at 3 time points of 10-min before anesthesia induction, and at stable 1 h of propofol 4 μg ml^{-1} anesthesia maintenance and at 20 min of anesthesia recovery, surgery side ipsilateral (75 ± 13, 25 ± 9, 72 ± 15) and contralateral (73 ± 15, 27 ± 9, 70 ± 16) were significantly lower than ipsilateral (95 ± 3, 46 ± 12, 83 ± 9) and contralateral (94 ± 3, 46 ± 12, 82 ± 8) mean ± SD Bispectral index values in matched control patients without hydrocephaly [96] (Fig. 5.10) [96].

Because our research group examined patients with severe intracranial hemorrhage and associated hydrocephaly compared with patients without intracranial pathology, differences in Bispectral index values could be accredited to numerous plausible explanations first and foremost to the patients' cerebral hemorrhage primary neurological intracranial pathology, as well

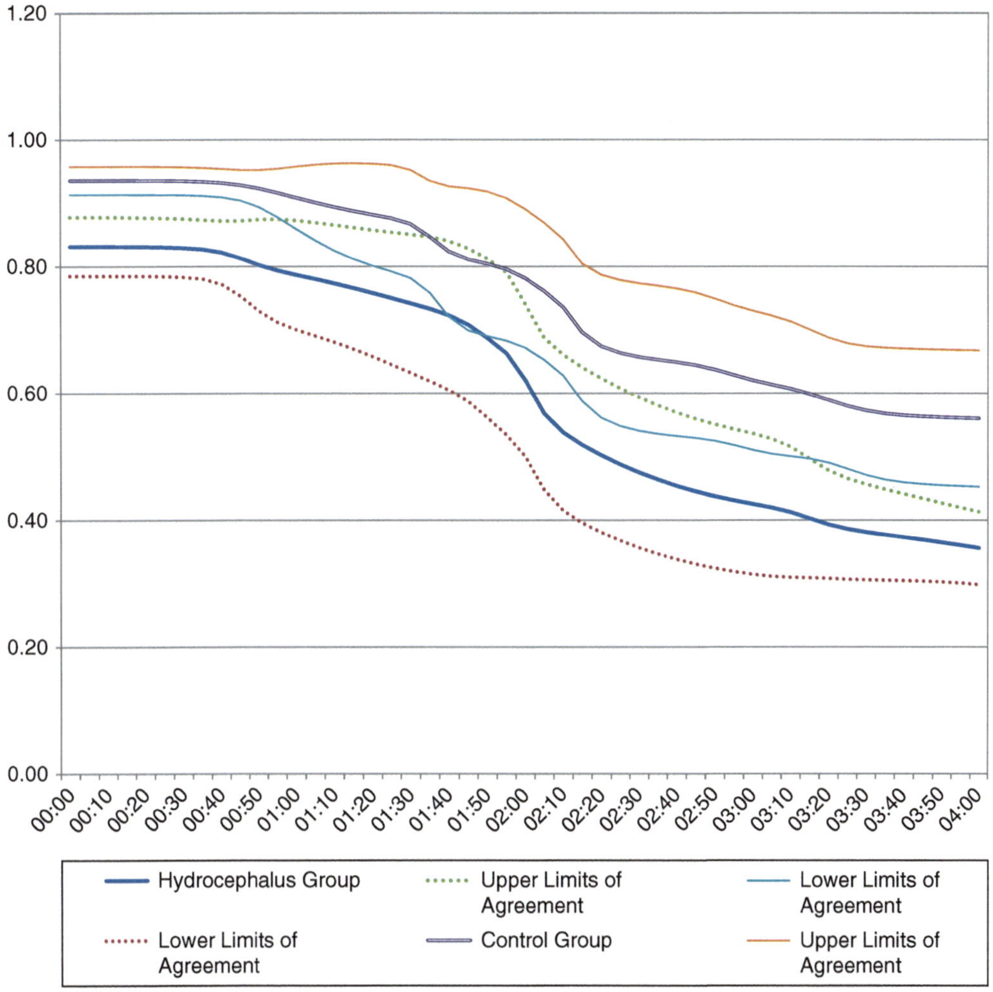

Fig. 5.9 Adult Emax Bispectral Index (BIS) versus propofol estimated plasma concentrations (C_p) in (μg ml^{-1}) ±95% Confidence Intervals Pharmacokinetic/Pharmacodynamic model in adult patients with and without hydrocephalus. (Reproduced with permission # 5885811037735 from Dahaba AA, et al. Propofol-Bispectral Index (BIS) Electroencephalography (EEG) Pharmacokinetic/Pharmacodynamic Model in patients with post-cerebral hemorrhage hydrocephalus. Clin EEG Neurosci. 2021;52:351–9)

as the widespread secondary injuries associated with their raised ICP and delayed cerebral ischemia [96].

Mean Bispectral index values significantly increased after we successfully placed a patent ventriculo-peritoneal shunt [96]. Such an increase in Bispectral index values after shunting could be attributed to many mechanisms that could include changes in direct pressure on certain brain areas, changes in oxygenation, or simply the shifting of the location of frontal cortex [96].

Interestingly in our Adult patients with hydrocephalus compared with controls study [96], patients with hydrocephaly exhibited considerably significantly larger decline in ipsi- and contra- lateral Bispectral index values during anesthesia maintenance (25 ± 9, 27 ± 9) [96] than the decline we previously encountered in our Pediatric patients with hydrocephaly (39 ± 13, 40 ± 12) study [95]. Despite the different nature of adults and children with completely different pathologies in the two studies [95, 96], we can speculate that this might be attributed, at least in

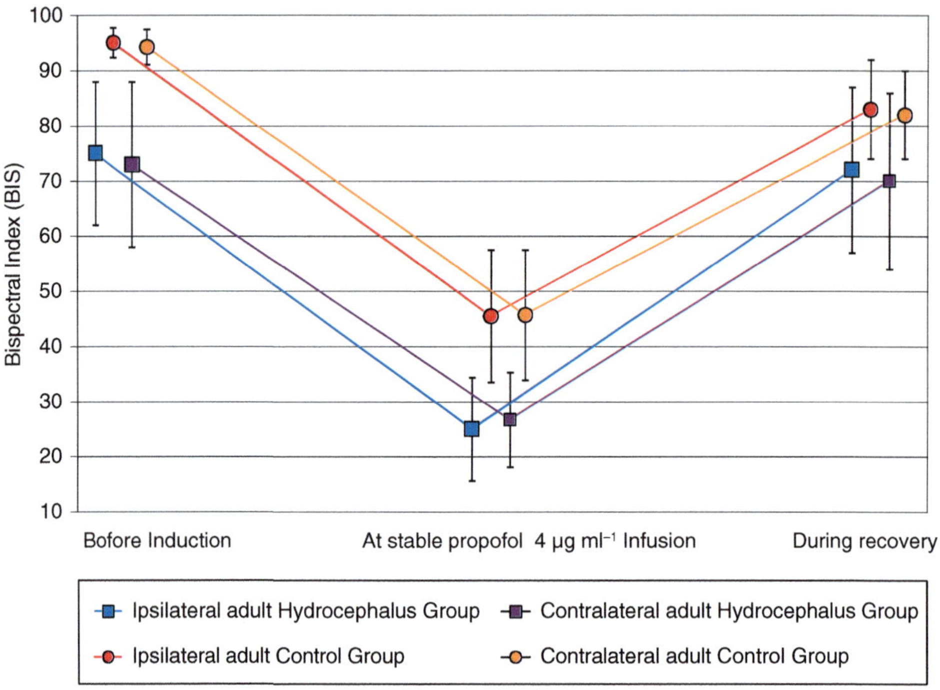

Fig. 5.10 Mean Ipsi– and contra–lateral Bispectral Index (BIS)-Vista values (SD) at 3 time points of 10-min before anesthesia induction, during 1 h of maintenance and at 20-min emergence from anesthesia in Adult patients with Hydrocephalus and matched control patients

part, to the acute nature in our adult trauma patients rapidly developing pathogenesis compared with the slow evolving hydrocephaly in mostly congenital hydrocephalus pediatric patients [95, 96].

5.6.4 Congenital Insensitivity to Pain Anhidrosis (CIPA) Syndrome

Congenital Insensitivity to Pain with Anhidrosis (CIPA) syndrome [98] is a rare neuropathy characterized by insensitivity to pain, impaired thermoregulation, anhidrosis, and mental retardation [98]. Urfalioglu and colleagues [98] present a rare case of a 9-year-old child with CIPA syndrome undergoing two operations for a calcaneal ulcer. In the first operation, in which anesthesia was induced with ketamine and midazolam during propofol Total Intravenous Anesthesia (TIVA) anesthesia maintenance despite pain insensitivity, extremity movements with surgical stimuli indicated tactile hyperesthesia [98]. In the second operation, Bispectral index monitoring was added, and temperature was monitored with an esophageal probe [98]. In the second operation despite pain insensitivity, at extremity movements with surgical stimuli, propofol infusion was adjusted [98]. Throughout the operation, the Bispectral Index values varied from 19 to 58 and body temperature was stable at 36.1–36.9 °C. Despite the absence of pain sensitivity in CIPA syndrome, there is an absolute need for the administration of anesthesia in surgical procedures because of tactile hyperesthesia [98].

5.6.5 Rett Syndrome

Rett Syndrome is a genetically inherited disorder [99] that almost exclusively affects females and is characterized by severe mental retardation after a period of apparent normal development for the first 6 to 18 months of life [99]. Kim and colleagues [100] present a case of a 12-year-old, 43-kg girl with confirmed Rett syndrome was scheduled for a T_3-L_4 posterior spinal fusion surgery to correct scoliosis. Her history included poorly controlled seizure disorder, episodes of irregular breathing, and reflux. On physical examination, the patient was noted to be cognitively impaired, doing purposeless movements [100]. She had a history of slow awakening after chloral hydrate anesthesia for a previous dental procedure 3 years prior to surgery [100]. It was decided to use propofol remifentanil anesthesia aiming at maintaining Bispectral index at 30–50 for 6.5 h. When the case ended, the patient had a delayed recovery and the Bispectral index was 60. A low Bispectral index trend was appropriate for the anesthetic management of a patient with Rett syndrome undergoing scoliosis surgery [100]. This case illustrates the difficulty of proper anesthetic management of a patient with Rett syndrome as these patients are known to possibly be excessively sensitive to both sedative drugs and volatile anesthetics [99–101].

5.6.6 Scimitar Syndrome

Zabala and colleagues [102] report a case of a 3-year-old girl with Scimitar Syndrome, a rare congenital disease of hypoplastic lung, right pulmonary artery and Dextrocardia, undergoing correction of the Atrial Septal Defect (ASD) and correction of anomalous pulmonary veins under Deep Hypothermic Circulatory Arrest (DHCA) [102]. The Bispectral index decline significantly correlated with the gradual hypothermia and circulatory arrest. However, the Bispectral index value remained low during the rewarming phase of the Cardiopulmonary Bypass (CPB) despite adequate pump flow and stable hemodynamics [102]. This observed significant lag in Bispectral index during the rewarming phase of DHCA may represent neuronal bewilderment or perhaps neuronal stunning and differs from previous reports that show significant increase in Bispectral index value during rewarming from mild hypothermia [102].

5.6.7 Alzheimer Disease and Dementia

The Bispectral Index was devised from healthy subjects with normal electroencephalography who volunteered to record their data; hence all neurologic conditions with unusual electroencephalographic patterns will have an impact on the Bispectral index values. Renna and colleagues [103] first using the Bispectral index A-2000 (version 4.0) monitor showed that the mean awake Bispectral index value of 89 in 36 elderly patients suffering from Alzheimer's disease or Dementia was statistically significantly lower than mean Bispectral index value of 95 recorded from 36 healthy elderly subjects [103]. When they reassessed, the same patients using a Bispectral index A-2000 (version 3.4) monitor, awake Bispectral index value was 93 in patients suffering from Alzheimer's disease or Dementia versus Bispectral Index values of 96 recorded from healthy elderly subjects [103]. This was attributed to the fact that patients with Alzheimer's disease or Dementia show reduced power in the 20–28 Hz β band [104], that could affect the Bispectral Index computation. This could be equally important in the interpretation of Bispectral Index values in other neurological disorders such as Schizophrenia that exhibits an increase in the slow electroencephalography (EEG) δ-wave and θ-wave activity [105], where the Bispectral Index may be falsely low [103].

Mini-Mental State Examination (MMSE) [106] is a quick and simple tool to identify elderly patients with Alzheimer disease, senile or vascular Dementia cognitive impairment that is associated with an increase in EEG slow-waves δ-wave and θ-wave activities, and a decrease in fast-waves α-wave and β-wave activities [106]. The loss of cholinergic neurons in the basal forebrain has been shown to correlate with the extent of the cognitive dysfunction of ageing in humans that may influence propofol's brain sensitivity in elderly patients. Laalou and colleagues [107] examined propofol's Target Controlled Infusion (TCI) anesthesia requirments, using Schnider's model, to maintain Bispectral Index between 40 and 60 in the 41 patients aged between 65 and 99 years elderly patients [107]. They demonstrated that the preoperative cognitive status, as assessed by MMSE, was the only covariate that may determine propofol's TCI requirements to maintain the Bispectral Index between 40 and 60, when age and gender had no significant effect on propofol TCI requirements in elderly patients [107]. A low MMSE <19 score in patients >65 year of age was associated with significant decrease in propofol requirement to maintain Bispectral Index within 40-60 [107].

Conversely Perez-Protto and colleagues [108] demonstrated that patients with Dementia were not more sensitive than patients without Dementia, to sevoflurane anesthesia regarding their Bispectral Index BIS-XP A-2000 (version (3.21) BIS-to-Minimal Alveolar Concentration (MAC) sevoflurane ratio, as they did not require fewer sevoflurane to maintain the same BIS-to-MAC ratio during the 5 minutes just before skin incision, hence not displaying deeper anesthesia levels at any given MAC fraction [108]. Each patient with dementia was matched with a maximum of five patients without dementia [108]. The difference in mean BIS-to-MAC ratio between 31 patients with dementia 85 (range 73–100) and 151 matched patients without dementia 78 (range 73–84) was not significantly different [108]. They concluded that patients with dementia are not more sensitive to sevoflurane volatile anesthetic than patients without dementia [108].

Erdogan and colleagues [109] evaluated the effects of cognitive impairment in elderly patients on anesthetic agents consumption during surgery and on emergence from anesthesia [109]. They compared induction of anesthesia propofol dose and Bispectral Index values in 47 elderly patients >65 year scheduled for elective orthopedic procedures suffering from cognitive impairment MMSE 16.6, 24 h before surgery and 15.6, 24 h after surgery, with 45 patients with normal cognitive function MMSE 26.8, 24 h before surgery [109]. Baseline Bispectral Index values were significantly lower in the awake state, during Loss of Consciousness (LOC) and before extubation, in patients with cognitive impairment (95 ± 4, 56 ± 6 and 84 ± 6) than age-matched healthy controls (96 ± 2, 58 ± 5 and 87 ± 5); but not significantly different during propofol remifentanil anesthesia maintenance or at termination of anesthesia in patients with cognitive impairment (49 ± 3 and 51 ± 5) than age-matched healthy controls (49 ± 3 and 53 ± 5) [109]. Propofol induction dose was significantly lower, eye-opening time was significantly longer, and propofol requirements to keep Bispectral Index between 45 and 60 were significantly less in patients with cognitive impairment than in those with normal cognitive function [109]. Based on their findings, it seems that the Bispectral Index values in awake state patients with cognitive impairment are lower than the general population [109].

5.6.8 Huntington Disease

Saeki and colleagues [110] present a case of a 41-year-old man with 10-year history of Huntington disease [110], suffering from choreiform movements, mis-swallowing and progressive mental deterioration. Because of repeated aspiration episodes, he was undergoing Percutaneous Endoscopic Gastrostomy (PEG), under sevoflurane 1.5% anesthesia [110]. Intravenous midazolam 2 mg was administered on transferring the patient to the operating room. On arrival, the patient was somnolent but responded to call; the Bispectral Index value of 55 was abnormally low before anesthesia induction. Anesthesia was maintained with sevoflurane 1.5% [110]. At the end of the procedure, return of spontaneous respiration and extubation, Bispectral Index was abnormally low after anesthesia recovery as the patient woke up at a Bispectral index value of 78. The postoperative course was uneventful. The authors concluded that anesthetists must keep in mind the phenomenon of low Bispectral Index values before anesthesia induction and at anesthesia recovery for avoiding any unintended adjustment of the depth of anesthesia [110].

5.7 Cerebral Injury

5.7.1 Brain Death Determination. Zero or Electrocardiography (ECG) Display

Brain death determination is essentially based on a full comprehensive clinical assessment. Such a comprehensive clinical examination, performed by a specialized skilled examiner, should yield precise diagnostic accuracy [111]. However, confirmatory tests in adult patients are not mandatory in the United States but are typically used as a safeguard or added complementary findings when clinical examination results are not conclusive [111]. In many countries, confirmatory Brain Death tests are mandatory, but in many countries they remain optional. These tests are divided into two main categories: Brain Electrical Function tests or Cerebral Blood Flow (CBF) tests [111]. A false-positive test result to determine brain death, i.e., the test result suggests brain death, but clinically the patient does not meet the clinical criteria of Brain Death, is rare [111]. On the other hand, a false-negative result to determine Brain Death, i.e., the test result suggests intact brain function, but clinically the patient meets the Brain Death clinical criteria, may result in no resolution. Confirmatory tests could help resolve such a condition [111].

Escudero and colleagues [112] evaluated the Bispectral Index BIS-XP as a method for Brain Death determination. They conducted an observational prospective study in a university hospital Critical Care Unit (CCU) in 19 patients with serious neurologic conditions progressing towards Brain Death [112]. In all cases, the Brain Death determination was made through neurological clinical examination and electroencephalography (EEG) [112]. Additionally, in 13 patients Transcranial Doppler (TCD) was used. Patients clinical condition deterioration corresponded with a gradual decrease in the Bispectral Index score, together with a rise in Burst Suppression Ratio (BSR) [112]. In all patients with a confirmed clinical Brain Death diagnosis, the Bispectral Index score showed values of Zero "0" and BSR of 100, except in one patient with high electromyography (EMG) activity, when a conventional EEG was being performed [112]. After a Neuromuscular Blocking Agent (NMBA) administration, the Bispectral Index score and BSR values became 0 and 100, respectively [112]. Their study results clearly demonstrated that Bispectral Index monitoring is a simple, noninvasive, and easy to interpret method for Brain Death determination [112]. Because all 19 patients with clinical Brain Death diagnosis, except for one, had a Bispectral Index value of Zero "0" and BSR of 100, this demonstrates a perfect correlation with the other clinical neurological diagnostic methods. They concluded that, albeit the Bispectral Index cannot be used on its own for Brain Death confirmation, it could represent a useful tool for Brain Death determination [112].

The diagnosis of Brain Death is a clinical condition in which it is difficult to perform confirmatory tests due to the sometimes-ineligible clinical status of the patient. Some reports have employed the Bispectral Index monitoring to detect "Brain Death" in comatose patients. The use of the Bispectral Index monitoring to determine Brain Death is important for organ transplants, cost-effectiveness, and reducing stressful wait of the family. Vivien and colleagues [113] using Bispectral Index A-2000 evaluated the accuracy of Bispectral Index monitoring for diagnosing Brain Death in 56 intracerebral hemorrhage, head injury, or post-anoxic severely comatose patients Glasgow Coma Scale (GCS) <5 [113]. Twelve patients were already confirmed brain-dead by EEG or Cerebral Angiography at CCU admission, and their individual Bispectral Index was Zero "0" [113]. Out of the remaining 44 not confirmed brain-dead patients on CCU admission, who's individual Bispectral Index values were between 20 and 79, 27 of these 44 patients became brain-dead and their Bispectral Index dropped to Zero "0" in few hours to few days. Their brain death was subsequently confirmed by EEG or Cerebral Angiography [113]. The remaining 17 patients, who never progressed to brain death during their CCU stay and had persistent electrocerebral activity on their EEG, their Bispectral Index values remained above 35. Thus specificity, sensitivity,

Positive Predictive Value (PPV), and Negative Predictive Value (NPV) were all 100% for Bispectral Index prediction of Brain Death [113].

When the EEG signal is absent as in extreme EEG suppression with Deep Hypothermic Circulatory Arrest (DHCA) or in Brain Death, the Bispectral Index algorithm becomes vulnerable to radio frequency noise from electrocardiography (ECG) signals from the heart, the only organ that is still sending strong electrical signals, and the Bispectral Index would then misinterpret these ECG signals from the heart as EEG activity. In 5 of the proven Brain Death subjects, Bispectral Index transitionally rose from Zero "0" to more than 90 and remained stable for few hours at a high level and in one patient to a Bispectral Index value of 98. In clinically brain-dead patients, this was accompanied by a significant increase in Bispectral Index monitor EMG activity [113]. Bispectral Index declined to Zero "0" with the administration of NMBAs. In one brain-dead patient, under epinephrine infusion, Bispectral Index increased from 0 to 60 before NMBA declined Bispectral Index value to only 38 as Bispectral Index remained stable at that level [113]. This hemodynamically unstable patient showed an important cardiovascular hyper-pulsation under epinephrine and the patient's head movement. Real time EEG waveform displayed tracing indicating oscillations that are exactly synchronized with the subjects' ECG Heart Rate (HR) frequency [113]. Immediately after switching epinephrine to norepinephrine, Bispectral Index declined to Zero "0" [113]. In their report, Bispectral Index score decreased to Zero "0" in 3 patients who were not initially clinically determined as brain-dead [113]. One of those patients had an intracranial hypertension that required treatment, and the other two patients had the Bispectral Index score decrease to Zero "0", 1-2 h before the criteria for clinical Brain Death were met [113].

Myles and Cairo [59] using Bispectral Index (version 3.4) monitor in another case of Brain Death following a pulmonary embolism cardiac arrest, in which following 60 min of Cardiopulmonary Resuscitation (CPR) patient's pupils were fixed and dilated while the Bispectral Index score was 0–5 with 100% Burst Suppression Ratio (BSR) indicating anoxic global cerebral injury [59]. During Normothermic Cardiopulmonary Bypass (CPB), epinephrine and norepinephrine infusions the Bispectral Index had an unexplained increase to 38 that was exactly synchronized with patient's ECG [59]. The pattern did not waver over 40 min. Cerebral Angiography confirmed Brain Death and life support was withdrawn. I must emphasize that the Bispectral Index was not designed, nor has it been validated as a surrogate monitor for detecting ischemic brain injuries, still one should always keep such possibilities in mind when interpreting the Bispectral Index data [59].

Puri and colleagues [114] observed similar findings to those previously reported by Myles and Cairo [59], and by Vivien and colleagues [113], in a patient following a cerebral aneurysm rupture, all the brain death tests were positive [114]. The authors have been routinely using the Bispectral Index as an adjuvant to determine brain death in their CCU. In the present case, Puri and colleagues [114] observed a Bispectral Index of 0–50 with 100-25% BSR without any BIS-EMG activity [114]. A careful analysis of the Bispectral Index tracings revealed that the Bispectral Index tracing was intermittently displaying the patient's own ECG waveform. Normally ECG and EMG electric artifacts are filtered by the Bispectral Index artifact-rejection algorithm: however in the present case, the Bispectral Index algorithm was selecting and displaying the patient's own ECG signal [114]. Despite Myles and Cairo, [59], Puri and Nakra, [114], and Vivien and colleagues [113] reported cases with severe brain injury where the ECG tracings were picked up by the Bispectral Index algorithm and not recognized /excluded as an artifact, still the interference of the patient's own ECG with the Bispectral Index reading phenomenon has been scarcely documented.

Çoban and colleague*s* [115], over a two-year period, used Bispectral Index monitoring for early detection of Brain Death in 12 CCU patients who were clinically determined brain-dead following brain injury. The Bispectral Index value was Zero "0" in all patients [115]. In one patient whose Bispectral index value was Zero "0," on the first brain Computerized Tomography (CT) Angiography, Cerebral Blood Flow (CBF) was still present [115]. Two days later, the second imaging revealed that the CBF was not present anymore. Bispectral Index monitoring could be a parameter for detection of brain death in patients with severe brain injury [115].

Dunham and colleagues [116] assessed the utility of using the Bispectral Index, a processed Electroencephalography-Depth Of Anesthesia (pEEG-DOA) parameter for Brain Death determination. Bispectral Index was assessed in Glasgow Coma Scale (GCS)-3 Traumatic Brain injury (TBI) brain-dead patients with no brain function including apnea, or Near Brain Death patients still breathing, but with negative CBF and no EEG activity [116]. In 23/37 (62%) patients, brain death was confirmed with Bispectral Index score of 3 ± 5, Bispectral Index score was <20 for 7 ± 6 h before the Brain Death determination. However, the remaining 14/37 patients (38%) showed "Near Brain Death" with Bispectral Index score of 36 ± 31 of which 9 patients had Bispectral Index ≥20 [116]. Bispectral Index score of <20 for predicting "Brain Death" had 100% Sensitivity (23/23), 84% PPV (23/28), and 100% NPV (9/9) [116]. Authors concluded that determining "Brain Death" and "Near Brain Death" in severely comatose Brain Trauma patients is complex; hence the Bispectral Index score represents a valuable tool for efficient evaluation of the likelihood of Brain Death [15, 116].

Smith and colleagues [117] reported a decline in the Bispectral Index value occurring before clinical manifestations indicating Brain Death in a patient being treated for severe Traumatic Brain Injury (TBI) in Neuro Intensive Care Unit (NICU) [117]. An 18-year-old male, who was not wearing a helmet at the time of his motorcycle accident, sustained severe closed TBI. Glasgow Coma Scale of 11 at the scene deteriorated to 3 on Emergency Department (ED) admission [117]. A Computerized Tomography (CT) scan revealed a large frontal bone fracture, right subdural hematoma, and sulcal effacement with edema of the right hemisphere indicating raised Intracranial Pressure (ICP). He was transferred to the NICU where therapy for raised ICP included deep sedation, neuromuscular block, and 35 °C hypothermia [117]. Bispectral Index monitoring was used to titrate his sedative agents. After 3 days, ICP stabilized and sedative agents were withdrawn to allow for clinical evaluation. However, several hours later while en-route to a repeat CT, one pupil became fixed dilated [117]. The CT scan revealed generalized severe cerebral swelling with loss of gray-white matter differentiation and profound cerebral sulci effacement [117]. Patient exhibited rapid and sustained decrease of isoelectric Bispectral Index score to Zero "0" and BSR 100% that preceded a clinical deterioration to brainstem death. Surgical intervention was discussed but was decided to be futile. Both pupils were now fixed dilated. Brainstem death was confirmed the next day ventilatory support was withdrawn [117].

Table 5.9 enumerates and describes all brain death that was successfully determined using the Bispectral Index (BIS) monitoring of Zero "0" or Electrocardiography (ECG) display.

Table 5.9 Brain death determination. Zero or electrocardiography (ECG) display

References	BIS Monitor	*n*	Change	BIS change	Explanation
Escudero et al. [112]	BIS-XP	19	Brain death No EEG activity	BIS 0 BSR 100	Isoelectricity
Vivien et al. [113]	BIS A-2000	39/56	Brain death No EEG activity	BIS 0 BSR 100	Isoelectricity
Vivien et al. [113]	BIS A-2000	5/56	Brain death EEG = ECG activity	BIS >90 = HR	ECG tracings
Myles and Cairo [59]	BIS (3.4)	1	Brain death EEG = ECG activity	BIS 0 BSR 100	ECG tracings
Puri et al. [114]	BIS	1	Brain death EEG = ECG activity	BIS 0 BSR 100	ECG tracings
Çoban et al. [115]	BIS	12	Brain death No EEG activity	BIS 0	Isoelectricity
Dunham et al. [116]	BIS	23/37	Brain death No EEG activity	BIS <20	Isoelectricity
Smith et al. [117]	BIS	1	Brain death TBI No EEG activity	BIS 0 BSR 100	Isoelectricity

n number of patients per group, *BIS* Bispectral Index, *BSR* Burst Suppression Ratio, *EEG* electroencephalography, *ECG* electrocardiography

5.7.2 Severe Brain Injury Neurologic Outcomes Predictions

The Glasgow Coma Scale (GCS) is regarded as a very useful tool for successfully predicting clinical outcomes in subjects with Traumatic Brain Injury (TBI). However, GCS estimation would not be possible when sedatives are used. Using processed Electroencephalography-Depth Of Anesthesia (pEEG-DOA) monitoring in TBI patients is an important tool for avoiding over- or under-sedation in subjects suffering from elevated Intra Cranial Pressure (ICP).

Dunham and colleagues [118] found that patients with Bispectral Index value of ≥60 were twenty folds prone to exhibit a GCS of more than 8 and to have 30-fold improved survival rate. They investigated whether noninvasive Transcranial Oxygen Saturation ($StcO_2$) measured with the INVOS 4100 system (Somanetics Corporation, Troy, MI, USA), and the Bispectral Index BIS-XP A-2000 correlate with severe TBI patients' survival outcomes in a prospective observational study [118]. Values of ICP, Bispectral Index, and $StcO_2$ were recorded hourly for the first 6 post-injury days in 18 Critical Care Unit (CCU) severe TBI patients [118]. The authors found that ICP ≤20, Cerebral Perfusion Pressure (CPP) ≥60, $StcO_2$ ≥70, and Bispectral Index score ≥60, are all independently associated with good neurological outcome and with improved survival rate ($P \leq 0.001$) [118]. With Bispectral Index ≥60, $StcO_2$ ≥70, CPP ≥60 the survival rate is 97.2% and the survival rate with ICP ≤25 is 97.1% [118]. The independent associations of Bispectral Index score, $StcO_2$, and ICP with survival outcomes suggest that noninvasive multi-modal monitoring may be beneficial in accurately predicting survival outcome in severe TBI patients [62, 118].

The Bispectral Index was similarly recommended for predicting outcome after severe Ischemic Hypoxic Brain Injury in patients rushed to the Operating Room (OR) for emergency surgery [119]. Myles and colleagues [119] retrospectively evaluated the prognostic ability of the Bispectral Index BIS-XP (version 4.1) in 25 CCU, unconscious patients with severe Ischemic Hypoxic Brain Injury who underwent time critical emergency surgery [119]. The Bispectral Index score and Burst Suppression Ratio (BSR) were recorded before and during surgery. Patients' Neurological Outcome was evaluated using GCS at 30 days after brain injury, where Poor Neurological Outcome was defined as Death or Severe Disability [119]. The incidence of Poor Neurological Outcome among the 25 study patients was in total 16/25 patients (64%) died and 1/25 patient (4%) had Severe Disability, giving an incidence of Poor Neurological Outcome of 17/25 patients (68%). Neither pupillary response ($P = 0.21$) nor clinical judgment ($P = 0.40$) was predictive of Neurological Outcome after surgery. An abnormal Bispectral Index score was strongly associated with Poor Neurological Outcome with 6.6 positive likelihood ratio (95% Confidence Interval CI 1.7–36.4; $P = 0.002$) [119]. A normal Bispectral Index score ($P < 0.0005$) but not clinical judgment ($P = 0.16$) could identify patients who are more likely to survive with a Good Neurological Outcome [119]. In patients with Poor Neurological Outcome, Bispectral Index score Receiver Operating Characteristic Area Under the Curve (ROC-AUC) of 0.80 (95% CI 0.62–0.98; $P = 0.017$) and BSR had a ROC-AUC of 0.84 (95% CI 0.68–0.99; $P = 0.007$), both were significantly different when compared with patients' Good Neurological Outcome [119]. The authors concluded that Bispectral Index score, but not clinical judgment, provides useful information to identify patients who are more likely to recover after Ischemic Hypoxic Brain Injury that required emergency surgery [119].

In 97 patients, Seder and colleague [120] evaluated the Bispectral Index BIS-XP A-2000 (version 3.1) or BIS- VISTA and BSR as early predictors of Neurologic Outcome after Therapeutic Hypothermia following Cardiac Arrest, compared with at discharge Cerebral

Performance Category (CPC) [120]. Receiver Operating Characteristic Area Under the Curve was constructed to predict Good Outcome (CPC 1–2) and Poor Outcome (CPC 3–5) Neurologic Outcome [120]. Fourteen patients were excluded from the final analysis, and 33/83 patients (40%) were classified as Good Outcome. The Bispectral Index score (range) in patients with Good Outcome of 37 (28–40) was significantly higher than the Bispectral Index score of 7 (3–15) in patients with Poor Outcome ($P < 0.001$) [120]. The Bispectral Index score of <22 predicted Poor Outcome with a Likelihood Ratio (LR) of 14.2 and AUC of 0.91 (95% Confidence Interval CI 0.85-0.98, $P < 0.001$) [120]. The BSR of ≥48 predicted Poor Outcome with a LR of 12.7 and an AUC of 0.90 (95% CI 0.83-0.98, $P < 0.001$). The authors concluded that the Bispectral Index and BSR were accurate very early predictors of Neurologic Outcome after Therapeutic Hypothermia following Cardiac Arrest [120].

Ebtehaj and colleagues [121] evaluated the correlation between GCS and Bispectral Index using the Bispectral Index A-2000 (version 3.4) as a prognostic outcome factor in 61 consecutive CCU patients suffering from TBI who received no sedative /hypnotics [121]. Bispectral Index scores and GCS, assessed for 20 min, correlated with the Glasgow Outcome Scale (GOS) at 6 months, revealed 12 Mild TBI, 14 Moderate TBI, and 35 Severe TBI. Mean Bispectral Index ± SD was significantly different ($P < 0.05$) in 96 ± 4 Mild TBI, 46 ± 1 Moderate TBI and 31 ± 2 Severe TBI [121]. There was a strong correlation ($r = 0.88$, $P < 0.05$) between GCS at admission and the mean Bispectral Index scores [121]. Bispectral Index score could be used together with the GCS to accurately predict TBI patients' Survival Outcome [121]. They concluded that an elevated Bispectral Index value is a prognostic outcome reassuring sign in TBI subjects who did not receive sedative /hypnotic agents [121].

Critical Care Unit treatment protocols for "Barbiturates Medically Induced Coma" would usually entail electroencephalography (EEG) "Targeted Burst Suppression" as their tertiary therapeutic parameter for patients with severe TBI [122]. During pentobarbital coma, EEG monitoring is used to document EEG Burst Suppression of 3–5 episodes/min. The Bispectral Index monitor displays cortical silence as Bispectral Index-Burst Suppression Ratio (BIS-BSR) of 100%. Even though Bispectral Index monitoring may be far easier to perform and requires no technical expertise to interpret, still Bispectral Index value of <15 corresponding to EEG Burst Suppression of 3–5 episodes min^{-1} is not with 100% Specificity and Sensitivity [122]. Thus, Bispectral Index Burst Suppression patterns must be verified on regular basis against real-time raw EEG [122, 123].

Tobias and colleagues [123], using the Bispectral Index A-2000 with a Smoothing Rate of 15 s, retrospectively assessed the relationship between the Bispectral Index score and Electroencephalography-Burst-Suppression (EEG-BS) in 7 patients, aged 2.9–14 years, on pentobarbital coma for elevated ICP [123]. When the EEG-BS exhibited ≤2, 3–5, and ≥6 EEG-BS episodes/min, the corresponding Bispectral Index value was 7 ± 5, 14 ± 3, and 37 ± 12 respectively, whereas the BIS-BSR was 93 ± 7%, 75 ± 6%, and 29 ± 18% respectively [123]. The authors demonstrated that with 3–5 EEG-BS episodes/min, the Sensitivity and Specificity of the Bispectral Index value of 10–20 were 96% and 92%, respectively, whereas the Sensitivity and Specificity of the BIS-BSR of 65–85% were 89% and 88%, respectively [123].

Cottenceau and colleague [122] investigated the relationship between Bispectral Index BIS-XP monitoring and EEG-BS patterns, in 11 TBI subjects with high ICP, under Barbiturate-induced Coma [122]. In daily recorded EEG, the number of EEG-BS episodes/min in cortical silence was compared to the Bispectral Index values and BIS-BSR. They considered EEG-BS 2–5 episodes /min as the Barbiturate Coma optimal level [122]. They estimated the Bispectral Index values that could predict EEG-BS from the EEG recordings [122]. There was strong correlation between EEG-BS and BIS-BSR ($r = 0.94$, CI

0.90–0.96) and a strong correlation between EEG-BS and Bispectral Index value (regression slope = –0.29, $P < 0.0001$), The optimal Barbiturate Coma level of EEG-BS 2–5 episodes /min was at Bispectral Index value of 6–15 [122]. The authors concluded that there was a strong correlation between Bispectral Index score and EEG-BS in TBI patients under Barbiturate Coma. Based upon their study results, the authors recommended that to maintain an optimal Barbiturate Coma level, the barbiturates infusion rate should be decreased if the Bispectral Index value is <6 and increased if the Bispectral Index value is >15 [122]. Concordance among the Bispectral Index value, BIS-BSR and EEG-BS should be regularly examined [122].

The probability of brain-injured patients recovering consciousness depends on the degree of brain damage and complications acquired during their CCU stay. Fàbregas and colleagues [124] examined the Bispectral Index and other variables derived from the EEG signal for prediction of patients' Return of Consciousness. Twenty-five critically ill comatose patients with extensive brain injuries were prospectively studied [124]. Sedative drugs were withdrawn for 24 h before recordings. Twenty-min Bispectral Index A-2000 (version 3.4), 95% Spectral Edge Frequency (SEF_{95}), and frontal electromyography (fEMG) were recorded [124]. The patients' GCS neurological condition was assessed for recovering consciousness 6 months after the injury. Variables were compared for their predictive power between 19/25 subjects who recovered consciousness and those 6/25 subjects who did not [124]. Mean (range) Bispectral Index value of 43 (13–49) was significantly lower in comatose subjects that never recovered from their extensive brain injuries compared with Bispectral Index value of 63 (35–97) in comatose subjects that regained consciousness [124]. There was a significant difference between patients who recovered consciousness median (range) 3 (1–10) fEMG and those who did not recover 0.5 (0–1). There was a significant difference between patients who recovered consciousness 8 (3–10) GCS and those who did not recover 3 (3–8), but not in SEF_{95} [124]. The authors showed that the Bispectral Index score was significantly different in patients who recovered consciousness and those with "Poor Neurological Outcome," Persistent Vegetative State or Brain Death. The Bispectral Index, other electrophysiological and clinical variables could be used to construct and cross-validate a Return of Consciousness probability model in severe brain injury comatose patients, after sedation has been withdrawn [124].

Table 5.10 enumerates and describes all conditions of Severe Brain Injury Neurologic Outcomes predictions that were successfully determined using the Bispectral Index (BIS) monitoring.

5.7.3 End-of-Life Care

Cardiac Death, or Non–Heart–Beating organ donation candidates, is another option for organ procurement when severely ill patients who do not meet "Brain Death Criteria" are declared

Table 5.10 Severe brain injury outcome prediction

References	BIS Monitor	*n*	Change	BIS change	Explanation
Dunham et al. [118]	BIS-XP A-2000	18	TBI	BIS ≥ 60	Good survival outcome
Myles et al. [119]	BIS-XP (4.1)	17/25	Ischemic brain injury	Abnormal BIS	Poor neurologic outcome
Seder et al. [120]	BIS-XP A-2000 (3.1), BIS-Vista	50/83	Ischemic brain injury	BIS < 22	Poor neurologic outcome
Fàbregas et al. [124]	BIS A-2000 (3.4)	6-19	Extensive brain injury	BIS 34 vs BIS 63	Poor neurologic outcome

n number of patients per group, *BIS* Bispectral Index, *TBI* Traumatic Brain Injury dead by "Cardiopulmonary Death Criteria" of absence of systemic blood flow without life-support, apnea, and unresponsiveness, before organ removal [125]. Non–Heart–Beating organ-donation candidates are dependent on life-support measures, and the removal of those life-sustaining treatments should result in rapid death [125]. Terminally ill, but not brain-dead, patients are brought to the Operating Room (OR) theater, where life-support medical care is terminated via halting artificial ventilation, intravenous infusions and all medications, other than measures to hasten their demise, are terminated [125]. Auyong and colleagues [126] presented a series of three case reports in which Bispectral Index values increased immediately after the termination of life-support process for organ-donation. The authors found that changes in Bispectral Index values after life-support termination were reproducible and happened in absence of electromyographic (EMG) or electrocardiographic (ECG) artifacts [126]. To validate the findings of their first case, Bispectral Index values were recorded in two other cases of life-support termination in patients who did not receive any sedatives /hypnotics. The first patient was monitored with electroencephalography (EEG) and Bispectral Index (BIS, A-2000) that showed, immediately after life-support termination, changes that are associated with "light planes of anesthesia" [126]. Within the first 5-min after life-support termination, the Bispectral Index value increased from 1 to 92, without any EMG activity on the Bispectral Index monitor. For the first 25-min after life-support termination, the Blood Pressure (BP) at the Radial Artery remained >100/60 mm Hg and the Heart Rate (HR), that was initially 74 beats /min, gradually increased to 108 beats/min. The Oxygen Saturation (SpO_2) oximetry was 100% and remained >90% for 8-min from life-support termination, after which SpO_2 gradually decreased to <85%, 8–20 min after life-support termination [126]. After 20-min, SpO_2 waveform disappeared with no noticeable spontaneous respiratory movements of the chest or abdomen. The Bispectral Index value remained elevated at 85–95 for 30-min after life-support termination. Twenty-five minutes after life-support termination, the patient's systolic BP decreased to <40 mm Hg and his HR decreased to Zero, while the Bispectral Index value declined to <4. The patient was pronounced dead by physical examination [126].

In their second case, subject's BP, just before life-support termination, was 142/59 mm Hg, HR was 98 beats/min, Respiratory Rate (RR) was 32 breaths/min, and SpO_2 was 100% on 30% Fractional Inspired Oxygen (FIO_2). The Bispectral Index values increased to 46–84 after life-support termination, whereas EMG values increased slightly from 31 to 42 decibel (dB) units before life-support termination to 36–55 dB at time of maximal Bispectral Index value increase [126]. The Burst Suppression Ratio (BSR) appropriately increased to 100% after life-support termination. The Mean Arterial Pressure (MAP) decreased to <50 mm Hg, 20-min after life-support termination. At 37-min HR decreased to zero with cardiopulmonary arrest.

Their third case arterial BP was 144 /78 mm Hg, HR was 104 beats/min, and SpO_2 was 98% on 40% FIO_2 before life-support termination [126]. The Bispectral Index value increased after life-support termination to 60–83 from a baseline of 28–46, EMG values decreased slightly from 43–58 dB before life-support termination to 40–54 dB at the time of maximal Bispectral Index value increase. As with their second case, BSR appropriately increased to 100% after life-support termination [126]. The Signal Quality Index (SQI) of the Bispectral Index monitor had a small decrease after life-support termination. Raw EEG recordings revealed patterns of high wave frequency activities with decreased amplitude after life-support termination [15, 126].

They concluded that these Bispectral Index changes deserve some consideration, if these Bispectral Index values are not due to artifact, then dosing of hypnotic or anesthetic agents might be warranted with the withdrawal of care process [126]. In the case of a patient in whom the brain stem is dead, it comes as no surprise that some electrical cortical activity is still present [126], this will allow processed Electroencephalography (pEEG) monitors like the Bispectral Index to process the patients' residual EEG activity for the pEEG monitor to calculate an index [15, 126].

Chawla and colleagues [127] demonstrated, in a very peculiar study, surges of EEG indicating level of consciousness at the End-Of-Life Near Death Experience (NDE) that was captured in a case series of seven neurologically intact Critical Care Unit (CCU) patients before the decision to withdraw life-support due to extensive systemic critical illness as part of End-Of-Life Care protocol [127]. In each case, loss of BP, as monitored by indwelling arterial line, was followed by a decline in SEDLine Patient State Index (PSI) (Hospira, Lake Forest, IL) and Bispectral Index, that was followed by short duration transient spikes in SEDLine PSI Patient State Index that was documented to go up as high as 97 in patient number 7 and Bispectral Index that was documented to go up as high as 85 in all patients except patients number 4 and number 6 when Bispectral Index went as high as 65; pEEG activities that approached levels normally associated with full consciousness [127]. This spike in pEEG activities had a short duration and the pEEG activity declined again this time to a level of activity associated with Burst Suppression (BS) [127]. The raw EEG signal confirmed that the EEG waveform was not artifact, and in fact showed that a high frequency waveform was present during the spike activity [127]. Authors speculate that since this increase in electrical activity occurred when there was no discernible blood pressure, patients who endure "Near Death Experience" may be recalling the aggregate memory of their synaptic activity [127].

Table 5.11 enumerates and describes all conditions of End-Of-Life Care that were successfully monitored using the Bispectral Index (BIS) and SEDLine Patient State Index (PSI).

5.7.4 Locked-in Syndrome

Cognitive abilities of patients with Amyotrophic Lateral Sclerosis (ALS) have been found to be impaired along with their neurodegeneration of their Motor Neurons. Electroencephalography (EEG) of end-stage ALS patients has reportedly shown specific features of neuronal network modulations, that quite differ from EEG of other patients with Brain Cognitive Dysfunctions or Dementia [128]. Hayashi and colleagues [128] report a case of 64-year-old male, an end-stage progressive ALS patient in Locked-in-State (LIS). He was supported by Tracheostomy Positive-Pressure Ventilation (TPPV) and Total Enteral Nourishment (TEN) for 9 years [128]. He could spontaneously blink and open his eyes, but seldom opened or closed his eyes on instruction. He underwent scheduled deep sevoflurane general anesthesia for repair of a fractured mandible under continuous Bispectral Index A-2000 (version 4.0) monitoring [128]. Prominent α-wave oscillations with small peak α-wave frequency (≈8 Hz) and high amplitude appeared during anesthesia that are usually associated with wakening, albeit in a manner quite different than the typical EEG characteristics seen in patients with Brain Cognitive Dysfunctions [128]. The prominent α-wave oscillations in this rare case may suggest modulation of the Functional Neuronal Network System during anesthesia in patients with progressive ALS, in a manner quite different than the typical Dementia patients [128]. Conversely, the Bispectral Index reflected sevoflurane concentration in this ALS patient that were like healthy individuals. This means that the

Table 5.11 End-of-life care

References	Monitor	*n*	Change	Index change	Explanation
Auyong et al. [126]	BIS A-2000	3	Life-support withdrawal	BIS ↑ 92, 83, 84	Cortical activities with brainstem death
Chawla et al. [127]	BIS, SEDLine PSI	7	End-of life-care no BP	BIS ↑ 65-85, PSI ↑ 95	Aggregate memory synaptic activity

n number of patients per group, *BIS* Bispectral Index, *PSI* SEDLine Patient State Index, *BP* Blood Pressure

EEG abnormalities in the ALS patient did not influence Bispectral Index monitoring, something that is quite useful for correctly estimating the depth of anesthesia in ALS patients [128].

5.7.5 Permanent Vegetative State (PVS)

Recent developments have provided several means to assess subjects who apparently seem to be in a "Persistent Vegetative State." Those methods aim at revealing when subjects are aware as well as some kind of primitive means of communicating with some of these subjects, who appear to be completely totally behaviorally nonresponsive [129]. Consequently, those new techniques could be easily extended far beyond dealing with the medical condition at hand, to even cover medicolegal aspects of such subjects supposedly in a "Persistent Vegetative State," as well as issues regarding patients' and their families' decision-making processes mainly withdrawal of care [129].

A "Persistent Vegetative State" is considered a medical condition of "Wakefulness" but no "Awareness" [130]. Prior to the year 1972, there was no unanimously acceptable definition that refers to subjects exhibiting any level of unconsciousness following brain injuries. Jennett and Plum [130], in the year 1972, coined the phrase "Persistent Vegetative State" in their publication under the title of "Persistent Vegetative State after Brain Damage. A Syndrome in Search of a Name" [130]. That distinctive terminology was chosen for stressing upon relative preservation to varying degrees of their "Vegetative" yet unconscious numerous brain mental functions including their core temperature regulations, breathing, digestion, and above all "wake-sleep cycles"; hence they are "awake" but not "aware" [130]. Whereas, when we go into normal physiological sleep, we are still "Aware" of our surroundings but not "Awake". At this time, the term "Persistent" was applied to delineate a condition that would last for a minimum of a whole month following brain damage [130]. Now, more than 52-years later, the term "Persistent Vegetative State" is still being the highest commonly used terminology describing this unique patients' state among both clinicians and the scientists' communities [129].

Patients in "Persistent Vegetative State" will frequently move spontaneously, open their eyes, and will often exhibit still preserved circadian "cycle of wake then sleep" [130]. This clearly demonstrates that the "Wakefulness" element of consciousness is still present. Normal physiological sleep is "Awareness" without "Wakefulness" then the "Persistent Vegetative State" could be considered a medical disturbance of "Wakefulness cycles with no Awareness" [130]. However, despite their circadian sleep and waking cycles, still meticulous repeated examinations of these patients reveal behavior that would not result in any indication of purposeful, maintained, voluntary "Awareness," nor repeated behavioral responses to noxious, visual, auditory, or tactile stimulations [130]. In other words, these subjects show no spontaneous purposefully initiated behavior and totally do not respond, in any shape or form, to any encouraging stimuli other than the most basic primitive reflex, that is why this is the ground for presuming that all the "Awareness" component is totally missing. However, highly experienced specialists in the field demonstrated that more than 43% of these patients exhibit, though not consistent but reproducible, behavioral indications of "Awareness" [131–133], and are currently placed in another category called "Minimally Conscious State" [134]. Unfortunately, it is still highly probable that several "Covertly Aware" subjects are not detected despite full efforts of specialists with vast experience using modern techniques, thus remaining wrongly identified as in "Persistent Vegetative State" [135].

Recent findings of Functional Magnetic Resonance Imaging (fMRI) and electroencephalography (EEG) trials have cast some doubt on considering patients being not "Aware" just because they do not show any noticeable behavioral responses to outside stimuli [135–138]. These recent trials showed that there is indeed a subgroup of subjects that are still "Aware" though totally incapable of responding in a fashion that can be easily elicited. These subjects, though they apparently meet the common standards for behavioral signs of being in a "Persistent Vegetative State", still they can clearly follow commands that the physicians can now establish and document via fMRI and EEG neuroimaging techniques, allowing these subjects, for the first time since they were brain injured, to communicate with their external environment [137].

Advanced high-tech fMRI response could provide strong verification proof that some "Persistent Vegetative State" patients are still "Aware." Bekinschtein and colleagues [139], in the year 2006, introduced a new pioneer method for eliciting covert command-following and hence detecting the "Awareness" component of consciousness [139]. Using fMRI, a patient who clearly fulfilled all the international clinical criteria for being in a "Persistent Vegetative State," turned out to be "covertly aware" and able to willfully respond to commands by simply "modulating" her fMRI activity [140]. Specifically, while in the fMRI scanner, the "Yes answer" was first established when the subject was instructed to envision striking a Tennis-ball forth and back as if they are playing a Tennis-match together with her fantasy Tennis-coach as soon as she hears the word "Tennis." The "No answer" was established when the patient was asked to visualize herself taking a walk passing through different rooms around her house as soon as she hears the word "House" [140]. The solid foundation for the experiment was rock-hard established when a group of control volunteers were asked to perform the exact same two mental imagery tasks [140]. The authors discovered that when you imagine that you are participating in a Tennis-match, this strongly generates vigorous activities in the region of "Supplementary Motor Area (SMA)" of the brain as seen in all participating volunteers [140]. Whereas volunteers who imagined that they are walking round their house resulted in activating the "Lateral Premotor Cortex," "Posterior Parietal Lobe," and "Parahippocampal Cortex," cortical regions all were shown strongly contributing to both the realistic and imaginary "Spatial Navigation" [140–145]. This robust and reliable fMRI response, demonstrated in every individual volunteer, was adequate to solidly establish such cerebral areas' activities as a firm response to specific verbal commands, as a "Neural Proxy" to bodily responses [135–137, 140, 146]. Essentially, fMRI response in apparent "Persistent Vegetative State" subjects was not different from responses seen in the subjects who volunteered to serve as controls when they were asked to perform the exact two tasks [136, 140]. Accordingly, the authors concluded that, contrary to her current diagnosis of being in "Persistent Vegetative State," the patient was still capable of understanding all the verbal orders and was still able to "coherently" respond to given commands just via her mental activities, that undoubtedly confirmed that she is "Consciously Aware." A full clinical trial followed [137], using a sample size of 23 subjects who were in a covert "Persistent Vegetative State" demonstrated that 4 /23 patients could generate similar consistent trustworthy fMRI responses to verbal commands, clearly showing that they were still "Aware."

Based on the same principle, numerous methods for detecting "covert awareness" in subjects who are not capable of physically responding, were devised and validated. Monti and colleagues [147], studied 12 healthy volunteers in addition to one "Minimally Conscious Subject (MCS)" because of brain anoxic injury following a cardiopulmonary arrest, whose Computerized Tomography (CT) revealed loss of grey-white matter differentiation indicating severe anoxic brain injury. Using fMRI, the 12 volunteers, and the one Minimally Conscious Subject, listened to a block of words and then they were asked the number of times a certain word was repeated among 4-seconds cues of numbers for the subjects to select when they hear the correct number the word was repeated. They demonstrated that the Frontal Cortex was activated bilaterally in the sub-lobar Inferior Frontal Gyrus and the Middle Frontal Gyrus areas when the subjects chose the correct number, that would be then considered as commands-following proxy [147].

Electroencephalographic responses could represent a viable alternative to fMRI as a proof of patient's "Awareness" [129]. Despite the astounding success of fMRI application for detecting covert "Awareness," still performing an fMRI on every subject that appears to be in an apparent Persistent Vegetative State is still not an easy task; due to the prohibitive costs of frequently performing such expensive procedure, or fMRI scanners being always available, as well as exposing patients to the repeated physical stressful experience of their transfer to the fMRI medical facilities. All this could be prohibitive for the wide usage for every subject that appears to be in an apparent Persistent Vegetative State [129]. Furthermore, patients' Movements Artefacts, that are a very common occurrence in restless subjects who are incapable of staying calm, as well as metallic implants that are very often placed in Traumatic Brain Injury (TBI) patients, could altogether exclude the use of fMRI. Thus, numerous neuroscientists explored whether cheap and portable to patients' bedside methods like EEG, could be utilized to reveal undetected covertly "Aware" patients who otherwise appear to be in Persistent Vegetative State [129].

This led to the next logical step in the efforts of styling EEG in a similar procedure that can detect the covertly "Aware" patients who otherwise appear to be in Persistent Vegetative State, using an electroencephalographic version of the "Motor Imagery Tasks" that was successfully used with fMRI [62, 148]. As subjects intend imaginary movements "Sensorimotor Cortical Activities," these clearly appear in their EEG as an Event-Related Desynchronization (ERD) or power reductions in the power of μ-wave and central β-wave (7-30 Hz) waveband rhythms, affected during imagery movements on the topographical contralateral primary Motor-Cortex area, i.e., on the "Lateral Premotor Cortex" when the hand is intended to move, together with a power increase in the α-wave (9–13 Hz and 10–12 Hz) and in the β-wave (22–29 Hz) indicating synchronization on the ipsilateral side, whereas the more "Medial Premotor Cortex" changes appear when the toe is intended to move [62, 148]. Thus, it seems that because these subjects with minimal consciousness could still possess some level of "Awareness," thus processed Electroencephalography (pEEG) could be a viable option for verifying their mental abilities [129]. The Bispectral Index can be used to yield insight into "consciousness levels" of the minimally conscious apparent Persistent Vegetative State subjects.

Davey and colleagues [149] examined EEG asymmetry and coherence of a Permanent Vegetative State patient with severe brain damage to subcortical gray matter structures mostly on the right side but relatively reserved function on the left side [149]. There were minimal differences in the frequencies power spectra between the two hemispheres as the damaged hemisphere showed a slight decline in the high frequency waves and a slight increase in the low frequency waves [149]. They showed that there was a dramatic difference in EEG Coherence between the two hemispheres (the extent to which the EEG of the different parts of the brain are correlated), as there was marked coherence reduction over the damaged hemisphere mainly at the frontal regions with the extensive damage, without a significant change in the overall EEG power spectrum [149]. At the TBI hemisphere, subcortical structures damage resulted in selective ipsilateral EEG coherence marked reduction because of damage to structures that mediate cortical areas communications, with only minimal power spectrum changes [149].

In a series of 8 subjects who received chronic Deep Brain Stimulation for the purpose of promoting recovery from the Permanent Vegetative State, compared with 11 patients who did not, Katayama and colleagues [150], showed that, in response to a painful Train Shock applied to a finger pad and recorded at the vertex, at the cortical level Pain-related P_{250} responsiveness to painful stimuli was depressed [150]. Pain-related P_{250} is unique in terms of its ability to assess cortical responsiveness to painful stimuli directly and quantitatively without involving functions of the motor system. The characteristics of cortical brain activity persist in some patients diagnosed with Permanent Vegetative State [150]. It was repeatedly demonstrated in four patients that the Pain-related P_{250} could be persistently increased by repeated sessions of Deep Brain Stimulation to the Mesencephalic Reticular Formation or

Nonspecific Thalamic Nuclei for more than 6 months, and this was correlated with their clinical improvements [150]. It appears that some part of the damage in Permanent Vegetative State patients may be functionally produced and could be reversible. It could also imply that the Bispectral Index as a technique might not be sufficiently strong enough to be able to discriminate non-integrated from integrated cerebral neuronal activities [150]. That is to say that an elevated Bispectral Index score might indicate an elevated general global cerebral activity indicating, in healthy subjects, high level of consciousness, but a high Bispectral Index score does not necessarily indicate that this cortical activity is integrated or meaningful in a subject with chronic brain damage like the Permanent Vegetative State patients [150].

Pandit and colleagues [151], using a Bispectral Index BIS A-1000 (version 3.2) monitor, reported a most interesting case of a patient in Permanent Vegetative State who required sevoflurane general anesthesia for dental surgery [151]. The Bispectral Index profile during surgery under sevoflurane anesthesia appeared to be like that of a normal subject. Before induction the patient appeared to be asleep, and the Bispectral Index value was 74–85 [151]. Venous cannulation and the change of the tracheostomy tube noxious stimuli provoked purposeless activity and an increase of Bispectral Index value to 98–100. Sevoflurane administration decreased the Bispectral Index value to 40–42 [151]. After the end of surgery and termination of sevoflurane anesthesia, the Bispectral Index increased to 98–100. This clearly indicates that some cortical activity persists in some patients currently diagnosed with Permanent Vegetative State [151].

5.8 Abnormal EEG Patterns

Table 5.12 enumerates and describes all conditions of abnormal electroencephalography patterns that were successfully monitored using the Bispectral Index (BIS) and State Entropy (SE).

Table 5.12 Abnormal EEG patterns

References	Monitor	*n*	Change	Index change	Explanation
Schnider et al. [154]	BIS A-1000 (3.11)	1	Low voltage EEG	Low BIS ↓ 40	Genetically determined
Muncaster et al. [155]	BIS A-1000 (3.31)	6/21	Low voltage EEG	Paradoxical Low BIS ↓	Drug induced Sevoflurane ↓ or remifentanil ↓
Hagihira et al. [156]	BIS A-1050 (3.4)	2/2	Low voltage EEG	Low BIS	Drug induced Sevoflurane ↓ or isoflurane ↓
Gunawardane et al. [157]	BIS A-1000 (3.31), SE	18	ECT Post-ictal	BIS ↓ 57, SE ↓ 58	ECT awake δ waves ↑, *SynchFastSlow* ↑
Nishihara et al. [159]	BIS	38	ECT Post-ictal	Awake low BIS ↓ 45	δ waves ↑
Chinzei et al. [161]	BIS A-1050 (3.12)	1	Seizures under anesthesia	BIS ↓40 ↑20	EEG fluctuations disappear with anticonvulsants
Tallach et al. [169]	BIS	1	Status Epilepticus	BIS ↑ 94 BIS ↓ 28	Propofol as anticonvulsant therapy
Musialowicz et al. [170]	BIS-XP A-2000	10	Status Epilepticus	BIS-BSR= EEG-BS	Propofol as anticonvulsant therapy
Dahaba et al. [172]	BIS-XP (3.4)	1	AERRPS	BIS ↑ 95 BIS ↓ 55	Diazepam /barbiturate coma
Dahaba et al. [180]	BIS-Vista (4.1)	11 + 11	BIS ↓	Epileptic Spikes ↑	Propofol ↑ /Sevoflurane ↑

n number of patients per group, *BIS* Bispectral Index, *BIS-BSR* Bispectral Index-Burst Suppression Ratio, *EEG-BS* Electroencephalography-Burst Suppression, *SE* State Entropy, *EEG* Electroencephalography, *ECT* Electroconvulsive Therapy, *AERRPS* Acute Encephalitis with Refractory Repetitive Partial Seizures

5.8.1 Genetically Determined Low-Voltage EEG

Genetically determined low-voltage electroencephalography (EEG) variant is defined as EEG amplitude of <20 mV over all head regions. This is a normal variant that occurs in 5–10% of the population and is not associated with any brain dysfunction [152]. The gene responsible for this variant was localized on the distal part of chromosome 20q [153]. Since the Bispectral Index algorithms' design is based upon healthy volunteers with ordinary electroencephalographic tracings, thus it would be presumed that the Bispectral Index algorithm would be "deceived" by any subject who would present with an unusual electroencephalographic tracing. This seems to be a potentially important source of error. If 5–10% of patients have a low amplitude EEG, and some fraction of those have a low Bispectral Index value, this could be gravely misleading as patients would be awake while the Bispectral Index would be giving an erroneous Bispectral Index value indicating deep anesthesia levels [152]. Hence it would be essential to ensure that all patients must exhibit a normal Bispectral Index when the patients are still awake, to prevent misleading interpretations of subsequent Bispectral Index values during anesthesia.

An interesting case was reported by Schnider and colleagues [154], in which a fully conscious volunteer was reported to exhibit, when he was completely awake, an irrationally decreased Bispectral Index A-1000 (version 3.11) value of around 40 on two different occasions repeated three days apart with the same result of Bispectral Index of around 40 [154]. Four days later, a 16-lead 10/20 international system electroencephalography (EEG) showed an inherently established "Low Voltage Electroencephalography" in which EEG is less than 20 mV amplitude globally all over the head area, a condition that occurs in 5–10% of the population, and a normal variant of the human EEG and is not associated with any brain dysfunction [154]. Since the Bispectral Index algorithms were derived from healthy subjects and the algorithms' design is based upon healthy volunteers with normal EEG tracings, thus it would be presumed that the Bispectral Index algorithm would be "deceived" by any subject who would present with an unusual EEG tracing. Therefore, the anesthesiologists should verify that all their patients exhibit ordinary Bispectral Index values before proceeding with the anesthesia induction [154].

5.8.2 Drug-Induced Low-Voltage EEG

The low-voltage electroencephalography (EEG) is not limited to the genetically determined variant, as it could also be "drug-induced." The low voltage EEG phenomenon was reported by Muncaster and colleagues [155], using Bispectral Index BIS A-1000 (version 3.31) monitor in 6/21 patients with a simultaneously recorded raw EEG during a two-stage awakening that consisted of first complete sevoflurane washout that was followed by remifentanil discontinuation [155]. They reported a mean Bispectral Index value paradoxical decline in 6 subjects, when remifentanil concentration decreased [155]. One of the Bispectral Index descriptors *SynchFastSlow* seems to have little influence on the Bispectral Index in this mode as it was trending higher, indicating awakening while at the same time the whole Bispectral Index value was decreasing. During this phase following complete sevoflurane washout but still during remifentanil withdrawal, the raw EEG in these six patients showed a very low amplitude EEG. This very low amplitude EEG was misinterpreted by the Bispectral Index algorithm as Burst Suppression (BS) and displayed a misleadingly low Bispectral Index value, which could represent a true shortcoming of the Bispectral Index monitoring [154]. The low amplitude EEG could be due to remifentanil withdrawal or sevoflurane washout [155].

The same effect was reported by Hagihira and colleagues [156], during monitoring of 2 inhalational anesthetic agents in 2 case reports on 2 separate occasions, using Bispectral Index A-1050 (version 3.4) and raw EEG downloaded and recorded from the Bispectral Index A-1050 monitor using the authors BSA software. The Bispectral Index A-1050 (version 3.4) value of 40–50 declined to 27–28 during the 1.5–1.6% sevoflurane "Washout Phase" in the first patient and during the 0.9% isoflurane "Washout Phase" in the second patient [156]. A low voltage EEG phenomenon was elicited during the "Washout Phases" of the two different inhalational anesthetic agents, that do not seem to be drug specific. During the "Washout Phase" of inhalational anesthesia at first the Bispectral Index value started to increase gradually, as few minutes later the raw EEG waveform displayed a spindle wave-dominant low voltage EEG pattern (≤3–4 μV) resembling rippled waves, as the Bispectral Index value now declined to 27–28. The Burst Suppression (BS) was 45–48 [156].

In all the above-mentioned patients, the simultaneously recorded raw EEG showed a very low-voltage EEG (≤15 μV) that was misinterpreted by the Bispectral Index monitor as Burst Suppression [155, 156]. Unfortunately, in these cases a normal Bispectral Index value before induction of anesthesia cannot guarantee a normal Bispectral Index performance with this form of drug-induced low-voltage EEG.

5.8.3 Seizures and Post-ictal EEG Patterns

Gunawardane and colleagues [157] studied the Bispectral Index A-1000 (version 3.31) and all its subparameters, as well as Spectral Entropy, in 18 patients undergoing Electro Convulsive Therapy (ECT) [157]. The Bispectral Index electroencephalography (EEG) data were downloaded to be later analyzed. The data were later analyzed using the Matlab software (Matlab 5.3, The Mathworks Inc., Natick, MA, USA). The subparameter descriptors components (*Relative BetaRatio* and *SynchFastSlow)* of the Bispectral Index were calculated using Ira Rampil analysis [158]. Patients awakened when the median (25–75 percentile) Bispectral Index was as low as 57 (47–78) and >25% of the patients woke up at a Bispectral Index score of <50. Similarly Spectral Entropy at awakening was 58 (31–85) [157]. At the awakening point, of all the significant independent predictors in their final model analysis, only the *SynchFastSlow* subparameter remained, reflecting a post-ictal electroencephalography (EEG) suppression state [157]. Acting via the *SynchFastSlow* subparameter, EEG post-ictal slow wave activities may result in low Bispectral Index value that does not reflect patient's real level of consciousness. This may be important in the interpretation of the Bispectral Index in other conditions with increased δ-band power in their EEG [157].

In a similar study Nishihara and colleagues [159], demonstrated that following ECT in 38 psychotically depressed patients [159], patients awakened when the mean ± SD Bispectral Index scores was as low as 45 ± 13 [159], reflecting the same post-ictal EEG suppression state in the first study [157]. In their study subjects were even capable of opening their eyes 7 ± 2 min following ECT [159]. Patients in the post-ictal state display very slow δ EEG waves even after regaining consciousness [160]. In Gunawardane and colleagues [157] study, in response to verbal command at the time of the first eye opening, the *BetaRatio* descriptor of the Bispectral Index algorithm recovered to pre-induction levels, while the *SynchFastSlow* descriptor remained low [157]. This clearly shows that, the low Bispectral Index value resulted from the low frequency δ waves strongly influencing the *SynchFastSlow* descriptor, that usually indicates deep anesthesia [157].

Epileptic seizures were shown to result in unusual Bispectral Index fluctuations during deep anesthesia in a subject with a known history of epilepsy [161]. The Bispectral Index may give very useful information, not only on the anesthetic depth, but also on the development of and recovery from abnormal epileptiform EEG activities. Chinzei and colleagues [161] reported abnormal fluctuation in Bispectral Index A-1050 (version 3.12) values during sevoflurane general anesthesia, appearing as repeated alternations between two different EEG waveform patterns in a patient with recent history of epileptic seizures [161]. The Bispectral Index value decreased to 40 with induction of anesthesia and remained stable at this value for about 1 h during which raw EEG baseline patterns before surgical stimulation displayed high-frequency (10–14 Hz) low-voltage waves [161]. About 40-min from the start of surgery, episodes of abrupt and transient reductions in Bispectral Index value from baseline value of 40 to 20 began to appear while raw EEG signals displayed low-frequency (1.5–2.5 Hz) high-voltage waves (slow δ-waves, with or without spikes), before abruptly the Bispectral Index recovers to 40 when high-frequency low-voltage raw EEG waves reappear [161]. The repeated Bispectral Index fluctuations between high-frequency low-voltage and low-frequency high-voltage EEG waves, at 1.5- to 5-min intervals, occurred for nearly 2 h, with the Bispectral Index monitor displaying error-messages of electrode impedances <2 kΩ, not due to high electromyography (EMG) signal contamination because vecuronium Neuromuscular Blocking Agent (NMBA) was administered [161]. Given the patient's epilepsy medical history, midazolam 1.5 mg and thiamylal 100 mg anticonvulsants were administered. Almost immediately, the low-frequency high-voltage EEG waves and the error messages disappeared as the Bispectral Index abruptly returned to 40 baseline values immediately after the anticonvulsants administration. The patient emerged uneventfully from anesthesia [161]. Although it is difficult to precisely identify epilepsy EEG patterns, the abrupt onset-offset low-frequency high-voltage EEG activities (slow δ-waves, with or without spikes) is widely recognized as a common epileptiform EEG pattern [162], that almost immediately disappeared with the administration of anticonvulsants [161]. Even though the patient received midazolam 4 mg as premedication before surgery, apparently this did not prevent epileptic seizures from occurring under anesthesia [161]. Although volatile anesthetics were reported to have some anticonvulsant effects [163, 164], conversely several reports of epileptoid EEG patterns were observed during sevoflurane general anesthesia [162, 165, 166], appearing as slow δ-wave EEG activity, with or without spikes, associated with jerky movements [162], whereas sevoflurane was shown to have stronger epileptogenic properties than those of isoflurane [167, 168].

A useful application of Bispectral Index is monitoring subclinical covert epileptic seizure activity that might go undetected in patients receiving NMBAs. The Bispectral Index changes, in that case, would greatly depend on the nature of the main ictal EEG waveform; thus the Bispectral Index values could increase or decrease during epileptic seizures. Furthermore, epileptic seizures' high EMG muscle activity may also affect Bispectral Index values [169]. Tallach and colleagues [169] reported the management of a patient who suffered from a refractory Status Epilepticus that did not respond to intravenous benzodiazepines and phenytoin anticonvulsive therapies. During the patient's epileptic seizure activity, authors observed an extremely high Bispectral Index value of 94 with an "ictal typical epileptiform" picture on the raw EEG tracings [169]. At no point was the patient awake or arousable, even when the Bispectral Index was extremely high. Because the patient's ictal seizure activity started in his left upper limb, the EEG remained unpolluted by the high EMG muscle activity in the head and neck, as the Bispectral Index monitor Signal Quality Index (SQI) remained consistently high ≥50 [169]. Control of seizures was achieved via propofol boluses administration, which resulted in the return of the raw EEG tracings to slow wave activity with the Bispectral Index value also declining from 94 during the patient's seizure to 28 when the seizure was under control [169]. Addition of phenobarbital therapy gradually controlled his seizure activity, that was evident in his Bispectral Index and raw EEG tracings [169]. The authors concluded that, in patients with medical history of epilepsy, any sudden unexplained change in the Bispectral Index value could indicate ictal seizure activity warranting further investigation and /or treatment [62, 169].

The aim of "Medically Induced Coma" for management of refractory Status Epilepticus or elevated Intra Cranial Pressure (ICP) patients is to reach Burst Suppression. Burst Suppression Ratio (BSR) of 15–30% on the Bispectral Index monitor corresponding to 3–5 bursts/min on the International 10/20 EEG System is considered by most neurophysiologists, as the optimal Burst

Suppression level to safely reduce Cerebral Metabolic Rate of Oxygen ($CMRO_2$) of patients in Medically Induced Coma [170]. Because the Bispectral Index values show strong correlation with Burst Suppression, this could be successfully utilized for Burst Suppression therapy for epileptic seizures of patients in Status Epilepticus, providing that the Bispectral Index is frequently checked against an International 10/20 EEG System when used for extended durations [170]. Furthermore, refractory Status Epilepticus sometimes could only be managed via administration of anesthetic agents in the Critical Care Unit (CCU) that must be continuously monitored. In that case, Bispectral Index monitoring would be very useful [62, 170].

Musialowicz and colleagues [170] used the Bispectral Index BIS-XP A-2000, as well as simultaneous EEG monitoring, for assessing Burst Suppression patterns when propofol was administered for management of 10 adult refractory Status Epilepticus patients in a CCU setting [170]. When propofol was titrated to Burst Suppression levels, authors compared Bispectral Index-Burst Suppression Ratio (BIS-BSR) percentage with Electroencephalography-Burst Suppression (EEG-BS) rate/minute [170]. They demonstrated significant $P < 0.001$ very strong correlation (r^2 −0.9) between Bispectral Index values and EEG-BS. The also demonstrated significant P <0.001 very strong correlation (r^2 −0.88) between BIS-BSR and EEG-BS [170]. Using a Bispectral Index cut-off value of 30, the Bispectral Index Sensitivity and Specificity in detecting EEG-BS were 0.99 and 0.98 respectively. However, the Bispectral Index could not detect local seizure activities or ictal discharges during EEG-Burst Suppression [170]. Although the authors considered EEG as the primary monitoring technique for assessment of Burst Suppression for treatment of refractory Status Epilepticus, Bispectral Index monitoring can still guide the level of anesthesia administration in CCU patients with refractory Status Epilepticus [62, 170].

International 10/20 EEG System, the Gold Standard for monitoring Epileptic activities, requires time and expertise for setting up and operating. Whereas Bispectral Index, an easy to set up and use processed Electroencephalography (pEEG) monitor, that was successfully used to monitor several EEG conditions, could be utilized to detect epileptic conditions. Acute Encephalitis with Refractory, Repetitive, Partial Seizures (AERRPS) is a refractory epileptic condition that was first described, in 2001 in Japan, by Sakuma and colleagues [171]. My research group Dahaba and colleagues used the Bispectral Index BIS-XP (version 3.4) for monitoring anticonvulsive therapy of a 14-year-old female patient with the very rare and peculiar AERRPS condition at the CCU of Peking Union Medical College Hospital, Beijing, People's Republic of China [172]. After Bispectral Index Quatro sensor placement, Bispectral Index, EMG in decibels (dB) as well as International 10/20 EEG System were continuously recorded [172]. Figure 5.11 demonstrates repeated 1–2 min epileptic seizures every 3–4 min Bispectral Index and Electromyography peaks and Burst Suppressions. Paired *t*-test revealed that at the beginning of epileptic seizure, mean Bispectral Index and EMG peak values (95 ± 4, 56 ± 3 dB) significantly declined at the end of seizures (55 ± 13, 35 ± 3 dB) respectively [172]. Electroencephalography showed that epileptic discharges were replaced by slow δ-waves and θ-waves at seizure remissions. As it was decided to induce pharmacological diazepam/barbiturate coma, Bispectral Index values declined to almost isoelectricity of 11.9 ± 2.5, as the BSR steadily increased to 66 ± 10 [172]. This characteristic Bispectral Index fluctuations of 1–2 min high EEG/EMG seizure activities, repeated every 3–4 min, probably is an indication of widespread cerebral ictal-seizure activities. Our report clearly shows that, after proper EEG confirmation of her diagnosis, the easy-to-use Bispectral Index monitor could be a useful tool for her diagnostic, prognostic as well as her therapy efficacy clinical monitoring [172].

Epilepsy is a chronic neurological condition characterized by epileptogenic activity recurring seizures. Although medical therapy is the corner-

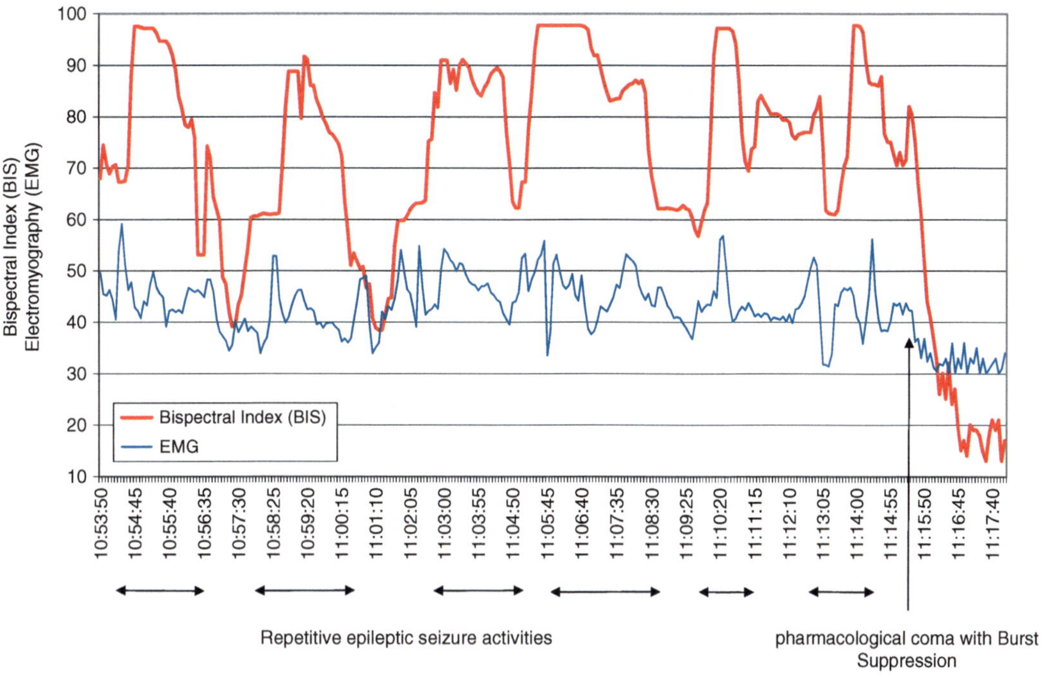

Fig. 5.11 Acute Encephalitis with Refractory, Repetitive, Partial Seizures (AERRPS) repeated 1-2 minutes epileptic seizures every 3–4 min Bispectral Index and Electromyography peaks and Burst Suppressions. (Reproduced with Minerva Anestesiologica permission from Dahaba AA, et al. Bispectral index (BIS) monitoring of acute encephalitis with refractory, repetitive partial seizures (AERRPS). Minerva Anestesiol. 2010;76:298–301)

stone of managing epilepsy, it still fails a substantial portion of patients who are refractory to therapy or intolerable to antiepileptic drugs [173]. Surgical treatment, in medically intractable patients, is currently an effective option for epileptogenic foci excision [174]. Because some epileptogenic foci are on proximity to motor, memory, speech, or sensory vital cortical areas [175], intraoperative electrocorticography (ECoG) from electrodes directly placed on the cerebral cortex [176] could help precisely locate the origin of epileptic activity and delineate the resection margins [174].

Different anesthetic agents and regimes could either obtund or provoke epileptiform activity. Many aspects regarding the choice of anesthetic regimes for their effect on ECoG epileptiform activity are still not guided by scientific evidence rather by local convention. The commonly used anesthetic agents for epilepsy surgery, sevoflurane [174, 177], and propofol [178, 179], have both raised controversies from their conflicting reports of pro- or anti-epileptogenic effects on ECoG spike activities.

Because we have not seen to date, any comparative studies answering this core issue, my research team Dahaba and colleagues at the Department of Neurosurgery, and the Epilepsy Center, Second Affiliated Hospital of Dalian Medical University, People's Republic of China, conducted a study to answer the core issue of whether increasing doses of anesthetic agents abolishes or decline the epileptogenic spikes, and for epileptic foci resection what is the optimal propofol/remifentanil or sevoflurane /remifentanil Bispectral Index levels for precisely identifying brain tissues that exhibit ECoG spikes [180]. We hypothesized that decreasing Bispectral Index levels would decline the spike frequency and amplitude of ECoG epileptiform activity. Twenty-two intractable epilepsy subjects were randomly allocated to the propofol /remifentanil or sevoflurane/remifentanil groups. We surgically placed an ECoG 28-Electrode grid directly on the

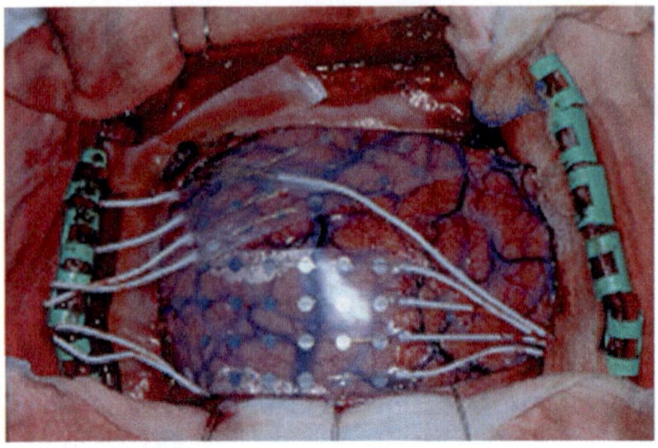

Fig. 5.12 Electrocorticography (ECoG) 28-Electrodes Grid directly placed on the Cerebral Cortex

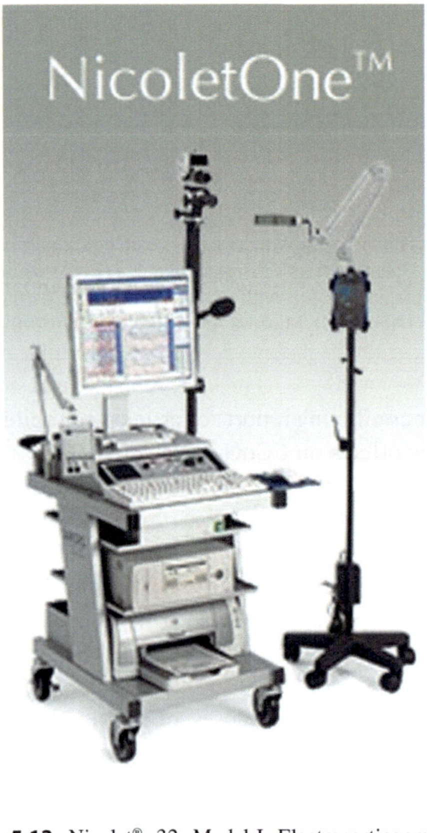

Fig. 5.13 Nicolet® 32 Model-I Electrocorticographic (ECoG) monitor

Cerebral Cortex for identifying the brain issue with the epileptogenic spike discharge (Fig. 5.12), that was connected to Nicolet® 32 Model-I Electrocorticographic monitor (Nicolet Biomedical Inc., Middleton, Wisconsin, USA) (Fig. 5.13) [180], that uses Cz as monopolar 70 Hz high frequency reference electrode, filtered at 0.01–1500 Hz bandwidth and a 0.03 s time constant 32-channel amplifier [180]. We synchronized our Bispectral Index BIS-Vista® (version 1.4) and Nicolet® monitors' clocks to the exact "h:min:s" time. Soon after the grid electrodes were placed, using sevoflurane Minimum Alveolar Concentration (MAC) or propofol Target Controlled Infusion (TCI) rate adjustments, subjects were randomly allocated to computer-generated four 15-min Bispectral Index levels sequence-order of BIS85 (BIS 71–85), BIS70 (BIS 56–70), BIS55 (BIS 41–55), and BIS40 (BIS < 40) [180].

To differentiate epileptiform spike discharges from other drug-induced, non-epileptiform waves such as K-Complexes, we considered the wave width as a reliable characteristic to easily distinguish our true epileptiform spike discharges of less than 70 ms from the K-Complexes that are more than 0.5 s [180]. We measured the spike amplitude using the automated capture "Nicolet monitoring software" of our ECoG monitoring window. Basically, the spike amplitude is expressed as "volatility" where the amplitude represents the "Nadir-Zenith vertical length of a wave" or the mean of the 2 Nadir-Zenith perpendicular lines for biphasic waves (Fig. 5.14) [180]. In addition, we further post hoc normalized/ extrapolated (smoothed) our raw Bispectral Index data to an arbitrary level of BIS85, as we "a-priori" hypothesized a decline in spike parameters from that Bispectral Index level, thus giving a percentage change within each patient [180].

Two-way ANOVA revealed no significant differences between the propofol or sevoflurane groups in spike frequency, spike amplitude, or number of spiking leads [180]. In the propofol group, decreasing Bispectral Index levels significantly increased mean spike frequency (Fig. 5.15) at BIS85 (10 ± 12), BIS70 (19 ± 17), BIS55 (22 ± 17) and BIS40 (25 ± 15) spike min^{-1}. Decreasing Bispectral Index levels increased spike amplitude (Fig. 5.16) at BIS85 (1100 ± 400), BIS70 (1200 ± 460), BIS55 (1300 ± 560) and BIS40 (1400 ± 570) spike μV [180]. Furthermore, in our post hoc normalized /extrapolated (smoothed) raw data to an arbitrary level of BIS85, compared with BIS85, spike frequency (Fig. 5.17) increased 100% at BIS70, 116% at

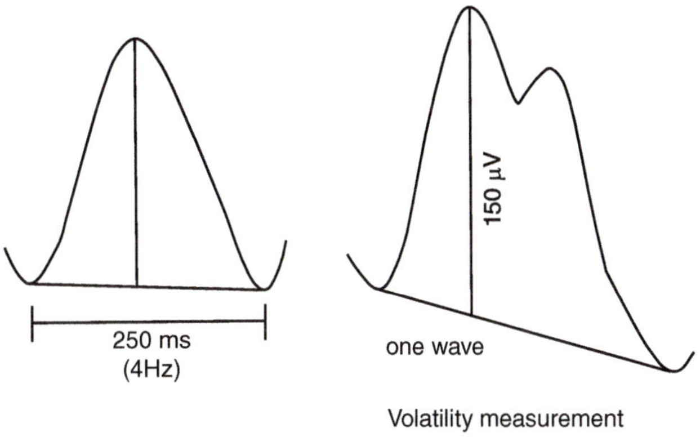

Fig. 5.14 The Spike amplitude measurement captured from our electrocorticography (ECoG) monitoring window using "Nicolet monitoring software". Spike amplitude is the "Nadir-Zenith vertical length of a wave" or the mean of the 2 nadir-zenith perpendicular lines for biphasic waves. (Reproduced with permission # 5895761372648 from Dahaba AA, et al. Different propofol-remifentanil or sevoflurane-remifentanil bispectral index levels for electrocorticographic spike identification during epilepsy surgery. Anesthesiology. 2013;119:582–92)

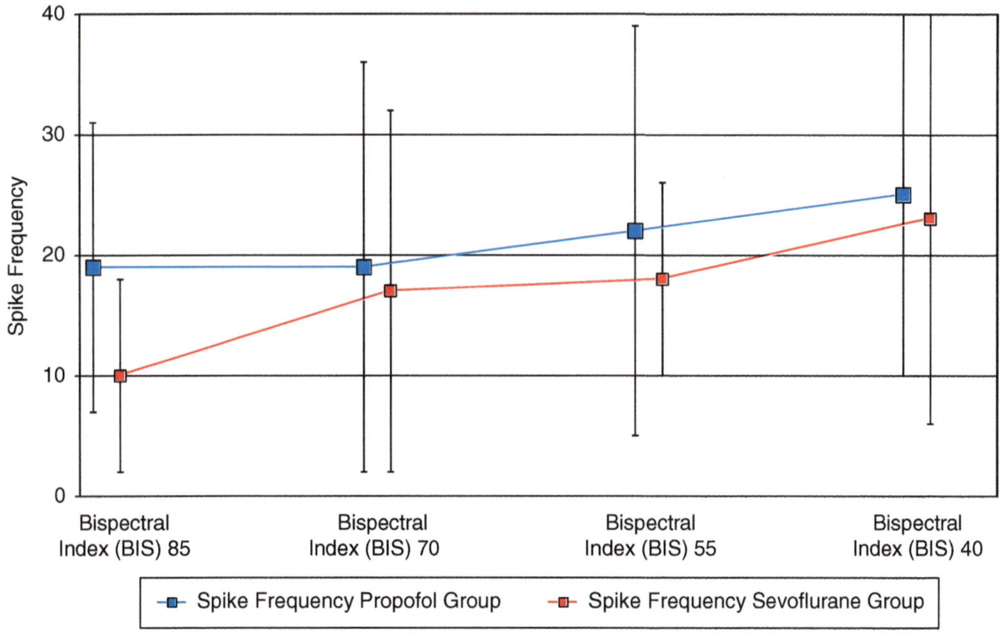

Fig. 5.15 Mean ± SD Spike frequency in the propofol and sevoflurane groups at Bispectral Index BIS85 (BIS 71–85), BIS70 (BIS 56–70), BIS55 (BIS 41–55) and BIS40 (BIS <40) levels. (Reproduced with permission # 5895761372648 from Dahaba AA, et al. Different propofol-remifentanil or sevoflurane-remifentanil bispectral index levels for electrocorticographic spike identification during epilepsy surgery. Anesthesiology. 2013;119:582–92)

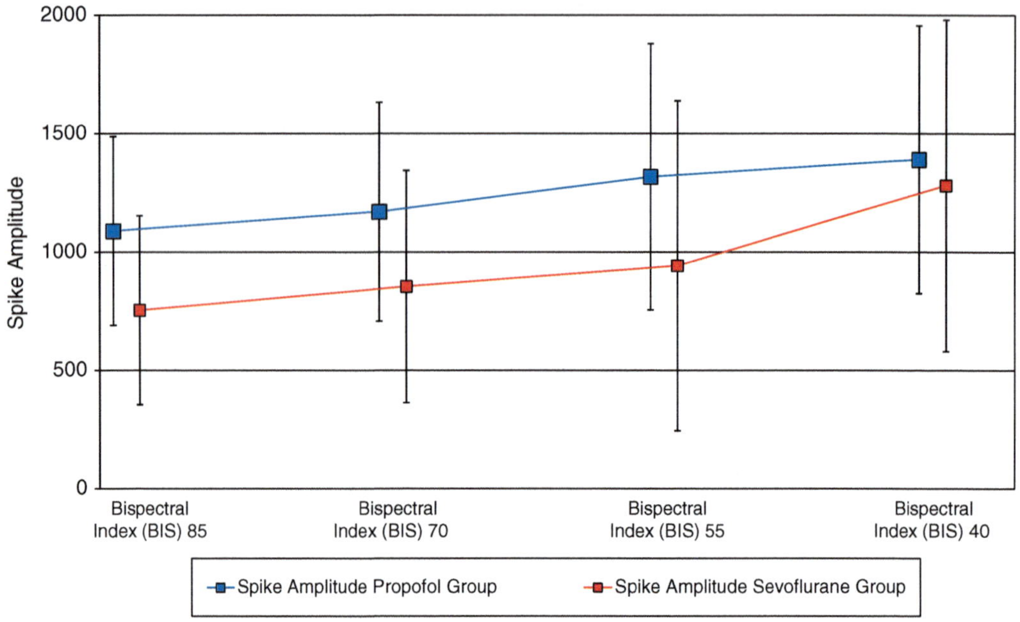

Fig. 5.16 Mean ± SD Spike amplitude in the propofol and sevoflurane groups at Bispectral Index BIS85 (BIS 71–85), BIS70 (BIS 56–70), BIS55 (BIS 41–55) and BIS40 (BIS <40) levels. (Reproduced with permission # 5895761372648 from Dahaba AA, et al. Different propofol-remifentanil or sevoflurane-remifentanil bispectral index levels for electrocorticographic spike identification during epilepsy surgery. Anesthesiology. 2013;119:582–92)

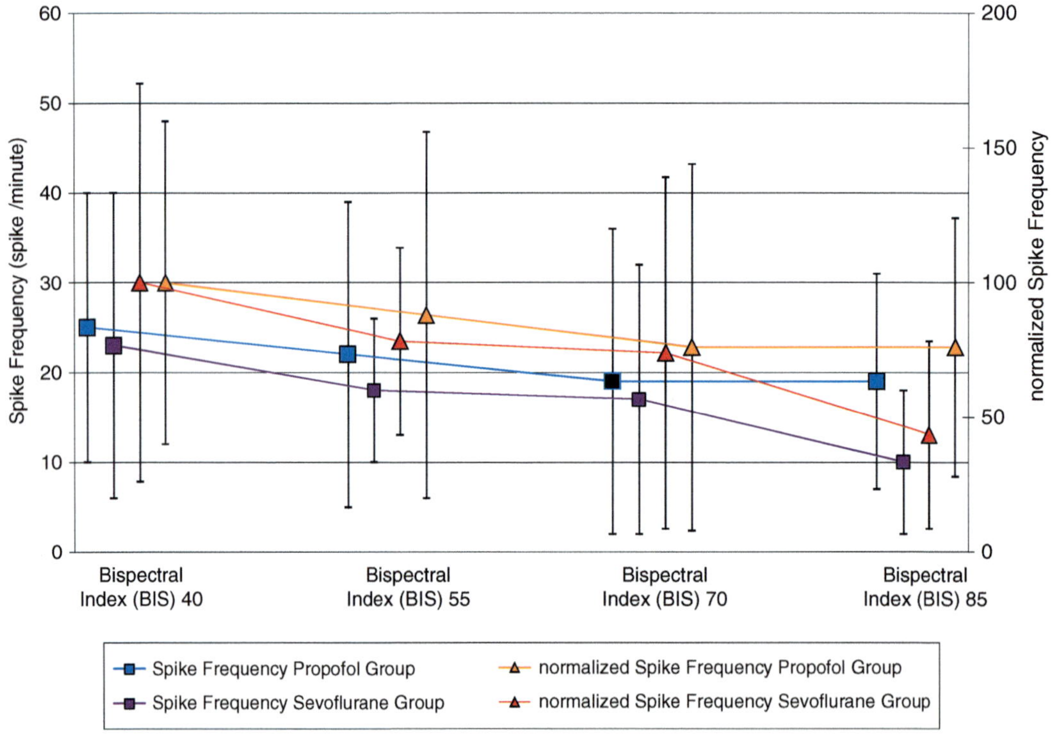

Fig. 5.17 Mean ± SD post hoc normalized/extrapolated (smoothed) raw data to an arbitrary level of BIS85; compared with BIS85 Spike frequency in the propofol and sevoflurane groups at Bispectral Index BIS85 (BIS 71–85), BIS70 (BIS 56–70), BIS55 (BIS 41–55) and BIS40 (BIS <40) levels

BIS55, and 132% at BIS40. Compared with BIS85, spike amplitude (Fig. 5.18) increased 108% at BIS70, 121% at BIS55, and 128% at BIS40 [180]. Decreasing Bispectral Index levels in the propofol group enhanced epileptogenic spike frequency and amplitude with no influence on location or number of spiking leads (Fig. 5.19) at the 4-BIS levels [180].

In the sevoflurane group, decreasing Bispectral Index levels significantly increased mean spike frequency (Fig. 5.15) at BIS85 (10 ± 8), BIS70 (17 ± 15), BIS55 (18 ± 8) and BIS40 (23 ± 17) spike min^{-1} [180]. Decreasing Bispectral Index levels increased spike amplitude (Fig. 5.16) at BIS85 (750 ± 400), BIS70 (850 ± 490), BIS55 (940 ± 700), and BIS40 (1300 ± 700) spike μV [180]. Furthermore, in our post hoc normalized/extrapolated (smoothed) raw data to an arbitrary level of BIS85, compared with BIS85, spike frequency (Fig. 5.17) increased 170% at BIS70, 180% at BIS55, and 230% at BIS40. Compared with BIS85 [180], spike amplitude (Fig. 5.18) increased 113% at BIS70, 125% at BIS55, and 170% at BIS40. Decreasing Bispectral Index levels in the sevoflurane group enhanced epileptogenic spike frequency and amplitude with no influence on location or number of spiking leads (Fig. 5.19) at the 4-BIS levels [180].

One of the major electroencephalographic "graphoelements" are the K-Complexes, first described by Loomis and colleagues [181]. The K-Complexes are defined using Rechtschaffen and Kales' criteria [182], as electroencephalographic waveforms with a well delineated negative sharp wave immediately followed by a positive component. The total duration of the K-Complex should exceed 0.5 s [182]. The K-Complexes are outlined as high-voltage, large amplitude 75–200 μV waveform, bi (tri) phasic slow waves, characterized by an initial short

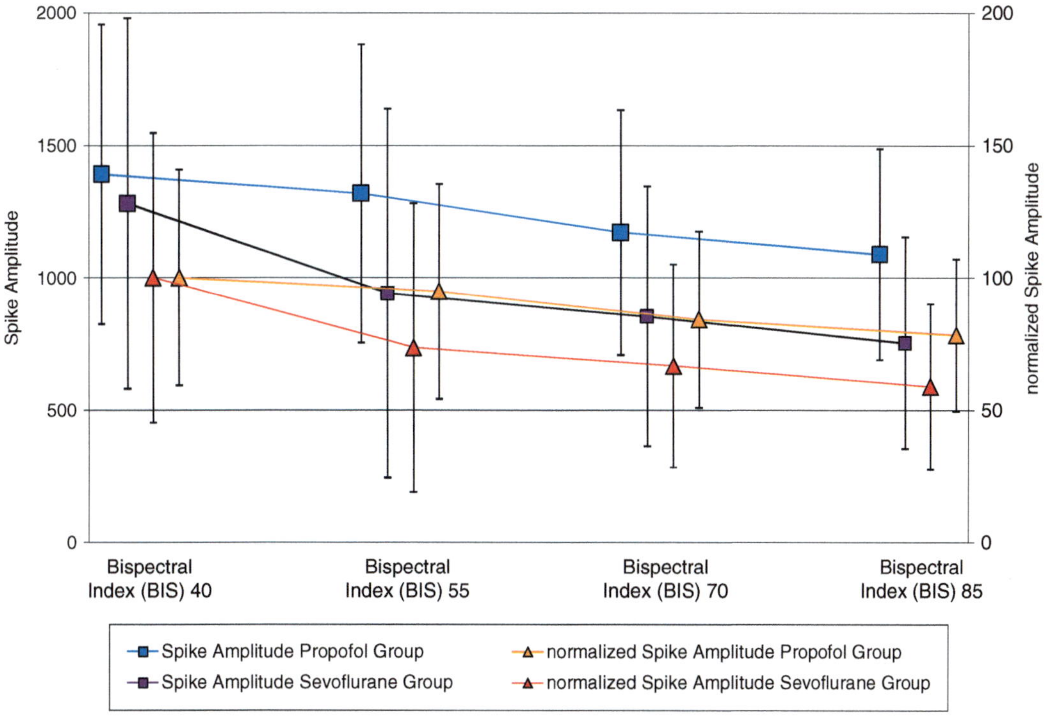

Fig. 5.18 Mean ± SD post hoc normalized/extrapolated (smoothed) raw data to an arbitrary level of BIS85; compared with BIS85 Spike amplitude in the propofol and sevoflurane groups at Bispectral Index BIS85 (BIS 71–85), BIS70 (BIS 56–70), BIS55 (BIS 41–55) and BIS40 (BIS <40) levels

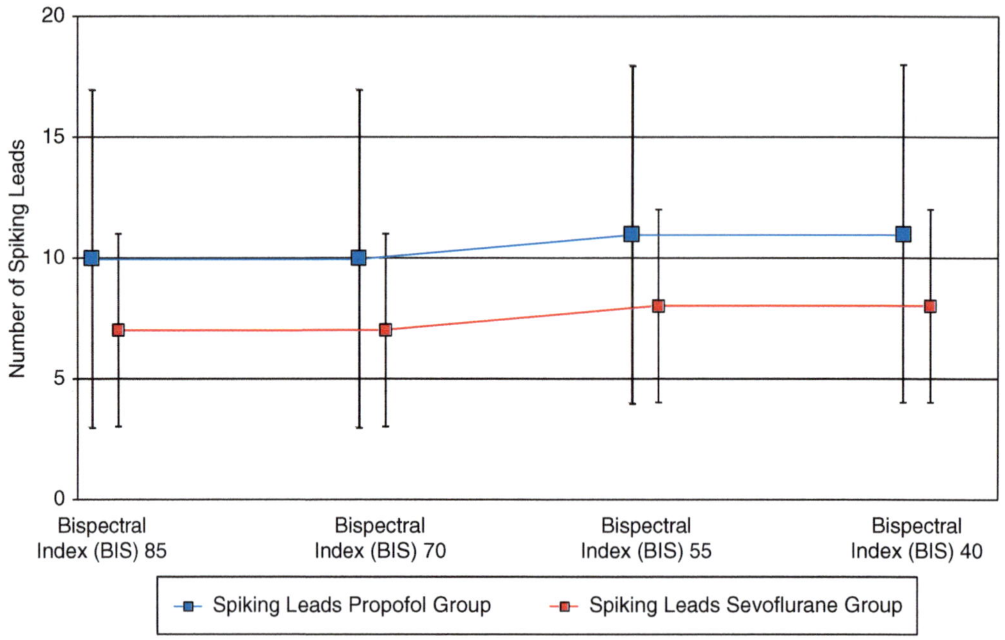

Fig. 5.19 Mean ± SD number of Spiking leads in the propofol and sevoflurane groups at Bispectral Index BIS85 (BIS 71–85), BIS70 (BIS 56–70), BIS55 (BIS 41–55) and BIS40 (BIS <40) levels. (Reproduced with permission # 5895761372648 from Dahaba AA, et al. Different propofol-remifentanil or sevoflurane-remifentanil bispectral index levels for electrocorticographic spike identification during epilepsy surgery. Anesthesiology. 2013;119:582–92)

surface-positive transient wave, followed by a slower larger surface-negative phase that peaks at 300–750 ms. This is immediately terminated by a final positive phase that peaks near 900–1300 ms [183]. The K-Complexes constitute a physiologic correlate of "Arousal." This notion is mostly based on the association of "spontaneous K-Complexes" with signs of Arousal such as those that occur with anesthesia lightning; the so-called "Arousal K-Complexes." The K-Complexes could also occur in response to surgical stimuli the so-called "Response K-Complexes" [184]. To differentiate epileptiform spike discharges from other drug-induced, non-epileptiform waves such as the K-Complexes, we considered the wave width as a reliable characteristic to easily distinguish true epileptiform spike discharges of less than 70 ms from the K-Complexes that are more than 0.5 s [184]. We captured what we considered EEG raw traces of inter-ictal K-Complexes discharges, distinguishable from EEG background, by their increase in frequency and amplitude before epileptic focii excision (Fig. 5.20), at 11: 50 (Fig. 5.21), at 12:14 (Fig. 5.22) and at 12 18 (Fig. 5.23).

Si and colleagues [185] compared the amplitude, frequency, and rising slopes changes of K-Complexes in refractory frontal lobe epilepsy patients versus control patients. K-Complexes' amplitude, frequency, and rising slopes were higher in refractory frontal lobe epilepsy patients than in control patients with no difference in K-Complexes durations [185].

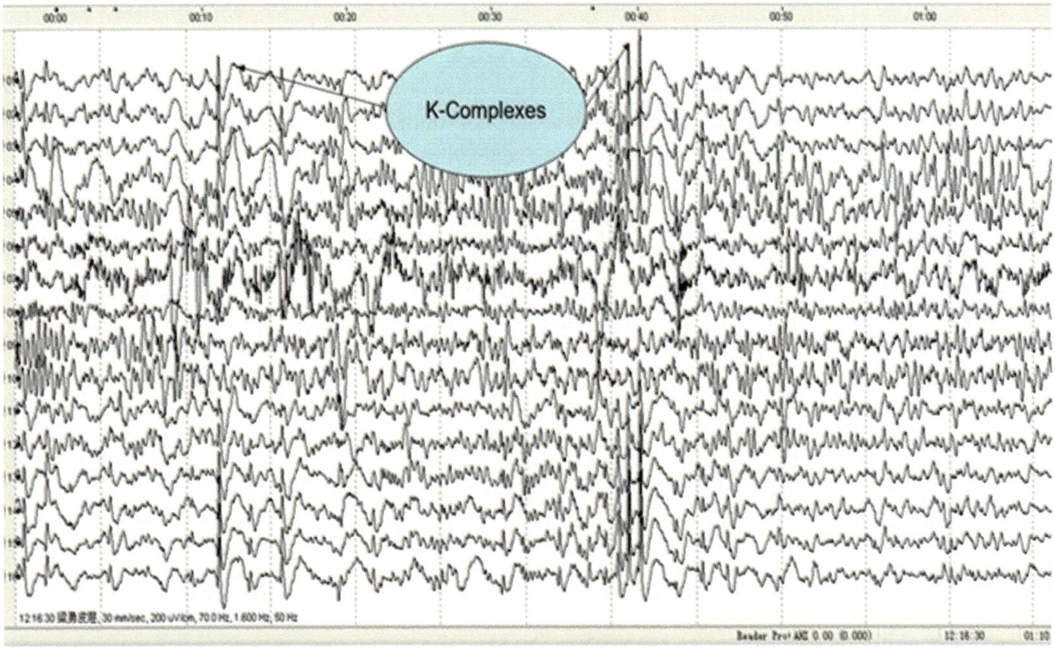

Fig. 5.20 Electrocorticography raw traces of inter-ictal K-Complexes discharges, distinguishable from EEG background by their increase in frequency and amplitude before epileptic focus excision

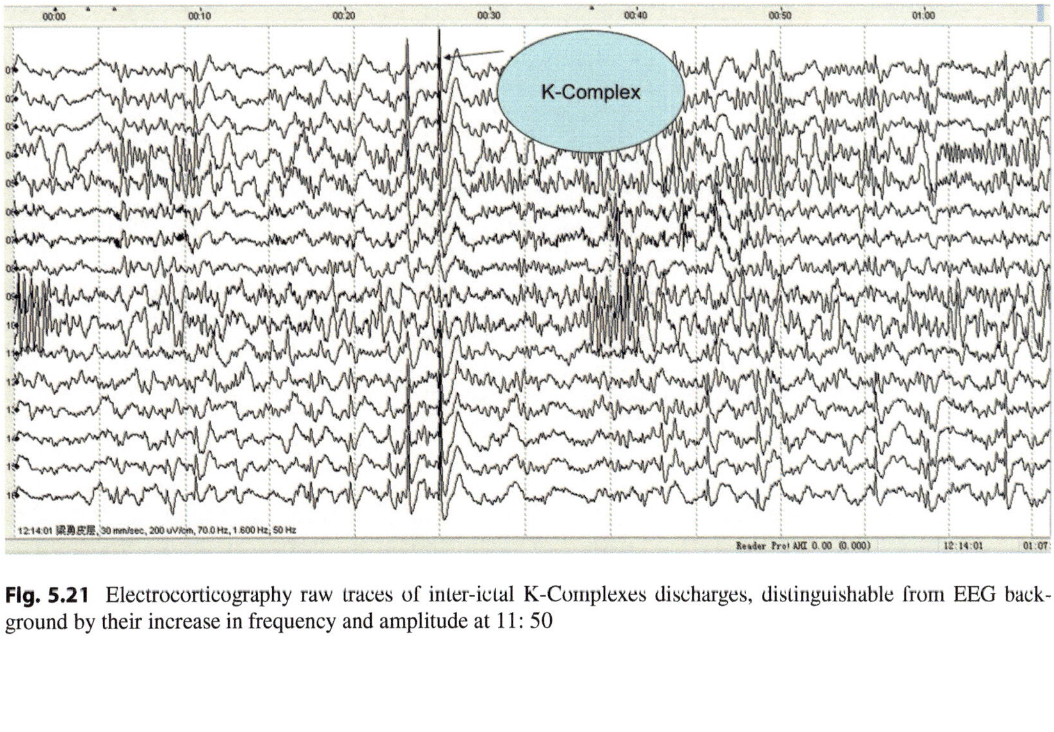

Fig. 5.21 Electrocorticography raw traces of inter-ictal K-Complexes discharges, distinguishable from EEG background by their increase in frequency and amplitude at 11: 50

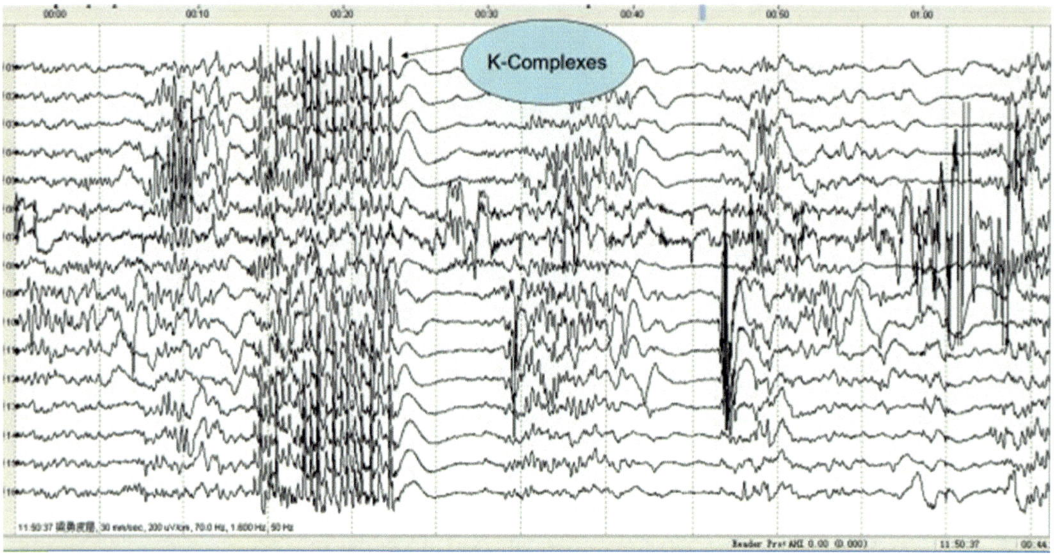

Fig. 5.22 Electrocorticography raw traces of inter-ictal K-Complexes discharges, distinguishable from EEG background by their increase in frequency and amplitude at 12:14

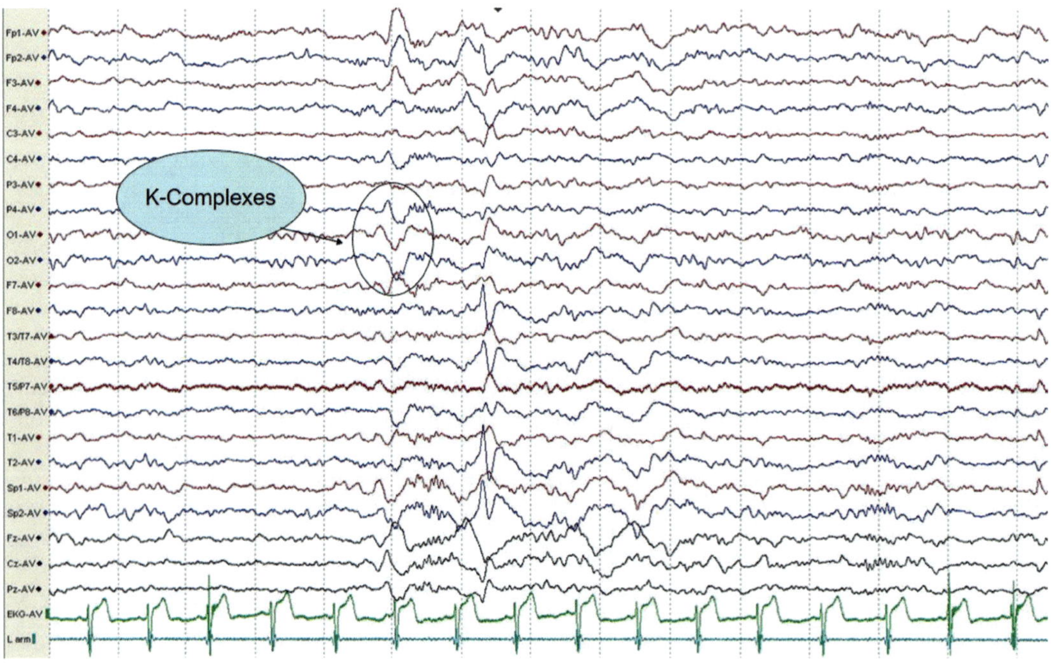

Fig. 5.23 Electrocorticography raw traces of inter-ictal K-Complexes discharges, distinguishable from EEG background by their increase in frequency and amplitude at 12 18

5.9 Effect of Acute Normovolemic Hemodilution (ANH)

The National Institute of Health Consensus Conference [186] recommended Acute Normovolemic Hemodilution (ANH) autologous blood procurement technique as one of the cost-effective blood conservation techniques [187], that aim at reducing or avoiding allogeneic blood transfusion during major surgical procedures when the expected blood loss is more than one liter [188]. The ANH technique, through preoperative dilution of the circulating blood, reduces the amount of red blood cells and other plasma constituents that could be lost during surgical bleeding [189]. The Bispectral Index monitoring was shown to be susceptible to numerous physiological and medical conditions that could alter cerebral function [2], whereas ANH used in clinical practice was shown to be associated with significant impairment of cognitive functions [190].

In a randomized observer-blinded study [189], my research team Dahaba and colleagues at the First Hospital of Peking University, Beijing, People's Republic of China, compared the effect of ANH on Bispectral Index changes prior to induction of anesthesia and after propofol Target Controlled Infusion (TCI) anesthesia induction and maintenance in 30 unmedicated patients whose Body Mass Index (BMI) was between 18 and 26 kg m^{-2} with an American Society of Anesthesiologists (ASA) classification physical status I-III undergoing general surgical or orthopedic procedures when the expected blood loss is more than 1 liter. Patients were randomly allocated and stratified for gender and ASA classification, to the ANH or control groups [189].

BIS "Quatro" sensors were placed on patients' forehead and connected to a Bispectral Index BIS-XP (version 3.4) monitor [189]. The ANH was carried out according to the published guidelines [188]. Fifteen to 20 ml kg^{-1} blood were procured from the cubital vein at an approximate rate of 75 ml min^{-1} and stored in an acid citrate-dextrose reservoir bag to be simultaneously replaced by an equal volume of warmed 6% Hydroxyethyl Starch (HES) 130/0.4 in 0.9% NaCl solution rapidly infused via a cubital vein cannula in the other arm aiming at reducing Hemoglobin (Hb) concentration to 9 g dl^{-1} [189]. Ideally, patients' blood was re-infused after the phase of major blood loss towards the end of the operation, or sooner if clinically indicated mainly for persistent hypotension or if Hb <6 g dl^{-1} threshold was reached [189]. Each subject was fitted with a face mask through which air was insufflated. After the conclusion of ANH in the ANH group and after a similar period of time in the control group, 100% oxygen was administered at a flow rate of 5 l min^{-1} and a 5-min equilibration period was allowed before induction of anesthesia [189] with propofol TCI set to reach a plasma concentration of 1.5 μg ml^{-1}, to be increased by 0.5 μg ml^{-1} once every 30 s until patients lost their eyelash reflex and no longer responded to verbal command. Loss of consciousness (LOC) time, propofol dose requirements for LOC, and the Bispectral Index values at LOC were recorded. Anesthesia was maintained at propofol TCI 4 μg ml^{-1} with no further rate adjustments until a stable Bispectral Index level was reached. The stable maintenance Bispectral Index level was recorded, and this marked the end of our study period [189].

Because Bispectral Index changes during ANH were not previously examined, our *a priori* power analysis was based upon a previous study that examined the effect of ANH on cognitive function tests. The study demonstrated that the mean (95% Confidence Intervals) Horizontal Addition Test of 2.811 s (2.405 – 3.218 s) increased to 3.345 s (2.923 – 3.767 s) with ANH to a Hb concentration of 6 g dl^{-1} [190]. Our power analysis ($\alpha = 0.05$) showed that a sample size of 7 patients would be required to reveal a statistically significant difference between the two groups with >90% power. To accommodate for the different nature of the Bispectral Index monitoring and cognitive function tests, we doubled our sample size to 15 patients [189].

Using paired *t*-test, mean ± SD Bispectral Index value significantly declined to 82 ± 4 with ANH to nadir Hb of 8.9 ± 1.7 g dl^{-1}, and was restored back to baseline value of 97 ± 2 with breathing 100% oxygen for 5-min [189].

Repeated measures ANOVA showed a significant difference in the mean Bispectral Index values over time between the ANH and control groups (Fig. 5.24). The LOC time was significantly shorter, with fewer propofol TCI dose requirements, and Bispectral Index value was significantly higher in the ANH group (1.3 ± 0.5 min, 2.41 ± 0.15 μg ml^{-1}, 73 ± 7) compared with the control group (1.7 ± 0.4 min, 2.75 ± 0.17 μg ml^{-1}, 61 ± 5) respectively [189]. There was no significant difference between the two groups in Bispectral Index values at stable propofol TCI 4 μg ml^{-1} anesthesia [189].

Two studies in healthy unmedicated volunteers demonstrated significant impairment of cognitive functions because of ANH reduction of "critical Hb threshold" to ≤6 g dl^{-1}, manifesting as slowing of reaction time to neuropsychological tests and Impaired Memory Free Recall tests [190, 191]. Because breathing 100% oxygen reversed the cognitive function deficits; they attributed the cerebral function deficit to decreased cerebral oxygenation [191]. However, that was not the case in our study simply because our study patients Hb concentrations declined to just a nadir of 8.9 g dl^{-1} considerably higher than the ≤6 g dl^{-1} "Critical Hb Threshold" necessary to produce cognitive functions impairment [189–191].

One possible explanation for the Bispectral Index decline with ANH [189] could be electrolyte shifts because of our "rapid" NaCl infusion in our administered HES in 0.9% NaCl. Rapid infusion of normal saline results in hyperchloremic metabolic acidosis [192] and cognitive mental confusion manifesting as electroencephalography (EEG) slowing. This self-limiting benign effect [192, 193] is probably compensated for or attenuated over time during the 100% oxygen administration [189].

In our study [189], ANH was associated with a significant increase in Heart Rate (HR) beats per minute (bpm) and Mean Arterial Pressure (MAP) after ANH (85.4 ± 20.4 bpm,

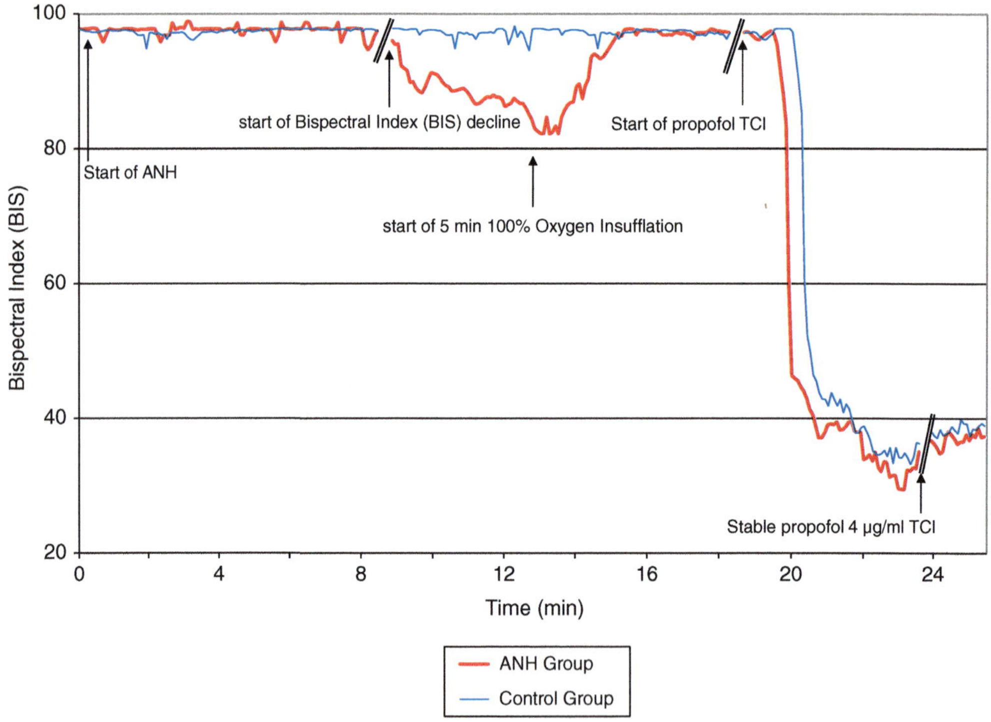

Fig. 5.24 Mean Bispectral Index (BIS) before and after propofol Target Controlled Infusion (TCI) in the Acute Normovolemic Hemodilution (ANH) and control groups

109 ± 19.4 mm Hg) compared with before ANH (70.3 ± 10.4 bpm, 98.6 ± 9.4 mm Hg) respectively. The ANH was associated with increases in cardiac output (CO) in humans as a study demonstrated 55% increase in cardiac index related to a 34% increase in stroke index with ANH [194]. In our study [189], the LOC time was significantly shorter with fewer propofol TCI dose requirements in the ANH group compared with the control group. This probably reflects a considerable change in propofol compartment volumes pharmacokinetics with increased CO, resulting in propofol sedative /hypnotic effect enhancement with the increased propofol delivery to the brain [12]. Interestingly, at LOC the mean Bispectral Index value was significantly higher in the ANH group compared with the control group of our study, reflecting the lower propofol dose requirements at LOC in the ANH group [189, 195].

We concluded that prior to induction of anesthesia, Bispectral Index values transiently declined with ANH that was restored to baseline with 5-min breathing 100% oxygen. Despite a transient ANH enhancement of propofol hypnotic /sedative effect at induction, during propofol maintenance of anesthesia, there was no significant difference in Bispectral Index values between the two groups [189].

5.10 Effect of Cardiovascular Medications

5.10.1 Adrenoreceptor Agonists

Shin and colleagues [196] investigated whether intravenous epinephrine administration affected the Bispectral Index A-2000 monitoring during propofol Target Controlled Infusion (TCI) anesthesia adjusted to maintain a modified Observer's Assessment of Alertness /Sedation (OAA/S) Scale of 2 (sedation period) or propofol adjusted to maintain general anesthesia Bispectral Index of 50 (anesthesia period) [196]. Intravenous 10 μg epinephrine was administered to 20 female patients undergoing elective lower abdominal surgery in the epinephrine group or normal saline was administered to 20 patients in the control group [196]. Epinephrine administration resulted in an arousal effect and an increase in mean ± SD Bispectral Index value from 65 ± 2 to 76 ± 4 in the sedation period but did not change during the anesthesia period. This suggests that the arousal effect of intravenous epinephrine administration during propofol anesthesia depends on anesthetic depth [196].

Andrzejowski and colleagues [197] using Bispectral Index BIS A-1000 (version 3.12) monitor in 8 patients undergoing surgery under propofol TCI general anesthesia using Graseby 3200 Diprofusor infusion pump (Graseby Medical, Herts., UK) adjusted to maintain a modified OAA/S score of 2, and a stable Bispectral Index score of peak-to-trough variation of <5 Bispectral Index points. Authors observed an arousal reaction following 15 μg intravenous epinephrine administration. The mean ± SD Bispectral Index increased from 63 ± 2 to 76 ± 4, which was associated with a rise in 6 of the 8 patients in OAA/S mean (range) score from 2 to 3.5 (2–4) [197]. Although epinephrine does not easily cross the blood brain barrier, it causes a direct analeptic arousal effect through directly stimulating the β-receptors in the Reticular Activating System (RAS). Another possible effect of epinephrine arousal reaction could be through the increased Cardiac Output (CO) enhancing anesthetic agents' clearance and a decrease in effect-site brain levels [197].

However, that is not true for all catecholamines [198], because not all catecholamines exhibit the same arousal effect. Ishiyama and colleagues [198], using Bispectral Index BIS A-1050, demonstrated in 17 patients that mean ± SD Bispectral Index values maintained at 42 ± 6 significantly increased to Bispectral Index value of 55 ± 7, 7–10 min following ephedrine 0.1 mg kg^{-1} administration. Whereas Bispectral Index values maintained at 43 ± 7 was not significantly different than Bispectral Index values of 45 ± 8 following phenylephrine 2 μg kg^{-1} administration in another 17 randomly allocated patients [198]. Similarly, Bispectral Index values of 45 ± 3 in 9 control patients were not significantly different than Bispectral Index value maintained at 42 ± 3. This was attributed to the fact that ephedrine crosses the blood brain barrier and exhibits a powerful stimulant effect [199], while phenylephrine does not cross the blood brain barrier [199], making it unlikely to influence arousal or the depth of anesthesia [200]. The arousal effect of ephedrine administration resulted in an increase in the Bispectral Index value leading to the authors increasing the sevoflurane concentrations [199].

Several reports appeared in literature regarding β-adrenergic agonists enhancing behavioral and electroencephalographic (EEG) arousal reactions. Studies exploring the effects of adrenoreceptor agonists on Bispectral Index monitoring reported that isoproterenol [201] may increase the Bispectral Index value. Matthews and colleagues [201] presented a case of an increase in Bispectral Index value and an arousal analeptic effect following the noradrenergic β_1- β_2-receptor agonist isoproterenol administration in a patient under general anesthesia. The rise in Bispectral Index values was not related to the surgical noxious stimulation [201]. Even though Bispectral Index values remained in the high 70s following isoproterenol administration, the patient had no intraoperative recall. This case report reinforces the belief that Bispectral Index values require thorough interpretation and can only be relevant to the circumstances of each individual case [201].

Table 5.13 enumerates and describes all Adrenoreceptor Agonists administrations that resulted in changes in the Bispectral Index (BIS) values.

5.10.2 Adrenoreceptor Antagonists

Johansen and colleagues [202], using Bispectral Index BIS A-1000 (version 3.3) monitor in 10 patients under stable Target Controlled Infusion (TCI) propofol /alfentanil anesthesia demonstrated that the administration of the ultra-short-acting β_1-adrenoceptor antagonist, esmolol 1 mg kg^{-1} bolus followed by a 30 min 250 μg kg^{-1} min^{-1} infusion, Bispectral Index values significantly decreased from 37 ± 6 to 22 ± 6 after a 12 min hysteresis [202]. This was accompanied by a 13.4-fold increase in the Burst Suppression Ratio (BSR)% from 5 ± 7 to 67 ± 23, after a 16 min hysteresis. These responses were reversed 30 min after esmolol discontinuation. There was no change in the plasma concentrations of propofol and alfentanil, ruling out a pharmacokinetic effect [202]. Esmolol does not cross the blood brain barrier [203]. However large doses esmolol 250 μg kg^{-1} min^{-1}, when combined with opioids, were shown to potentiate reduction of isoflurane Minimum Alveolar Concentration (MAC) requirements [204] and a significant reduction in propofol anesthetic required to suppress the response to skin incision [205].

Berkenstadt and colleagues [206], using a Bispectral Index BIS A-1000 (version 3.12) monitor, demonstrated that administration of 80 mg esmolol bolus in 30 patients did not affect the Bispectral Index value 3 min from administration. Hysteresis may explain why they failed to demonstrate such an effect, since no data on 12 min following their administration where available [206].

Table 5.13 Effect of adrenoreceptor agonists cardiovascular drugs

References	BIS Monitor	*n*	Change	BIS change	Explanation
Shin et al. [196]	BIS A-2000	20 + 20	Epinephrine 10 μg	BIS 65 ↑ BIS 76	cross the bbb, arousal analeptic effect
Andrzejowski et al. [197]	BIS A-1000 (3.12)	8	Epinephrine 15 μg	BIS 63 ↑ BIS 76	cross the bbb, arousal analeptic effect
Ishiyama et al. [198]	BIS A-1050	17	Ephedrine 0.1 mg /kg	BIS 42 ↑ BIS 55	cross the bbb, arousal analeptic effect
Ishiyama et al. [198]	BIS A-1050	17	Phenylephrine 2 μg/kg	BIS 43 = BIS 45	does not cross the bbb, no arousal effect
Matthews et al. [201]	BIS	1	Isoproterenol	BIS ↑ 70	noradrenergic β_1- β_2-receptor agonist arousal effect

n number of patients per group, *BIS* Bispectral Index, *bbb* blood brain barrier

Menigaux and colleagues [207] explored whether the addition of esmolol ultra-short-acting β_1-adrenoceptor antagonist, to propofol anesthesia modified the Bispectral Index BIS A-1000 (version 3.3) monitoring during anesthesia induction and endotracheal intubation [207]. Fifty patients were randomly assigned, in a double-blind study, to receive either esmolol 1 mg kg^{-1} followed by 250 µg kg^{-1} min^{-1} infusion or saline, 6-min after a propofol TCI effect-site concentration (C_e) 4 µg ml^{-1} was started [207]. Mean arterial pressure (MAP), Heart rate (HR), and Bispectral Index values were continuously recorded. After Loss of Consciousness (LOC), vecuronium 0.1 mg kg^{-1} was administered. The trachea was intubated 11-min after the start of TCI, and gross movements after endotracheal intubation were recorded [207]. There were no significant intergroup differences in Bispectral Index values before laryngoscopy. However, compared with pre-laryngoscopy mean ± SD Bispectral Index values significantly increased after endotracheal intubation in the control group to a maximum Bispectral Index percentage increase of 40 ± 18% but not in the esmolol group with just a maximum percentage increase of 8 ± 11% ($P < 0.01$) [207]. Similarly, the maximum percentage increase in HR of 45 ± 19% in the control group was significantly higher than the 23 ± 14% increase in the esmolol group. The maximum percentage increase in MAP of 62 ± 24% was significantly higher than the 45 ± 23% increase in the esmolol patients. More control subjects, than in the esmolol group, moved after endotracheal intubation (23 versus 12 respectively, $P < 0.01$) [207]. The authors concluded that esmolol not only attenuated hemodynamic/somatic responses to laryngoscopy and endotracheal intubation, but also prevented Bispectral Index arousal reactions in patients anesthetized with propofol [207].

In another randomized, double-blind, controlled study, Oda and colleagues [208] tested the hypothesis that the ultra-short-acting β_1-adrenoceptor antagonists esmolol and landiolol suppress the hemodynamic and Bispectral Index BIS A-2000 (version 3.0) increases, both of which supposedly induced by endotracheal intubation under general anesthesia with sevoflurane [208]. Forty-five patients were randomly allocated to the Control-, Esmolol-, or Landiolol-Groups (n = 15). Anesthesia was induced with sevoflurane in oxygen (O_2) and then maintained with 1-MAC end-tidal concentration. Five minutes following induction of anesthesia normal saline infusion was administered to patients allocated to the Control-Group. Esmolol 1 mg kg^{-1} bolus followed by esmolol 250 µg kg^{-1} min^{-1} infusion were administered to patients allocated to the Esmolol-Group. Landiolol 0.125 mg kg^{-1} bolus followed by landiolol 40 µg kg^{-1} min^{-1} infusion were administered to patients allocated to the Landiolol-Group [208]. Twelve minutes from induction of anesthesia, tracheal intubation was attempted. There was no significant difference in MAP between the three groups. One and a half minute following endotracheal-intubation, HR was elevated in the three groups albeit at a statistically significantly slower rates ($P < 0.05$) in the Esmolol- and the Landiolol-Groups than in the Control-Group [208]. Baseline Bispectral Index values of 96–98 decreased following anesthesia induction with no significant differences in the Bispectral Index values of 38 ± 4, 39 ± 5 and 39 ± 5, preceding laryngoscopy in the Landiolol-, Esmolol- and Control-Groups respectively [208]. One minute following endotracheal intubation, there was a statistically significant ($P < 0.05$) increase in the Bispectral Index value of 54 ± 10 in the Control-Group, with no change in the Bispectral Index value of 45 ± 10 in the Esmolol-Group or the Bispectral Index value of 41 ± 6 in the Landiolol-Group. The authors concluded that increases in both HR and Bispectral Index following endotracheal intubation under one-MAC sevoflurane anesthesia were both significantly suppressed via either esmolol or landiolol administrations [208]. There have been conflicting reports in literature regarding the effects of the adrenoreceptor antagonists, esmolol, and landiolol on Bispectral Index values [207–209]. This is because β-adrenoceptor antagonists not only block cardiovascular /hemodynamic stress responses after noxious stimulation, but also have an added effect of increasing the antinociceptive component of anesthesia [209].

The mechanism for this effect remains not completely clear.

On the other hand, the Bispectral Index may decline following single intravenous administration of the selective β_1-Adrenoceptor antagonist metoprolol [210]. Sinha and colleagues [210] report a transient decrease in the Bispectral Index value, in a 28-year-old male 60-kg patient, following intravenous administration of a metoprolol bolus during craniotomy and decompression excision of a large left frontal glioma under general anesthesia [210]. Patient was American Society of Anesthesiologists (ASA) classification physical status III, with medical history of mild motor dysphasia, decreased memory, and generalized tonic–clonic seizures [210]. For premedication, the patient received intramuscular glycopyrrolate 0.2 mg and oral diazepam 5 mg 1 h before surgery. Before induction, the Bispectral Index BIS A–2000 Index was 96. Anesthesia, induced with fentanyl 200 μg, propofol 110 mg and pancuronium for orotracheal intubation, was maintained with 1% isoflurane in 1:1 nitrous oxide (N_2O) in O_2 [210]. During the scalp flap retraction, the patient experienced tachycardia with hypertension episode that could have resulted in surgical field excessive bleeding; thus isoflurane was briefly increased to 3% and 50 mg propofol was administered. When Blood Pressure (BP) normalized to 105/49 mm Hg, albeit the HR remaining at 126 beats/min, isoflurane was decreased to 1%, during which the Bispectral Index value was 35–45 [210]. To alleviate the tachycardia, metoprolol 3 mg intravenous bolus dose was administered, and within 45 s the Bispectral Index value decreased to 3–5. At this time, Signal Quality Index (SQI), BP and HR were 88%, 100/45 mm Hg, and 102 beats/min, respectively. This extremely low Bispectral Index value persisted for about 3-min and returned to 45 over 3-min. With an unremarkable intraoperative course, the patient's trachea was extubated after regaining full consciousness with no sign of neurological damage [210]. The rapid decrease in Bispectral Index values following the metoprolol administration might be explained by the fact that metoprolol readily crosses the blood–brain barrier (bbb) to a greater extent than the more hydrophilic agents esmolol and atenolol [207] and hence is more likely to result in more abrupt decline in Bispectral Index value than other β-adrenoreceptor antagonists.

Table 5.14 enumerates and describes all Adrenoreceptor Antagonists administrations that resulted in changes in the Bispectral Index (BIS) values.

Table 5.14 Effect of Adrenoreceptor Antagonists cardiovascular drugs

References	BIS Monitor	*n*	Change	BIS change	Explanation
Johansen et al. [202]	BIS A-1000 (3.3)	10	Esmolol 1 mg /kg +infusion	BIS 37 ↓ BIS 22	Does not cross bbb, enhance propofol effect
Berkenstadt et al. [206]	BIS A-1000 (3.12)	30	Esmolol 80 mg bolus	BIS no change	No data after 12 min hysteresis
Menigaux et al. [207]	BIS A-1000 (3.3)	25/50	Esmolol 1 mg /kg +infusion	BIS ↑ 8%	Attenuated BIS ↑ to laryngoscopy
Oda et al. [208]	BIS A-2000 (3.0)	15/45	Esmolol 1 mg /kg +infusion	BIS 39 = BIS 45	Attenuated BIS ↑ to laryngoscopy
Oda et al. [208]	BIS A-2000 (3.0)	15/45	Landiolol 0.125 mg / kg +infusion	BIS 38 = BIS 41	Attenuated BIS ↑ to laryngoscopy
Sinha et al. [210]	BIS A-2000	1	Metoprolol 3 mg	BIS 35-45 ↓ BIS 3-5	Rapid BIS decline rapid cross bbb

n number of patients per group, *BIS* Bispectral Index, *bbb* blood brain barrier

5.11 Extremes of Age and Age Progression

Patients' age is a factor that has been least studied as one of the factors that can affect the processed Electroencephalography-Depth Of Anesthesia (pEEG-DOA) indices. Even though Bispectral Index and Entropy have been promoted as pEEG-DOA indices, effects of changes with brain maturation were not adequately studied. Davidson and colleagues [211] compared the differences in anesthetic monitoring of the Bispectral Index BIS-XP A–1000 and Entropy module of the S/5 GE Monitor (General Electric GE Healthcare, Chicago, Illinois, USA) in different age groups [211]. Forty-six children with complete data set, divided into 4 age groups: 0–1 (n = 7), 1–2 (n = 9), 2–4 (n = 16) and 4–12 (n = 14) years, underwent cardiac catheterization under sevoflurane anesthesia. Entropy and Bispectral Index were recorded pre-awakening and at 1.5%, 2% and 2.5% steady-state end-tidal sevoflurane concentrations [211]. In 46 children, median pre-awakening values in the youngest 0–1 year age group Response Entropy (RE) of 45 compared with 84, 87, and 89, $P = 0.003$; State Entropy (SE) of 36 compared with 78, 74, and 77, $P = 0.009$; and Bispectral Index value of 56 compared with 78, 76.5, and 72, $P = 0.02$ were significantly lower than in the 1–2, 2–4, and 4–12 years age groups respectively. The 0–1, 1–2, 2–4, and 4–12 years age groups showed significant difference in Entropy or Bispectral Index at the 3 sevoflurane concentrations [211]. Sevoflurane concentrations required to achieve mid-scale values of Entropy and Bispectral Index were highest in the 1–2 years age group, lower in the 0–1 year age group and then progressively declining in the 2–4 and 4–12 years age groups. Their study results suggest that the Bispectral Index and Entropy are significantly different among young children different age groups [211].

Bernard Baruch once said: "To me, old age is always 15 years older than I am." Elderly patients constitute society's fastest growing sector, that comprises one third of all performed surgeries. Half of all elderly patients over 65 years will require at least one surgery in their lifetime [212]. These surgical patients' demographics have such a tremendous impact on medical practice. Elderly surgical patients tend to have multiple comorbidities for which they receive poly-pharmacotherapies. Because of their altered physiological, pharmacological, and pathological aspects, titrating general anesthesia in the elderly is a challenging task as these patients are at particular risk of incurring undesirable adverse effects from our commonly used anesthetic agents. Not all pEEG-DOA monitors have software designed for the elderly [213]. The pEEG-DOA monitors seem to be particularly sensitive to extremes of age populations, for their altered anesthetic agents' pharmacokinetic and pharmacodynamics, which brings to question the reliability of the pEEG-DOA monitors in the elderly. Nowadays Population Pharmacokinetic - Pharmacodynamic (PopPK-PD) modeling has made an enormous contribution to intravenous anesthesia [213].

Folstein and colleagues [106], Mini-Mental State Examination (MMSE) is a simple quick tool to identify elderly patients with Alzheimer cognitive impairment, vascular dementia, or senile dementia, all associated with an increase in the electroencephalography (EEG) slow waves and a decrease in the EEG fast waves activities, that would eventually lead to a significant Bispectral Index value decline in the awake state and during anesthesia maintenance [103]. My research team Dahaba and colleagues at the Department of Anesthesiology, Xijing Hospital of Fourth Military Medical University, Xi'an, Shaanxi, People's Republic of China, evaluated all study candidate patients' mental function using MMSE. We excluded patients with poor mental cognitive MMSE function of ≤23 [107]. Eighty subjects with Body Mass Index (BMI) between 16 $kg{\cdot}m^{-2}$ and 28 $kg{\cdot}m^{-2}$ who were undergoing general anesthesia for scheduled elective general surgical or orthopedic procedures were enrolled in our study and divided by decades to eight age groups. We used both numerical age values as well as age decade groups: vicenarians (20–29 years), tricenarian (30–39 years), quadragenarian (40–49 years), quinquagenarian (50–59 years), sexagenarian (60–69 years), septuagenarian (70–79 years), octogenarian (80–89 years), and nonagenarian

(90–99 years) as continuous variables. Each group consisted of five male and five female subjects [213]. Using Target Controlled Infusion (TCI) and Bispectral Index, we set propofol plasma concentration (C_p) to gradually reach 3.5 μg·ml^{-1} over a period of 3.5-min. In each patient, we constructed a Pharmacokinetic / Pharmacodynamic (PK/PD) model and conducted a PopPK-PD covariate analysis [213].

Because there are no previous publications that we could have used for an *a-priori* sample size calculation, we decided to use our clinically relevant primary parameter, in an interim sample size analysis of a significant "age effect" on Bispectral Index monitoring [213]. Thus, our interim power analysis for *t-test* ($\alpha = 0.05$) of mean ± SD 1 h Bispectral Index maintenance of 55.3 ± 10.2 in our first 5 sexagenarians (60–69 year) patients and mean ± SD 1 h Bispectral Index maintenance of 87.4 ± 7.2 in our first 5 tricenarians (30–39 year) showed that a group size of 10 patients (n = 10) would be necessary to reveal a statistically significant difference with 90% power [213]. However, because all statistical results are subject to random variation; this might not have been the most appropriate way in determining our sample size for such a study, particularly when the difference used in our sample size calculation was consequently used in our final study analysis [213].

We estimated all our covariate-free "base model" meaningful PK/PD parameters namely, E_0 is the Bispectral Index baseline value before propofol administration, E_{max} is the Bispectral Index nadir maximum depression, C_{50} is inhibitory propofol concentration associated with 50% maximum Bispectral Index depression, and Hill-coefficient is the coefficient that influences the slope and Sigmoidicity of the Concentration–Response curve (γ) [97]. We then conducted a full PopPK-PD analysis using an Individual Pharmacokinetic Parameters (IPP) sequential approach [214] of all available biological and demographic characteristics namely; gender, age, height, weight, and BMI, as formal covariates [215]. Using an IPP sequential approach with C_p historical post-hoc values provided us with a far better fit than our measured arterial blood sampling (C_m, 4 samples per patient at 1-min, 2-min, 3-min 4-min time points) [213]. To assess the Goodness of Fit describing deviation of the model points from the raw data points, we used patient's Root Squared Error (RSE), where the lower the RSE the better the fit of the model. We applied "Constraints" on the model, as Bispectral Index E_0 value was constrained to BIS of ≤100, and Bispectral Index E_{max} value was constrained to BIS of ≥1 [213].

We did not premedicate our study patients. We placed a BIS "Quatro" sensor on patients' foreheads that was attached to a Bispectral Index BIS-Vista (version 4.1) monitor [213]. Raw real time EEG tracings were all band passed 2–70 Hz filtered. Immediately before our anesthesia induction, one of two assigned anesthesiologists, who was blinded to the patients' date of birth, recorded the Bispectral Index for an artefact-free period of 10 min [213]. Because the Bispectral Index was shown to decline with sleep [216], and spuriously increase with high Electromyography (EMG) activity [217], we instructed our study patients to remain calm with their eyes closed but not to fall asleep and to try to keep their facial muscles completely relaxed. All tactile or verbal stimulation as well as noise were all completely avoided; thus, we averted any external stimuli from provoking a transient Bispectral Index response [213]. We started Bispectral Index recordings after we verified a Signal Quality Index (SQI) of >95% and electrodes impedance of <5 kΩ. We set the Smoothing Rate (SR, signal averaging time that is derived from the previous artifact-free EEG) to 15 s. The Bispectral Index monitor displays frontal Electromyography (fEMG) in decibel (dB) units relative to 0.01 μV, in which the use of the logarithmic units is particularly useful in detecting changes at low levels of EMG activity [213]. We chose the level of ≤35 dB, equivalent to 0.56 μV, suggested by Mathews and colleagues [218], as an indicator of adequate frontal EMG suppression [213].

After we entered our study patients' anthropometric data [213], we programmed an Orchestra® (Fresenius Kabi, Brezins, France) propofol TCI pump, using Schnider's pharmacokinetic model [219], C_p to reach 3.5 μg·mL^{-1} over a set period of 3.5-min and this marked the end of our statistical model data collection period [213]. With a gentle prod on the patients' shoulders, once every 30 s for the first 4 min, we asked our patients whether they were awake, until they reached time of Loss-Of-Behavioral-Response (T_{LOBR}) defined as an Observer's Assessment of Alertness/Sedation (OAA/S) Scale [220], of ≤2, i.e., lost their eyelash reflex and did not respond to verbal commands or mild prodding [221]. We recorded the Bispectral Index value at LOBR (BIS_{LOBR}) [213].

After our 3.5-min model data collection period, we started remifentanil 0.1–0.3 μg kg^{-1} min^{-1} infusion and we administered rocuronium 600 μg kg^{-1} for tracheal intubation. We then maintained anesthesia at propofol C_p 3.5 μg mL^{-1} [213]. After anesthesia induction, and at the anesthesiologists' discretion, they adjusted remifentanil infusion to maintain hemodynamic stability for their surgical anesthesia maintenance. Anesthesiologists adjusted rocuronium 30 μg kg^{-1} min^{-1} continuous infusion to maintain first twitch (T_1%) of Train-Of-Four (TOF) at 25% [222], via Neuromuscular Transmission Module (GE Healthcare, Madison, Wisconsen, USA) kinemyography monitoring [223]. During surgical anesthesia maintenance period, we kept propofol C_p infusion constant at 3.5 μg ml^{-1} with no further rate adjustments for an artefact-free 1 h anesthesia maintenance study-recording period [213]. After end of surgical procedures, we recorded Bispectral Index values at 20 min from the termination of propofol and remifentanil infusions. We also recorded time of Return-Of-Behavioral-Response (T_{ROBR}) and Bispectral Index values at patients' Return-Of-Behavioral-Response (BIS_{ROBR}) with return of their eyelash reflex and patients responding to verbal commands [213, 221].

In our 3-Dimensional (3D) diagram (Fig. 5.25) with the Bispectral Index 20–100 scale

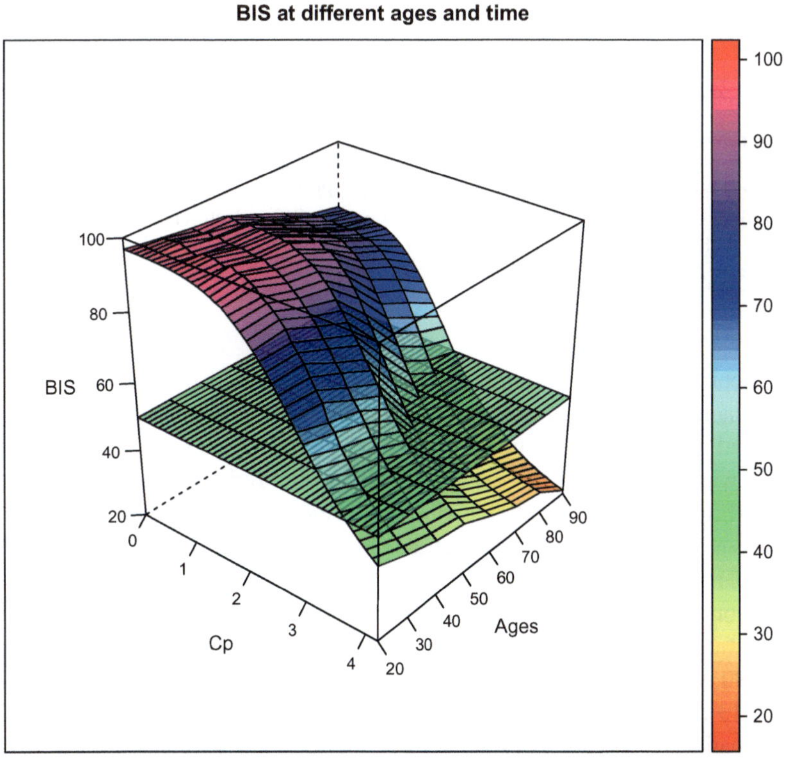

Fig. 5.25 3-Dimensional (3D) diagram with the Bispectral Index 20-100 scale depicting the relationship between the 3 variables; Bispectral Index (BIS), Propofol estimated plasma concentration (C_p) and Age (n = 10). Compared with vicenarians (20–29 years), the cut-off in light blue of BIS 50 was descriptively shifted by 25% in elderly patients above 60 years, and by 33% in patients above 90 years nonagenarians (90–99 years). (Reproduced with permission # 5898220646529 from Dahaba AA, et al. Age progression from vicenarians (20–29 year) to nonagenarians (90–99 year) among a Population Pharmacokinetic-Pharmacodynamic (PopPk-PD) covariate analysis of propofol-bispectral index (BIS) electroencephalography. J Pharmacokinet Pharmacodyn. 2020;47:145–61)

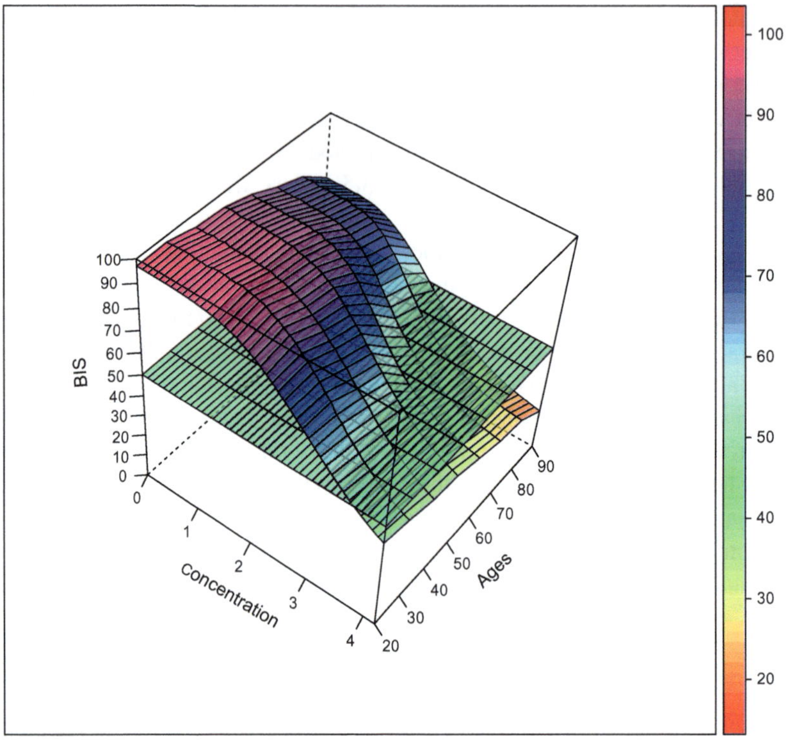

Fig. 5.26 3-Dimensional (3D) Top-down view diagram with the Bispectral Index 20–100 scale depicting the relationship between the three variables; Bispectral Index (BIS), propofol estimated plasma concentration (C_p) and age. (Reproduced with permission # 5898220646529 from Dahaba AA, et al. Age progression from vicenarians (20–29 year) to nonagenarians (90–99 year) among a Population Pharmacokinetic-Pharmacodynamic (PopPk-PD) covariate analysis of propofol-bispectral index (BIS) electroencephalography. J Pharmacokinet Pharmacodyn. 2020;47:145–61)

(Fig. 5.25), we depicted the relationship between all 3 variables namely, Bispectral Index BIS, Propofol estimated plasma concentration C_p and Age. Compared with vicenarians (20–29 years) Top-down view (Fig. 5.26), and Front-side view (Fig. 5.27), we observed that the cut-off Bispectral Index value of 50, in light blue, was descriptively 25% shifted in our elderly patients above 60 years of age, and further 33% shifted in our study patients above 90 years of age nonagenarians (90–99 years) [213].

We depicted patient 16 (Fig. 5.28) as one of the best model fits of Bispectral Index versus propofol estimated plasma concentration (C_p) raw data and logistic model, as well as patient 62 (Fig. 5.29) as one of the worst model fits of Bispectral Index versus propofol C_p raw data and logistic model who did not meet our Goodness of Fit criteria, thus required utilization of the optimizer "Solver" function.

We depicted mean Bispectral Index BIS (95% Confidence Intervals) raw data presentation of two groups: our vicenarians (20–29 years) compared with nonagenarians (90–99 years) (Fig. 5.30). We depicted our raw data presentation of our 8 age groups from our vicenarians

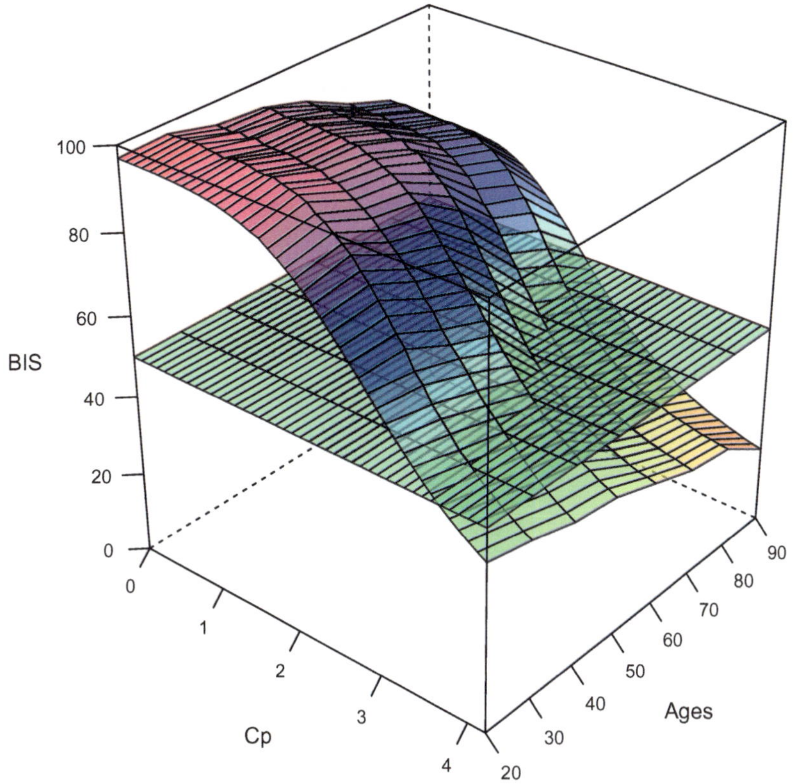

Fig. 5.27 3-Dimensional (3D) Front-side view diagram with the Bispectral Index 20–100 scale depicting the relationship between the three variables; Bispectral Index (BIS), Propofol estimated plasma concentration (C_p) and age. (Reproduced with permission # 5898220646529 from Dahaba AA, et al. Age progression from vicenarians (20–29 year) to nonagenarians (90–99 year) among a Population Pharmacokinetic-Pharmacodynamic (PopPk-PD) covariate analysis of propofol-bispectral index (BIS) electroencephalography. J Pharmacokinet Pharmacodyn. 2020;47:145–61)

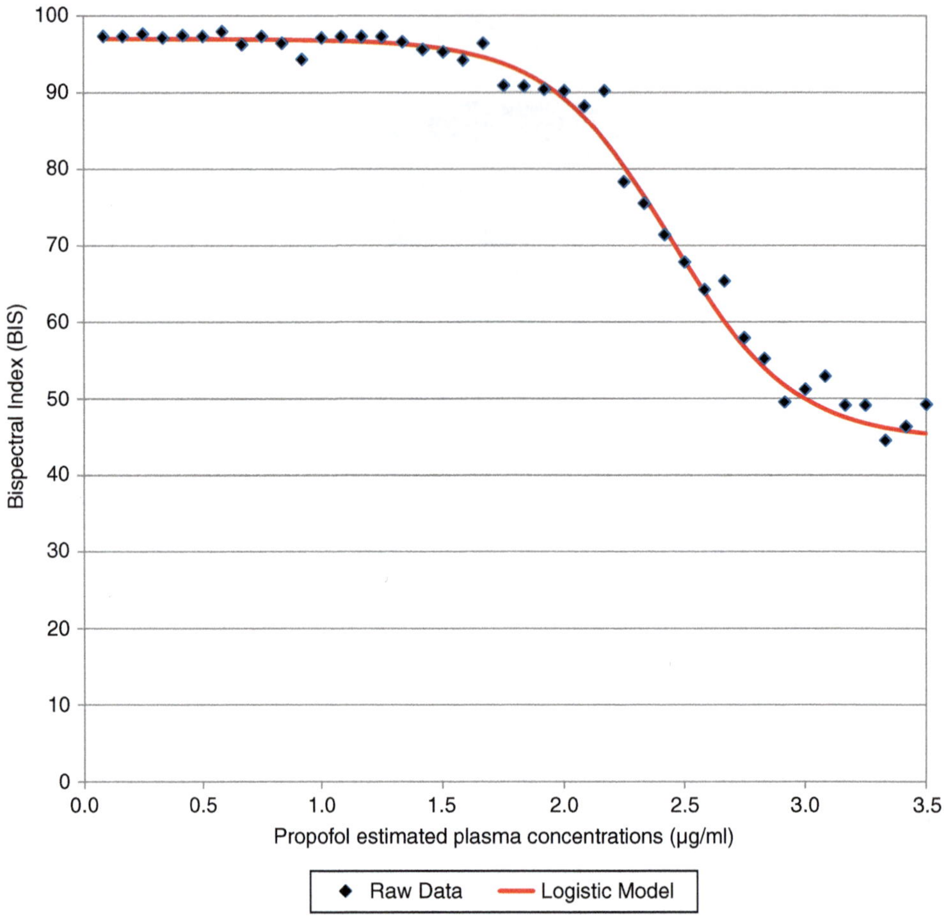

Fig. 5.28 Patient 16 is one of the best model fits. Bispectral Index (BIS) versus Propofol estimated plasma concentration (C_p) raw data and logistic model

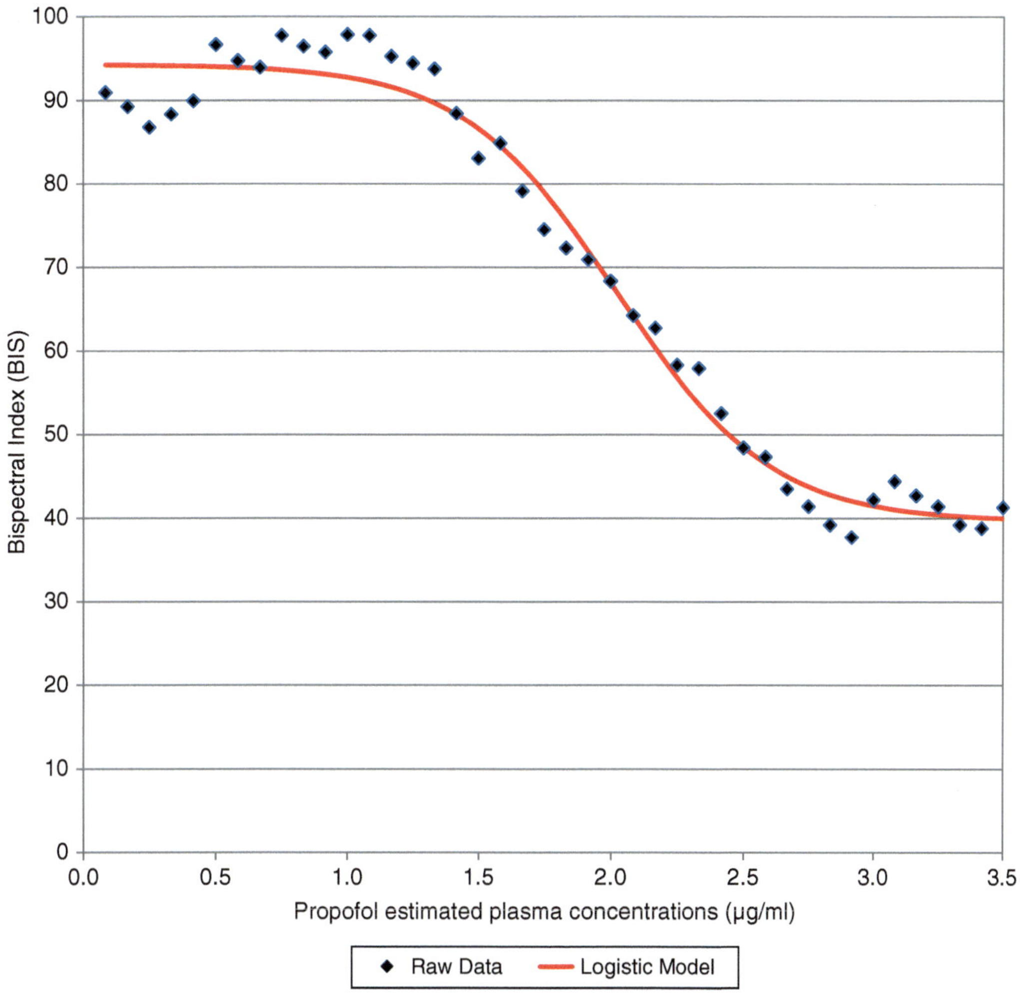

Fig. 5.29 Patient 62 is one of the worst model fits. Bispectral Index (BIS) versus propofol estimated plasma concentration (C_p) raw data and logistic model. Patient 62 did not meet our Goodness of Fit criteria, thus required utilization of the optimizer "Solver" function

Fig. 5.30 Our raw data presentation of two groups. Our vicenarians (20–29 years) group compared with nonagenarians (90–99 years) group

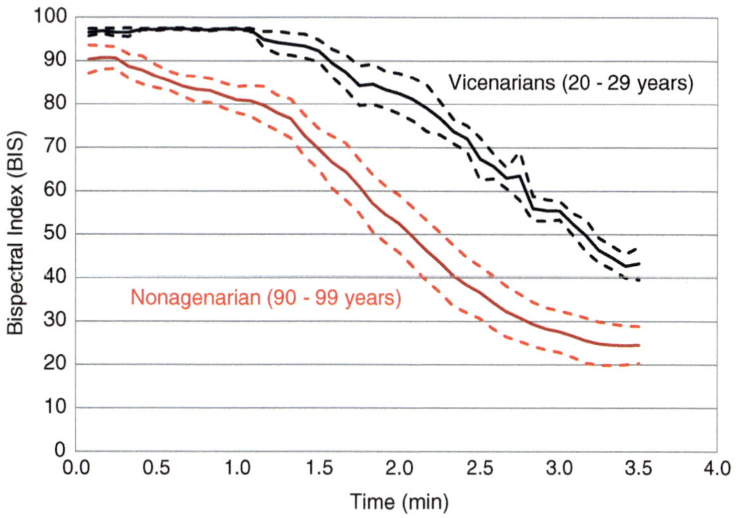

(20–29 years) up to our nonagenarians (90–99 years) (Fig. 5.31). In addition to our raw data presentation, we further post hoc normalized / extrapolated (smoothed) our 8 age groups raw data to an arbitrary level of BIS 20–29 as we a-priori hypothesized a decline with age from that level, thus, giving a percentage change within each group (Fig. 5.32). Separately analyzed and unrelated to any model, mean ± SD Bispectral Index value before induction, for 1 h anesthesia maintenance, and at 20 min recovery after termination of propofol infusion in different age groups (Fig. 5.33) [213].

Regarding our study patients' time to Loss-of-Behavioral-Responses (LOBR) and Return-of-Behavioral-Responses (ROBR), Bispectral Index levels, propofol plasma concentrations, and propofol effect-site concentrations at LOBR in our 20s, 30s, 40s, 50s, 60s, 70s, 80s, and 90s eight study groups (n = 10), mean ± SD Time to Loss-of-Behavioral-Responses (T_{LOBR}) was significantly faster with age progression (6.9 ± 3.5,

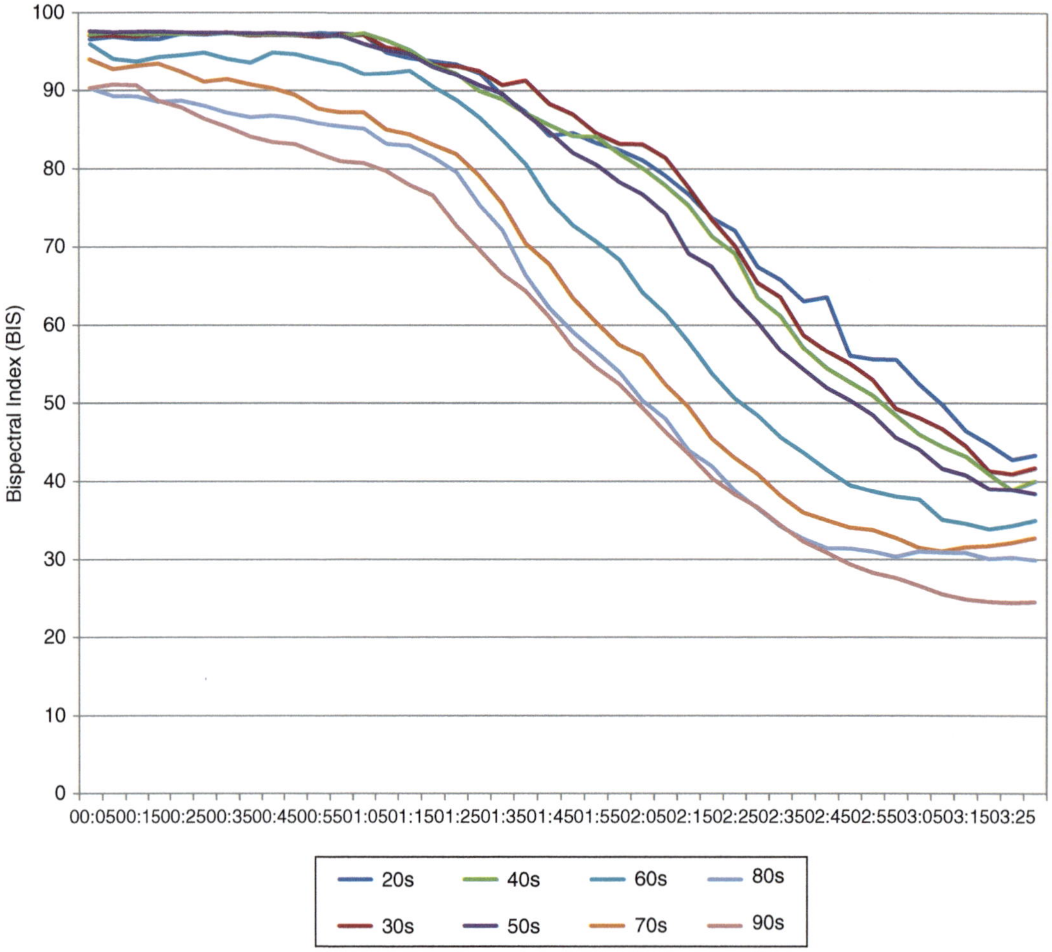

Fig. 5.31 Raw data presentation of our 8 age groups from our vicenarians (20–29 years) up to our nonagenarians (90–99 years). (Reproduced with permission # 5898220646529 from Dahaba AA, et al. Age progression from vicenarians (20–29 year) to nonagenarians (90–99 year) among a Population Pharmacokinetic-Pharmacodynamic (PopPk-PD) covariate analysis of propofol-bispectral index (BIS) electroencephalography. J Pharmacokinet Pharmacodyn. 2020;47:145–61)

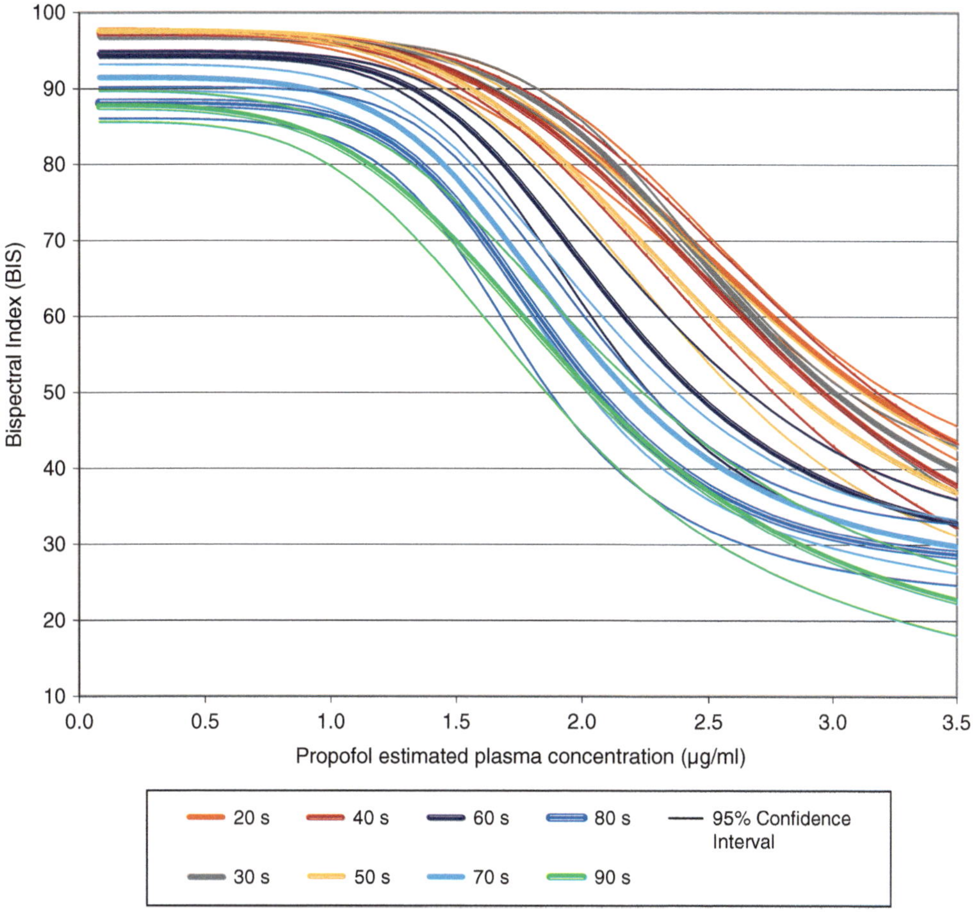

Fig. 5.32 Our eight age groups normalized /extrapolated (smoothed) to an arbitrary level of BIS 20–29 as we a-priori hypothesized a decline with age from that level, thus giving a percentage change within each group. (Reproduced with permission # 5898220646529 from Dahaba AA, et al. Age progression from vicenarians (20–29 year) to nonagenarians (90–99 year) among a Population Pharmacokinetic-Pharmacodynamic (PopPk-PD) covariate analysis of propofol-bispectral index (BIS) electroencephalography. J Pharmacokinet Pharmacodyn. 2020;47:145–61)

6.6 ± 3.7, 5.9 ± 4.6, 6.2 ± 3.3, 4.8 ± 5.6, 3.9 ± 4.3, 3.8 ± 6.6, 2.8 ± 1.6 min). Propofol estimated plasma concentration C_p at LOBR was gradually declining with age progression (2.9 ± 1.8, 2.8 ± 2, 2.8 ± 1.1, 2.4 ± 0.5, 2 ± 0.9, 1.8 ± 1.6, 1.8 ± 1.2, 1.6 ± 2.1 μg·mL^{-1}). Similarly, propofol predicted effect-site concentration C_e at LOBR was gradually declining with age progression (1.9 ± 0.8, 1.9 ± 0.8, 1.8 ± 0.6, 1.5 ± 0.2, 1.2 ± 0.4, 1.1 ± 0.5, 0.9 ± 0.3, 0.6 ± 0.2 μg·mL^{-1}) [213]. Interestingly, we observed a new phenomenon in our study, as the significantly faster T_{LOBR} was accomplished with similar Bispectral Index values at Loss-of-Behavioral-Responses (BIS_{LOBR}) levels (60 ± 7, 60 ± 4, 57 ± 6, 61 ± 4, 63 ± 7, 65 ± 7, 64 ± 8, 68 ± 14) [213]. Similarly, the significantly longer Time to Return-of-Behavioral-Responses (T_{ROBR}) with age progression (15.3 ± 4, 15.8 ± 4.8, 14.8 ± 5, 16.7 ± 6, 29.3 ± 10, 33 ± 19.5, 33.3 ± 12.3, 45 ± 10 min) was accomplished with similar Bispectral Index values at Return-of-Behavioral-Responses (BIS_{ROBR}) levels (81 ± 5, 81 ± 5, 81 ± 8, 82 ± 5, 84 ± 6, 83 ± 6, 86 ± 7, 84 ± 7) [213].

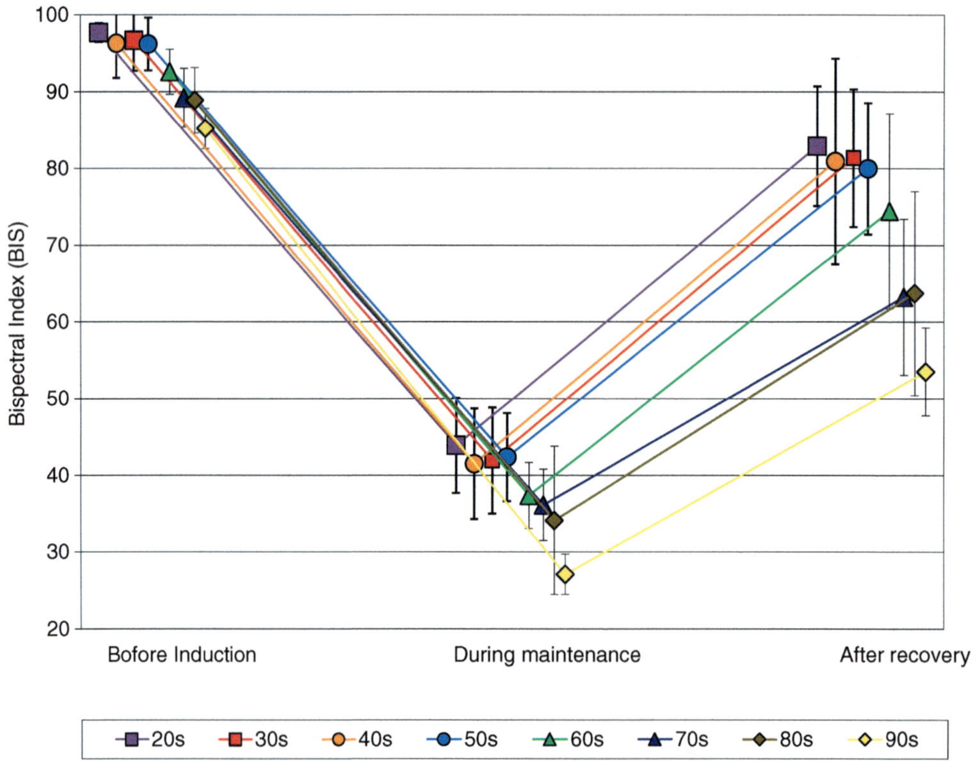

Fig. 5.33 Mean ± SD Bispectral Index (BIS) before induction, for 1 h maintenance of anesthesia and at 20 min after termination of propofol target controlled infusion (TCI) in different age groups

We have very precisely quantified age progression from vicenarians (20–29 years) to nonagenarians (90–99 years) among a Population Pharmacokinetic -Pharmacodynamic (PopPk-PD) covariate analysis of propofol-Bispectral Index model [213]. Age was a significant covariate for baseline BIS effect (E_0), inhibitory propofol concentration at 50% BIS decline (IC_{50}) and maximum BIS decline (E_{max}). First-order rate constant ($\boldsymbol{K}_{e0}$) of 0.47 min^{-1} in vicenarians (20–29 year) gradually increased with age-progression to 1.85 min^{-1} in nonagenarians (90–99 year) [213]. Our simulation modelling showed that the clinically recommended C_p of 3.5 μg·ml^{-1} in vicenarians (20–29 years) for a Bispectral Index value of 50 (BIS_{50}) should be reduced to 3.0 in 30–49 years of age (tricenarian and quadragenarian), further reduced to 2.5 for 50–69 years of age (quinquagenarian and sexagenarian), and 2.0 for octogenarian (80–89 years) [213]. We have successfully quantified and graded Bispectral Index age-progression among different age groups divided by decades. We demonstrated deeper Bispectral Index values with decades' age progression. Our data has important implications for propofol dosing. The practical information for physicians in their daily clinical practice is using propofol C_p of 3.5 μg·ml^{-1} might not yield Bispectral Index value of 50 in elderly patients [213].

As I have previously mentioned, the Bispectral Index monitor uses a proprietary algorithm that was never published before. Even though in our study [213], T_{LOBR} was faster with increasing age, interestingly we demonstrated a new phenomenon of all our age groups losing their behavioral response at the same narrow range of BIS_{LOBR} between 57 and 68 Bispectral Index value [213]. Even though in our study, T_{ROBR} was longer with increasing age, interestingly we demonstrated a new phenomenon of all our age groups regaining their behavioral response at another very narrow range of BIS_{ROBR} between 81 and 86 Bispectral Index value [213]. We can speculate an explanation of these "phenomena" being the fact that LOBR (or ROBR) appears to occur only when the algorithm of the Bispectral Index monitor "reads" a very "specific BIS_{LOBR} configuration" or a "specific BIS_{ROBR} configuration" within very narrow range configurations that is "immune" to other effects as we have previously demonstrated [224]. This phenomenon that we demonstrated, as far as I know, for the first time with propofol total intravenous anesthesia Bispectral Index depth of anesthesia monitoring was previously reported with sevoflurane inhalational anesthesia as Katoh and colleagues [225], using a Bispectral Index BIS A-1000 monitor, reported a similar BIS_{LOBR} occurring in their 3 age groups, despite their elderly age group (65–85 years) needed significantly lower Minimum Alveolar Concentration (MAC) end-tidal sevoflurane concentrations to suppress response-to-verbal-command (MAC_{awake}) than their young age group (18–39 years) or middle age group (40–64 years) [225].

If BIS_{50} (BIS 40–60 range) is the supposed Bispectral Index monitoring level associated with a deep hypnotic state for all age groups, I believe that our study revealed a "fundamental flaw" when we try to estimate the correct recommended dose of propofol that would yield the desired BIS_{50} among different age groups [213]. Our simulation model clearly revealed that the widely recommended regimen of propofol C_p 3.5 $\mu g{\cdot}mL^{-1}$ for the 20–29 years that would yield the desired BIS_{50} should be gradually decreased till propofol C_p 2.0 $\mu g{\cdot}mL^{-1}$ for the 80–89 years [213].

The first "plausible" explanation of our observed findings is a "pharmacodynamic assertion" of assumed linearly increasing "intrinsic brain sensitivity" to propofol that is accredited to the evolution of the raw EEG with age. This means that the elderly patients respond "more" to the same C_p propofol plasma concentration, i.e., a true difference in "potency," as potency is *stricto sensu* a pharmacodynamic parameter. Published animal studies have revealed that age related loss of forebrain hippocampal basal cholinergic frontal cortex neurons correlated to propofol potency in rodent animal models [226]. Whereas in another animal model, young rats needed larger propofol induction dose than old rats [227]. This is not restricted to animal models as it was also evident in humans. Under propofol constant infusion anesthesia maintenance, elderly patients more than 70 years of age exhibit smaller EEG power amplitudes when compared with younger patients under 50 years. Elderly patients reach deeper EEG stages during the same propofol induction compared with younger patients [228]. In elderly patients, EEG patterns appearing with advanced age manifest as a steady decrease in the α wave band dominant frequency [229]. However, the biologic basis for these age dependent brain sensitivity EEG changes remains unclear, although reduction in the cortical synaptic density and decrease in cortical cells synchronization are all possible explanations [230].

On the other hand, a "pharmacokinetic assertion" could also contribute to our observed results; as in animal models, older animals exhibit higher propofol brain concentrations at EEG loss of brain response [227]. Volume of distribution (V_d) and Clearance (Cl) of certain extensively plasma protein bound drugs such as propofol characterized by their very large V_d, would then progressively decline with age [231]. Propofol is extensively bound to the plasma proteins (>98%), predominantly albumin and erythrocytes. The free fraction of propofol is just 1.2–1.7%, as up to 50% of propofol is extensively bound to erythrocytes [231]. Propofol's redistribution to and from a slow compartment is due to propofol's high lipid solubility. This slow compartment has large capacity to absorb propofol thus resulting in very large Volume of distribution at Steady State (Vd_{ss}), numerically representing 3–4 multiples of the total body volume [232]. Eleveld and colleagues [233] demonstrated a strong relationship between age and volumes of the peripheral compartments and with metabolic clearance. The volume of peripheral compartment V_2 declines with age, while peripheral compartment V_3 and metabolic clearance decline only when other medications are concurrently administered [233]. Eleveld and colleagues [233] findings confirm our current study simulation model results, that is also widely shared by many practicing anesthetists [233], that the recommended regimen of C_p 3.5 $\mu g{\cdot}ml^{-1}$ for the 20-29 years that would yield BIS_{50} should be gradually decreased till C_p of 2.0 $\mu g{\cdot}ml^{-1}$ for the 80–89 years [213].

5.12 Transgeographic Ethnic Differences in Propofol–BIS Response Curves

Total intravenous anesthesia (TIVA) is less popular in North America than in other parts of the world [234] because the Food and Drug Administration (FDA) did not deem Target Controlled Infusion (TCI) devices sufficiently advantageous to deliver anesthetic agents and opioids such as propofol and remifentanil to approve their commercial use [235]. The principle of how the commercially available TCI devices operate is rather simple [234], as anesthesiologists select a target plasma concentration of an anesthetic agent such as propofol required for a specific procedure, and the TCI device will then automatically administer a small bolus loading dose followed by a kinetically driven infusion rate to rapidly attain and then maintain that specific drug concentration until a new drug concentration is selected [48, 236]. The TCI device displays in real time the theoretical population-based attained drug plasma concentration and the quantity of drug delivered. Alternatively, the anesthesiologist can set the TCI device to attain target effect-site concentrations [234], where the infusion rate algorithm considers the time lag between the plasma and the site of action of hypnotics and opioids, the brain or subcortical structures. These TCI devices could be truly a useful tool to improve delivery of intravenous anesthetics in the clinical setting, or in the research arena [234].

The concept of the TCI devices is clinically appealing as anesthesiologists enter patient's age, weight, and height anthropometric data, with a population pharmacokinetic data set from a selection menu; the TCI device then calculates and delivers in real time the correct amount of drug that will rapidly achieve the desired calculated target plasma or the predicted effect-site drug concentrations [234]. Plasma drug concentration is a useful measure that could be regarded as a surrogate measure of its effect-site concentration at the site of action. In the case of propofol, the site of action is the brain, so TCI plasma concentrations are useful if they have a predictable relationship with the brain concentrations [234]. Whenever a new drug concentration is selected, the TCI calculations achieve the new target concentration as soon as possible. There is no need to adjust the infusion rate to take into consideration complex pharmacokinetic redistribution and elimination processes or effect site time lag–the machine does all that [234]. When we use TCI devices, we do not think in terms of bolus doses and infusion rates, rather just in terms of target concentrations. My research group Dahaba and colleagues pushed the concept further as we used TCI devices in our research to precisely quantify the relationship between concentrations and desired effects such as Loss of Consciousness (LOC) or a specific processed Electroencephalography-Depth Of Anesthesia (pEEG-DOA) indices [234]. My Austrian research group at Medical University of Graz, Austria, and my Chinese research group at Zhe Jiang University Hospitals, Department of Anaesthesiology, Sir Run Run Shaw Hospital, Zhe Jiang University, Women's Hospital School of Medicine, Zhe Jiang University, Hang Zhou, People's Republic of China, collaborated to examine the responses to propofol using the TCI technology in two different patient populations on two continents [224]. Our findings strongly suggest that Chinese patients were more sensitive to propofol than Austrian patients. Our study can be considered as an example of the advantages and limitations of TCI devices [224].

The Bispectral Index (BIS) is a pEEG-DOA monitor whose Bispectral Index was devised from experimental dataset that used complicated proprietary algorithms combining three descriptors to form a dimensionless metric score [158]. The Bispectral Index quantifies electroencephalographic responses to anesthetic agents [224]. Whereas the Diprifusor™ TCI device (AstraZeneca Pharmaceuticals, Macclesfield, United Kingdom) utilizes pharmacokinetic models to achieve and maintain pre-set plasma or effect-site concentrations [237]. Because plasma is not the site of drugs' effect; hence the Diprifusor can simultaneously display predicted effect-site concentration (C_e) as well as estimated plasma concentration (C_p) [237]. Although the measured plasma concentration might overshoot or undershoot the selected TCI target concentration [236], in our study for both Austrian or Chinese patients, the dosing regimen required to attain the TCI target drug concentration was the same for patients of the same sex, with similar height, weight, and age [224]. Thus, we can assume, for our study research purposes, that our study subjects with similar demographics on both continents received comparable propofol doses [224].

Abruptly changing TCI plasma concentrations will create significant discrepancies between plasma and effect-site concentrations because of the lag period required for the changes in plasma concentrations to reach the brain effect-site concentrations [234]. This lag time might be relatively short, but clinically apparent as anesthesiologists must wait for 1–2 min before a bolus dose of propofol induces anesthesia [234]. This lag time is incorporated into the pharmacokinetic model, and the effect-site concentrations are accordingly calculated [48]. The effect-site concentrations depend on the kinetic data selected to run the device normally represented by the parameter "plasma-effect site equilibration rate (k_{eo})". Since propofol concentration cannot be measured in the brain effect-site organ, effect compartment models are constructed as to add a "virtual compartment" that does not affect drug distribution. This "virtual effect compartment" [238] is structurally linked to the central compartment by a first-order rate constant (k_{e0}). With the central compartment (V_1) and peripheral compartments (V_2, V_3) rate constants (K_{12}, K21, k_{13}, K_{31}) and Clearance (Cl) only from the central compartment rate constant (K_{10}), the effect-site compartment is linked to the central compartment by a first-order rate constant (k_{e0}) Pharmacokinetic/Pharmacodynamic models with lag periods (Fig. 5.34) [234, 237].

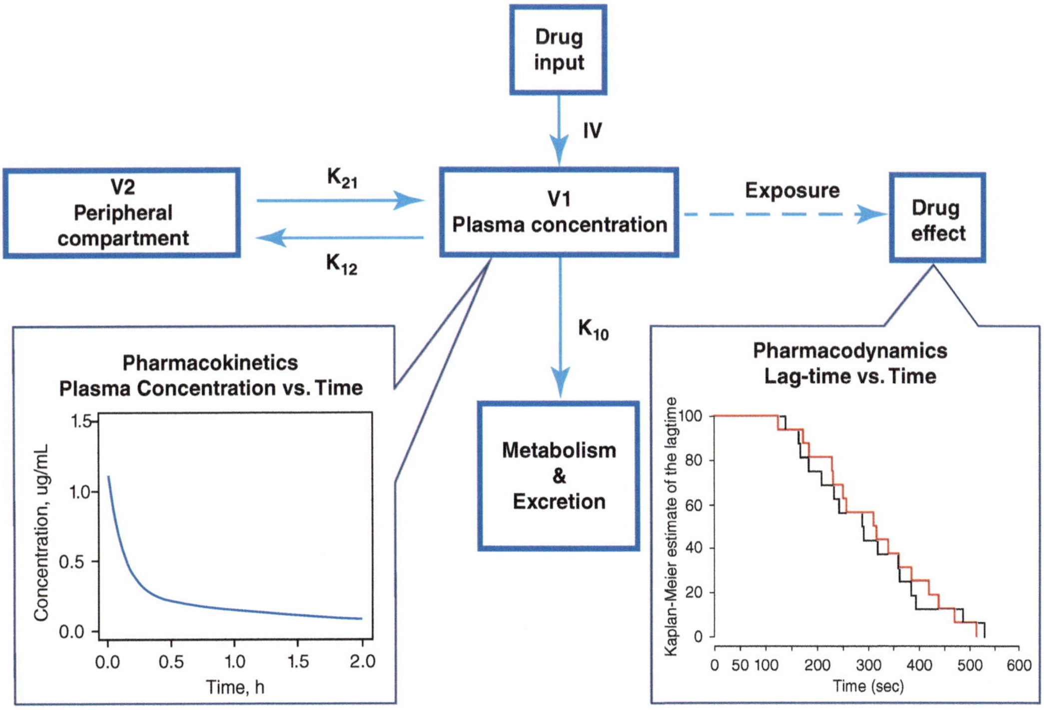

Fig. 5.34 Central compartment (V_1) and peripheral compartments (V_2, V_3), rate constants (K_{12}, K_{21}, k_{13}, K_{31}) and Clearance (Cl) only from the central compartment rate constant (K_{10}) Pharmacokinetic/Pharmacodynamic models with lag periods. (Reproduced with permission # 5900111352009 from Dahaba AA, et al. Location matters: Overlooked ethnic-geographic effect in China and Austria on propofol/cisatracurium sex differences among a Population Pharmacokinetic-Pharmacodynamic (PopPK/PD) covariate analysis in men, women, and one transgender subject. Fundam Clin Pharmacol. 2022;36:182–98)

Ethnic geographic variability in drug response could arise from both genetics "Nature" as well as an array of environmentally influenced factors that are hard to specify or measure, i.e., "Nourish or Nurture" [224]; hence drug effect could differ with ethnic geographic locations for both reasons. Despite published studies showing propofol dose recommendations differ among ethnic groups of subjects who reside in various geographical regions [239, 240], ethnic geographic locations variability is generally not yet recognized among the stratifying factors which influences drugs effects [224]. That was always an overlooked topic of investigation. For instance, propofol TCI pharmacokinetic set, developed in a mostly Caucasian white European population, is totally different in Chinese patients living in People's Republic of China [224]. To date there are no trials comparing propofol estimated plasma concentrations or predicted effect-site concentration-Bispectral Index responses in European white Caucasians living in Europe and ethnic Han-Chinese subjects living in People's Republic of China. Even though responsible authorities of certain countries like People's Republic of China, Korea, and Japan demand material data from their home population prior to issuing drugs registrations, still anesthetic agents all over the world are still issued licenses with the exact dose recommendations that will be used in different populations living in various geographical regions [224]. Responsible authorities of certain states are still accepting anesthetic agents' information material obtained from a small sample size subjects of white Caucasian European or North America patients and are still not demanding detailed explicit information material on the exact dosing toxicity and efficacy obtained from subjects of their ethnic geographical regions [224]. The different results between Chinese and European patients could be due to either pharmacokinetic or pharmacodynamic differences, or a combination of both [224, 234]. Although an efficient and rapid change in blood concentration is better accomplished using a TCI device to maintain a reasonably stable plasma concentration, still when using the wrong recommended doses for ethnic-geographic reasons, the anesthesiologist who uses a TCI device might overdose the administered anesthetic when the ethnic-geographic patients are more sensitive or underdose the administered anesthetic when the ethnic-geographic patients are less sensitive [234].

None of the published trials have accurately quantified differences between white Caucasian [241] and Chinese [242, 243] populations. Using logistic regression of Sigmoid E_{max} curves, Milne and colleagues [241] in white Caucasian patients' population, compared with Irwin and colleagues [242], and Xu and colleagues [243] in Chinese patients' population, assessed propofol-Bispectral Index relationship using the same study design protocols. Three separate studies implementing the same study design protocol in 40 Caucasian patients in the United Kingdom by Milne and colleagues [241], and 60 Chinese patients in Hong Kong, People's Republic of China by Irwin and colleagues [242], and 405 Chinese patients in 5 Beijing centers, People's Republic of China multicenter study, used propofol TCI Diprifusor incorporated into a Graesby 3500 syringe pump version 2.0 software (AstraZeneca Pharmaceuticals, Macclesfield, UK) in the 3 centers [241–243], estimated concentrations at Bispectral Index BIS A-1000 (version 3.0 revision 0.5) in United Kingdom and Hong Kong [241, 242], and Bispectral Index BIS-XP A-2000 (version 3.2) in Beijing [243] peak effect. The 3 studies individually assessed propofol predicted Effective-Concentration$_{50}$ (EC_{50}) effect-site concentrations (C_e) when 50% of patients lost consciousness and become nonresponsive to painful stimulus. An initial propofol target concentration of 1.5 μg ml^{-1} was increased by 0.5 μg ml^{-1} every 2 minutes in United Kingdom and Hong Kong [241, 242], or an initial propofol target concentration of 1.2 μg ml^{-1} was increased by 0.3 μg ml^{-1} every 0.5 minutes until loss of consciousness in the 5 Beijing centers multicenter study [243]. Two teams [241, 242] reported similar estimated EC_{50} at Loss of Consciousness (LOC) of 2.8 (range 2.7–2.9) μg·mL^{-1} in Caucasian patients in the United Kingdom [241], and 2.66 (range 2.59–2.72) μg·mL^{-1} in Chinese patients in Hong Kong [242].

However, the third multicenter team [243] reported considerably lower 2.2 (range 2.2–2.3) $\mu g \cdot mL^{-1}$ estimated EC_{50} at LOC in Chinese patients in the five centers in Beijing, People's Republic of China [243]. To the best of my knowledge, our study [224] is the first direct comparison in the same study, of propofol TCI at peak Bispectral Index effect between Han Chinese patients in mainland People's Republic of China and White Caucasians in mainland Europe [224].

My research team Dahaba and colleagues conducted a multicenter prospective study in two continents in Zhe Jiang University, Hang Zhou, People's Republic of China and Medical University of Graz, Austria, hospitals [224]. To properly delineate the health status among the two patients' populations, we only included American Society of Anesthesiologists (ASA) classification physical status I–II patients, aged 30–50 years with Body Mass Index (BMI) 18–26 $kg \cdot m^{-2}$ [224]. At Zhe Jiang University hospitals, we first recruited 30 consecutive, 15 female/15 male patients, belonging to the second generation Chinese of Han ethnicity who lived for 10 years in People's Republic of China and were undergoing general anesthesia for scheduled orthopedic or general surgery [224]. Each ethnic Han Chinese patient was then matched for gender, ASA classification, approximate age, and weight in Medical University of Graz with 30 second-generation White Caucasians living in Austria for 10 years [224]. Caucasian subjects were defined as member of Indo-European white-ethnicity who supposedly originated from the Caucasus Mountain Regions [224].

A Bispectral Index BIS "Quatro" sensor was placed on patients' forehead [224], according to manufacturer's guidelines and connected to Bispectral Index BIS-XP monitor. The raw EEG signal was band-pass filtered to 2–70 Hz and processed in real time using (version 3.4) of the Bispectral Index algorithm. Bispectral Index recordings were started after verifying a Signal Quality Index (SQI) >95%, electrodes impedance <5 kΩ and electromyography (EMG) values below 35 dB [224].

Patients were not premedicated [224]. Because of possible different quality controls of pharmaceutical production lines, the propofol we used on the two continents in the two centers of our study was produced by the same AstraZeneca Pharmaceuticals European branch. We followed extremely strict trial protocols in the two centers on the two continents [224]. After entering patients' anthropometric data, Diprifusor™ incorporated into a Graesby 3500 syringe pump version 2.0 of the software (AstraZeneca Pharmaceuticals, Macclesfield, United Kingdom) TCI was started using Marsh and colleagues [244] pharmacokinetic model. To maintain the same "C_p variation versus time" for all patients, we chose to set the Diprifusor estimated C_p to reach 5 $\mu g \cdot mL^{-1}$ over 5-min, after that we maintained propofol TCI at 5 $\mu g \cdot mL^{-1}$ infusion rate for maintenance of anesthesia with no further changes [224]. To remove differences in investigators' approach for LOC assessments, our investigators implemented a standard form of similar oral commands in the patients' own native languages. Patients were asked every 10 s, with a gentle prod on the shoulder, whether they are still awake until they reached an Observer's Assessment of Alertness and Sedation (OAA/S) Scale of 1 which is defined as loss of their eyelashes reflexes and stopped responding to verbal commands and this was considered the definition of "Loss of Consciousness" [243]. Subjects were ventilated via a face mask. Estimated propofol plasma concentrations C_p–BIS Bispectral Index and Predicted propofol effect-site concentrations C_e–BIS Bispectral Index data were constantly collected and digitally stored to be statistically analyzed. We defined Steady-State propofol concentrations as Bispectral Index values that did not decline and was within the Bispectral Index values ±5 points of the preceding Bispectral Index value. [224]. We started remifentanil 0.1–0.3 $\mu g\ kg^{-1}\ min^{-1}$ infusion and administered rocuronium 600 $\mu g\ kg^{-1}$ for endotracheal intubation and that was taken as a demarcation of the end of data collection of our trial. Bispectral Index values that showed artifacts on the offline recordings were deleted after being positively identified [224].

Using the program NONMEM Scientist® (MicroMath Scientific Software Inc., Saint Louis, Missouri, US), estimated propofol plasma concentrations C_p–BIS Bispectral Index and predicted propofol effect-site concentrations C_e–BIS Bispectral Index relations were modeled by means of Non-Linear Regression analysis (NLR) by Non-Linear Mixed Effects Modeling (NONMEM) Sigmoid E_{max} Pharmacokinetic–Pharmacodynamic Model [245]. Goodness Of Fit (GOF) for every subject was determined according to Akaike's Information Criterion (AIC) [246]. The estimated propofol plasma concentrations C_p–BIS Bispectral Index and predicted propofol effect-site concentrations C_e–BIS Bispectral Index entries of each patient were used to form the Sigmoid E_{max} curve mathematical model:

$$E = \left(E_{baseline} - E_{max}\right)\frac{C^{\gamma}}{C_{50}^{\ \gamma} + C^{\gamma}}$$

The $E_{baseline}$ was defined as the baseline Bispectral Index score of 100 prior to propofol administration, E_{max} is the Bispectral Index score reached at the maximal propofol effect (nadir Bispectral Index value at steady state C_p of 5 μg mL^{-1}), C_{50} is the estimated concentration at 50% of maximum Bispectral Index effect, and γ is Hill's coefficient Sigmoidicity factor [97], which describes the Sigmoidicity relationship and the influence of the slope and steepness of the curve [97]. Over the 5-min period, we obtained all patients' standard data of Bispectral Index score versus estimated C_p plasma concentrations and C_e effect-site concentrations entries in the constructed Sigmoid curves, from which we constructed a sigmoid curve for each Chinese and Austrian study patient [224].

Because our primary outcome parameter was E_{max} (nadir Bispectral Index value at steady state C_p of 5 μg·mL^{-1}) [224], we performed an interim power analysis *t*-test ($\alpha = 0.05$) based upon the first 10 Chinese pilot patients, who were included in the final analysis, in whom the Bispectral Index BIS of 43.2 ± 2.6 compared with 54.8 ± 6.4 in 10 Austrian pilot patients, revealed that a study group size of 25 subjects is required to reveal a statistically significant difference, with more than 90% power, between the 2 groups [224].

Bispectral Index (BIS) versus propofol estimated plasma concentration (C_p) Sigmoid E_{max} Cp–BIS curve model (95% CI Confidence Intervals) revealed a statistically significant shift to left in the Chinese group patients compared with the Austrian group patients (Fig. 5.35). Bispectral Index (BIS) versus propofol predicted effect-site concentration (C_e) Sigmoid E_{max} Ce–BIS curve model revealed a statistically significant shift to left in the Chinese group patients compared with the Austrian group patients, (Fig. 5.36) [224]. Maximal propofol E_{max} effect (nadir Bispectral Index score at the propofol Estimated C_p steady-state plasma concentration of 5 μg mL^{-1}) showed a statistically significant decline in the Chinese patients (47.2 ± 3.6) than in Austrian patients (63.6 ± 5.4) [224].

Time to loss of consciousness, predicted effect-site C_e at loss of consciousness and propofol estimated C_p at loss of consciousness, were all significantly less in the Chinese patients (2.8 ± 0.6 min, 1.6 ± 0.4 μg mL^{-1} and 3.3 ± 0.8 μg mL^{-1}) than in the Austrian patients (3.9 ± 0.5 min, 2.4 ± 1.5 μg mL^{-1} and 4.6 ± 2.8 μg mL^{-1}) respectively [224]. Using two-way Analysis of Variance (ANOVA, Group × Time) showed a statistically significant difference in Bispectral Index scores over time [224], whereas using Dunnett's *post hoc* two-sided multiple-comparison's analysis showed a statistically significant difference between the two groups in Bispectral Index scores starting from propofol estimated plasma concentration of C_p 2.0 μg mL^{-1} [224]. What I find most interesting is the fact that subjects of both group ethnicities experienced their loss of consciousness at almost the same Bispectral Index scores albeit at statistically significant different propofol predicted effect-site C_e at loss of consciousness and different propofol estimated C_p at loss of consciousness [224].

We demonstrated that for the same propofol estimated C_p, Chinese patients lost consciousness 1.1 min earlier and required 0.8 μg mL^{-1} lower C_p to lose consciousness. The first possible explanation for that is that subjects of the Chinese ethnic-

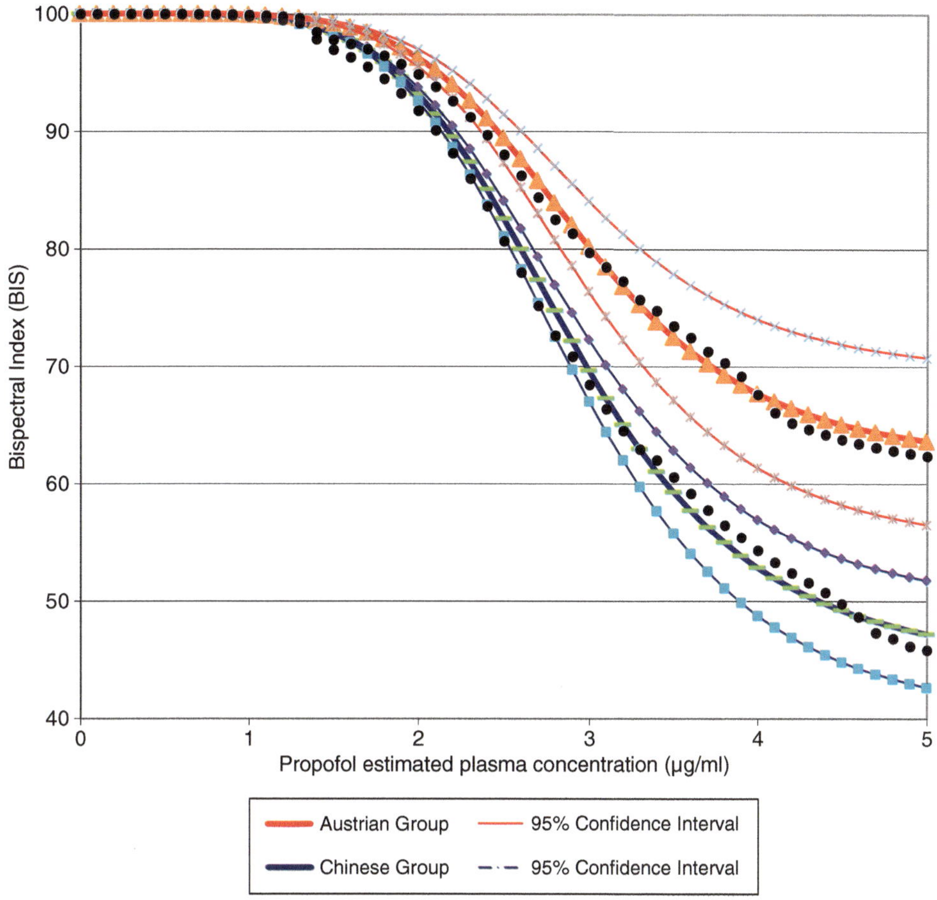

Fig. 5.35 Bispectral Index (BIS) versus propofol estimated plasma concentration (C_p) Sigmoid E_{max} curve model (95% confidence intervals) in the Austrian group patients compared with the Chinese group patients. (Reproduced with permission # 5900140805427 from Dahaba AA, et al. Geographic differences in the target-controlled infusion estimated concentration of propofol: bispectral index response curves. Can J Anaesth. 2011;58:364–70)

ity living in the People's Republic of China respond to a greater extent to equal estimated propofol C_p plasma concentrations (Pharmacodynamic clarification), or Chinese subjects have higher propofol plasma concentrations at similar Diprifusor's C_p (Pharmacokinetic clarification), or a mixture of the two explanations [224]. We concluded that Diprifusor's recommended concentration settings must be reduced in ethnic Chinese subjects living in People's Republic of China to attain the same desired Bispectral Index effect.

In our study we revealed statistically significant differences in propofol's plasma concentrations C_p–BIS Bispectral Index as well as statistically significant differences in propofol's effect-site concentrations C_e–BIS Bispectral Index response-relationships between Chinese and Austrian subjects [224]. Of these, propofol's effect-site concentrations C_e–BIS Bispectral Index is probably more clinically important, and not the plasma, this is because propofol's site of action monitoring is indeed via the Bispectral Index, a processed electroencephalographic index of the cerebral cortex [224]. Conversely, our study conclusion did not seem to agree with a previous study by Ortolani and colleagues [239], were following termination of propofol anesthe-

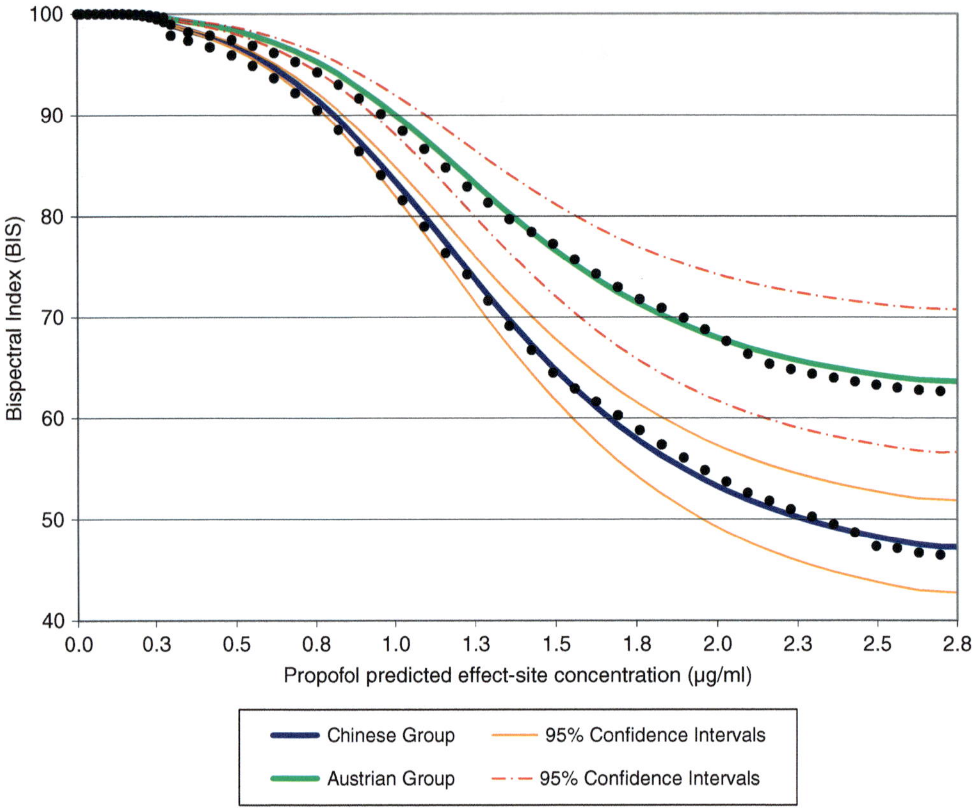

Fig. 5.36 Bispectral Index (BIS) versus propofol predicted effect-site concentration (C_e) Sigmoid E_{max} curve model (95% confidence intervals) in the Austrian group patients compared with the Chinese group patients. (Reproduced with permission # 5900140805427 from Dahaba AA, et al. Geographic differences in the target-controlled infusion estimated concentration of propofol: bispectral index response curves. Can J Anaesth. 2011;58:364–70)

sia, maintained at the same infusion rate, patient's Bispectral Index (BIS-XP A-2000, Software Revision 1.06) time of Bispectral Index return to baseline of 11 ± 4.2 min in 40 ethnic white Caucasian patients living in Italy was not significantly different than time of Bispectral Index return to baseline of 12.5 ± 5.1 min in 40 ethnic Chinese subjects living in Malysia [239]. Time to eye-opening after propofol discontinuation of 11.63 ± 4.2 min in Caucasian patients living in Italy was not significantly different than 13.23 ± 4.9 min in Chinese subjects living in Malysia. Time to respond to verbal commands recovery after propofol termination of 12.9 ± 5.5 min in Chinese subjects living in Malysia was not significantly different than 11 ± 5.1 min in Caucasian patients living in Italy [239].

In 1969, Ronald Katz [247] described how, during his one-year sabbatical, he was surprised to notice clinically that the dose of the Neuromuscular Blocking Agent (NMBA) d-tubocurarine that regularly abolished the neuromuscular twitch response to ulnar nerve stimulations in white Caucasian patients in New York, USA, did not do the same in white Caucasian patients in London, United Kingdom [247]. He was fascinated by the fact that the North American anesthesiologists routinely used d-tubocurarine doses that appeared to be "homeopathic" to their English anesthesiologists' colleagues [247]. What he found even more intriguing was the fact that the same d-tubocurarine manufacturer, in response to anesthesiologists' demand on both sides of the Atlantic Ocean, prepared d-tubocurarine in two different concentrations, one for each country on both sides of the Atlantic Ocean [247].

In his so called "The English First Transatlantic study," Katz and colleagues [247], using Grass force-displacement transducer Mechanomyograph (model FT-03), measured the adductor pollocis muscle single twitch response to 0.1 mg kg^{-1} d-tubocurarine neuromuscular block, in 11 white Caucasian patients in London, United Kingdom compared with 15 white Caucasian patients in New York, USA, anesthetized with thiopental, halothane, and Nitrous Oxide (N_2O) [247]. The mean ± SD percentage block magnitude and duration till 90% neuromuscular block recovery in the London patients (43 ± 21% block, 14.6 ± 8.8 min) were both significantly ($P < 0.01$) less than in a similar group of patients in New York (75 ± 14.1% block, 31.2 ± 12.4 min) [247]. They also studied an additional group of 11 United States Air Force (USAF) white Caucasian American citizens who lived in London and its environs for periods of 13–36 months, who were partially integrated in the local non-American community as some of them bought food at the USAF base while others shopped at local markets and were operated upon at the USAF hospital in London. The magnitude and duration of action of d-tubocurarine (62.4 ± 26.7% block, 26.6 ± 16.0 min) was like that seen in New York patients ($P > 0.05$) [247].

This was followed in 1991 by the so called The French Second Transatlantic study by Fiset and colleagues [248]. They performed a second transatlantic study comparing the dose response curves of vecuronium NMBA in 19 patients in Montreal, Canada, with 18 patients in Paris, France, anesthetized with thiopental, N_2O and fentanyl [248]. Using the Grass force-displacement transducer Mechanomyograph (model FT-10), they measured the adductor pollocis muscle response to Train-Of-Four (TOF) stimulation of the ulnar nerve. They demonstrated the same transatlantic significant difference in vecuronium potency between patients in Montreal and patients in Paris [248], as that previously demonstrated for d-tubocurarine by Katz and colleagues [247]. They created dose response curves using linear regression of the first TOF response (T_1) logit versus log dose of randomly allocated vecuronium 20, 30, and 40 μg kg^{-1} dose administrations [248]. Despite similar demographic data in both centers and identical anesthetic and neuromuscular techniques, the Paris dose response curve was significantly shifted to the right by 27% difference as the dose response curve of patients in Montreal. Patients in Paris required 27% (Confidence Interval CI 5–53%, $P = 0.01$) more vecuronium for the same "degree" of neuromuscular blockade than patients in Montreal [248]. Furthermore, when the dose was expressed as ideal Body Mass Index (BMI) of 22.5 kg m^{-2}, the difference between patients in Paris and patients in Montreal increased to 41% (CI 18–81%, $P = 0.01$) [248]. In Montreal the mean ± SEM ED_{50} and Ed_{90} (Effective Dose required for 50% and 90% first twitch depression respectively) were 26.0 ± 1.4 and 44.2 ± 2.5 μg kg^{-1} compared with 33.0 ± 3.3 and 71.9 ± 7.2 μg kg^{-1} in Paris [248]. They concluded that vecuronium was more potent in Montreal than in Paris, and such difference could not be explained by differences in age, sex, weight, or habits [248]. Their results confirmed for vecuronium the same transatlantic difference in potency of d-tubocurarine NMBA [248], previously observed between London and New York [247].

In our so called Global Third Transcontinental study, my research team Dahaba and colleagues similarly conducted a study in which ethnic geographic differences in the NMBA rocuronium potency was statistically significantly higher in white Caucasian patients living in New Jersey, USA, than in ethnic Han Chinese subjects living in Dalian, People's Republic of China, or in white Caucasian patients living in Graz, Austria [222]. My research team Dahaba and colleagues conducted a prospective controlled clinical multicenter study in 9 males and 9 females, North American white Caucasian patients, 18 matched for gender and approximate age and body weight Han Chinese patients and 18 matched Austrian white Caucasian patients, in 3 countries on 3 continents in 3 centers namely Englewood Medical Center (Englewood, New Jersey, USA); Dalian Medical University, Dalian, People's Republic of China, and Medical University of Graz, Austria [222]. Neuromuscular block of six consecutive rocuronium 50 μg kg^{-1} incremental doses was evaluated using the same force-displacement transducer Relaxometer Mechanomyograph (Groningen University, Groningen, The Netherlands) monitoring of the adductor pollocis muscle response to TOF stimulation of the ulnar nerve [249]. I carried the same Relaxometer Mechanomyograph [249], to be used first to monitor patients in Dalian, People's Republic of China, then I carried the same Relaxometer Mechanomyograph to monitor patients in New Jersey, USA, before returning with the same Relaxometer Mechanomyograph to monitor patients in Graz, Austria [249]. In that regard, Professor Sandor Agoston, the renowned scientist pioneer of neuromuscular monitoring and my research mentor at Groningen University, Groningen, The Netherlands, once told me that almost all neuromuscular monitoring devices monitor the NMBAs paralysis of the "adductor pollocis muscle," a muscle he knows of almost no surgeon who wants to see that muscle paralyzed. We created rocuronium Dose-response curves using log-dose-probit transformation [222]. We compared rocuronium bromide ED_{50}, ED_{90} and ED_{95} (Effective Dose required for 50%, 90%, 95% first twitch depression respectively) as well as Dur_{25} and $Dur_{0.8}$ (time from rocuronium last incremental dose administration until 25% first twitch T_1%, and 0.8 Train-Of-Four (TOF) ratio recoveries, in patients of the three countries [222]. Rocuronium ED_{50}, ED_{90} and ED_{95} were significantly higher in Austrian patients (258 ± 68, 530 ± 159, 598 ± 189 μg kg^{-1}) and in Chinese patients (201 ± 59, 413 ± 107, 475 ± 155 μg kg^{-1}) than in North American patients (148 ± 48, 316 ± 116, 362 ± 149 μg kg^{-1}) respectively [222]. Like the Montreal/Paris trans-atlantic study's dose response curves of patients in Paris shift to the right compared with the dose response curves of patients in Montreal [248], in our study, our white Caucasian North American patients in New Jersey, USA, rocuronium dose response curves were shifted to the right compared with our Han Chinese patients in Dalian, People's Republic of China, dose response curves and further shifted to the right compared with our white Caucasian patients in Graz, Austria, dose response curves (Fig. 5.37) [222]. The Dur_{25} and $Dur_{0.8}$ were significantly shorter in Austrian patients (22.3 ± 5.5, 36.9 ± 12.8 min) and in Chinese patients (30.4 ± 7.5, 45.7 ± 15.9 min) than in North American patients (36.7 ± 8.5, 56.2 ± 16.7 min) respectively [222]. We demonstrated a significant difference in rocuronium potency and duration of action between patients in the three countries for determining dosage recommendations for different geographic regions [222].

Although subjects in one ethnic-geographic region have distinct shared characteristics in genetics as well as environment, however in nowadays' global multi-cultural societies, groups of people who have dissimilar ancestries or even mixed ethnicities might still be residing at far geographical areas [224]. In our Trans-geographic multicenter study that my research group Dahaba and colleagues conducted of differences in propofol-Bispectral Index response curves between patients in Medical University of Graz, Austria and patients in Zhe Jiang University Hospitals, Hang Zhou, People's Republic of China, our aim was identifying two racial ethnic groups of inhabitants who share very clear described features, such as dietary habits, environmental pollution, and physical exercise; all contributing to their body composition, body fat distribution, as well as the proportions of skeletal muscular fiber types [224]. Thus, in our propofol-Bispectral Index response curves investigation, we attempted to evaluate the overall "collective effects" of all these ethnic /environmental influencing factors combinations. Our study did not determine which individual variables might have contributed to our results as the differences between white Caucasian and Chinese patients is likely to be multifactorial [224].

As a rule, propofol plasma concentrations is different than propofol effect-site concentrations, because propofol protein binding takes place only in plasma and just the free drug interacts with the receptors in the brain effect-site [224]. A possible explanation of our study's [224] ethnic geographic differences in propofol-Bispectral Index response curves could be attributed to the differences in propofol plasma protein binding between the two ethnic-geographic race groups

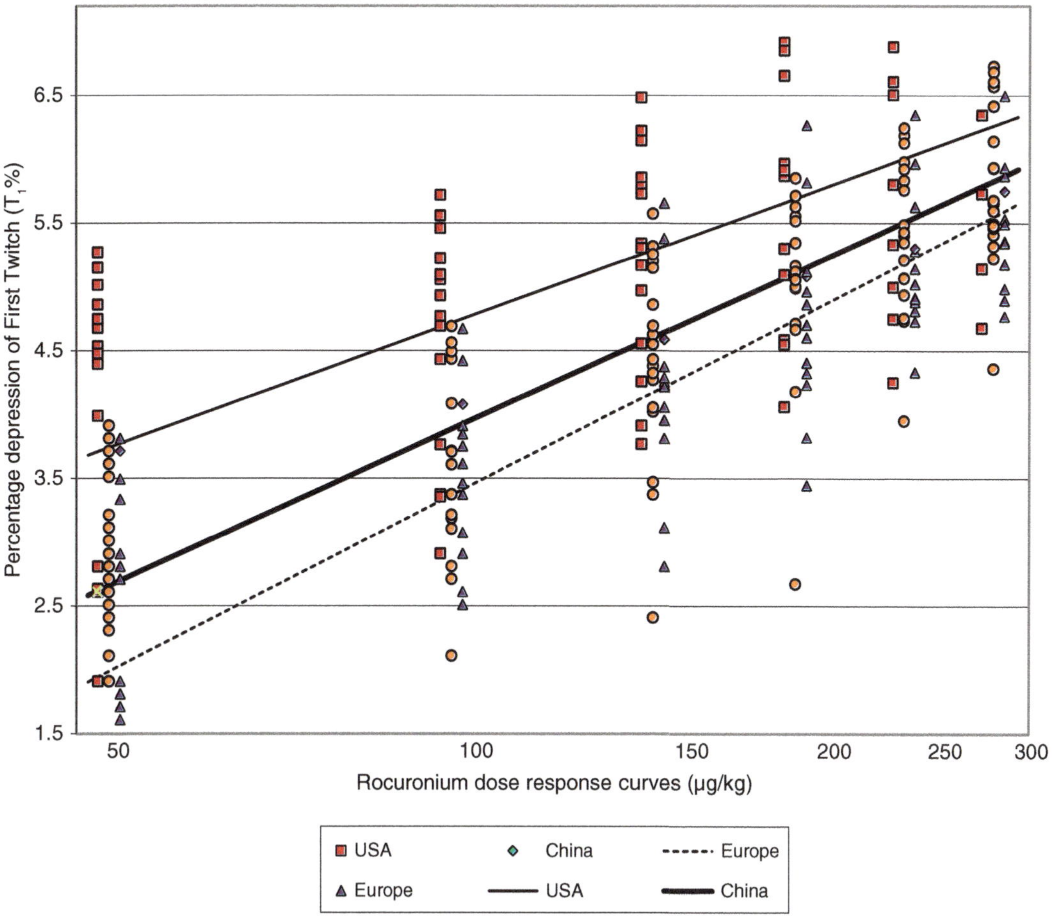

Fig. 5.37 Our white Caucasian patients in New Jersey, USA, rocuronium Dose-response curves were shifted to the right compared with our Han Chinese patients in Dalian, People's Republic of China, Dose-response curves, and further shifted to the right compared with our white Caucasian patients in Graz, Austria. (Reproduced with permission # 5902540430962 from Dahaba AA, et al. Geographic regional differences in rocuronium bromide dose-response relation and time course of action: an overlooked factor in determining recommended dosage. Anesthesiology. 2006;104:950–3)

[224]. Because the propofol unbound free fraction is the only portion that can exert a pharmacologically active effect [224], therefore, aside from very few classes of medications, any increase in plasma protein-binding would result in a considerable decline in the clinical effects of medications that are extensively protein bound [250, 251], including propofol that is >96% protein bound [231, 252]. A study by Takizawa and colleagues [253] has clearly shown that abrupt Cardiopulmonary Bypass (CBP) hemodilution lead to a significant 2-fold increase in propofol unbound portion that resulted in a significant instant decline in Bispectral Index BIS A-2000 scores and a significant rise in Burst Suppression Ratio (BSR), although propofol's total plasma concentrations remained the same [253]. Our results [224] suggest that lower propofol serum protein binding in the Chinese population might explain the enhanced effect of propofol in the Chinese population compared with the white Caucasian Europeans [254]. Compared with white Caucasians, Chinese patients have totally different drugs' plasma protein bindings, simply because Chinese subjects have less plasma concentrations of α-1-acid Glycoprotein, a glycoprotein responsible for accelerating drugs' recovery via increasing their protein binding [254].

The significant differences in propofol-Bispectral Index response curves between white Caucasians and Chinese populations to the same anesthetic regime that we demonstrated might require a TCI pharmacokinetic model especially devised for different trans-geographic regions' inhabitants, or different recommended TCI infusions when utilizing similar pharmacokinetic models [224]. Having said that, because substantial interindividual differences and variations still frequently exist even among the same ethnic race and geographic location, this greatly emphasizes the importance of TCI drug titration to Bispectral index effect and gives credence to the current clinical practice of "dosing to clinical Bispectral index effect" [224].

References

1. Smith D, Andrzejowski J, Smith A. Certainty and uncertainty: NICE guidance on "depth of anaesthesia" monitoring. Anaesthesia. 2013;68:1000–5.
2. Dahaba AA. Different conditions that could result in the bispectral index indicating an incorrect hypnotic state. Anesth Analg. 2005;101:765–73.
3. Bjørgaas M, Sand T, Vik T, Jorde R. Quantitative EEG during controlled hypoglycaemia in diabetic and non-diabetic children. Diabet Med. 1998;15:30–7.
4. Tribl G, Howorka K, Heger G, Anderer P, Thoma H, Zeitlhofer J. EEG topography during insulin-induced hypoglycemia in patients with insulin-dependent diabetes mellitus. Eur Neurol. 1996;36:303–9.
5. Pramming S, Thorsteinsson B, Stigsby B, Binder C. Glycaemic threshold for changes in electroencephalograms during hypoglycaemia in patients with insulin dependent diabetes. Br Med J (Clin Res Ed). 1988;296:665–7.
6. Wu C-C, Lin C-S, Mok MS. Bispectral index monitoring during hypoglycemic coma. J Clin Anesth. 2002;14:305–6.
7. Vivien B, Langeron O, Riou B. Increase in bispectral index (BIS) while correcting a severe hypoglycemia. Anesth Analg. 2002;95:1824–5.
8. Alkire MT. Quantitative EEG correlations with brain glucose metabolic rate during anesthesia in volunteers. Anesthesiology. 1998;89:323–33.
9. Narayanaswamy M. Decrease in Bispectral Index while correcting hyperglycemia and an increase in Bispectral Index with correction of hypoglycemia. Anesth Analg. 2009;109:995.
10. Mahaldar DAC, Gadhinglajkar S, Sreedhar R. Decrease in bispectral index in a patient undergoing coronary artery bypass graft surgery. J Cardiothorac Vasc Anesth. 2011;25:e39–40.
11. Chakravarthy M, Holla S, Jawali V. Index of consciousness and bispectral index values are interchangeable during normotension and hypotension but not during non pulsatile flow state during cardiac surgical procedures: a prospective study. J Clin Monit Comput. 2010;24:83–91.
12. Honan DM, Breen PJ, Boylan JF, McDonald NJ, Egan TD. Decrease in bispectral index preceding intraoperative hemodynamic crisis: evidence of acute alteration of propofol pharmacokinetics. Anesthesiology. 2002;97:1303–5.
13. De Paepe P, Belpaire FM, Rosseel MT, Van Hoey G, Boon PA, Buylaert WA. Influence of hypovolemia on the pharmacokinetics and the electroencephalographic effect of propofol in the rat. Anesthesiology. 2000;93:1482–90.
14. Meybohm P, Gruenewald M, Höcker J, Renner J, Graesner J-T, Ilies C, et al. Correlation and agreement between the bispectral index vs. state entropy during hypothermic cardio-pulmonary bypass. Acta Anaesthesiol Scand. 2010;54:169–75.
15. Lobo FA, Schraag S. Limitations of anaesthesia depth monitoring. Curr Opin Anaesthesiol. 2011;24:657–64.
16. Musialowicz T, Lahtinen P, Pitkänen O, Kurola J, Parviainen I. Comparison of spectral entropy and BIS VISTA™ monitor during general anesthesia for cardiac surgery. J Clin Monit Comput. 2011;25:95–103.
17. Hayashi K, Mita K, Sawa T. Electroencephalographic changes in the late cardiopulmonary bypass period are not reflected in the bispectral index. Clin Neurophysiol. 2010;121:1198–204.
18. Levy WJ, Pantin E, Mehta S, McGarvey M. Hypothermia and the approximate entropy of the electroencephalogram. Anesthesiology. 2003;98:53–7.
19. Honan D, Doherty D, Frizelle H. A comparison of the effects on bispectral index of mild vs. moderate hypothermia during cardiopulmonary bypass. Eur J Anaesthesiol. 2006;23:385–90.
20. Hayashida M, Sekiyama H, Orii R, Chinzei M, Ogawa M, Arita H, et al. Effects of deep hypothermic circulatory arrest with retrograde cerebral perfusion on electroencephalographic bispectral index and suppression ratio. J Cardiothorac Vasc Anesth. 2007;21:61–7.
21. Mychaskiw G, Heath BJ, Eichhorn JH. Falsely elevated bispectral index during deep hypothermic circulatory arrest. Br J Anaesth. 2000;85:798–800.
22. Sebel PS. Central nervous system monitoring during open heart surgery: an update. J Cardiothorac Vasc Anesth. 1998;12(2 Suppl 1):3–8.
23. Vretzakis G, Dragoumanis C, Argiriadou H, Papaziogas B, Konstantinou F, Tsakiridis K. Inaccuracy of BIS values produced by the cardiopulmonary bypass machine during the operative repair of an aortic dissection. J Cardiothorac Vasc Anesth. 2006;20:68–70.
24. Doi M, Gajraj RJ, Mantzaridis H, Kenny GN. Effects of cardiopulmonary bypass and hypothermia on

electroencephalographic variables. Anaesthesia. 1997;52:1048–55.
25. Russell GN, Wright EL, Fox MA, Douglas EJ, Cockshott ID. Propofol-fentanyl anaesthesia for coronary artery surgery and cardiopulmonary bypass. Anaesthesia. 1989;44:205–8.
26. Schmidlin D, Hager P, Schmid ER. Monitoring level of sedation with bispectral EEG analysis: comparison between hypothermic and normothermic cardiopulmonary bypass. Br J Anaesth. 2001;86:769–76.
27. Mathew JP, Weatherwax KJ, East CJ, White WD, Reves JG. Bispectral analysis during cardiopulmonary bypass: the effect of hypothermia on the hypnotic state. J Clin Anesth. 2001;13:301–5.
28. Antognini JF. Hypothermia eliminates isoflurane requirements at 20 degrees C. Anesthesiology. 1993;78:1152–6.
29. Thompson C. Henry Hill Hickman. A forgotten pioneer of anesthesia. Br Med J. 1912;1:843–5.
30. Sprigge JS. Henry Hill Hickman. Br J Anaesth. 2005;95:842.
31. Martoft L, Lomholt L, Kolthoff C, Rodriguez BE, Jensen EW, Jørgensen PF, et al. Effects of CO2 anaesthesia on central nervous system activity in swine. Lab Anim. 2002;36:115–26.
32. Turnbull D, Furlonger A, Andrzejowski J. The influence of changes in end-tidal carbon dioxide upon the Bispectral Index. Anaesthesia. 2008;63:458–62.
33. Yoshida H, Kushikata T, Kabara S, Takase H, Ishihara H, Hirota K. Flat electroencephalogram caused by carbon dioxide pneumoperitoneum. Anesth Analg. 2007;105:1749–52. table of contents
34. Lee YW, Chang CC. The bispectral index in a patient with carbon dioxide narcosis. Anaesth Intensive Care. 2007;35:453–4.
35. Fothergill DM, Hedges D, Morrison JB. Effects of CO2 and N2 partial pressures on cognitive and psychomotor performance. Undersea Biomed Res. 1991;18:1–19.
36. Takahashi K, Nosaka S. Carbon dioxide narcosis caused by midazolam in a patient with myotonic dystrophy. Anaesthesia. 2000;55:97.
37. Smyth E, Egan TD. Apneic oxygenation associated with patient-controlled analgesia. J Clin Anesth. 1998;10:499–501.
38. Jain V, Singh D, Prabhakar H. Sudden decrease in bispectral index value during emergence from general anaesthesia. Anaesth Intensive Care. 2008;36:460–1.
39. Messner M, Beese U, Romstöck J, Dinkel M, Tschaikowsky K. The bispectral index declines during neuromuscular block in fully awake persons. Anesth Analg. 2003;97:488–91.
40. Hayashi K, Fujikawa M, Sawa T. Hyperventilation-induced hypocapnia changes the pattern of electroencephalographic bicoherence growth during sevoflurane anaesthesia. Br J Anaesth. 2008;101:666–72.
41. Kakinohana M, Miyata Y, Kawabata T, Kawashima S, Tokumine J, Sugahara K. Bispectral index decreased to "0" in propofol anesthesia after a cross-clamping of descending thoracic aorta. Anesthesiology. 2003;99:1223–5.
42. Veroli P, O'Kelly B, Bertrand F, Trouvin JH, Farinotti R, Ecoffey C. Extrahepatic metabolism of propofol in man during the anhepatic phase of orthotopic liver transplantation. Br J Anaesth. 1992;68:183–6.
43. Billard V. Brain injury under general anesthesia: is monitoring of the EEG helpful? Can J Anaesth. 2001;48:1055–60.
44. Mérat S, Lévecque JP, Le Gulluche Y, Diraison Y, Brinquin L, Hoffmann JJ. BIS monitoring may allow the detection of severe cerebral ischemia. Can J Anaesth. 2001;48:1066–9.
45. Ackerstaff RG, Moons KG, van de Vlasakker CJ, Moll FL, Vermeulen FE, Algra A, et al. Association of intraoperative transcranial doppler monitoring variables with stroke from carotid endarterectomy. Stroke. 2000;31:1817–23.
46. Halsey JH. Risks and benefits of shunting in carotid endarterectomy. The International Transcranial Doppler Collaborators. Stroke. 1992;23:1583–7.
47. Estruch-Pérez MJ, Barberá-Alacreu M, Ausina-Aguilar A, Soliveres-Ripoll J, Solaz-Roldán C, Morales-Suárez-Varela MM. Bispectral index variations in patients with neurological deficits during awake carotid endarterectomy. Eur J Anaesthesiol. 2010;27:359–63.
48. Absalom AR, Mani V, De Smet T, Struys MMRF. Pharmacokinetic models for propofol–defining and illuminating the devil in the detail. Br J Anaesth. 2009;103:26–37.
49. Dahaba AA, Xue JX, Hua Y, Liu QH, Xu GX, Liu YM, et al. The utility of using the bispectral index-Vista for detecting cross-clamping decline in cerebral blood flow velocity. Neurosurgery. 2010;67(3 Suppl Operative):ons102–7; discussion ons107.
50. Gosling RG, King DH. Arterial assessment by Doppler-shift ultrasound. Proc R Soc Med. 1974;67(6 Pt 1):447–9.
51. England MR. The changes in bispectral index during a hypovolemic cardiac arrest. Anesthesiology. 1999;91:1947–9.
52. Azim N, Wang CY. The use of bispectral index during a cardiopulmonary arrest: a potential predictor of cerebral perfusion. Anaesthesia. 2004;59:610–2.
53. Szekely B, Saint-Marc T, Degrémont AC, Castelain MH, Fischler M. Value of bispectral index monitoring during cardiopulmonary resuscitation. Br J Anaesth. 2002;88:443–4.
54. Losasso TJ, Muzzi DA, Meyer FB, Sharbrough FW. Electroencephalographic monitoring of cerebral function during asystole and successful cardiopulmonary resuscitation. Anesth Analg. 1992;75:1021–4.
55. Goodman PG, Mehta AR, Castresana MR. Predicting ischemic brain injury after intraoperative cardiac arrest during cardiac surgery using the BIS monitor. J Clin Anesth. 2009;21:609–12.
56. Chakravarthy M, Patil T, Jayaprakash K, Shivananda N, Jawali V. Bispectral index is an indicator of adequate cerebral perfusion during cardiopulmo-

nary resuscitation. J Cardiothorac Vasc Anesth. 2003;17:506–8.
57. Morimoto Y, Monden Y, Ohtake K, Sakabe T, Hagihira S. The detection of cerebral hypoperfusion with bispectral index monitoring during general anesthesia. Anesth Analg. 2005;100:158–61.
58. Puri GD, Kumar A, Thignam SKS. BIS as an indicator of cerebral perfusion during closed mitral commissurotomy. Can J Anaesth. 2003;50:971–2.
59. Myles PS, Cairo S. Artifact in the bispectral index in a patient with severe ischemic brain injury. Anesth Analg. 2004;98:706–7. table of contents
60. Fatovich DM, Jacobs IG, Celenza A, Paech MJ. An observational study of bispectral index monitoring for out of hospital cardiac arrest. Resuscitation. 2006;69:207–12.
61. Burjek NE, Wagner CE, Hollenbeck RD, Wang L, Yu C, McPherson JA, et al. Early bispectral index and sedation requirements during therapeutic hypothermia predict neurologic recovery following cardiac arrest. Crit Care Med. 2014;42:1204–12.
62. Hajat Z, Ahmad N, Andrzejowski J. The role and limitations of EEG-based depth of anaesthesia monitoring in theatres and intensive care. Anaesthesia. 2017;72(Suppl 1):38–47.
63. Stammet P, Collignon O, Werer C, Sertznig C, Devaux Y. Bispectral index to predict neurological outcome early after cardiac arrest. Resuscitation. 2014;85:1674–80.
64. Seder DB, Dziodzio J, Smith KA, Hickey P, Bolduc B, Stone P, et al. Feasibility of bispectral index monitoring to guide early post-resuscitation cardiac arrest triage. Resuscitation. 2014;85:1030–6.
65. Khan J, Mariappan R, Venkatraghavan L. Entropy as an indicator of cerebral perfusion in patients with increased intracranial pressure. J Anaesthesiol Clin Pharmacol. 2014;30:409–11.
66. Harclerode Z, Andrzejowski J, Coley S, Dyde R. Bispectral index detects intraoperative cerebral ischaemia during balloon assisted cerebral aneurysm coiling. F1000Res. 2013;2:225.
67. Win NN, Kohase H, Miyamoto T, Umino M. Decreased bispectral index as an indicator of syncope before hypotension and bradycardia in two patients with needle phobia. Br J Anaesth. 2003;91:749–52.
68. Mercader MA, Varghese PJ, Potolicchio SJ, Venkatraman GK, Lee SW. New insights into the mechanism of neurally mediated syncope. Heart. 2002;88:217–21.
69. Bleeker CP, Smits B, Vos PE, Mourisse JMJ. Bispectral index changes during acute brainstem TIA/ischemia. Case Rep Med. 2010;2010:697185.
70. Hashimoto H, Nakamura H, Hirota K. Marked reduction in bispectral index with severe bradycardia without hypotension in a diabetic patient undergoing ophthalmic surgery. J Anesth. 2008;22:300–3.
71. Chauhan S, Ingole P, Kale S, Kiran U. Correlation of BIS with Tet spells. J Cardiothorac Vasc Anesth. 2004;18:682–3.
72. Welsby IJ, Ryan JM, Booth JV, Flanagan E, Messier RH, Borel CO. The bispectral index in the diagnosis of perioperative stroke: a case report and discussion. Anesth Analg. 2003;96:435–7. table of contents
73. Roach GW, Kanchuger M, Mangano CM, Newman M, Nussmeier N, Wolman R, et al. Adverse cerebral outcomes after coronary bypass surgery. Multicenter Study of Perioperative Ischemia Research Group and the Ischemia Research and Education Foundation Investigators. N Engl J Med. 1996;335:1857–63.
74. Villacorta J, Kerbaul F, Collart F, Guidon C, Bonnet M, Guillen JC, et al. Perioperative cerebral ischaemia in cardiac surgery and BIS. Anaesth Intensive Care. 2005;33:514–7.
75. Kin N, Konstadt SN, Sato K, Hanaoka K. Reduction of bispectral index value associated with clinically significant cerebral air embolism. J Cardiothorac Vasc Anesth. 2004;18:82–4.
76. Ellenberger C, Diaper J, Licker M, Panos A. Bispectral index and detection of acute brain injury during cardiac surgery. Eur J Anaesthesiol. 2007;24:807–9.
77. Uhrig L, Devys J-M, Piotin M, Taylor G. Bispectral index transiently decreased to "0" during per-embolization rupture of an intracranial aneurysm. Ann Fr Anesth Reanim. 2014;33:e15–7.
78. Prabhakar H, Ali Z, Rath GP, Singh D. Zero bispectral index during coil embolization of an intracranial aneurysm. Anesth Analg. 2007;105:887–8.
79. Umegaki N, Hirota K, Kitayama M, Yatsu Y, Ishihara H, Mtasuki A. A marked decrease in bispectral index with elevation of suppression ratio by cervical haematoma reducing cerebral perfusion pressure. J Clin Neurosci. 2003;10:694–6.
80. Lauwick S, English M, Hemmerling TM. An unusual case of cerebral hypoperfusion detected by bispectral index monitoring. Can J Anaesth. 2007;54:680–1.
81. Kaki AM, Almarakbi WA. Does patient position influence the reading of the bispectral index monitor? Anesth Analg. 2009;109:1843–6.
82. Soto SL, Tellería AM, Pacheco Adamuz MJ. Bispectral index (BIS) as a fast indicator of brain stem compression in posterior cervical spine surgery in the sitting position. J Neurosurg Anesthesiol. 2010;22:270–1.
83. Mourisse J, Booij L. Bispectral index detects period of cerebral hypoperfusion during cardiopulmonary bypass. J Cardiothorac Vasc Anesth. 2003;17:76–8.
84. Hemmerling TM, Olivier J-F, Basile F, Le N, Prieto I. Bispectral index as an indicator of cerebral hypoperfusion during off-pump coronary artery bypass grafting. Anesth Analg. 2005;100:354–6.
85. Hayashida M, Kin N, Tomioka T, Orii R, Sekiyama H, Usui H, et al. Cerebral ischaemia during cardiac surgery in children detected by combined monitoring of BIS and near-infrared spectroscopy. Br J Anaesth. 2004;92:662–9.
86. Hayashida M, Chinzei M, Komatsu K, Yamamoto H, Tamai H, Orii R, et al. Detection of cerebral hypo-

perfusion with bispectral index during paediatric cardiac surgery. Br J Anaesth. 2003;90:694–8.
87. Choudhry DK, Brenn BR. Bispectral index monitoring: a comparison between normal children and children with quadriplegic cerebral palsy. Anesth Analg. 2002;95:1582–5. table of contents
88. Katoh T, Suzuki A, Ikeda K. Electroencephalographic derivatives as a tool for predicting the depth of sedation and anesthesia induced by sevoflurane. Anesthesiology. 1998;88:642–50.
89. Valkenburg AJ, de Leeuw TG, Tibboel D, Weber F. Lower bispectral index values in children who are intellectually disabled. Anesth Analg. 2009;109:1428–33.
90. Ponnudurai RN, Clarke-Moore A, Ekulide I, Sant M, Choi K, Stone J, et al. A prospective study of bispectral index scoring in mentally retarded patients receiving general anesthesia. J Clin Anesth. 2010;22:432–6.
91. Faraoni D, Ghoundiwal D, Barvais L. Artefact in the electroencephalographic monitoring in a patient with brain metastasis. Eur J Anaesthesiol. 2010;27:921–2.
92. Ferreira DA, Nunes CS, Lobo F, Casal M, Antunes LM, Amorim P. Brain tumors may alter the relationship between bispectral index values and propofol concentrations during induction of anesthesia. J Clin Anesth. 2008;20:116–21.
93. Fudickar A, Jacobsen JH, Weiler N, Scholz J, Bein B. Bilateral measurement of bispectral index and mid-latency auditory evoked potentials in patients with unilateral brain lesions. J Crit Care. 2009;24:545–50.
94. Martin Bland J, Altman Douglas G. Statistical methods for assessing agreement between two methods of clinical measurement. The Lancet. 1986;327:307–10.
95. Dahaba AA, Lin H, Ye XF, Lin J, Zhang N, Lian QQ. Bispectral index monitoring of propofol anesthesia in pediatric patients with hydrocephalus. A prospective observational study. Paediatr Anaesth. 2014;24:1190–2.
96. Dahaba AA, Lin H, Ye XF, Zhang N, Wang K, Reibnegger G, et al. Propofol-Bispectral Index (BIS) Electroencephalography (EEG) pharmacokinetic-pharmacodynamic model in patients with post-cerebral hemorrhage hydrocephalus. Clin EEG Neurosci. 2021;52:351–9.
97. Hill AV. The possible effects of the aggregation of the molecules of haemoglobin on its disociation (dissociation) curves. J Physiol. 1910;40:iv–vii.
98. Urfalioglu A, Arslan M, Duman Y, Gisi G, Oksuz G, Yildiz H, et al. Anesthesia procedure for congenital insensitivity to pain in a child with anhidrosis syndrome: a rare case. J Nippon Med Sch. 2017;84:237–40.
99. Khalil SN, Hanna E, Farag A, Armendartz G. Rett syndrome: anaesthesia management. Paediatr Anaesth. 2002;12:375.
100. Kim JT, Muntyan I, Bashkirov IL, Bloom M, Hartmannsgruber MWB. The use of bispectral index monitoring in the anesthetic management of a patient with Rett syndrome undergoing scoliosis surgery. J Clin Anesth. 2006;18:161–2.
101. Konarzewski WH, Misso S. Rett syndrome and delayed recovery from anaesthesia. Anaesthesia. 1994;49:357.
102. Zabala L, Ahmed MI, Denman WT. Bispectral index in a 3-year old undergoing deep hypothermia and circulatory arrest. Paediatr Anaesth. 2003;13:355–9.
103. Renna M, Handy J, Shah A. Low baseline Bispectral Index of the electroencephalogram in patients with dementia. Anesth Analg. 2003;96:1380–5.
104. Holschneider DP, Leuchter AF, Uijtdehaage SH, Abrams M, Rosenberg-Thompson S. Loss of high-frequency brain electrical response to thiopental administration in Alzheimer's-type dementia. Neuropsychopharmacology. 1997;16:269–75.
105. Veiga H, Deslandes A, Cagy M, Fiszman A, Piedade RAM, Ribeiro P. Neurocortical electrical activity tomography in chronic schizophrenics. Arq Neuropsiquiatr. 2003;61:712–7.
106. Folstein MF, Folstein SE, McHugh PR. "Mini-mental state". A practical method for grading the cognitive state of patients for the clinician. J Psychiatr Res. 1975;12:189–98.
107. Laalou F-Z, Egard M, Guillot M, Noll E, Taglang G, Pain L. Influence of preoperative cognitive status on propofol requirement to maintain hypnosis in the elderly. Br J Anaesth. 2010;105:342–6.
108. Perez-Protto S, Geube M, Ontaneda D, Dalton JE, Kurz A, Sessler DI. Sensitivity to volatile anesthetics in patients with dementia: a case-control analysis. Can J Anaesth. 2014;61:611–8.
109. Erdogan MA, Demirbilek S, Erdil F, Aydogan MS, Ozturk E, Togal T, et al. The effects of cognitive impairment on anaesthetic requirement in the elderly. Eur J Anaesthesiol. 2012;29:326–31.
110. Saeki H, Shirasawa Y, Nagamizo D, Morimoto Y, Matsumoto M, Sakabe T. Anesthetic management for a patient with Huntington disease. Masui. 2007;56:1358–61.
111. Wijdicks EFM. The case against confirmatory tests for determining brain death in adults. Neurology. 2010;75:77–83.
112. Escudero D, Otero J, Muñiz G, Gonzalo JA, Calleja C, González A, et al. The Bispectral Index Scale: its use in the detection of brain death. Transplant Proc. 2005;37:3661–3.
113. Vivien B, Paqueron X, Le Cosquer P, Langeron O, Coriat P, Riou B. Detection of brain death onset using the bispectral index in severely comatose patients. Intensive Care Med. 2002;28:419–25.
114. Puri GD, Nakra D. ECG artifact and BIS in severe brain injury. Anesth Analg. 2005;101:1566–7.
115. Çoban Y, Yildizdas D, Horoz OO, Aslan N, Herguner O. Can bispectral index be an early marker in the diagnosis of brain death? Acta Neurol Belg. 2023;123:513–7.
116. Dunham CM, Katradis DA, Williams MD. The bispectral index, a useful adjunct for the timely diag-

nosis of brain death in the comatose trauma patient. Am J Surg. 2009;198:846–51.
117. Smith MMJ, Andrzejowski JC. Decrease in bispectral index preceding signs of impending brain death in traumatic brain injury. J Neurosurg Anesthesiol. 2010;22:268–9.
118. Dunham CM, Ransom KJ, McAuley CE, Gruber BS, Mangalat D, Flowers LL. Severe brain injury ICU outcomes are associated with Cranial-Arterial Pressure Index and noninvasive Bispectral Index and transcranial oxygen saturation: a prospective, preliminary study. Crit Care. 2006;10:R159.
119. Myles PS, Daly D, Silvers A, Cairo S. Prediction of neurological outcome using bispectral index monitoring in patients with severe ischemic-hypoxic brain injury undergoing emergency surgery. Anesthesiology. 2009;110:1106–15.
120. Seder DB, Fraser GL, Robbins T, Libby L, Riker RR. The bispectral index and suppression ratio are very early predictors of neurological outcome during therapeutic hypothermia after cardiac arrest. Intensive Care Med. 2010;36:281–8.
121. Ebtehaj M, Yaqubi S, Seddighi AS, Seddighi A, Yazdi Z. Correlation between BIS and GCS in patients suffering from head injury. Ir J Med Sci. 2012;181:77–80.
122. Cottenceau V, Petit L, Masson F, Guehl D, Asselineau J, Cochard J-F, et al. The use of bispectral index to monitor barbiturate coma in severely brain-injured patients with refractory intracranial hypertension. Anesth Analg. 2008;107:1676–82.
123. Tobias JD. Bispectral index monitoring documents burst suppression during pentobarbital coma. J Intensive Care Med. 2008;23:258–62.
124. Fàbregas N, Gambús PL, Valero R, Carrero EJ, Salvador L, Zavala E, et al. Can bispectral index monitoring predict recovery of consciousness in patients with severe brain injury? Anesthesiology. 2004;101:43–51.
125. Van Norman GA. Another matter of life and death: what every anesthesiologist should know about the ethical, legal, and policy implications of the non-heart-beating cadaver organ donor. Anesthesiology. 2003;98:763–73.
126. Auyong DB, Klein SM, Gan TJ, Roche AM, Olson D, Habib AS. Processed electroencephalogram during donation after cardiac death. Anesth Analg. 2010;110:1428–32.
127. Chawla LS, Akst S, Junker C, Jacobs B, Seneff MG. Surges of electroencephalogram activity at the time of death: a case series. J Palliat Med. 2009;12:1095–100.
128. Hayashi K, Araki R, Tanaka A. Electroencephalographic monitoring during sevoflurane anaesthesia in an amyotrophic lateral sclerosis patient with locked-in state. J Clin Neurosci. 2017;43:126–8.
129. Fernández-Espejo D, Owen AM. Detecting awareness after severe brain injury. Nat Rev Neurosci. 2013;14:801–9.
130. Jennett B, Plum F. Persistent vegetative state after brain damage. A syndrome in search of a name. Lancet. 1972;1:734–7.
131. Andrews K, Murphy L, Munday R, Littlewood C. Misdiagnosis of the vegetative state: retrospective study in a rehabilitation unit. BMJ. 1996;313:13–6.
132. Childs NL, Mercer WN, Childs HW. Accuracy of diagnosis of persistent vegetative state. Neurology. 1993;43:1465–7.
133. Schnakers C, Vanhaudenhuyse A, Giacino J, Ventura M, Boly M, Majerus S, et al. Diagnostic accuracy of the vegetative and minimally conscious state: clinical consensus versus standardized neurobehavioral assessment. BMC Neurol. 2009;9:35.
134. Giacino JT, Ashwal S, Childs N, Cranford R, Jennett B, Katz DI, et al. The minimally conscious state: definition and diagnostic criteria. Neurology. 2002;58:349–53.
135. Owen AM. Detecting consciousness: a unique role for neuroimaging. Annu Rev Psychol. 2013;64:109–33.
136. Owen AM, Coleman MR, Boly M, Davis MH, Laureys S, Pickard JD. Detecting awareness in the vegetative state. Science. 2006;313:1402.
137. Monti MM, Vanhaudenhuyse A, Coleman MR, Boly M, Pickard JD, Tshibanda L, et al. Willful modulation of brain activity in disorders of consciousness. N Engl J Med. 2010;362:579–89.
138. Cruse D, Chennu S, Chatelle C, Bekinschtein TA, Fernández-Espejo D, Pickard JD, et al. Bedside detection of awareness in the vegetative state: a cohort study. Lancet. 2011;378:2088–94.
139. Bekinschtein TA, Manes FF, Villarreal M, Owen AM, Della-Maggiore V. Functional imaging reveals movement preparatory activity in the vegetative state. Front Hum Neurosci. 2011;5:5.
140. Boly M, Coleman MR, Davis MH, Hampshire A, Bor D, Moonen G, et al. When thoughts become action: an fMRI paradigm to study volitional brain activity in non-communicative brain injured patients. Neuroimage. 2007;36:979–92.
141. Aguirre GK, Detre JA, Alsop DC, D'Esposito M. The parahippocampus subserves topographical learning in man. Cereb Cortex. 1996;6:823–9.
142. Ghaem O, Mellet E, Crivello F, Tzourio N, Mazoyer B, Berthoz A, et al. Mental navigation along memorized routes activates the hippocampus, precuneus, and insula. Neuroreport. 1997;8:739–44.
143. Mellet E, Briscogne S, Tzourio-Mazoyer N, Ghaëm O, Petit L, Zago L, et al. Neural correlates of topographic mental exploration: the impact of route versus survey perspective learning. Neuroimage. 2000;12:588–600.
144. O'Craven KM, Kanwisher N. Mental imagery of faces and places activates corresponding stiimulus-specific brain regions. J Cogn Neurosci. 2000;12:1013–23.
145. Ino T, Inoue Y, Kage M, Hirose S, Kimura T, Fukuyama H. Mental navigation in humans is pro-

cessed in the anterior bank of the parieto-occipital sulcus. Neurosci Lett. 2002;322:182–6.
146. Owen AM, Coleman MR. Functional neuroimaging of the vegetative state. Nat Rev Neurosci. 2008;9:235–43.
147. Monti MM, Coleman MR, Owen AM. Executive functions in the absence of behavior: functional imaging of the minimally conscious state. Prog Brain Res. 2009;177:249–60.
148. Pfurtscheller G, Neuper C. Motor imagery activates primary sensorimotor area in humans. Neurosci Lett. 1997;239:65–8.
149. Davey MP, Victor JD, Schiff ND. Power spectra and coherence in the EEG of a vegetative patient with severe asymmetric brain damage. Clin Neurophysiol. 2000;111:1949–54.
150. Katayama Y, Tsubokawa T, Yamamoto T, Hirayama T, Miyazaki S, Koyama S. Characterization and modification of brain activity with deep brain stimulation in patients in a persistent vegetative state: pain-related late positive component of cerebral evoked potential. Pacing Clin Electrophysiol. 1991;14:116–21.
151. Pandit JJ, Schmelzle-Lubiecki B, Goodwin M, Saeed N. Bispectral index-guided management of anaesthesia in permanent vegetative state. Anaesthesia. 2002;57:1190–4.
152. Da Niedermeyer SEL, editor. The normal EEG of the waking adults In: Niedermeyer's electroencephalography: basic principles, clinical applications, and related fields. 4th edition. Lippincott Williams and Wilkins; 1999.
153. Steinlein O, Anokhin A, Yping M, Schalt E, Vogel F. Localization of a gene for the human low-voltage EEG on 20q and genetic heterogeneity. Genomics. 1992;12:69–73.
154. Schnider TW, Luginbühl M, Petersen-Felix S, Mathis J. Unreasonably low bispectral index values in a volunteer with genetically determined low-voltage electroencephalographic signal. Anesthesiology. 1998;89:1607–8.
155. Muncaster ARG, Sleigh JW, Williams M. Changes in consciousness, conceptual memory, and quantitative electroencephalographical measures during recovery from sevoflurane- and remifentanil-based anesthesia. Anesth Analg. 2003;96:720–5.
156. Hagihira S, Okitsu K, Kawaguchi M. Unusually low bispectral index values during emergence from anesthesia. Anesth Analg. 2004;98:1036–8.
157. Gunawardane PO, Murphy PA, Sleigh JW. Bispectral index monitoring during electroconvulsive therapy under propofol anaesthesia. Br J Anaesth. 2002;88:184–7.
158. Rampil IJ. A primer for EEG signal processing in anesthesia. Anesthesiology. 1998;89:980–1002.
159. Nishihara F, Saito S. Pre-ictal bispectral index has a positive correlation with seizure duration during electroconvulsive therapy. Anesth Analg. 2002;94:1249–52. table of contents
160. Small JG. Psychiatric disorders and the EEG. In: Electroencephalography: basic principles, clinical applications and related fields. 4th ed. Lippincott Williams and Wilkins; 1999.
161. Chinzei M, Sawamura S, Hayashida M, Kitamura T, Tamai H, Hanaoka K. Change in bispectral index during epileptiform electrical activity under sevoflurane anesthesia in a patient with epilepsy. Anesth Analg. 2004;98:1734–6.
162. Yli-Hankala A, Vakkuri A, Särkelä M, Lindgren L, Korttila K, Jäntti V. Epileptiform electroencephalogram during mask induction of anesthesia with sevoflurane. Anesthesiology. 1999;91:1596–603.
163. Franks NP, Lieb WR. Molecular and cellular mechanisms of general anaesthesia. Nature. 1994;367:607–14.
164. Murao K, Shingu K, Tsushima K, Takahira K, Ikeda S, Nakao S. The anticonvulsant effects of volatile anesthetics on lidocaine-induced seizures in cats. Anesth Analg. 2000;90:148–55.
165. Komatsu H, Taie S, Endo S, Fukuda K, Ueki M, Nogaya J, et al. Electrical seizures during sevoflurane anesthesia in two pediatric patients with epilepsy. Anesthesiology. 1994;81:1535–7.
166. Kaisti KK, Jääskeläinen SK, Rinne JO, Metsähonkala L, Scheinin H. Epileptiform discharges during 2 MAC sevoflurane anesthesia in two healthy volunteers. Anesthesiology. 1999;91:1952–5.
167. Iijima T, Nakamura Z, Iwao Y, Sankawa H. The epileptogenic properties of the volatile anesthetics sevoflurane and isoflurane in patients with epilepsy. Anesth Analg. 2000;91:989–95.
168. Watts AD, Herrick IA, McLachlan RS, Craen RA, Gelb AW. The effect of sevoflurane and isoflurane anesthesia on interictal spike activity among patients with refractory epilepsy. Anesth Analg. 1999;89:1275–81.
169. Tallach RE, Ball DR, Jefferson P. Monitoring seizures with the Bispectral index. Anaesthesia. 2004;59:1033–4.
170. Musialowicz T, Mervaala E, Kälviäinen R, Uusaro A, Ruokonen E, Parviainen I. Can BIS monitoring be used to assess the depth of propofol anesthesia in the treatment of refractory status epilepticus? Epilepsia. 2010;51:1580–6.
171. Sakuma H, Fukumizu M, Kohyama J. Efficacy of anticonvulsants on acute encephalitis with refractory, repetitive partial seizures (AERRPS). No To Hattatsu. 2001;33:385–90.
172. Dahaba AA, Liu DW, Metzler H. Bispectral index (BIS) monitoring of acute encephalitis with refractory, repetitive partial seizures (AERRPS). Minerva Anestesiol. 2010;76:298–301.
173. Wass CT, Grady RE, Fessler AJ, Cascino GD, Lozada L, Bechtle PS, et al. The effects of remifentanil on epileptiform discharges during intraoperative electrocorticography in patients undergoing epilepsy surgery. Epilepsia. 2001;42:1340–4.
174. Endo T, Sato K, Shamoto H, Yoshimoto T. Effects of sevoflurane on electrocorticography in patients

with intractable temporal lobe epilepsy. J Neurosurg Anesthesiol. 2002;14:59–62.
175. Asano E, Benedek K, Shah A, Juhász C, Shah J, Chugani DC, et al. Is intraoperative electrocorticography reliable in children with intractable neocortical epilepsy? Epilepsia. 2004;45:1091–9.
176. Ogawa S, Okutani R, Nakada K, Suehiro K, Shigemoto T. Spike-monitoring of anaesthesia for corpus callosotomy using bilateral bispectral index. Anaesthesia. 2009;64:776–80.
177. Kurita N, Kawaguchi M, Hoshida T, Nakase H, Sakaki T, Furuya H. The effects of sevoflurane and hyperventilation on electrocorticogram spike activity in patients with refractory epilepsy. Anesth Analg. 2005;101:517–23.
178. Soriano SG, Eldredge EA, Wang FK, Kull L, Madsen JR, Black PM, et al. The effect of propofol on intraoperative electrocorticography and cortical stimulation during awake craniotomies in children. Paediatr Anaesth. 2000;10:29–34.
179. Samra SK, Sneyd JR, Ross DA, Henry TR. Effects of propofol sedation on seizures and intracranially recorded epileptiform activity in patients with partial epilepsy. Anesthesiology. 1995;82:843–51.
180. Dahaba AA, Yin J, Xiao Z, Su J, Bornemann H, Dong H, et al. Different propofol-remifentanil or sevoflurane-remifentanil bispectral index levels for electrocorticographic spike identification during epilepsy surgery. Anesthesiology. 2013;119:582–92.
181. Loomis AL, HEN, HGA. Distribution of disturbance-patterns in the human electroencephalogram with special reference to sleep. J Neurophysiol. 1938;1:413–30.
182. Rechtschaffen Allan KA. A manual of standardized terminology, techniques and scoring system for sleep stages of human subjects. Washington DC: Public Health Service, US Government Printing Office; 1968.
183. Cash SS, Halgren E, Dehghani N, Rossetti AO, Thesen T, Wang C, et al. The human K-complex represents an isolated cortical down-state. Science. 2009;324:1084–7.
184. Roth M, Shaw J, Green J. The form voltage distribution and physiological significance of the K-complex. Electroencephalogr Clin Neurophysiol. 1956;8:385–402.
185. Si Y, Liu L, Li Q, Mu J, Tian L-Y, Chen J-N, et al. Features of the K-complex waves in refractory nocturnal frontal lobe epilepsy. Epilepsy Res. 2010;92:219–25.
186. Consensus conference. Perioperative red blood cell transfusion. JAMA. 1988;260:2700–3.
187. Monk TG, Goodnough LT, Brecher ME, Colberg JW, Andriole GL, Catalona WJ. A prospective randomized comparison of three blood conservation strategies for radical prostatectomy. Anesthesiology. 1999;91:24–33.
188. Napier JA, Bruce M, Chapman J, Duguid JK, Kelsey PR, Knowles SM, et al. Guidelines for autologous transfusion. II. Perioperative haemodilution and cell salvage. British Committee for Standards in Haematology Blood Transfusion Task Force. Autologous Transfusion Working Party. Br J Anaesth. 1997;78:768–71.
189. Dahaba AA, Rinnhofer S, Wang G, Xu X, Liu XY, Wu XM, et al. Influence of acute normovolaemic haemodilution on bispectral index monitoring and propofol dose requirements. Acta Anaesthesiol Scand. 2008;52:815–20.
190. Weiskopf RB, Kramer JH, Viele M, Neumann M, Feiner JR, Watson JJ, et al. Acute severe isovolemic anemia impairs cognitive function and memory in humans. Anesthesiology. 2000;92:1646–52.
191. Weiskopf RB, Feiner J, Hopf HW, Viele MK, Watson JJ, Kramer JH, et al. Oxygen reverses deficits of cognitive function and memory and increased heart rate induced by acute severe isovolemic anemia. Anesthesiology. 2002;96:871–7.
192. Wilkes NJ, Woolf R, Mutch M, Mallett SV, Peachey T, Stephens R, et al. The effects of balanced versus saline-based hetastarch and crystalloid solutions on acid-base and electrolyte status and gastric mucosal perfusion in elderly surgical patients. Anesth Analg. 2001;93:811–6.
193. Scheingraber S, Rehm M, Sehmisch C, Finsterer U. Rapid saline infusion produces hyperchloremic acidosis in patients undergoing gynecologic surgery. Anesthesiology. 1999;90:1265–70.
194. Ickx BE, Rigolet M, Van Der Linden PJ. Cardiovascular and metabolic response to acute normovolemic anemia. Effects of anesthesia. Anesthesiology. 2000;93:1011–6.
195. Lysakowski C, Dumont L, Pellegrini M, Clergue F, Tassonyi E. Effects of fentanyl, alfentanil, remifentanil and sufentanil on loss of consciousness and bispectral index during propofol induction of anaesthesia. Br J Anaesth. 2001;86:523–7.
196. Shin H-W, Ban Y-J, Lee H-W, Lim H-J, Yoon S-M, Chang S-H. Arousal with iv epinephrine depends on the depth of anesthesia. Can J Anaesth. 2004;51:880–5.
197. Andrzejowski J, Sleigh JW, Johnson IA, Sikiotis L. The effect of intravenous epinephrine on the bispectral index and sedation. Anaesthesia. 2000;55:761–3.
198. Ishiyama T, Oguchi T, Iijima T, Matsukawa T, Kashimoto S, Kumazawa T. Ephedrine, but not phenylephrine, increases bispectral index values during combined general and epidural anesthesia. Anesth Analg. 2003;97:780–4.
199. Steffey EP, Eger EI. The effect of seven vasopressors of halothane MAC in dogs. Br J Anaesth. 1975;47:435–8.
200. Guo TZ, Tinklenberg J, Oliker R, Maze M. Central alpha 1-adrenoceptor stimulation functionally antagonizes the hypnotic response to dexmedetomidine, an alpha 2-adrenoceptor agonist. Anesthesiology. 1991;75:252–6.

201. Matthews R. Isoproterenol-induced elevated bispectral indexes while undergoing radiofrequency ablation: a case report. AANA J. 2006;74:193–5.
202. Johansen JW. Esmolol promotes electroencephalographic burst suppression during propofol/alfentanil anesthesia. Anesth Analg. 2001;93:1526–31. table of contents
203. Sum CY, Yacobi A, Kartzinel R, Stampfli H, Davis CS, Lai CM. Kinetics of esmolol, an ultra-short-acting beta blocker, and of its major metabolite. Clin Pharmacol Ther. 1983;34:427–34.
204. Johansen JW, Schneider G, Windsor AM, Sebel PS. Esmolol potentiates reduction of minimum alveolar isoflurane concentration by alfentanil. Anesth Analg. 1998;87:671–6.
205. Johansen JW, Flaishon R, Sebel PS. Esmolol reduces anesthetic requirement for skin incision during propofol/nitrous oxide/morphine anesthesia. Anesthesiology. 1997;86:364–71.
206. Berkenstadt H, Loebstein R, Faibishenko I, Halkin H, Keidan I, Perel A. Effect of a single dose of esmolol on the bispectral index scale (BIS) during propofol/fentanyl anaesthesia. Br J Anaesth. 2002;89:509–11.
207. Menigaux C, Guignard B, Adam F, Sessler DI, Joly V, Chauvin M. Esmolol prevents movement and attenuates the BIS response to orotracheal intubation. Br J Anaesth. 2002;89:857–62.
208. Oda Y, Nishikawa K, Hase I, Asada A. The short-acting beta1-adrenoceptor antagonists esmolol and landiolol suppress the bispectral index response to tracheal intubation during sevoflurane anesthesia. Anesth Analg. 2005;100:733–7.
209. Coloma M, Chiu JW, White PF, Armbruster SC. The use of esmolol as an alternative to remifentanil during desflurane anesthesia for fast-track outpatient gynecologic laparoscopic surgery. Anesth Analg. 2001;92:352–7.
210. Sinha PK, Jacob D, Koshy T. Rapid decrease in bispectral index following a single intravenous injection of metoprolol. Anesth Analg. 2007;105:548–9.
211. Davidson AJ, Huang GH, Rebmann CS, Ellery C. Performance of entropy and Bispectral Index as measures of anaesthesia effect in children of different ages. Br J Anaesth. 2005;95:674–9.
212. Arnold G, Kluger M, Voss L, Sleigh J. BIS and Entropy in the elderly. Anaesthesia. 2007;62:907–12.
213. Dahaba AA, Xiao Z, Zhu X, Dong H, Xiong L, Rehak P, et al. Age progression from vicenarians (20-29 year) to nonagenarians (90-99 year) among a population pharmacokinetic/pharmacodynamic (PopPk-PD) covariate analysis of propofol-bispectral index (BIS) electroencephalography. J Pharmacokinet Pharmacodyn. 2020;47:145–61.
214. Zhang L, Beal SL, Sheiner LB. Simultaneous vs. sequential analysis for population PK/PD data I: best-case performance. J Pharmacokinet Pharmacodyn. 2003;30:387–404.
215. Mandema JW, Verotta D, Sheiner LB. Building population pharmacokinetic--pharmacodynamic models. I. Models for covariate effects. J Pharmacokinet Biopharm. 1992;20:511–528.
216. Dahaba AA, Xue JX, Xu GX, Liu QH, Metzler H. Bilateral Bispectral Index (BIS)-Vista as a measure of physiologic sleep in sleep-deprived anesthesiologists. Minerva Anestesiol. 2011;77:388–93.
217. Dahaba AA, Bornemann H, Hopfgartner E, Ohran M, Kocher K, Liebmann M, et al. Effect of sugammadex or neostigmine neuromuscular block reversal on bispectral index monitoring of propofol/remifentanil anaesthesia. Br J Anaesth. 2012;108:602–6.
218. Mathews DM, Kumaran KR, Neuman GG. Bispectral index-derived facial electromyography-guided fentanyl titration in the opiate-exposed patient. Anesth Analg. 2003;96:1062–4.
219. Schnider TW, Minto CF, Shafer SL, Gambus PL, Andresen C, Goodale DB, et al. The influence of age on propofol pharmacodynamics. Anesthesiology. 1999;90:1502–16.
220. Chernik DA, Gillings D, Laine H, Hendler J, Silver JM, Davidson AB, et al. Validity and reliability of the Observer's Assessment of Alertness/Sedation Scale: study with intravenous midazolam. J Clin Psychopharmacol. 1990;10:244–51.
221. Dahaba AA, Lischnig U, Kronthaler R, Bornemann H, Georgiev V, Rehak PH, et al. Bispectral-index-guided versus clinically guided remifentanil/propofol analgesia/sedation for interventional radiological procedures: an observer-blinded randomized study. Anesth Analg. 2006;103:378–84. table of contents
222. Dahaba AA, Perelman SI, Moskowitz DM, Bennett HL, Shander A, Xiao Z, et al. Geographic regional differences in rocuronium bromide dose-response relation and time course of action: an overlooked factor in determining recommended dosage. Anesthesiology. 2006;104:950–3.
223. Dahaba AA, von Klobucar F, Rehak PH, List WF. The neuromuscular transmission module versus the relaxometer mechanomyograph for neuromuscular block monitoring. Anesth Analg. 2002;94:591–6. table of contents
224. Dahaba AA, Zhong T, Lu HS, Bornemann H, Liebmann M, Wilfinger G, et al. Geographic differences in the target-controlled infusion estimated concentration of propofol: bispectral index response curves. Can J Anaesth. 2011;58:364–70.
225. Katoh T, Bito H, Sato S. Influence of age on hypnotic requirement, bispectral index, and 95% spectral edge frequency associated with sedation induced by sevoflurane. Anesthesiology. 2000;92:55–61.
226. Wang Y, Kikuchi T, Sakai M, Wu JL, Sato K, Okumura F. Age-related modifications of effects of ketamine and propofol on rat hippocampal acetylcholine release studied by in vivo brain microdialysis. Acta Anaesthesiol Scand. 2000;44:112–7.
227. Larsson JE, Wahlström G. The influence of age and administration rate on the brain sensitiv-

ity to propofol in rats. Acta Anaesthesiol Scand. 1998;42:987–94.
228. Schultz A, Grouven U, Zander I, Beger FA, Siedenberg M, Schultz B. Age-related effects in the EEG during propofol anaesthesia. Acta Anaesthesiol Scand. 2004;48:27–34.
229. Roubicek J. The electroencephalogram in the middle-aged and the elderly. J Am Geriatr Soc. 1977;25:145–52.
230. Lysakowski C, Elia N, Czarnetzki C, Dumont L, Haller G, Combescure C, et al. Bispectral and spectral entropy indices at propofol-induced loss of consciousness in young and elderly patients. Br J Anaesth. 2009;103:387–93.
231. Mazoit JX, Samii K. Binding of propofol to blood components: implications for pharmacokinetics and for pharmacodynamics. Br J Clin Pharmacol. 1999;47:35–42.
232. Simons PJ, Cockshott ID, Douglas EJ, Gordon EA, Hopkins K, Rowland M. Disposition in male volunteers of a subanaesthetic intravenous dose of an oil in water emulsion of 14C-propofol. Xenobiotica. 1988;18:429–40.
233. Eleveld DJ, Colin P, Absalom AR, Struys MMRF. Pharmacokinetic-pharmacodynamic model for propofol for broad application in anaesthesia and sedation. Br J Anaesth. 2018;120:942–59.
234. Donati F, Miller DR, Fiset P. Target-controlled infusion devices: are we missing much? Can J Anaesth. 2011;58:349–53.
235. Egan TD, Shafer SL. Target-controlled infusions for intravenous anesthetics: surfing USA not! Anesthesiology. 2003;99:1039–41.
236. Masui K, Upton RN, Doufas AG, Coetzee JF, Kazama T, Mortier EP, et al. The performance of compartmental and physiologically based recirculatory pharmacokinetic models for propofol: a comparison using bolus, continuous, and target-controlled infusion data. Anesth Analg. 2010;111:368–79.
237. White M, Schenkels MJ, Engbers FH, Vletter A, Burm AG, Bovill JG, et al. Effect-site modelling of propofol using auditory evoked potentials. Br J Anaesth. 1999;82:333–9.
238. Dahaba AA, Xiao Z, Zhu X, Oettl K, Dong H, Xiong L, et al. Location matters: Overlooked ethnic-geographic effect in China and Austria on propofol/cisatracurium sex differences among a population pharmacokinetic/pharmacodynamic (PopPK/PD) covariate analysis in men, women, and one transgender subject. Fundam Clin Pharmacol. 2022;36:182–98.
239. Ortolani O, Conti A, Chan YK, Sie MY, Ong GSY. Comparison of propofol consumption and recovery time in Caucasians from Italy, with Chinese, Malays and Indians from Malaysia. Anaesth Intensive Care. 2004;32:250–5.
240. Ortolani O, Conti A, Ngumi ZW, Texeira L, Olang P, Amani I, et al. Ethnic differences in propofol and fentanyl response: a comparison among Caucasians, Kenyan Africans and Brazilians. Eur J Anaesthesiol. 2004;21:314–9.
241. Milne SE, Troy A, Irwin MG, Kenny GNC. Relationship between bispectral index, auditory evoked potential index and effect-site EC50 for propofol at two clinical end-points. Br J Anaesth. 2003;90:127–31.
242. Irwin MG, Hui TWC, Milne SE, Kenny GNC. Propofol effective concentration 50 and its relationship to bispectral index. Anaesthesia. 2002;57:242–8.
243. Xu Z, Liu F, Yue Y, Ye T, Zhang B, Zuo M, et al. C50 for propofol-remifentanil target-controlled infusion and bispectral index at loss of consciousness and response to painful stimulus in Chinese patients: a multicenter clinical trial. Anesth Analg. 2009;108:478–83.
244. Marsh B, White M, Morton N, Kenny GN. Pharmacokinetic model driven infusion of propofol in children. Br J Anaesth. 1991;67:41–8.
245. Minto CF, Schnider TW, Short TG, Gregg KM, Gentilini A, Shafer SL. Response surface model for anesthetic drug interactions. Anesthesiology. 2000;92:1603–16.
246. Dahaba AA, Oettl K, von Klobucar F, Reibnegger G, List WF. End-stage renal failure reduces central clearance and prolongs the elimination half life of remifentanil. Can J Anaesth. 2002;49:369–74.
247. Katz RL, Norman J, Seed RF, Conrad L. A comparison of the effects of suxamethonium and tubocurarine in patients in London and New York. Br J Anaesth. 1969;41:1041–7.
248. Fiset P, Donati F, Balendran P, Meistelman C, Lira E, Bevan DR. Vecuronium is more potent in Montreal than in Paris. Can J Anaesth. 1991;38:717–21.
249. Rowaan CJ, Vandenbrom RH, Wierda JM. The Relaxometer: a complete and comprehensive computer-controlled neuromuscular transmission measurement system developed for clinical research on muscle relaxants. J Clin Monit. 1993;9:38–44.
250. Benet LZ, Hoener B. Changes in plasma protein binding have little clinical relevance. Clin Pharmacol Ther. 2002;71:115–21.
251. MacKichan JJ. Protein binding drug displacement interactions fact or fiction? Clin Pharmacokinet. 1989;16:65–73.
252. Hiraoka H, Yamamoto K, Okano N, Morita T, Goto F, Horiuchi R. Changes in drug plasma concentrations of an extensively bound and highly extracted drug, propofol, in response to altered plasma binding. Clin Pharmacol Ther. 2004;75:324–30.
253. Takizawa E, Hiraoka H, Takizawa D, Goto F. Changes in the effect of propofol in response to altered plasma protein binding during normothermic cardiopulmonary bypass. Br J Anaesth. 2006;96:179–85.
254. Zhou HH, Adedoyin A, Wilkinson GR. Differences in plasma binding of drugs between Caucasians and Chinese subjects. Clin Pharmacol Ther. 1990;48:10–7.

6 Effect of Electromyographic (EMG) Activity and Neuromuscular Blocking Agents (NMBAs) on Depth of Anesthesia Monitoring: Achilles First Heel

6.1 Effect of High Electromyographic (EMG) Activity on Depth of Anesthesia Monitoring

In their review article, Jensen and colleagues [1] assessed the present methods used for validating the Depth of Anesthesia (DOA) using the processed Electroencephalography-Depth Of Anesthesia (pEEG-DOA) monitors [1]. They reviewed the present methods used for validating DOA, introducing three concepts: the real DOA (DOA-real) is the state of hypnosis that the patient has in any given moment during the general anesthesia, the observed DOA (DOA-obs) is the subjective assessment of the anesthesiologist based on clinical signs, and the electronic indices DOA (DOA el-ind) is any estimation of the DOA given by an electronic device [1]. The three entities should, in the ideal situation, be identical. However, this is rarely the case. The correlation between the three entities can be affected by several factors such as the administration of analgesic agents, neuromuscular blocking agents (NMBAs), or the presence of facial muscular activity [1]. In conclusion, although several monitors and clinical scoring scales are available to assess the DOA during general anesthesia, care should be taken when interpreting their results [1]. Jensen and colleagues [1] clarified that both the Bispectral Index (BIS) and Narcotrend algorithms are based upon frequency range between 0.5 and 47 Hz used to compute the raw electroencephalography (EEG) hypnotic stages. However, Electromyographic (EMG) activity occurs at the 25–300 Hz range frequency [1]. This is a source of interference in the two monitors Narcotrend or Bispectral Index systems and the EMG muscular activity, as the frequencies generated by increased muscular activity and the frequencies originating from the EEG are overlapping in that specific frequency range [1]. The BIS-XP platform contains an EMG activity bar that indicates, in real time, high EMG muscle activity. The BIS-XP manufacturers added improvements in the software of the Bispectral Index in the BIS Quattro® electrode that has a newly added fourth EMG electrode indicating EMG activity and filtering EMG artifacts [1].

High EMG activities are artifact signals that occur within the frequency "range of interest" of the bispectrum, because EMG frequencies overlap the Bispectral Index algorithm's *Relative BetaRatio* in the 30–47 Hz range [2]. Although the EMG power spectrum is quite small in relation to the total EEG power spectrum (0.5%) [2], Sleigh and colleagues [3] in an interesting study estimated the electromyography $EMG_{30–47\ Hz}$ portion to be one-tenth the magnitude of the electroencephalography $EEG_{30–47\ Hz}$ signal in the Bispectral Index BIS A-1000 (version 3.12) algorithm [3]. Contamination of pEEG-DOA indices in general and Bispectral Index in particular by EMG activity is therefore "inherent" in the

A. A. Dahaba, *Understanding the Numbers in Depth of Anesthesia Monitors*,
https://doi.org/10.1007/978-3-031-88260-9_6

calculation of the Bispectral Index value, because frontal Electromyography (fEMG) frequencies could simulate EEG frequencies, in particular the 30–47 Hz component of the *Relative BetaRatio* descriptor that is typically associated with awake or a light level of anesthesia [4]. Thus, EMG contamination [5], during deep anesthesia in patients who generate little EEG power in the 30–47 Hz range, may falsely mimic the EEG of awake subjects; this will make patients under deep anesthesia appear more awake than they really are that would be misinterpreted by the pEEG-DOA algorithms as EEG activity and assigned a spuriously elevated Bispectral Index value [3]. The administration of NMBAs, in that case, would then decrease the Bispectral Index value by alleviating the artifact of high EMG activity from muscles lying near the electrodes and "unveiling" the true calculated Bispectral Index value [6].

The most common artifactual interference with the pEEG-DOA indices signal acquisition is from fEMG [6]. Because most pEEG-DOA sensors are placed on the frontal region, the temporal and facial spontaneous high-frequency muscular activities will frequently compromise the EEG signal acquisition and signal processing and artifactually change the pEEG-DOA indices such as the Bispectral Index, Narcotrend, Index of Consciousness (IoC), or Entropy [7]. As a matter of fact, when NMBAs are not used, the fEMG could also serve as a DOA indicator per se, as the fEMG is selectively processed and displayed in almost all currently commercially available DOA devices. Whatever the nature of the pEEG-DOA device used, anesthesiologists must expect a certain degree of fEMG biased effect in their displayed pEEG-DOA indices. Assessing the Bispectral Index value when NMBAs are not used, it is essential for anesthesiologists to take into consideration and always keep in mind the effect of fEMG on the Bispectral Index value before deciding on DOA change [8, 9].

Because the great majority of pEEG-DOA monitors are reliant, in part, on EMG for their calculations, administration of NMBAs would then decrease their values [6]. The $EMG_{32\text{–}47\ Hz}$ signal overlaps Entropy's indices Response Entropy (RE) $EEG_{0.8\text{–}47\ Hz}$ but not State Entropy (SE) $EEG_{0.8\text{–}32\ Hz}$ [10]. The Entropy algorithm uses spectral analysis to produce two main Entropy parameters for assessing depth of anesthesia: the SE indicates the depth of hypnosis with an index scale from 0 to 91. The RE indicates nociception/responsiveness to stimuli (derived from the fEMG) with an index scale from 0 to 100. When the difference between RE and SE widens, this indicates inadequate anesthesia.

Bruhn and colleagues [6] reported two case reports of an anomalous Bispectral Index readings. In the first case, when propofol target controlled infusion (TCI) increased from 1 to 3 μg ml^{-1}, paradoxically the Bispectral Index BIS A-1000 (version 3.22) value increased from 54 to 84 [6]. This was attributed to the fact that drug-induced EMG muscle rigidity resulted in the Bispectral Index misclassifying the depth of the propofol anesthetic [6].

Baulig and colleagues [11], using the Bispectral Index and Entropy S/5 GE modules (General Electric GE Healthcare, Chicago, Illinois, USA), compared the Entropy indices SE and RE with Bispectral Index in 30 patients undergoing coronary artery bypass graft surgery [11]. A strong correlation $r^2 = 0.62$ existed between fEMG and Bispectral Index, in contrast to a moderate $r^2 = 0.45$ correlation between fEMG and RE a weak correlation and between fEMG $r^2 = 0.39$ and SE [11]. They demonstrated RE and SE showed significantly less interference from fEMG and superior resistance against artifacts [11].

Wennervirta and colleagues [12], using the Bispectral Index (version 4.0) and Entropy S/5 GE modules, compared the Entropy indices SE and RE with Bispectral Index in 16 brain-dead organ donors [12]. After excluding the results from the four subjects with persistent residual EEG activity, SE, RE, and Bispectral Index values differed from 0 in 17%, 18%, and 62% of the recorded time, respectively [12]. Both Entropy and Bispectral Index showed nonzero values due to artifacts after brain death diagnosis. The Bispectral Index was more liable to artifacts than Entropy. The authors demonstrated that, compared with Bispectral Index, the Entropy indices are more resistant to fEMG artifacts [11, 12].

Table 6.1 Effect of high Electromyography (EMG) activity on depth of anesthesia monitoring

Reference	Monitor	*n*	Change	Index change	Explanation
Bruhn et al. [6]	BIS A-1000 (3.22)	1	EMG ↑	BIS 54 ↑ BIS 84	EMG ↑ muscle rigidity
Baulig et al. [11]	BIS (S/5), RE, SE	30	EMG ↑	BIS ↑, RE, SE less increase	RE, SE less EMG effect
Wennervirta et al. [12]	BIS (S/5), RE, SE	12	Brain dead EMG ↑	BIS 62%, RE 18%, SE 17% of recorded time	RE, SE less EMG effect

n number of patients per group, *EMG* electromyography, *BIS* Bispectral Index, *SE* State Entropy, *RE* Response Entropy

Table 6.1 enumerates and describes the effect of high Electromyography (EMG) activity on depth of anesthesia monitoring.

6.2 Effect of Neuromuscular Blocking Agents (NMBAs) on Depth of Anesthesia Monitoring

It appears that, in adult patients, Neuromuscular Blocking Agents (NMBAs) may decrease the processed Electroencephalography-Depth Of Anesthesia (pEEG-DOA) indices. Conversely, Weber and colleagues [13] assessed in a prospective randomized trial in 40 pediatric patients undergoing elective minor surgical procedures under propofol general anesthesia adjusted to maintain Bispectral Index values between 45 and 55, the effect of NMBA administration on the Bispectral Index and the Cerebral State Index (CSI). Patients were randomly allocated to receive either 0.25 mg kg^{-1} mivacurium or NaCl 0.9% [13]. Pediatric BIS sensors were placed on patients' forehead and attached to a BIS-XP monitor, and silver-silver chloride electrodes (Medicotest, Olstykke, Denmark) were placed on patients' forehead and attached to a Cerebral State Monitor (Danmeter, Odense, Denmark). There were no significant differences between changes in Bispectral Index and CSI values following mivacurium administration. Mean ± SD baseline Bispectral Index values (48.0 ± 4.8) and baseline CSI (54.0 ± 7.7) values did not decrease, rather conversely significantly increased 5 min following mivacurium administration to Bispectral Index value of 53.0 ± 6.6, but not CSI value of 57.4 ± 7.1 [13]. Their study results showed that in pediatric patients under propofol anesthesia, mivacurium does not induce a decline in pEEG-DOA indices rather a significant increase in Bispectral Index but not CSI values that appear 5 min following mivacurium administration as a delayed effect [13].

Bruhn and colleagues [6], in the second case, report that the administration of 2 bolus doses each of vecuronium 1 mg NMBA decreased the Bispectral Index BIS A-1000 (version 3.22) value from 85 to 58 despite constant propofol and remifentanil Target Controlled Infusions (TCIs), as vecuronium NMBA administration alleviated the EMG artifact, hence "unveiling" the truly calculated Bispectral Index value [6].

However, it seems that not all pEEG-DOA monitors are susceptible to high Electromyography (EMG) activity. This was demonstrated by Panousis and colleagues [14], as they enrolled 33 patients undergoing major urological procedures under combined thoracic epidural analgesia and deep surgical anesthesia maintained at Bispectral Index BIS-XP 40–55 [14]. Patients were simultaneously assessed using the Bispectral index BIS-XP and Narcotrend monitors, using BIS Quatro and silver-silver chloride Narcotrend electrodes (Medicotest, Olstykke, Denmark). Authors investigated the susceptibility of the Bispectral Index and Narcotrend monitoring to EMG activity. Narcotrend is a pEEG-DOA monitor [14]. In their study, the Bispectral Index, but not Narcotrend, values significantly correlated with EMG activity. Within mean ± SD 40 ± 10 minutes from 0.5 mg kg^{-1} rocuronium administration, Bispectral Index values increased in 16/33 (48.4%) of their study patients to >70 with high >35 dB EMG activity; the Narcotrend and the clinical signs of DOA remained unchanged during the period of elevated Bispectral Index values [14]. None of the patients reported intraoperative awareness. They concluded that under combined anesthesia, the Narcotrend monitor is more reliable when compared with the Bispectral Index regarding susceptibility to increased EMG activity [14]. However, the results of Panousis and colleagues' [14] investigation did not confirm the expectations of these EMG improvements [14]. On the other hand, none of the Narcotrend versions contain an indicator of EMG activity [14]. Considering that the Narcotrend algorithm includes frequencies in the 0.5–47 Hz EMG activity range, one would assume that increased EMG activity would also interfere with the Narcotrend EEG recordings and thereby affect the accuracy of the Narcotrend values [14]. However, in the previous study, Narcotrend patients' recordings, Panousis and colleagues [14] did not observe any increase in the Narcotrend values like the increase in the Bispectral Index values with enhanced EMG activity. The detailed algorithms of both monitor systems have not been published [14]. Regarding the Bispectral Index monitor, the possible explanation of the sudden and/or periodical increases of the Bispectral Index values in the presence of high EMG activity could be due to the overlapping frequencies of EMG and the Bispectral Index EEG [14]. Although the Narcotrend algorithm and EMG activity "share" the same EMG frequency band, increased EMG activity did not lead to any change in the Narcotrend values, possibly because the Narcotrend algorithm does not process EMG activity [14].

Vivien and colleagues [4] investigated 45 critically ill patients under midazolam sufentanil sedation to achieve a Sedation-Agitation Scale (SAS) equal to 1, who required administration of NMBA. The administration of NMBAs caused a significant $P < 0.0001$ decline in the Bispectral Index BIS A-2000 (version 2.10) value from 67 ± 19 to 43 ± 10 and significant $P < 0.001$ decline in the BIS-EMG (dB) from 36.8 ± 9 to 27.5 ± 3.3 in 32/45 (71%) lightly sedated patients with high EMG activity [4]. Furthermore, the Bispectral Index did not significantly change in 13/45 (29%) patients deeply sedated who showed minimal EMG activity. They reported similar results when 16/45 of their study patients were simultaneously studied with the BIS-XP (version 3.12) as NMBAs administration caused a significant decline in the Bispectral Index BIS-XP value from 62 ± 16 to 42 ± 12 and significant $P < 0.001$ decline in the BIS-EMG (dB) from 35.3 ± 8.3 to 26.2 ± 1.8 [4].

Doufas and colleagues [15] investigated 24 volunteers who sequentially received in a randomized order the NMBAs mivacurium, succinylcholine, or saline, as they assessed the simultaneous effect of neuromuscular block and the effect of a strong noxious stimulation on Bispectral Index BIS A-2000 (version 3.21) cortical activation under near-Minimal Alveolar Concentration desflurane general anesthesia. They found that noxious stimulation increased the Bispectral Index after succinylcholine or mivacurium neuromuscular block as well as after the control "saline" administration [15]. Before the application of a strong noxious stimulus, succinylcholine resulted in a small but significant increase in the Bispectral Index value that was not correlated to EMG activity and was totally independent of noxious stimulation or EMG of the Bispectral Index monitor [15].

In a prospective, randomized, double-blind study by Liu and colleagues [10], 39 patients received remifentanil and propofol TCI titrated till loss of consciousness (LOC). Two minutes after LOC, 19 patients in the study group received 0.5 mg kg^{-1} atracurium, while the 20 patients in the control group received a placebo [10]. At LOC and 2 min following LOC Bispectral Index A-2000 BIS-XP (version 3.11), Spectral Edge Frequency (SEF), EMG, Response Entropy (RE), and State Entropy (SE) were recorded. Bispectral Index, SEF, EMG, RE, and SE values before atracurium administration (78 ± 8, 22 ± 4, 39 ± 9, 80 ± 15, and 74 ± 13, respectively) significantly decreased following atracurium administration in Bispectral Index, SEF, EMG, and RE (58 ± 12, 18 ± 4, 27 ± 2, 65 ± 12, respectively) but not SE (62 ± 12) [10]. Declines in the BIS, EMG, and RE values were significantly larger in the atracurium group than in the placebo group. They concluded that in lightly anesthetized patients, NMBA administration decreases Bispectral Index, SEF, EMG, and RE, but not SE values [10].

Table 6.2 Effect of Neuromuscular Blocking Agents (NMBAs) on depth of anesthesia monitoring

Reference	Monitor	*n*	Change	Index change	Explanation
Weber et al. [13]	BIS-XP, CSI	40	mivacurium 5-min	5-min BIS ↑, CSI no change	CSI no change
Bruhn et al. [6]	BIS A-1000 (3.22)	1	vecuronium EMG ↓	BIS 85 BIS 54 ↓	Alleviating EMG artifact
Panousis et al. [14]	BIS-XP, Narcotrend	16/33	rocuronium recovery EMG ↑	BIS >70 ↑, Narcotrend no change	EMG not in Narcotrend algorithm
Vivien et al. [4]	BIS A-2000 (2.1)	32/45	NMBAs EMG ↓	BIS 67 BIS 43 ↓	Alleviating EMG artifact
Doufas et al. [15]	BIS A-2000 (3.21)	24	succinylcholine, mivacurium	succinylcholine BIS ↑	succinylcholine Afferentation Theory
Liu et al. [10]	BIS A-2000 (3.11), RE, SE	19 + 20	atracurium EMG ↓	BIS 54 ↓, RE 65 ↓	No change in SE 63

n number of patients per group, *EMG* electromyography, *BIS* Bispectral Index, *SE* State Entropy, *RE* Response Entropy, *CSI* Cerebral State Index, *NMBAs* neuromuscular blocking agents

Table 6.2 enumerates and describes the effect of Neuromuscular Blocking Agents (NMBAs) on depth of anesthesia monitoring.

6.3 Effect of Sugammadex Neuromuscular Block Reversal on Depth of Anesthesia Monitoring

Sugammadex, a modified γ-cyclodextrin, possesses a novel mechanism of action for the reversal of the steroidal Neuromuscular Blocking Agents (NMBAs) vecuronium or rocuronium [16]. The Bispectral Index, an Electroencephalography (EEG)-derived parameter, shown to be susceptible to frontal Electromyography (fEMG) artifacts [2]. Entropy, another processed Electroencephalography-Depth Of Anesthesia (pEEG-DOA) monitor, exhibited similar results with neuromuscular blockade, while lacking improvements in EMG artifacts filtering when compared with Bispectral Index monitoring [17]. Two prospective, randomized, double-blind studies clearly demonstrated that, under light propofol-remifentanil anesthesia, the induced high EMG activity reappearance in the biosignal caused a rise in Bispectral Index or Entropy following sugammadex or neostigmine neuromuscular blockade reversal, which was only noticeable in patients with preexisting high EMG activity [16, 17].

Illman and colleagues [18] investigated, in 29 patients undergoing oto-rhino-laryngological surgery, the effect of intermediate level of rocuronium neuromuscular block on propofol 4 μg ml^{-1}/remifentanil 2 ηg ml^{-1} Target Controlled Infusions (TCIs) pEEG-DOA monitoring with Bispectral Index, Response Entropy (RE), and State Entropy (SE) [18]. They also evaluated the effect of reversal of rocuronium neuromuscular block with sugammadex on Bispectral Index and Entropy, two pEEG-DOA indices. After surgery, while keeping propofol and remifentanil TCI unchanged, neuromuscular block was antagonized via sugammadex 2 mg kg^{-1}. Bispectral Index and Entropy indices stayed unchanged despite the neuromuscular block reversal as the mean ± SD Bispectral Index, RE, and SE values before neuromuscular block reversal 31.7 ± 9.9, 36.8 ± 13.3, and 35.3 ± 12.9 were not significantly changed 32.0 ± 11.9, 38.4 ± 18.0, and 36.3 ± 15.9 after sugammadex administration [18]. In summary, it seems that based upon published material as well as anesthesiologists' clinical experience, the EMG artifactual alteration of the pEEG-DOA values greatly depends on both the depth of anesthesia and the noxious stimuli [9]. Thus, anesthesiologists should always keep in mind that high EMG activity could always erroneously increase the pEEG-DOA indices [9].

Sugammadex only reverses the neuromuscular block of amino-steroidal NMBAs and does not reverse the neuromuscular block of benzylisoquinoline NMBAs. Fassoulaki and colleagues [19], in an interesting study with 60 patients, investigated the effect of three doses of sugammadex alone on Depth of Anesthesia (DOA) in patients who did not receive amino-steroidal NMBAs. They demonstrated that sugammadex in three randomly allocated doses 2 mg kg^{-1}, 4 mg kg^{-1}, or 16 mg kg^{-1} (n = 20) has no effect on the Bispectral Index or SE values when given to patients receiving cisatracurium 0.2 mg kg^{-1} benzylisoquinoline NMBA for neuromuscular blockade [19], under anesthesia maintained with nitrous oxide (N_2O) and sevoflurane adjusted to maintain Entropy and Bispectral Index (BIS) A-2000 values between 40 and 50. There was no significant difference $P = 0.812$ in the overall SE values, the primary outcome of the study of 44 ± 11, 43 ± 10, and 43 ± 11 in patients who received sugammadex 2 mg kg^{-1}, 4 mg kg^{-1}, or 16 mg kg^{-1}, respectively. Their results underline the likelihood that high EMG activity does indeed contribute to ≈20% of the Bispectral Index value in the higher ranges [19]. Sugammadex alone at low, medium, or high clinical doses has no effect on DOA as assessed by State Entropy and Bispectral Index. This is important to remember when dealing with patients at emergence from anesthesia whose neuromuscular block is inadequately reversed, as probably these patients with unrecognized residual postoperative curarization will spuriously appear completely awake because of their high EMG activity [20].

It seems that most pEEG-DOA indices are susceptible to high EMG activity artifactual interference. Christ and colleagues [21] investigated the influence of 0.07 mg kg^{-1} rocuronium-induced neuromuscular block and the effect of 2 mg kg^{-1} sugammadex reversal on Bispectral Index (BIS) A-2000 (version 3.31) and NeuroSENSE (NS-701, version 1.12.0 NeuroWave Systems Inc., Ohio, USA) in 27 patients undergoing elective surgical procedures under propofol 2.4 μg ml^{-1} remifentanil 2.0 μg ml^{-1} TCI [21]. With rocuronium-induced neuromuscular block, both indices significantly $P < 0.0001$ declined as Bispectral Index declined from 60.1 ± 6.8 to 51.7 ± 8.9 and NeuroSENSE declined from 61.4 ± 8.8 to 53.9 ± 7.4. The Bispectral Index 42.1 ± 8.9 nor NeuroSENSE 45.3 ± 7.4 before reversal of neuromuscular block did not significantly change after sugammadex 2 mg kg^{-1} administration and remained at Bispectral Index 43.6 ± 8.4 or NeuroSENSE 47.6 ± 11.5 after reversal of neuromuscular block [21]. In their study, the NeuroSENSE index was equally susceptible to EMG influence as the Bispectral Index. During stable, but light level of anesthesia, neuromuscular blockade resulted in a decrease in the Bispectral Index and NeuroSENSE values. Thus, the EMG signal may similarly be misinterpreted by the NeuroSENSE monitor as the BIS monitor would [21], although, due to the NeuroSENSE algorithm that is based on the analysis of wavelets, the NeuroSENSE monitor should have been more resistant to EMG activity than the Bispectral Index, which proved not to be the case [22], as it appears that the WAV_{CNS} analysis is also altered by the presence of EMG activity [21].

In the first study, Aho and colleagues [17] assessed 30 patients undergoing gynecological procedures under propofol remifentanil TCIs and rocuronium neuromuscular block. After termination of surgery, patients were randomly allocated for neuromuscular block reversal using either sugammadex or neostigmine. Using the Bispectral Index and Entropy S/5 GE modules (General Electric GE Healthcare, Chicago, Illinois, USA), the authors recorded the Bispectral Index values, the Electromyography of the Bispectral Index monitor (BIS-EMG), and Entropy indices for 5 minutes, as well as the biosignal recorded from the Entropy module [17]. Both time and frequency domain of the Entropy biosignal was studied offline to see if neuromuscular block reversal causes EEG changes [17]. The Entropy algorithm uses spectral analysis to produce two main Entropy parameters for assessing depth of anesthesia: the SE indicates the depth of hypnosis with an index scale from 0 to 91. The RE indicates nociception/responsiveness to stimuli (derived from the fEMG with an index scale from 0 to 100). When the difference between RE and SE widens, this indicates inadequate anesthesia. The EMG artifactual EEG signal contamination results in an increase in both Entropy and Bispectral Index with rocuronium neuromuscular block reversal under light propofol remifentanil anesthesia. The authors demonstrated that there was a statistically significant strong increase from the baseline Bispectral Index, BIS-EMG (dB), SE, and RE indices (45 ± 12, 34 ± 8, 41 ± 8, and 50 ± 10) to (85 ± 5, 53 ± 4, 87 ± 3, and 99 ± 1), respectively, in five patients following sugammadex administration and five patients following neostigmine administration. There was a weak increase response in four patients following sugammadex administration and one patient following neostigmine administration. There was no response in six patients following sugammadex administration and no response in nine patients following neostigmine administration. The authors attributed this phenomenon to an increase in EMG activity, even though sugammadex or neostigmine administrations appear to have just minimal EEG effects [17].

In the second study [16], my research team Dahaba and colleagues at the Department of Anesthesiology and Intensive Care Medicine, Medical University of Graz, Austria, compared the Bispectral Index values before and after neostigmine or sugammadex reversal of neuromuscular block in patients with or without high EMG activity. According to manufacturer's guidelines, we mounted a Bispectral Index BIS Quatro sensor on patients' forehead that we connected to a Bispectral Index BIS-Vista monitor. The raw EEG signals were band-pass filtered to 2–70 Hz band and was processed in real time using Bispectral Index algorithm version 1.4 for bilateral Bispectral Index sensors which is the Bispectral Index algorithm version 4.1 for the single-sided BIS Quatro sensor [16]. We started Bispectral Index recording after we verified Signal Quality Index (SQI) of >95% and electrodes impedance of <5 kΩ. In the once every 5 s mode, we digitally collected all EEG variables that we stored on a laptop computer. In the off-line analysis, we identified and eliminated Bispectral Index values that showed artifacts [16].

The Bispectral Index monitor displays fEMG logarithmic transformation of the 70–110 Hz frequency band in decibels (dB) units relative to 0.01 μV [16]. Although 30 dB is the lowest displayed dB value by the Bispectral Index monitor, the lowest dB value measured and recorded by the Bispectral Index monitor to be retrieved in the off-line mode is 0 dB [16]. To discriminate patients with no EMG, we elected the cutoff value of 35 dB to clearly indicate adequate fEMG suppression when EMG values were consistently <35 dB. We considered patients as exhibiting high EMG when their fEMG values were constantly ≥35 dB [16].

Because my research team did not find previously published studies for an *a*-priori sample size assessment, using our first 10 consecutive patients interim sample size assessment, whose data are included in the final analysis, in which mean ± SD Bispectral Index value of 48.7 ± 4.4 increased following sugammadex administration to 54.9 ± 5.7, our (α =0.05) *t*-test interim power analysis revealed that we need a group size of 11 patients to reveal a statistically significant difference with 80% power [16]. We increased the sample size to 24 patients, to accommodate dropout patients and to ensure adequate number of patients with or without EMG activity. We randomly allocated 48 patients to receive neostigmine 50 μg kg^{-1}/glycopyrrolate 10 μg kg^{-1} or sugammadex 4 mg kg^{-1} [16].

We evaluated neuromuscular block at the adductor pollicis muscle using the Relaxometer Mechanomyograph (Groningen University, The Netherlands), the gold standard for precisely quantifying neuromuscular block [23]. We attached the Relaxometer's force-transducer to the patients' thumb, and using surface electrodes, we supramaximally stimulated the ulnar nerve at the wrist with rectangular wave of pulse width 200 μs, with Train-Of-Four (TOF) stimuli (2 Hz for 2 sec) at 12 s intervals [16]. We used the first twitch of the TOF (T_1) expressed as percentage of control response and the TOF ratio (T_4:T_1) to evaluate neuromuscular block. We digitally collected and stored the T_1% and TOF ratio of all patients on a laptop computer for the duration of our study [16].

The only premedication we gave 1 hour preoperatively was oral midazolam 7.5 mg [16]. For induction, after entering patients' anthropometric data, we started propofol 4 μg ml^{-1} TCI using Diprifusor™ infusion pump (AstraZeneca Pharmaceuticals, Macclesfield, United Kingdom) [16]. We started remifentanil 0.1–0.3 μg kg^{-1} min^{-1} infusion, and we adjusted propofol ±0.2 μg ml^{-1} TCI rate to maintain a stable Bispectral Index value of 40–60 [16].

We administered rocuronium 600 μg kg^{-1} for tracheal intubation. We mechanically ventilated the lungs with 40% oxygen (O_2) in air, adjusted to maintain end-tidal carbon dioxide at 30–40 mm Hg [16]. We warmed our patients using a Forced-Hot-Air-Blanket to maintain core temperature at >36 °C and skin temperature at >32 °C. We closely observed our study patients for clinical signs of recovery such as spontaneous breathing, eyes opening, or spontaneous movement [16].

After termination of surgical procedures and all noxious stimulation, we maintained and recorded remifentanil and propofol infusion rates for a stabilization period of 10 min [16]. For reversal of neuromuscular block, our study patients randomly received either neostigmine 50 μg kg^{-1}/glycopyrrolate 10 μg kg^{-1} or sugammadex 4 mg kg^{-1} when patients regained 1–2 TOF responses as patients were allowed to recover from neuromuscular block up to 1.0 TOF ratio [16].

Five min following sugammadex administration, mean ± SD Bispectral Index value of 50.1 ± 10.3 significantly increased $P = 0.018$ to 61.7 ± 7.9 in 11 patients with reappearance of high EMG activity [16]. Conversely, the Bispectral Index value of 49.3 ± 4.9 remained unchanged at 51.9 ± 5.4 in 13 patients who did not have EMG activity. On the other hand, 15 min following neostigmine administration, mean Bispectral Index value of 51.9 ± 8.1 significantly increased $P = 0.007$ to 63.9 ± 8.1 in 13 patients with reappearance of high EMG activity [16]. Whereas in 11 patients who had no EMG activity, Bispectral Index value of 52.3 ± 7.4 remained unchanged at 53.3 ± 6.8 (Fig. 6.1) [16].

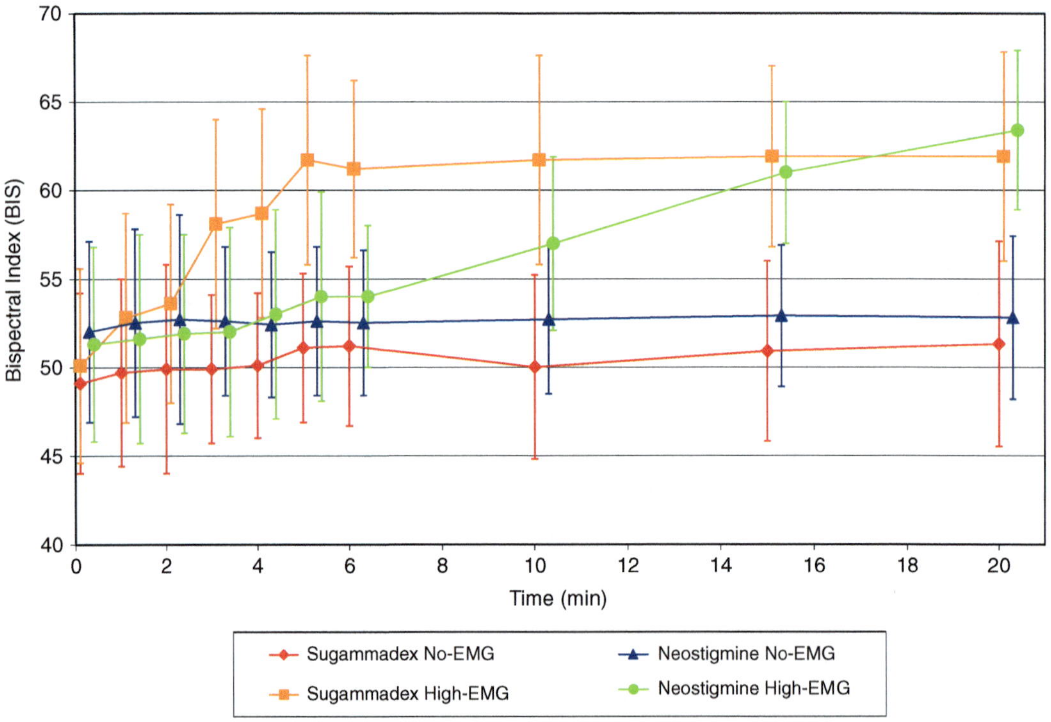

Fig. 6.1 Mean ± SD Bispectral Index (BIS) following sugammadex and neostigmine neuromuscular block reversal in the no EMG activity and in the high EMG activity patients. Reproduced with permission # 5906490423983 from Dahaba et al. Effect of sugammadex or neostigmine neuromuscular block reversal on bispectral index monitoring of propofol/remifentanil anaesthesia. Br J Anaesth. 2012;108:602–6

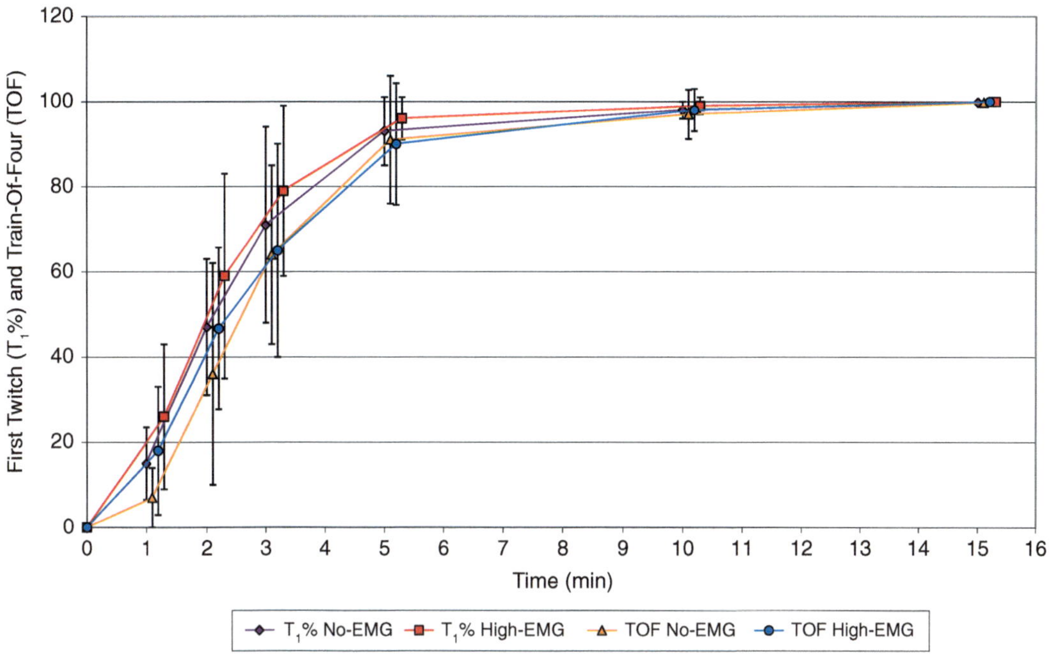

Fig. 6.2 Mean ± SD first twitch (T_1%) of the Train-Of-Four (TOF) ratio following sugammadex administration in the no EMG and high EMG activity patients

We clearly demonstrated in our study [16] that the reversal of neuromuscular block with sugammadex or neostigmine resulted in an increase in Bispectral Index values dependent on the presence of high EMG activity [16]. Our results revealed that following sugammadex administration, mean ± SD time to reach 0.8 TOF ratio of 3.1 ± 0.4 min was significantly shorter than time to reach 0.8 TOF ratio of 15.4 ± 5.6 min following neostigmine administration [16]. Five min following sugammadex administration, mean ± SD T_1% and TOF ratio increased to >95% T_1 and 0.9 TOF ratio. In all patients, full reversal of the neuromuscular blockade to 1.0 TOF ratio was subsequently reached (Fig. 6.2) [16].

In our study, we demonstrated that the reappearance of muscular activities because of rocuronium's neuromuscular block reversal erroneously increased the Bispectral Index value [16]. We recommend that this should be taken into consideration when using the Bispectral Index value during recovery from propofol remifentanil anesthesia. After neuromuscular block recovery and termination of anesthesia, we extubated our patients, and we transferred them to the Post-Anesthesia Care Unit (PACU) [16].

Table 6.3 enumerates and describes the effect of sugammadex neuromuscular block reversal on depth of anesthesia monitoring.

6.3.1 The Facial Affective Consciousness Electromyograph (FACE™) Monitor

Historically, my research group investigated in the 1990s the "FACE™" (Patient Comfort Inc. LLC, 293 Greenwood Avenue Florham Park, NJ 07932, USA), at that time a novel monitoring method based on the relative electromyographic (EMG) intensity of four muscles of the face. The basic theory of my research group was that the relative activation of muscles responsible for "smiling" and "frowning" indicates whether patients' anesthesia or antinociception is sufficient or not. The electrodes for

Table 6.3 Effect of sugammadex neuromuscular block reversal on depth of anesthesia monitoring

Reference	Monitor	*n*	Change	Index change	Explanation
Illman et al. [18]	BIS, RE, SE	19 + 20	rocuronium reversal	BIS, RE, SE no change	sugammadex no effect
Fassoulaki et al. [19]	BIS A-2000, SE	60	cisatracurium reversal sugammadex 2, 4, 16 mg kg^{-1}	BIS, SE no change	sugammadex no effect
Christ et al. [21]	BIS A-2000 (3.31), NeuroSense	27	rocuronium reversal	BIS, NeuroSense no change	sugammadex no effect
Aho et al. [17]	BIS (S/5), RE, SE	15 + 15	rocuronium reversal EMG ↑	BIS ↑, EMG ↑, RE ↑, SE ↑	Only in 5 EMG ↑ patients
Dahaba et al. [16]	BIS Vista	24 + 24	rocuronium reversal EMG ↑	BIS 50 BIS 62 ↑	Only in 11 EMG ↑ patients

n number of patients per group, *EMG* electromyography, *BIS* Bispectral Index, *SE* State Entropy, *RE* Response Entropy

the FACE™ system were designed to be specifically utilized for the facial monitoring application. The arrangement of the electrodes is such that they specifically record from the muscles values that are used in the proprietary algorithms to aid in determining the SUM and RATIO algorithms. This assures that the pattern of muscle activity utilized for analytic purposes will always come from the frontalis, corrugator, zygomaticus, and orbicularis oculi, the four facial muscles. The FACE™ system includes a four-channel EMG, an integrated laptop computer running the FACE™ software, amplifiers, and initial signal processors integrated in the FACE box and connected to a four-cable harness ending with two electrode sensors to be placed at the patient face as shown in Fig. 6.3. The front panel of the FACE™ contains the four patient cable connectors and an Aux connector for the patient cable labeled "Ch 1," "Ch 2," "Ch 3," and "Ch 4″ for data collection from the four facial muscle electrodes (Fig. 6.4). Once the FACE™ software is running, the available information includes the current values for each muscle area being monitored and the current values of the SUM, RATIO1, and RATIO2 algorithms (Fig. 6.5). Each muscle area is monitored, and data is collected ten times per second and converted into digital values. A file on the hard disk stores data every 5 s as the average values for each of the four muscles and the SUM and RATIO1 and RATIO2 algorithms.

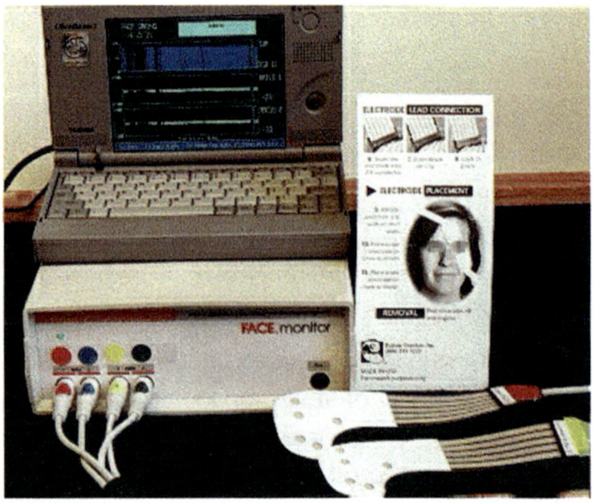

Fig. 6.3 The Facial Affective Consciousness Electromyograph (FACE™) system includes a four channel electromyograph (EMG), an integrated laptop computer running the FACE™ software, amplifiers, and processors in the FACE box which are connected to a four-cable harness and two electrode sensors to be placed at the patient face as shown

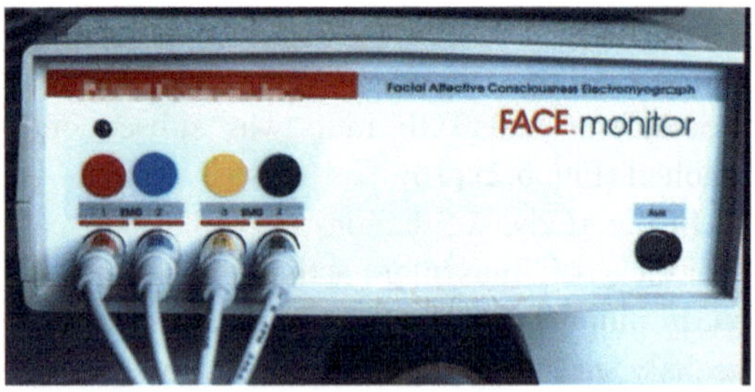

Fig. 6.4 The front panel of the FACE™ contains the four-patient cable connectors and an Aux connector for the patient cable labeled "Ch 1," "Ch 2," "Ch 3," and "Ch 4" for data collection from the four facial muscle electrodes. (Patient Comfort Inc.)

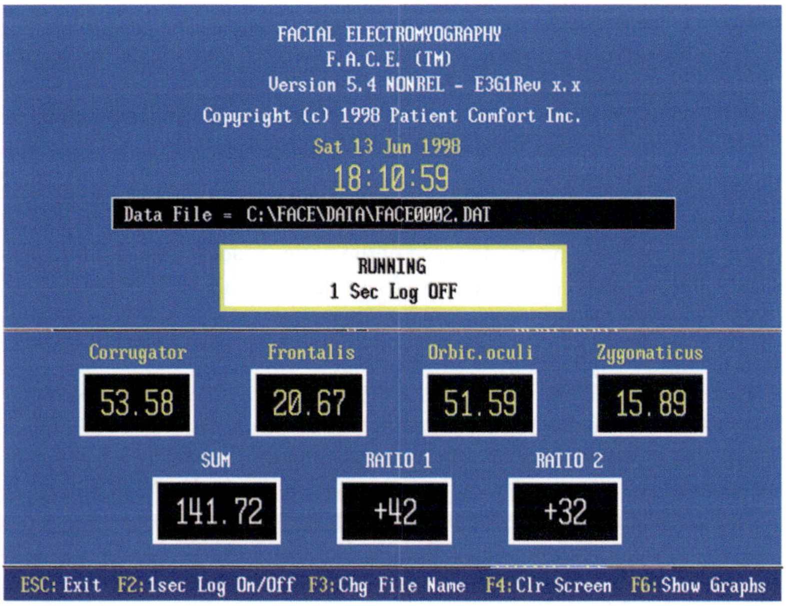

Fig. 6.5 Once the FACE™ software is running, the available information includes the current values for each muscle area being monitored and the current values of the SUM, RATIO1, and RATIO2 algorithms

6.3.2 Rationale of Facial Electromyography

The term "depth of anesthesia" is of historical interest, having been developed for ether anesthesia based on *Guedel's stages* [24], of ether anesthesia classification (Fig. 6.6) [24], for use in primitive anesthetizing locations such as battlefield surgeries [24]. Ether has an analgesic antinociception pain control effect before it became anesthetic unconsciousness agent and thus led to early observations of a consciousness, analgesic and later amnestic states [24]. Current potent inhalation agents have minimal antinociception analgesic properties, so narcotics are often given in the hope of "blunting" or eliminating surgical nociception registration. Outside of cardiovascular stability, recognizing the components of anesthesia as unconsciousness, immobility, and antinociception, one could say each of these has "depth," but "adequacy" may be a better term. "Depth" implies one dimension, whereas "adequacy" allows for more than one dimension to achieve sufficient anesthesia. A monitor of the "Adequacy of Anesthesia" requires both sensitivity and specificity, when the measures available are subjective and indirect, by the clinician's assessment of vital signs and the pharmacology of the anesthetic agents employed. Consequences of inadequate anesthesia include patient movement, catabolism and its consequences, nociception, and/or recall. An overly deep level of anesthesia may result in cardiovascular compromise and prolonged time to awakening.

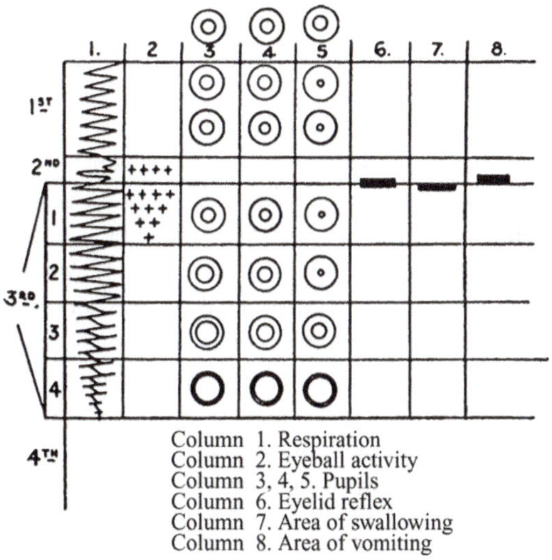

Fig. 6.6 Guedel's stages of ether anesthesia classification

6.3.3 Neurophysiology of the Facial Affective Consciousness Electromyograph (FACE™) Monitor

Facial muscles have been classified by anatomists as autonomic viscera. The face has been called an "inside out brain", because of the brainstem, limbic, and cortical control over facial expressions. The FACE™ monitor uses the facial spontaneous Electromyography (EMG) from four muscles to record and display two components of anesthesia, namely, the degree of unconsciousness and level of antinociception pain registration.

The facial muscles have long been noted clinically to register responses under surgical anesthesia. However, the sensitivity and specificity of facial responses is a concern when developing a monitor for clinical use. In a discussion of our research team using the facial EMG as a monitor of "Adequacy of Anesthesia," of those considered, the tone of the frontalis muscle seems to be the only one which looks as though this is a real possibility. The seventh cranial nerve is primarily an autonomic nerve, and the frontalis muscle is almost a "window" on the autonomic system. The "tears of frustration," the "frown of worry," and the "agonized face of pain" all register emotions which are not under conscious control. With any other processed Electroencephalography (pEEG) signal indices, there does not seem to be a clear cutoff, without overlap, between "consciousness and unconsciousness".

Having the facial muscles serve as a consciousness monitor is also attractive because, compared to other groups of muscles, they seem to possess higher resistance to Neuromuscular Blocking Agents (NMBAs). Except for the facial muscles, the musculoskeletal system depends on the number of synapses firing (recruitment) for the strength of muscular contraction. My research group's rationale was that, in the face, the strength of contraction in the facial muscles is about 70% determined by the rate of firing across the myoneural junction. The result of this difference in function is that with up to 95% neuromuscular blockade at the ulnar/hypothenar recording site, my research team hypothesized that facial muscles could remain, to some degree, responsive and that such an EMG response could then be detected given sufficient amplification. Thus, our rationale was that, despite neuromuscular blockade, with sensitive enough amplifiers, spontaneous EMG from the muscles of facial expression may be used to monitor responses under anesthesia.

Increased frontalis muscle activity alone may indicate any number of facial responses, and so one must look at more than just the frontalis muscle. Emotionally, the frontalis is involved in only surprise and fear emotions, whereas other primary muscles of facial expression, the corrugator, zygomaticus, and orbicularis oculi can, by their relative levels of activity, indicate the pattern of expression from limbic nuclei for each primary emotion. Conceptually, by utilizing more than just the frontalis muscle, it is possible to observe at the same time both the degree of unconsciousness from the tonic level across multiple facial muscles and independently the relative pattern of activity among other facial muscles which indicate valence or emotional specificity as in a "grimace" or "wince" with a nociceptive response.

6.4 Fully Awake Volunteers Receiving NMBAs

A study by Schuller and colleagues [25] assessed the effect of succinylcholine or rocuronium administration on Bispectral Index monitoring in the absence of anesthesia, in 11 fully awake volunteer anesthesiologists [25]. Two Bispectral Index BIS Quatro sensors were simultaneously placed on each side of volunteers' forehead and connected to two Bispectral Index monitors BIS-Vista (version 4.1) and BIS A-2000 (version 3.3). The onset of neuromuscular block resulted in a decline in the Bispectral Index values that returned to original values upon recovery of volunteer anesthesiologists' muscular activity [25]. The Bispectral Index showed a two-stage decline in the form of an immediate decline to values around 80, followed several minutes later by a sharp decrease to 44 with succinylcholine and to 47 with rocuronium administrations that returned to original Bispectral Index values with the return of neuromuscular activity. Sugammadex was used to antagonize the rocuronium neuromuscular block [25]. The two Bispectral Index monitors agreed closely; however, there were some periods were the two monitors differed for up to 15 BIS points for several minutes [25]. The authors argued that the algorithm upon which the Bispectral Index is derived must incorporate information from the Electromyography (EMG) to derive the Bispectral Index proprietary algorithm [25]. The authors further argued that a strong, painful noxious stimulus would have increased the Bispectral Index value in these volunteer anesthesiologists, which confirmed that the Bispectral Index was not wholly reliant on EMG activity [25].

Messner and colleagues [26], three volunteer anesthesiologists, who were the three coauthors of their published article [26], went as far as volunteering to receive succinylcholine or alcuronium Neuromuscular Blocking Agents (NMBAs) while fully awake [26]. During the entire study, the three volunteers were fully conscious. The response to succinylcholine administration, in one volunteer, was a peculiar transient decline of the Bispectral Index BIS A-1000 (version 3.31) value of 33 and a decline to 9 on repeated administration when total neuromuscular block was reached [26], whereas Bispectral Index value decreased to just 64 and 57 because total neuromuscular block was not achieved in the other two volunteers [26]. With reappearance of spontaneous EMG activity, Bispectral Index value fully recovered [26]. Because NMBAs administration without sedative/hypnotic in fully conscious three volunteer anesthesiologists was found to decrease Bispectral Index value to deep sedation/anesthesia levels, EMG activity seems to be an integral part of the "high end" BIS >80 algorithm that is indicative of an awake state or light anesthesia, rather than EMG activity artifacts simply distorting otherwise appropriate Bispectral Index values at that level [25, 26]. However, the two studies raised serious ethical reservations of fully awake volunteers receiving NMBAs. There were very serious concerns raised in subsequent discussions about the administration of succinylcholine to an awake volunteer, as something ethics committees should never support, even though both studies were conducted in fully equipped operating rooms, and after obtaining appropriate ethics committee approvals [25, 26].

Because succinylcholine administration before skin incision in adult patients was accredited with directly provoking certain EEG patterns, namely, low voltage Electroencephalography (EEG) patterns (δ and θ waves) [27, 28], that could have provoked a transient Bispectral Index artifact. This could also be a possible clarification for the previously reported Bispectral Index value decline to 33 and a further decline to a Bispectral Index value of 9 on repeated succinylcholine administrations in a fully awake volunteer [26], and the Bispectral Index two-stage decline to values around 80 that was followed several minutes later by a sharp decrease to 44 with succinylcholine administration in the second study [25].

Neuromuscular block may also reduce signals from muscle stretch receptors that normally contribute to arousal. Some authors have stated that Bispectral Index values increased after reversal of neuromuscular blockade and have explained their results with the "Afferent Muscle Spindle Theory" [29]; conversely, others showed no significant changes [5, 30]. The "Afferentation Theory" or the "Afferent Muscle Spindle Theory" that could be a possible explanation states that signals from muscle stretch receptors, i.e., the muscle afferents, hence the name, produce certain EEG stimulations [29]. This is because the slowly contracting facial muscle fibers sustain a high level of muscle tone activities [31].

Table 6.4 enumerates and describes fully awake volunteers receiving Neuromuscular Blocking Agents (NMBAs).

Table 6.4 Fully awake volunteers receiving Neuromuscular Blocking Agents (NMBAs)

Reference	BIS Monitor	*n*	Change	BIS change	Explanation
Schuller et al. [25]	BIS Vista (4.0), A-2000 (3.3)	11	Awake NMBAs succinylcholine, rocuronium	BIS 44 ↓ succinylcholine, BIS 47 ↓ rocuronium	succinylcholine δ, θ waves ↑
Messner et al. [26]	BIS A-1000 (3.31)	3	Awake NMBAs succinylcholine, alcuronium	BIS 9 ↓, BIS 64 ↓, BIS 57 ↓	Alleviating EMG artifact

n number of patients per group, *EMG* electromyography, *BIS* Bispectral Index, *NMBAs* neuromuscular blocking agents

6.5 Effect of Different Neuromuscular Block Levels on Bispectral Index Monitoring

Frontal Electromyography (fEMG) could greatly influence the Bispectral Index algorithm calculations. My research team Dahaba and colleagues in the Department of Anesthesiology and Intensive Care Medicine, Medical University of Graz, Austria, simultaneously used a conventional Bispectral Index A-2000 BIS™ ($BIS_{3.4}$, version 3.4) and a Bispectral Index BIS-XP™ (BIS_{XP}, version 4.0) to study the effect of different neuromuscular block levels on Bispectral Index monitoring [32]. We maintained surgical deep anesthesia and profound antinociception to completely abolish fEMG activity that could falsely elevate the Bispectral Index value. We then examined the effect of different neuromuscular block levels on the Bispectral Index values [32]. In 30 patients undergoing general selective surgical procedures expected to last 1–2 hours, propofol 3.6 μg ml^{-1} Target Controlled Infusion (TCI) and remifentanil 0.3 μg kg^{-1} min^{-1} continuous infusion maintained a stable Bispectral Index value of around 40 deep surgical anesthesia [32]. Mivacurium 0.15 mg kg^{-1} was administered, and neuromuscular block at the adductor pollicis muscle was evaluated using the Relaxometer® (Groningen University, Holland) Mechanomyograph (MMG) that could precisely quantify the neuromuscular block [23]. We attached the force-transducer of the Relaxometer to the thumb, and we supramaximally stimulated the ulnar nerve at the wrist (pulse width 200 μs, rectangular wave) via surface electrodes with Train-Of-Four (TOF) stimuli (2 Hz for 2 s) at 12 s intervals [23]. We used the First Twitch of

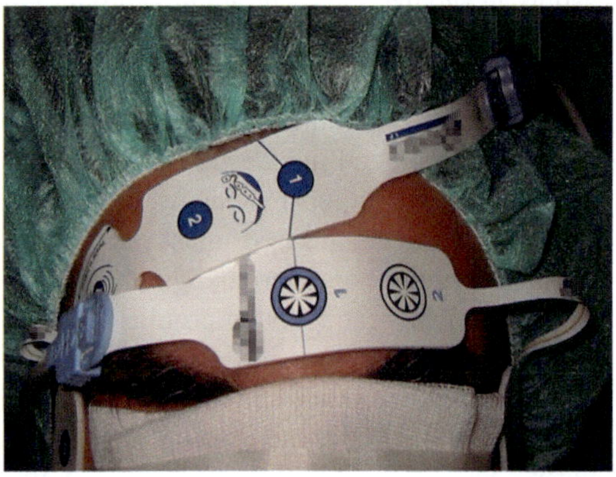

Fig. 6.7 Bispectral Index (BIS) BIS-3.4 conventional sensor 3 electrodes (1–2–3) and BIS-XP Quatro 4 electrodes (1–2–4–3) sensor simultaneous montage. The BIS-3.4 conventional sensor strip is transversely placed on the forehead with electrode-1 in the midline and electrode-3 adjacent to the angle of the eye. The BIS-XP Quatro sensor strip is inclined on the forehead with electrode-4 slanting just above the eyebrow, electrode-1 placed slightly higher on the forehead, and electrode-3 adjacent to the angle of the eye. (Reproduced with permission # 5861390605923 from Dahaba et al. The effect of different stages of neuromuscular block on the bispectral index and the bispectral index-XP under remifentanil/propofol anesthesia. Anesth Analg. 2004;99:781–7)

the TOF (T_1) expressed as percentage of control response and the TOF ratio (T_4:T_1) to evaluate neuromuscular block. We digitally collected and stored the T_1% and TOF ratio values on a laptop computer for the duration of our study [23].

On each side of patients' forehead, a conventional Bispectral Index ($BIS_{3.4}$) 3 electrodes (electrode-1, electrode-2, and electrode-3) sensor and a Bispectral Index BIS-XP (BIS_{XP}) newly introduced at the time Quatro™ 4 electrodes (electrode-1, electrode-2, electrode-4, and electrode-3) sensor were simultaneously placed (Fig. 6.7) [32]. Connected to a conventional Bispectral Index monitor, the Bispectral Index

$BIS_{3.4}$ sensor was horizontally placed on the patients' forehead with electrode-1 in the midline and electrode-3 placed adjacent to the angle of the eye. Connected to a BIS_{XP} monitor, on the other side, the XP sensor tangential montage allowed the $BIS_{3.4}$ and BIS_{XP} midline electrodes to be simultaneously placed with the BIS_{XP} electrode-4 placed just above the eyebrow [32]. Displayed on the $BIS_{3.4}$ and BIS_{XP} monitors, Channel 12 (CH12) is the combination of the lowest value of Channel 1 (CH1, electrodes 1 + 2 impedance) and Channel 2 (CH2, electrodes 1 + 3 impedance) [32]. To level out any dominance of right-handed subjects' cerebral hemisphere effect, the $BIS_{3.4}$ and BIS_{XP} monitors were alternatively placed on the right and left sides in consecutive patients. The Bispectral Index and 95% Spectral Edge Frequency (SEF_{95}) recordings were started after verifying absence of Electromyography (EMG) activity, Signal Quality Index (SQI) >95%, and electrodes impedance <5 kΩ [32].

In our study patients, we simultaneously recorded the Bispectral Index values from the $BIS_{3.4}$ and BIS_{XP} monitors, EMG in decibels (dB), and SEF_{95} in Hz at onset, maintenance, and recovery of TOF and T_1% neuromuscular block. We demonstrated an increase in EMG of the Bispectral Index ($BIS_{3.4}$) activity before neuromuscular block [32], whereas the Bispectral Index $BIS_{3.4}$ and BIS_{XP} values did not change at the full myorelaxation or the myorelaxation recovery phases [32]. However, in 70% of patients, during mivacurium onset, there was a significant, albeit transient, Bispectral Index $BIS_{3.4}$ increase to 49 ± 7 at 1 ± 0.2 min and a simultaneous significant BIS_{XP} decline to 35 ± 3 at 1.4 ± 0.5 min. After 2.5 ± 1 min from mivacurium administration, both monitors returned to baseline values (Fig. 6.8) [32].

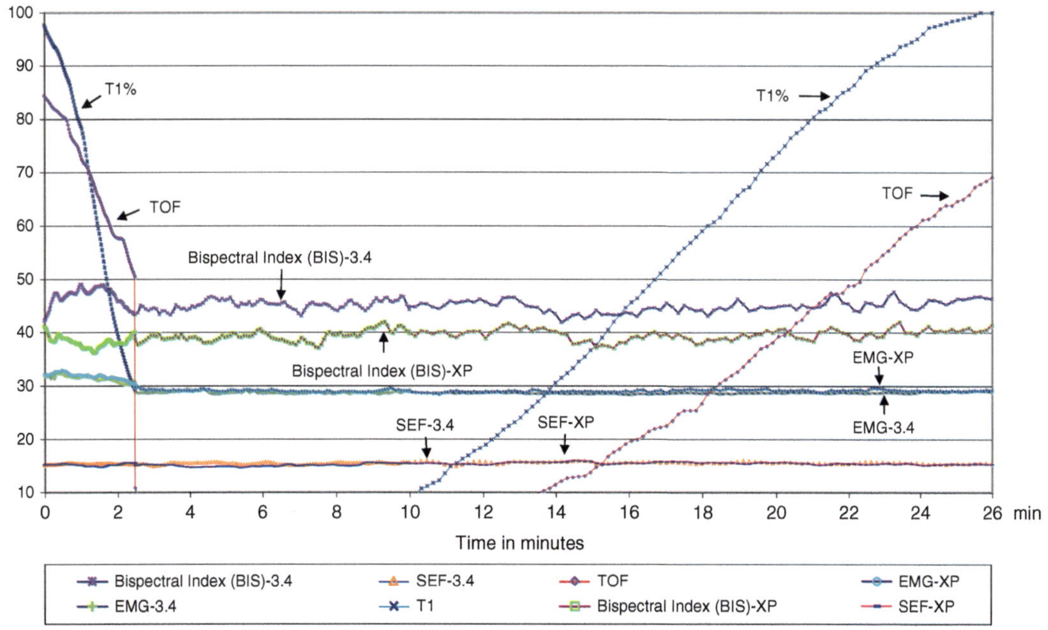

Fig. 6.8 Bispectral Index (BIS) $BIS_{3.4}$ and BIS_{XP} monitor simultaneous recordings, Electromyography (EMG) in decibels (dB) and 95% Spectral Edge Frequency (SEF_{95}) in Hz at onset, maintenance, and recovery of Train-Of-Four (TOF) and first twitch of the TOF (T_1%) neuromuscular block. (Reproduced with permission # 5861390605923 from Dahaba et al. The effect of different stages of neuromuscular block on the bispectral index and the bispectral index-XP under remifentanil/propofol anesthesia. Anesth Analg. 2004;99:781–7)

Our study clearly demonstrated that at Bispectral Index 40, remifentanil profound antinociception is effectively preventing EMG activity from erroneously elevating the Bispectral Index value rendering no effect of mivacurium on Bispectral Index monitoring [32]. However, with the onset of neuromuscular block, the change of muscle tone status trigger $BIS_{3.4}$ and BIS_{XP} artifact recognition algorithms to manifest as mirror-image opposite transient-artifacts, lasting till patients were fully relaxed [32]. We demonstrated a very peculiar response to mivacurium Neuromuscular Blocking Agent (NMBA) administration manifesting as a transient increase in the Bispectral Index value of the $BIS_{3.4}$ model and conversely at the same time a transient decline in the Bispectral Index value of the BIS_{XP} model. The change in the EMG muscle tone was instantly detected by the two monitors the $BIS_{3.4}$ model and BIS_{XP} model artifact recognition algorithms but differently interpreted in opposite direction "artifacts" [32]. However, in our now fully relaxed subjects, as soon as the new stretch receptor muscle tone status of a fully relaxed muscle was established, the $BIS_{3.4}$ Bispectral Index value and BIS_{XP} Bispectral Index value both returned to their original values before neuromuscular block. After termination of anesthesia propofol TCI and remifentanil infusion discontinuation, we simultaneously recorded Bispectral Index $BIS_{3.4}$ and BIS_{XP} values increase at the emergence from anesthesia phase starting from termination of anesthesia till full recovery. We recorded EMG of BIS-3.4 (EMG-3.4), BIS-XP (EMG-XP), SEF_{95} of BIS-3.4 (SEF-3.4), and BIS-XP (SEF-XP) (Fig. 6.9).. According to Bland and Altman statistical method [33], when data simultaneously collected from the two monitors were analyzed, at all phases, the BIS_{XP} data was constantly lower than the $BIS_{3.4}$. The $BIS_{3.4}-BIS_{XP}$ bias and Limits of Agreement (LOA) [33] were + 7 (+22, −6) with a Bispectral Index difference range LOA of +22 and − 6 BIS value between the two monitors [32].

Greif and colleagues [5] specifically designed a study to precisely quantify the fEMG contribution in the Bispectral Index. In their study, the degree of neuromuscular block did not significantly change the Bispectral Index value under propofol anesthesia, neither by decreasing artifact resulting from fEMG nor by reducing the afferent neuronal input [5]. There were no statistically significant differences in Bispectral Index values in paralyzed patients, partially paralyzed patients, or patients who did not receive a NMBA [5]. The authors tested the hypothesis that non-depolarizing NMBAs could decrease the Bispectral Index BIS A-1000 (version 3.3) value in 10 volunteers under TCI propofol 3.8 $\mu g\ ml^{-1}$ anesthesia [5]. A mivacurium infusion was adjusted to randomly maintain the $T_1\%$ in the TOF at 80%, 30%, 20%, 15%, 10%, 5%, or 2% while recording the Bispectral Index value and BIS-EMG [5]. Mean Bispectral Index value of 95 ± 4 before anesthesia induction significantly decreased to 40 ± 5 following propofol administration. However, there were no significant differences between Bispectral Index values at the designated neuromuscular block levels [5]. Mean frontal BIS-EMG of 47 ± 3 dB before anesthesia induction significantly decreased to 27 ± 1 dB following propofol administration. However, there were no significant differences between frontal BIS-EMG values at the designated neuromuscular block levels [5]. The authors demonstrated that in the absence of noxious stimulation and absence of EMG activity, a Bispectral Index value of 40 was unaltered by randomly allocated different mivacurium neuromuscular block levels [5]. Different levels of neuromuscular block did not alter Bispectral Index values during propofol anesthesia, either by reducing EMG artifact or by decreasing afferent neuronal input. The Bispectral Index will

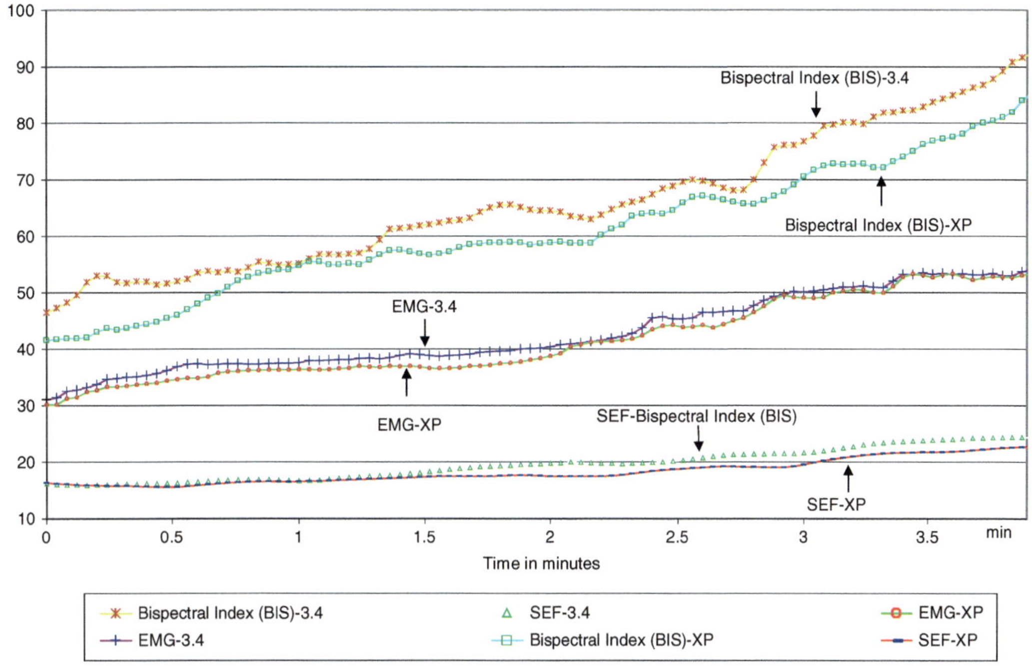

Fig. 6.9 Bispectral Index (BIS) $BIS_{3.4}$ and BIS_{XP} monitor simultaneous recordings in 30 patients at emergence from anesthesia phase starting from termination of anesthesia till full recovery. Electromyography (EMG) in decibels (dB) of BIS-3.4 (EMG-3.4) and BIS-XP (EMG-XP) and 95% Spectral Edge Frequency (SEF_{95}) in Hz of BIS-3.4 (SEF-3.4) and BIS-XP (SEF-XP). (Reproduced with permission # 5861390605923 from Dahaba et al. The effect of different stages of neuromuscular block on the bispectral index and the bispectral index-XP under remifentanil/propofol anesthesia. Anesth Analg. 2004;99:781–7)

Table 6.5 Effect of different neuromuscular block levels on Bispectral Index monitoring

Reference	BIS Monitor	*n*	Change	BIS change	Explanation
Dahaba et al. [32]	BIS-XP (4.0), A-2000 (3.4)	30	mivacurium no EMG	BIS deep TCI anesthesia	No EMG BIS change
Greif et al. [5]	BIS A-1000 (3.3)	10	mivacurium no EMG	BIS deep TCI anesthesia	No EMG BIS change

n number of patients per group, *EMG* electromyography, *BIS* Bispectral Index, *TCI* target controlled infusion

thus comparably estimate depth of anesthesia in deeply unconscious patients who are paralyzed, partially paralyzed, or unparalyzed in the absence of EMG activity [5].

Table 6.5 enumerates and describes the effect of different neuromuscular block levels on Bispectral Index monitoring.

References

1. Jensen EW, Litvan H, Struys M, Vazquez PM. Pitfalls and challenges when assessing the depth of hypnosis during general anaesthesia by clinical signs and electronic indices. Acta Anaesthesiol Scand. 2004;48:1260–7.
2. Rampil IJ. A Primer for EEG Signal Processing in Anesthesia. Anesthesiology. 1998;89:980–1002.

3. Sleigh JW, Steyn-Ross DA, Steyn-Ross ML, Williams ML, Smith P. Comparison of changes in electroencephalographic measures during induction of general anaesthesia: influence of the gamma frequency band and electromyogram signal. Br J Anaesth. 2001;86:50–8.
4. Vivien B, Di Maria S, Ouattara A, Langeron O, Coriat P, Riou B. Overestimation of Bispectral Index in sedated intensive care unit patients revealed by administration of muscle relaxant. Anesthesiology. 2003;99:9–17.
5. Greif R, Greenwald S, Schweitzer E, Laciny S, Rajek A, Caldwell JE, et al. Muscle relaxation does not alter hypnotic level during propofol anesthesia. Anesth Analg. 2002;94:604–8.
6. Bruhn J, Bouillon TW, Shafer SL. Electromyographic activity falsely elevates the bispectral index. Anesthesiology. 2000;92:1485.
7. Bowdle TA. Depth of anesthesia monitoring. Anesthesiol Clin North Am. 2006;24:793–822.
8. Renna M, Wigmore T, Mofeez A, Gillbe C. Biasing effect of the electromyogram on BIS: a controlled study during high-dose fentanyl induction. J Clin Monit Comput. 2002;17:377–81.
9. Lobo FA, Schraag S. Limitations of anaesthesia depth monitoring. Curr Opin Anaesthesiol. 2011;24:657–64.
10. Liu N, Chazot T, Huybrechts I, Law-Koune J-D, Barvais L, Fischler M. The influence of a muscle relaxant bolus on bispectral and datex-ohmeda entropy values during propofol-remifentanil induced loss of consciousness. Anesth Analg. 2005:1713–8.
11. Baulig W, Seifert B, Schmid ER, Schwarz U. Comparison of spectral entropy and bispectral index electroencephalography in coronary artery bypass graft surgery. J Cardiothorac Vasc Anesth. 2010;24:544–9.
12. Wennervirta J, Salmi T, Hynynen M, Yli-Hankala A, Koivusalo A-M, Van Gils M, et al. Entropy is more resistant to artifacts than bispectral index in brain-dead organ donors. Intensive Care Med. 2007;33:133–6.
13. Weber F, Kriek N, Blussé van Oud-ALBLAS HJ. The effects of mivacurium-induced neuromuscular block on Bispectral Index and Cerebral State Index in children under propofol anesthesia—a prospective randomized clinical trial. Pediatr Anesth. 2010;20:697–703.
14. Panousis P, Heller AR, Burghardt M, Bleyl JU, Koch T. The effects of electromyographic activity on the accuracy of the Narcotrend ® monitor compared with the Bispectral Index during combined anaesthesia. Anaesthesia. 2007;62:868–74.
15. Doufas AG, Komatsu R, Orhan-Sungur M, Sengupta P, Wadhwa A, Mascha E, et al. Neuromuscular block differentially affects immobility and cortical activation at near-minimum alveolar concentration anesthesia. Anesth Analg. 2009;109:1097–104.
16. Dahaba AA, Bornemann H, Hopfgartner E, Ohran M, Kocher K, Liebmann M, et al. Effect of sugammadex or neostigmine neuromuscular block reversal on bispectral index monitoring of propofol/remifentanil anaesthesia. Br J Anaesth. 2012;108:602–6.
17. Aho AJ, Kamata K, Yli-Hankala A, Lyytikäinen L-P, Kulkas A, Jäntti V. Elevated BIS and E ntropy values after sugammadex or neostigmine: an electroencephalographic or electromyographic phenomenon? Acta Anaesthesiol Scand. 2012;56:465–73.
18. Illman H, Antila H, Olkkola KT. Reversal of neuromuscular blockade by sugammadex does not affect EEG derived indices of depth of anesthesia. J Clin Monit Comput. 2010;24:371–6.
19. Fassoulaki A, Chondrogiannis K, Staikou C. Sugammadex at both high and low doses does not affect the depth of anesthesia or hemodynamics: a randomized double blind trial. J Clin Monit Comput. 2017;31:297–302.
20. Hajat Z, Ahmad N, Andrzejowski J. The role and limitations of EEG -based depth of anaesthesia monitoring in theatres and intensive care. Anaesthesia. 2017;72:38–47.
21. Christ B, Guerci P, Baumann C, Meistelman C, Schmartz D. Influence of neuromuscular block and reversal on bispectral index and NeuroSense values. Eur J Anaesthesiol. 2014;31:437–9.
22. Zikov T, Bibian S, Dumont GA, Huzmezan M, Ries CR. Quantifying cortical activity during general anesthesia using wavelet analysis. IEEE Trans Biomed Eng. 2006;53:617–32.
23. Rowaan CJ, Vandenbrom RHG, Wierda JMKH. The Relaxometer: A complete and comprehensive computer-controlled neuromuscular transmission measurement system developed for clinical research on muscle relaxants. J Clin Monit. 1993;9:38–44.
24. Guedel AE. Inhalation Anestheia. A Fundamental Guide. 2nd ed. New York: The MacMillan Company; 1947.
25. Schuller PJ, Newell S, Strickland PA, Barry JJ. Response of bispectral index to neuromuscular block in awake volunteers. Br J Anaesth. 2015;115:i95–103.
26. Messner M, Beese U, Romstöck J, Dinkel M, Tschaikowsky and K. The bispectral index declines during neuromuscular block in fully awake persons. Anesth Analg. 2003;97:488–91.
27. Oshima E, Shingu K, Mori KEEG. Activity during halothane anaesthesia in man. Br J Anaesth. 1981;53:65–72.
28. Lanier WL, Milde JH, Michenfelder JD. Cerebral stimulation following succinylcholine in dogs. Anesthesiology. 1986;64:551–9.
29. Lanier WL, Laizzo PA, Milde JH, Sharbrough FW. The cerebral and systemic effects of movement in response to a noxious stimulus in lightly

anesthetized dogs possible modulation of cerebral function by muscle afferents. Anesthesiology. 1994;80:392–401.

30. Chazot T, Dumont G, Le Guen M, Hausser-Hauw C, Liu N, Fischler M. Sugammadex administration results in arousal from intravenous anaesthesia: a clinical and electroencephalographic observation. Br J Anaesth. 2011;106:914–6.
31. Tammisto TTO. Spontaneous EMG activity for detection of arousal during general anaesthesia--comparison between recordings from frontal and neck musculature. Eur J Anaesthesiol. 1991;8:109–14.
32. Dahaba AA, Mattweber M, Fuchs A, Zenz W, Rehak PH, List WF, et al. The effect of different stages of neuromuscular block on the bispectral index and the bispectral index-XP under remifentanil/propofol anesthesia. Anesth Analg. 2004;99:781–7.
33. Bland JM, Altman DG. Statistical methods for assessing agreement between two methods of clinical measurement. Lancet. 1986;1:307–10.

7 Effect of Noxious Stimulation and Antinociception Opioids on Depth of Anesthesia Monitoring: Achilles Second Heel (!)

Basically, the Bispectral Index has two Achilles heels. When I said that in an anesthesia congress presentation in Thessaloniki, Greece, this almost started a riot against me—how can Achilles have two heels?

In their review article, Jensen and colleagues [1] assessed processed Electroencephalography-Depth Of Anesthesia (pEEG-DOA) monitors when using opioids. They concluded that there is a poor correlation between pEEG-DOA indices and administered opioids [1]. They review the present methods used for validating the Depth of Anesthesia (DOA) and introduce three concepts: the real DOA (DOA real) state of hypnosis that the patient has in a given moment during the general anesthesia, the observed DOA (DOA obs) that is the subjective assessment of the anesthesiologist based on clinical signs, and the electronic indices DOA (DOA el-ind) that is any estimation of the DOA given by an electronic device [1]. The three entities should, in the ideal situation, be identical. However, this is rarely the case [1]. The correlation between the three entities can be affected by several factors such as the administration of opioids, the administration of Neuromuscular Blocking Agents (NMBAs), or the presence of facial muscular activity [1]. Opioids can block the response to tactile and noxious stimuli, and even the response to verbal command could vanish, hence deeming the patient in a lower depth of hypnosis than the real patient state. They concluded that although several monitors and clinical scoring scales are available to assess the depth of hypnosis during general anesthesia, care should be taken when interpreting their results [1].

7.1 Effect of Opioids, Without Noxious Stimulation, on Depth of Anesthesia Monitoring: Frontal Sensors Montage Limitations

Because all processed Electroencephalography-Depth Of Anesthesia (pEEG-DOA) monitors data acquisition is limited to the frontal lobe activity where most pEEG-DOA sensors are placed, this basically denies anesthesiologists the ability to judge events happening in deep cortical regions. A major limitation of the pEEG-DOA devices is the fact that monitoring is confined to the frontal lobe area and does not penetrate deep subcortical structures that could detect noxious stimulation. Furthermore, each anesthetic agent is uniquely characterized by its own spatial electroencephalography (EEG) distribution and evolution. The β activity "frontal predominance" is eventually superseded by an increase in δ and θ low-frequency activities migrating anteriorly in the "anteriorization" phenomena topographic pattern as the anesthesia deepens.

Ferreira and colleagues [2] examined in 25 patients the effect of administration of the uncon-

A. A. Dahaba, *Understanding the Numbers in Depth of Anesthesia Monitors*,
https://doi.org/10.1007/978-3-031-88260-9_7

ventional 2 μg kg^{-1} "remifentanil bolus" under propofol target controlled infusion (TCI) when there is no noxious stimulation of any kind, on Bispectral Index monitoring using the BIS A-2000XP (Bispectral Index model as reported), a practice that was prohibited by the remifentanil manufacturer very shortly after remifentanil introduction [2]. They reported no significant decline in Bispectral Index values, but a significant decline in blood pressure (BP) and heart rate (HR) beats per minute (bpm) hemodynamic parameters following remifentanil bolus administration [2]. Mean ± SD Bispectral Index value of 42 ± 6 (range 36–54) prior to the remifentanil bolus was not significantly different $P = 0.7$ than the Bispectral Index value of 41.1 ± 6 (range 32–57) 90–120 s after the remifentanil bolus administration [2]. Mean BP and HR were significantly $P = 0.001$ reduced from 96.4 ± 19.9 mm Hg and 70 ± 16.4 bpm at the time of the bolus administration to 74.2 ± 16.6 mm Hg and 61 ± 13.6 bpm, respectively, at 90–120 s after remifentanil bolus administration [2]. They concluded that remifentanil bolus given under propofol total intravenous anesthesia did not decrease Bispectral Index values, an effect independent of hemodynamic changes, endotracheal intubation, or surgical stimuli [2].

Unlike inhalational or intravenous anesthetic agents, opioids in analgesic concentrations produce minimal or no electrophysiological alterations on the cerebral cortex EEG. Opioids seem to be incapable of exerting an influence on the Bispectral Index in the absence of noxious stimulation [3]. Guignard and colleagues [3], using a Bispectral Index BIS A-1000 (version 3.12) monitor, in 50 patients randomly allocated to five groups of 10 patients each to receive one of five randomly allocated remifentanil effect-site (C_e) concentrations of 0, 2, 4, 8, or 16 ηg ml^{-1}, showed that before laryngoscopy and orotracheal intubation, at constant propofol TCI C_e of 4 μg ml^{-1} and in the absence of painful stimuli, remifentanil in all 5 C_e concentrations of 0, 2, 4, 8, or 16 ηg ml^{-1} groups, including the highest TCI rate of 16 ηg ml^{-1}, had no influence on Bispectral Index values [3].

Schmidt and colleagues [4], in 25 patients, in the absence of painful stimuli, demonstrated that the Bispectral Index BIS A-1000 (version 2.0) values recorded during remifentanil 0.3 μg kg^{-1} min^{-1} two-stage recovery during step-by-step reduction till remifentanil was completely discontinued, followed by propofol TCI 3.0 μg ml^{-1} reduction till propofol was also discontinued [4], and in the absence of painful stimuli, remifentanil reductions, at all-time points, were not associated with any significant change in Bispectral Index values. Conversely, propofol concentration reductions were associated with significant increases in Bispectral Index values. They concluded that Bispectral Index could clearly indicate the hypnotic/sedative component of propofol general anesthesia but does not reflect the analgesic component of remifentanil as the Bispectral Index algorithm is not designed to detect the direct effect of opioids on the cerebral cortex [4].

Shafer and colleagues [5] demonstrated that, in clinical doses, opioids have minimal direct effect on the cerebral cortex; rather an increase in opioids, clinical dose up to approximately fivefold is required to elicit any perceptible effect [5]. Furthermore, opioids would require even higher concentrations of up to eightfold the clinical dose to produce a discernible alteration of cortical EEG [5]. This is because opioids' analgesic effect is mediated through neuromodulation of subcortical structures such as locus coeruleus-noradrenergic system pathway involved in the mechanism of opioids' drug effect, which are undetectable by the pEEG-DOA monitors [6]. Because hypnotics and opioids have totally different sites of action [3], this would result in a totally different Bispectral Index effect following hypnotics or opioid administrations. Hypnotics exert their EEG effect directly on cerebral cortical structures, whereas opioids exert an analgesic effect via inhibition of subcortical structures including those in the spinal cord [3]. Because the classic 10/20 international system Fast Fourier Transform EEG does not penetrate deep subcortical structure, this confines most of the DOA EEG-derived parameter to the frontal lobe activities [6].

Spatial EEG distribution is an area of great implications, because disease pathology loca-

Table 7.1 Effect of opioids, without noxious stimulation, on depth of anesthesia

References	BIS Monitor	*n*	Agent	BIS change	Explanation
Ferreira et al. [2]	BIS A-1000XP (as reported)	25	Remifentanil 2 μg kg^{-1} bolus	BIS 42–41 no change	Lack of cortical effect
Guignard et al. [3]	BIS A-1000 (3.12)	10 × 5	Remifentanil C_e 0, 2, 4, 8, or 16 ηg ml^{-1}	BIS no change	Lack of cortical effect
Schmidt et al. [4]	BIS A-1000 (2.0)	25	Remifentanil 0.3 μg kg^{-1} min^{-1} ↓	BIS no change	Lack of cortical effect

n number of patients per group, C_e predicted effect-site concentrations, *BIS* Bispectral Index

tions such as cerebral ischemia as well as effect sites of anesthetic agents' mechanism of action are mostly deep cortical events. Furthermore, the site of cerebral events such as cerebral ischemia is potentially very relevant to pEEG-DOA monitoring. Here we have an inherent structural flaw, because the classic EEG 10/20 international system Fast Fourier Transform from which almost all pEEG-DOA indices are derived is completely separated from deep subcortical structures that detect noxious stimulation, thus rendering pEEG-DOA monitors placed on patients' foreheads incapable of detecting deep cortical signal events.

The concept of Minimum Alveolar Concentration (MAC) [7], which was first introduced in 1963 in the infamous "Anesthesiology" publication by Merkel G and Eger EI II [7], in a comparative study of halothane and halopropane anesthesia as a pioneer method for determining equipotency of inhalational anesthetics, is universally used to quantify the potency of inhalational anesthetics [7]. The MAC [7], initially created to compare volatile anesthetics and often used as a parameter to measure reactivity to adrenergic stimuli, is not an appropriate indicator of the "anesthetic hypnotic state," as it is greatly modified by the opioids on board. Thus, the MAC as displayed on anesthesia machines cannot be accurately used, most of the time, to guide anesthesia delivery [7]. This led to the creation of "Interaction Display Devices" such as the Navigator (GE General Electric Healthcare, Chicago, Illinois, USA) or the Smart Pilot View (Dräger, Lübeck, Germany) that take into consideration pharmacokinetics of all used anesthetics to "predict" a hypnotic state to better guide anesthetic agent delivery.

Table 7.1 enumerates and describes the effect of opioids, without noxious stimulation, on depth of anesthesia monitoring.

7.2 Effect of Noxious Stimulation on Depth of Anesthesia Monitoring

The relationship between processed Electroencephalography-Depth Of Anesthesia (pEEG-DOA) indices and noxious stimulation seems to be another huge pitfall. Sandin and colleagues [8] demonstrated in six male and four female healthy, young volunteers, American Society of Anesthesiologists (ASA) classification physical status I, applying Transcutaneous Electrical Nerve Stimulation (TENS) and Ice Water Pain test (IWP) noxious stimulation, that Bispectral Index values BIS-XP 2000 (BIS model as reported) and hemodynamic parameters significantly increased at the lower 1.0-Minimum Alveolar Concentration (MAC) sevoflurane concentration compared with no significant change in Bispectral Index values or hemodynamic parameters at the higher 1.5-MAC sevoflurane concentration [8]. At 1.0 MAC, there was a significant increase in Bispectral Index values during pain stimulation both with IWP ($P < 0.03$) and with TENS ($P < 0.005$), but at 1.5 MAC there were no significant changes [8].

Takamatsu and colleagues [9] studied response of the Bispectral Index BIS A-2000 (version 3.4) monitor and simultaneously placed State Entropy (SE) and Response Entropy (RE) S/5 module (GE General Electric Healthcare, Chicago, Illinois, USA) to noxious increasing intensity electric stimuli of 20, 40, 60, and 80 mA, 100 Hz for 5 s and after skin incision, in 40 female patients randomly allocated to one of four end-tidal sevoflurane concentrations (ET_{sev}) groups of 1.3%, 1.7%, 2.1%, or 2.5% [9]. All pEEG-DOA indices increased significantly ($P < 0.0001$) as electric stimuli intensity increased [9]. Median (range) RE of 29 (15–41) and SE of 24 (14–41) before skin incision in patients who did not move were significantly lower than RE of 38 (24–53) and SE of 37 (24–52) in those patients that did move, with no significant difference in Bispectral Index values [9]. The differences between RE and SE significantly increased as the intensity of electric stimuli increased. However, pEEG-DOA showed wide variability depending on exposure to noxious stimuli, as none of the pEEG-DOA indices were able to accurately distinguish the differences in intensity of the electric stimuli intensities [9].

Yoo and colleagues [10] demonstrated that Bispectral Index BIS-XP A-2000 (version 3.31) values in 20 women undergoing Cesarean Section under general anesthesia show a totally different pattern depending on whether women experienced "labor pain" beforehand or not [10]. Intraoperative Bispectral Index values under 1% sevoflurane together with 1:1 Nitrous Oxide (N_2O) general anesthesia during the period between intubation and delivery was significantly $P < 0.001$ lower in women who did experience "labor pain" prior to undergoing cesarean section delivery compared with 20 women who did not experience "labor pain" prior to their Cesarean Section delivery under similar 1% sevoflurane together with 1:1 N_2O general anesthesia [10]. The authors concluded that prior labor pain increased "pain tolerance" and hence was associated with significantly lower intraoperative Bispectral Index values during sevoflurane/N_2O general anesthesia and reduced postoperative analgesic consumption in women undergoing Cesarean Section delivery that experienced prior labor compared with women without prior labor pain [10].

This was strongly confirmed by Erden and colleagues [11] in a study showing that anesthetic requirements for sevoflurane, at Bispectral Index values strictly maintained at similar levels, were significantly lower in 25 women in labor pain versus 25 women not in labor pain all undergoing Cesarean Section delivery [11, 12]. A total of 50 women were enrolled, of whom 25 had uterine activity less than 30 Montevideo units in the Prelabor group and 25 women had uterine activity greater than 80 Montevideo units in the Labor group [11]. Anesthesia was maintained with 50% oxygen in N_2O with sevoflurane [11]. Sevoflurane concentrations were adjusted to sustain a constant Bispectral Index BIS A-2000 value of 47–50. The ET_{sev} in the Prelabor group were $P < 0.05$ significantly higher than in the Labor group at similar Bispectral Index values [11]. This confirms the previous study results [10] that prior labor pain increased "pain tolerance" and hence was associated with significantly lower intraoperative sevoflurane concentrations at strictly similar Bispectral Index values compared with women not in labor pain [11]. Because labor results in the release of sensitizing substances such as progesterone, prostaglandins, cytokines, and cortisol, some of which have been observed to participate in sleep regulation [11], the authors speculated that these substances could affect the amount of volatile agent required to provide general anesthesia for cesarean delivery [11]. However, no significant differences between the two groups were found in prolactin, progesterone, and cortisol levels immediately before cesarean delivery; hence, prolactin, progesterone, and cortisol do not appear to be responsible for this observation [11].

Table 7.2 enumerates and describes the effect of noxious stimulation on depth of anesthesia monitoring.

Table 7.2 Effect of noxious stimulation on depth of anesthesia monitoring

References	Monitor	*n*	Agent/stimulus	Index change	Explanation
Sandin et al. [8]	BIS-XP 2000 (as reported)	10	Sevoflurane 1–1.5-MAC	BIS ↑ 1-MAC, BIS no change 1.5-MAC	Blunting noxious stimuli
Takamatsu et al. [9]	BIS A-2000 (3.4), RE, SE	10 × 4	20, 40, 60, and 80 mA	BIS ↑, RE ↑, SE ↑ mA ↑	Noxious stimuli↑
Yoo et al. [10]	BIS-XP A-2000 (3.31)	20 + 20	Labor pain/no labor pain CS	BIS ↓ with labor pain CS	Labor pain ↑ pain tolerance
Erden et al. [11]	BIS A-2000	25 + 25	Labor pain/no labor pain CS	at similar BIS sevoflurane ↓ with labor pain CS	Labor pain ↑ pain tolerance

n number of patients per group, *MAC* minimum alveolar concentrations, *BIS* Bispectral Index, *SE* state entropy, *RE* response entropy, *CS* cesarean section

7.3 Opioids First Effect

Even though opioids do not significantly change the cortical Electroencephalography (EEG) when given in the usual clinical doses because hypnotics and opioids exert their actions at totally different sites, opioids could still influence Depth of Anesthesia (DOA) indices in another peculiar fashion known as the "Opioids First Effect." Several studies have demonstrated that the addition of a μ-agonist opioid to Target Controlled Infusion (TCI) propofol results in loss of consciousness (LOC) at significantly lower propofol predicted effect-site concentrations (C_e) and as a result at higher Bispectral Index BIS A-1000 (version 3.12) [13], BIS A-1050 (version 3.2) [14], BIS A-1050 [15], and BIS A-2000 (version 3.4) [16] Bispectral Index values.

Lysakowski and colleagues [13], using a Bispectral Index BIS A-1000 (version 3.12) monitor in a most illuminating study, demonstrated a possible effect of opioids on Bispectral Index monitoring. The authors noticed that the addition of an opioid to propofol TCI using Diprifusor™ infusion pump (AstraZeneca Pharmaceuticals, Macclesfield, United Kingdom) gradually increasing concentrations, to achieve predicted C_e of 1, 2, and 4 μg ml^{-1} for anesthesia induction, results in Loss Of Consciousness (LOC) at lower propofol C_e and consequently at higher Bispectral Index BIS_{50} (Bispectral Index BIS value associated with 50% probability of LOC) [13]. They demonstrated in 75 patients randomly allocated to five groups of 15 patients each to receive one of four different μ-agonist opioids fentanyl, sufentanil, remifentanil, alfentanil, or placebo that LOC occurred at a lower TCI propofol C_e and at a higher Bispectral Index BIS_{50} values of 70, 66, 74, and 75, respectively, compared with a placebo group Bispectral Index value of 61 [13]. This clearly showed that the propofol hypnotic / anesthetic clinical effect is enhanced with the addition of antinociception doses of μ-agonist opioids [13]. However, the Bispectral Index values cannot "directly detect" this "opioids' enhanced hypnotic /anesthetic effect" and in that case would then just display the lower propofol C_e concentrations and consequently only displays the higher Bispectral Index values which could mislead the anesthesiologists in administering an inadvertent anesthetic agent over dosage [13]. The addition of μ-opioids to γ-amino butyric acid (GABA)ergic anesthetic agents would greatly enhance the "anesthetic /hypnotic effect" of GABAergic anesthetic agents, thus requiring less anesthetic dosage to achieve the same anesthetic state, as μ-opioids possess a synergistic anesthetic /hypnotic effect to anesthetic agents, an effect that cannot be directly detected by these processed Electroencephalography-Depth Of Anesthesia (pEEG-DOA) cortical devices that would then just indicate the GABAergic anesthetics' lower dose requirements and not the actual "anesthetic /hypnotic state" [13].

Mi and colleagues [14] used Bispectral Index BIS A-1050 (version 3.2) monitor in 42 patients randomly allocated to receive propofol 30 mg kg^{-1} h^{-1} infusion with 2 μg kg^{-1} fentanyl

2 min before propofol administration in 20 patients or propofol 30 mg kg^{-1} h^{-1} infusion without fentanyl administration in 22 patients. They found that the addition of fentanyl resulted in reaching the hypnotic endpoints of "no response to voice commands," "loss of eyelash reflexes," and "loss of nasal stimulation responses" at mean ± SD Bispectral Index values of 75 ± 11, 73 ± 11, and 47 ± 9 that were significantly higher in the propofol with fentanyl group compared with Bispectral Index values of 66 ± 10, 60 ± 10, and 34 ± 6, respectively, in the propofol without fentanyl group [14]. The addition of fentanyl resulted in reduction of propofol dose requirement by 48%, with a corresponding 32% reduction in mean ± SD measured propofol plasma concentrations at "inhibition of nasal body movement response" that was 9.2 ± 2.0 μg mL^{-1} significantly lower in the propofol with fentanyl group compared with 14.1 ± 4.2 μg mL^{-1} in the propofol without fentanyl group [14]. Their results suggest that fentanyl pretreatment potentiates the effects of propofol and achieves the hypnotic endpoints at higher Bispectral Index values and lower propofol doses and propofol plasma concentrations measured at inhibition of nasal body movement response.

Kodaka and colleagues [15] used Bispectral Index BIS A-1050 monitor in 64 patients randomly assigned to one of four groups of 16 patients to receive placebo or one of three doses of fentanyl 0.5, 1 and 2 μg kg^{-1} [15]. Their study sought to determine whether fentanyl influenced the Bispectral Index values or propofol predicted effective concentration for 50% of the attempts to secure Laryngeal Mask Airway (LMA) insertion (EC_{50LMA}) using a TCI Diprifusor™ infusion pump (Graseby 3500; SIMS Graseby, Watford, United Kingdom) [15]. Movement was defined as presence of bucking or gross purposeful muscular movement within 1 min after LMA insertion at the EC_{50LMA} [15]. Predicted EC_{50LMA} of the control, fentanyl 0.5, 1 and 2 μg kg^{-1} groups were 3.25 ± 0.20, 2.06 ± 0.55, 1.69 ± 0.38, and 1.50 ± 0.54 μg ml^{-1} respectively. Predicted EC_{50LMA} in all fentanyl groups were significantly lower than the predicted EC_{50LMA} in the control group [15]. Mean ± SD Bispectral Index values of 60 ± 13 and 70 ± 15 following fentanyl 1 and 2 μg kg^{-1} respectively, were significantly higher than Bispectral Index values of 45 ± 16 and 47 ± 9 in the control and 0.5 μg kg^{-1} groups [15]. They found that administration of fentanyl 0.5, 1 and 2 μg kg^{-1} before propofol TCI results in the suppression of movements in response to LMA insertion achieved at significantly decreasing TCI propofol EC_{50LMA} concentrations and increasing Bispectral Index values [15].

Struys and colleagues [16], using a Bispectral Index BIS A-2000 (version 3.4) and Middle Latency Auditory Evoked Potential (MLAEP) A-Line Autoregressive index (AAI) monitors in 45 patients randomly allocated to one of three groups to receive placebo or TCI remifentanil C_e infusion rates of 2 or 4 ɳg ml^{-1}, using Fresenius Modular DPS Infusion Pump (Fresenius Vial Infusion Systems, Brézins, France) [16], demonstrated that increasing the remifentanil TCI infusion rates from 0 to 2 and to 4 ng ml^{-1} resulted in a significant decrease in propofol TCI C_e requirements and a significant increase in Bispectral Index and AAI values at "loss of response to verbal command" using the Observer's Assessment of Alertness/Sedation (OAA/S) Scale, "loss of response to eyelash reflex," and "loss of response to noxious stimulus" [16]. At TCI remifentanil 0, 2 and 4 ɳg ml^{-1}, Bispectral Index values (CI, 95% confidence interval) at loss of response to verbal were 61 (60–62), 66 (65–68), 77 (75–78); and at loss of response to eyelash reflex were 67 (65–68), 80 (79–82), and 84 (82–85); and at loss of response to noxious stimulus were 41 (40–43), 72 (70–73), and 85 (84–87) [16]. At TCI remifentanil 0, 2, and 4 ng ml^{-1}, AAI at loss of response to verbal command were 26 (25–28), 33 (31–34), and 40 (39–42); at loss of response to eyelash reflex were 34 (33–36), 56 (55–58), and 56 (54–58); and at loss of response to noxious stimulus were 16 (15–17), 42 (40–44), and 55 (52–57) [16]. At TCI remifentanil 0, 2 and 4 ɳg ml^{-1}, propofol C_e at loss of response to verbal command were 2.9 (2.7–3.1), 2.4 (2.2–2.6), 2.0 and (1.8–2.2); at loss of response to eyelash reflex were 2.8 (2.5–3.0), 1.8 (1.7–2.0), and 1.7 (1.6–1.9); and at loss of response to noxious stimulus were 4.1 (3.8–4.4), 1.8 (1.4–2.2), and 1.3 (1.1–1.5) respectively [16].

Table 7.3 Opioids first effect

References	Monitor	*n*	Agent	Index change	Explanation
Lysakowski et al. [13]	BIS A-1000 (3.12)	15 × 5	Propofol TCI +1/4 opioids	BIS ↑, propofol C_e ↓	↓ propofol requirements
Mi et al. [14]	BIS A-1050 (3.2)	20 + 22	Propofol +fentanyl	BIS ↑, propofol plasma concentration ↓	↓ propofol requirements
Kodaka et al. [15]	BIS A-1050	16 x 4	Propofol +fentanyl 0, 0.5, 1, 2 μg kg^{-1}	BIS ↑, propofol C_e ↓	↓ propofol requirements
Struys et al. [16]	BIS A-2000 (3.4), AAI	15 × 3	Propofol +remifentanil TCI 0, 2, 4 ηg ml^{-1}	BIS ↑, AAI ↑, propofol C_e ↓	↓ propofol requirements

n number of patients per group, *TCI* target controlled infusion, C_e predicted effect-site concentrations, *BIS* Bispectral Index, *AAI* A-Line Autoregressive index auditory evoked potential

Table 7.3 enumerates and describes the "Opioids First Effect" on depth of anesthesia monitoring.

7.4 Opioids Second Effect

In many instances, Bispectral Index changes could be a sensitive indicator of painful stimuli, because when Bispectral Index values suddenly rise in response to noxious stimuli such as endotracheal intubation and laryngoscopy, which are far more painful than surgical skin incision, this is a "cortical arousal" reflex as a result of insufficient antinociception component of the general anesthesia that would require increasing the component opioid and not the hypnotic anesthetic [3]. In that case, the Bispectral Index response to noxious stimulation could be modulated by antinociception opioid analgesic regimes [17]. Blunting the noxious stimulation is another mechanism by which opioids could directly influence processed Electroencephalography-Depth Of Anesthesia (pEEG-DOA) monitoring, through blunting or attenuation of the pEEG-DOA "cortical arousal" responses to noxious stimuli, the so-called "Opioids Second Effect". The addition of Nitous Oxide (N_2O) [18] or remifentanil [3] to constant propofol Target Controlled Infusion (TCI) would attenuate or abolish, in a dose-dependent fashion, increases in Bispectral Index after laryngoscopy and orotracheal intubation [3, 18]. When TCI propofol infusion is kept constant, μ-agonist opioids administration would then attenuate increases in Bispectral Index BIS (A-1000) in response to endotracheal intubation and laryngoscopy [3], endotracheal suction [19], as well as to pinning-fixation of the Mayfield Head- holder [17]. The so-called "Opioids Second Effect" could mean that the Bispectral Index value when kept constant does not just reflect the degree of hypnotic anesthesia unconsciousness but rather simultaneously reflect the degree of opioid-induced blunting or even total inhibition of response to noxious stimuli.

Guignard and colleagues [3], using Bispectral Index BIS A-1000 (version 3.12) monitor in 50 patients randomly allocated to five groups of 10 patients each to receive a placebo or one of four randomly allocated remifentanil predicted effect-site concentrations (C_e) of 2, 4, 8 or 16 ηg ml^{-1}, showed that remifentanil negatively correlated to the laryngoscopy and orotracheal intubation increases in Bispectral Index value, i.e., remifentanil resulted in the attenuation or abolishing of any increase in the Bispectral Index value in response to laryngoscopy and endotracheal intubation [3]. Maximal Bispectral Index values were significantly ($P < 0.01$) only increased by laryngoscopy and orotracheal intubation in the remifentanil 0 and 2 ηg mL^{-1} groups but not in the other three remifentanil 4, 8 or 16 ηg ml^{-1} C_e groups [3]. The authors showed that the addition of remifentanil to constant TCI propofol 4 μg ml^{-1} in 50 patients randomly allocated to receive one of five remifentanil C_e of 0, 2, 4, 8 or 16 ηg ml^{-1} concentrations affects the Bispectral Index value only in the presence of painful stimuli [3]. Moreover, remifentanil attenuated or abolished increases in Bispectral Index values after laryngoscopy and tracheal intubation in a comparable dose-dependent fashion [3].

Hans and colleagues [17], using a Bispectral Index BIS A-1000 monitor in 20 patients under propofol constant TCI estimated plasma concentrations (C_p) of 3 μg ml^{-1}, randomly allocated to receive sufentanil TCI C_p of 0.5 ɳg ml^{-1} or 1 ɳg ml^{-1}, showed that at the extremely painful peak point at 1 min from pinning of the Mayfield head holder fixator, Bispectral Index values increase over the baseline value was significantly lower when sufentanil dose was doubled from TCI C_p of 0.5 ɳg ml^{-1} to 1 ɳg ml^{-1} [17]. The mean ± SD of Bispectral Index value was 41.4 ± 7.0 (range 30–52) before Mayfield head holder fixator and reached a peak Bispectral Index value 1 min after Mayfield head holder fixator of 56.4 ± 9.7 (range 44–71) in the 0.5 ɳg ml^{-1} group [17]. The mean ± SD of Bispectral Index value was 45.8 ± 6.9 (range 39–60) before Mayfield head holder fixator and reached a peak Bispectral Index value 1 min after Mayfield head holder fixator of 52.1 ± 8.1 (range 41–65) in the 1 ɳg ml^{-1} group [17]. The increase in Bispectral Index value of 14.6 ± 7.9 over the baseline value was significantly higher $P = 0.036$ in the 0.5 ɳg ml^{-1} group than the increase in Bispectral Index value of 6.7 ± 6.5 in the 1 ɳg ml^{-1} group, i.e., at the peak point pinning of the Mayfield head holder fixator; the administration of adequate sufentanil dose abolished increases in Bispectral Index over baseline value [17].

Furthermore, Brocas and colleagues [19], using a Bispectral Index BIS A-1000 (version 3.03) monitor in 11 Critical Care Unit (CCU) sedated patients with each patient serving as their own control, Bispectral Index was continuously recorded before, during and after tracheal suction without opioid administration (control period). Fifteen min after Bispectral Index level returned to baseline value an intravenous bolus of alfentanil 15 μg kg^{-1} was administrated 2-min before twice tracheal suction for 30 s (alfentanil period) [19]. They demonstrated that, before endotracheal suction, at a stable steady-state baseline level of Bispectral Index value of 61 ± 8 in the control period and 59 ± 7 in the alfentanil period; the increase in Bispectral Index value to 88 ± 14 during the control period 1 min after endotracheal suction without opioid was $P = 0.02$ significantly higher than the Bispectral Index value of 65 ± 9 during the alfentanil period 1 min after endotracheal suction following the administration of alfentanil 15 μg kg^{-1}, which remained significantly higher for 10 min after suction [19]. They concluded that an alfentanil bolus of 15 μg kg^{-1} markedly reduced the increase in Bispectral Index values observed immediately after tracheal suction [19]. Therefore, Bispectral Index monitoring in CCU might help optimize analgesia during invasive events. This study suggests that Bispectral Index variations reflect cortical reactivity to tracheal suction in critically ill sedated and ventilated patients [19].

This clearly demonstrates that the addition of μ agonist opioids markedly blunted, attenuated, or abolished increases in Bispectral Index values in response to nociceptive stimulation cortical reactivity "cortical arousal" due to stimulation of central noradrenergic neurons [19]. Therefore, the Bispectral Index reflects the pain-induced cortical arousal as well as cortical arousal without pain emergence from hypnotic anesthetic agents' effect [19]. This provides evidence of a distinction between the analgesic and hypnotic effects on the pEEG-DOA monitoring. This means that Bispectral Index correlates best with the effect of the primary hypnotic-based anesthetic techniques, and to avoid confounding the Bispectral Index values when they are used as a measure of anesthetic adequacy the adjunctive use of μ agonist opioid analgesics "Opioids Second Effect" must always be considered [20].

Table 7.4 enumerates and describes the "Opioids Second Effect" on depth of anesthesia monitoring.

Table 7.4 Opioids second effect

References	BIS Monitor	*n*	Agent/stimulus	BIS change	Explanation
Guignard et al. [3]	A-1000 (3.12)	10 × 5	Pain + remifentanil C_e 0, 2, 4, 8 or 16 ɳg ml^{-1}	Blunted BIS increase	Blunting noxious response
Hans et al. [17]	A-1000	10 × 2	Pain + sufentanil C_p 0.5 or 1 ɳg kg^{-1}	Blunted BIS increase	Blunting noxious response
Brocas et al. [19]	A-1000 (3.03)	11	Pain + alfentanil 15 μg kg^{-1}	Blunted BIS increase	Blunting noxious response

n number of patients per group, C_e predicted effect-site concentrations, C_p estimated plasma concentrations, *BIS* Bispectral Index, *AAI* A-Line Autoregressive index auditory evoked potential

References

1. Jensen EW, Litvan H, Struys M, Vazquez PM. Pitfalls and challenges when assessing the depth of hypnosis during general anaesthesia by clinical signs and electronic indices. Acta Anaesthesiol Scand. 2004;48:1260–7.
2. Ferreira DA, Nunes CS, Antunes LM, Santos IA, Lobo F, Casal M, et al. The effect of a remifentanil bolus on the bispectral index of the EEG (BIS) in anaesthetized patients independently from intubation and surgical stimuli. Eur J Anaesthesiol. 2006;23:305–10.
3. Guignard B, Menigaux C, Dupont X, Fletcher D, Chauvin M. The effect of remifentanil on the bispectral index change and hemodynamic responses after orotracheal intubation. Anesth Analg. 2000;90:161–7.
4. Schmidt GN, Bischoff P, Standl T, Voigt M, Papavero L, Schulte EJ. Narcotrend, bispectral index, and classical electroencephalogram variables during emergence from propofol/remifentanil anesthesia. Anesth Analg. 2002;95:1324–30.
5. Shafer SL, Varvel JR. Pharmacokinetics, pharmacodynamics, and rational opioid selection. Anesthesiology. 1991;74:53–63.
6. Pan Y-Z, Li D-P, Chen S-R, Pan H-L. Activation of μ-opioid receptors excites a population of locus coeruleus-spinal neurons through presynaptic disinhibition. Brain Res. 2004;997:67–78.
7. Merkel G, Eger EI. A comparative study of halothane and Halopropane anesthesia. Anesthesiology. 1963;24:346–57.
8. Sandin M, et al. Effects of pain stimulation on bispectral index, heart rate and blood pressure at different minimal alveolar concentration values of sevoflurane. Acta Anaesthesiol Scand. 2008;52:420–6.
9. Takamatsu I, Ozaki M, Kazama T. Entropy indices vs the bispectral index™ for estimating nociception during sevoflurane anaesthesia. Br J Anaesth. 2006;96:620–6.
10. Yoo KY, Jeong CW, Kang MW, Kim SJ, Chung ST, Shin MH, et al. Bispectral index values during sevoflurane-nitrous oxide general anesthesia in women undergoing cesarean delivery: a comparison between women with and without prior labor. Anesth Analg. 2008;106:1827–32.
11. Erden V, Erkalp K, Yangin Z, Delatioglu H, Kiroglu S, Ortaküz S, et al. The effect of labor on sevoflurane requirements during cesarean delivery. Int J Obstet Anesth. 2011;20:17–21.
12. Lobo FA, Schraag S. Limitations of anaesthesia depth monitoring. Curr Opin Anaesthesiol. 2011;24:657–64.
13. Lysakowski C, Dumont L, Pellégrini M, Clergue F, Tassonyi E. Effects of fentanyl, alfentanil, remifentanil and sufentanil on loss of consciousness and bispectral index during propofol induction of anaesthesia. Br J Anaesth. 2001;86:523–7.
14. Mi WD, Sakai T, Singh H, Kudo T, Kudo M, Matsuki A. Hypnotic endpoints vs. the bispectral index,95% spectral edge frequency and median frequency during propofol infusion with or without fentanyl. Eur J Anaesthesiol. 1999;16:47–52.
15. Kodaka M, Okamoto Y, Handa F, Kawasaki J, Miyao H. Relation between fentanyl dose and predicted EC 50 of propofol for laryngeal mask insertion. Br J Anaesth. 2004;92:238–41.
16. Struys MMRF, Vereecke H, Moerman A, Jensen EW, Verhaeghen D, De Neve N, et al. Ability of the bispectral index, autoregressive modelling with exogenous input-derived auditory evoked potentials, and predicted propofol concentrations to measure patient responsiveness during anesthesia with propofol and remifentanil. Anesthesiology. 2003;99:802–12.
17. Hans P, Brichant JF, Dewandre PY, Born JD, Lamy M. Effects of two calculated plasma sufentanil concentrations on the hemodynamic and bispectral index responses to mayfield head holder application. J Neurosurg Anesthesiol. 1999;11:81–5.
18. Coste C, Guignard B, Menigaux C, Chauvin M. Nitrous oxide prevents movement during orotracheal intubation without affecting BIS value. Anesth Analg. 2000;91:130–5.
19. Brocas E, Dupont H, Paugam-Burtz C, Servin F, Mantz J, Desmonts J-M. Bispectral index variations during tracheal suction in mechanically ventilated critically ill patients: effect of an alfentanil bolus. Intensive Care Med. 2002;28:211–3.
20. Sebel PS, Lang E, Rampil IJ, White PF, Cork R, Jopling M, et al. A multicenter study of bispectral electroencephalogram analysis for monitoring anesthetic effect. Anesth Analg. 1997;84:891–9.

8 Off-Label Use of Bispectral Index to Grade Conditions Other Than Depth of Anesthesia

Numerous articles appeared in literature using different processed Electroencephalography-Depth Of Anesthesia (pEEG-DOA) monitors, such as the Bispectral Index (BIS) to assess cerebral cognitive conditions other than Depth of Anesthesia (DOA) [1]. The pEEG-DOA algorithms were retrospectively derived from volunteers' Electroencephalography (EEG) changes with increasing incremental doses of certain γ-amino butyric acid (GABA)ergic anesthetic agents. Keeping in mind that these pEEG-DOA indices were specifically derived as the dose effect of these (GABA)ergic anesthetic agents, many of us would feel curiosity, angst, and cautious optimism when we read such articles promoting other pEEG-DOA monitors' uses outside anesthesia and the operating room [1].

However, such pEEG-DOA devices cannot be considered a "true" reflection of the EEG signal nor an independent measure of cerebral or cognitive brain functions [1]. Apparently, authors of these publications treat EEG-derived devices as monitors that somehow simplified, in a single numerical, the complex 10/20 Fast Fourier Transform (FFT) international system derived from 20 EEG electrodes within the complete EEG spectrum of δ, θ, β, and α band amplitudes [1]. These pEEG-DOA devices are certainly not. Even when using the pEEG-DOA devices in the intended target population of anesthetized patients, we often encounter a disparate clinical picture deviating from what these pEEG-DOA devices display. Here I present a balanced exposition of the intrinsic problems we will encounter when using pEEG-DOA devices outside their intended DOA monitoring, while focusing on light and shadow of the pEEG-DOA device use other than DOA [1]. Could such monitors be off-label utilized outside the scope of DOA, to "grade" and assess other EEG conditions like hepatic encephalopathy? The answer to that question seems to be a very cautious "yes", but only in very limited clinical conditions when the EEG changes of those clinical conditions' grades would somehow resemble EEG changes of increasing concentrations of the original anesthetic agents from which these pEEG-DOA algorithms were initially derived [1]. Because the Bispectral Index monitor has a rather appealing scale from 100 to 0, it is very tempting to find numerical cutoff values for already very well clinically graded conditions like hepatic encephalopathy West Haven criteria [2]. This is only because EEG changes with hepatic encephalopathy resemble, to a great extent, those EEG changes of increasing anesthetic agents' concentrations from which the pEEG-DOA algorithms were initially derived.

A. A. Dahaba, *Understanding the Numbers in Depth of Anesthesia Monitors*,
https://doi.org/10.1007/978-3-031-88260-9_8

8.1 Hepatic Encephalopathy

Severity of hepatic encephalopathy is currently clinically graded via West Haven Criteria (Table 8.1) [2], as well as various psychometric tests. In a consecutive, observer-blinded, validation, multicenter study, my research team Dahaba and colleagues assessed the discriminative power of the Bispectral Index monitor specificity and sensitivity for classification of overt hepatic encephalopathy [1]. We also evaluated the utility of the Bispectral Index monitor for grading the degree of deterioration or improvement progression of hepatic encephalopathy at three different centers in three countries on three continents, namely, Medical University of Graz, Austria; Zhe Jiang First University Hospital, Hang Zhou, People's Republic of China; and Cairo University, Egypt [1].

Over 18 months, our study's "Test-Set" was at the Medical University of Graz, which included eight healthy adult volunteers as control subjects and 28 consecutive patients with alcohol abuse liver cirrhosis suffering from episodic overt hepatic encephalopathy or acute deterioration of a persisting hepatic encephalopathy displaying changes or deterioration in personality/behavior, alterations in intellectual cognitive functions, or changes in consciousness level [1]. We first established Bispectral Index cutoff values for discriminating between the control subjects and patients with West Haven grades from 1 to 4 as

Table 8.1 West Haven criteria [2]

Grade 0	Normal mental state (one or more quantifiable abnormalities on psychometric testing)
Grade 1	Trivial lack of awareness, euphoria or anxiety, shortened attention span, impaired performance of addition
Grade 2	Lethargy or apathy, minimal disorientation for time or place, subtle personality change, inappropriate behavior, impaired performance of subtraction
Grade 3	Somnolence to semi-stupor, but responsive to verbal stimuli, confusion, gross disorientation
Grade 4	Coma (unresponsive to verbal or noxious stimuli)

"Test-Set" at Graz Medical University. We subsequently tested the Bispectral Index cutoff "Test-Set" values in 26 consecutive patients at Kasr El-Aini Cairo University Hospital and 31 consecutive patients at Zhe Jiang First University Hospital as the two "Validation-Sets" [1]. We meticulously chose these three centers because patients in the Medical University of Graz mostly suffered from liver cirrhosis because of alcohol abuse, whereas patients in Cairo University Hospital suffered mostly from liver cirrhosis because of chronic hepatitis C, and patients in Zhe Jiang First University Hospital suffered mostly from liver cirrhosis because of hepatitis B [1]. This provided a wide spectrum of liver disease etiologies to include in our study. Our study design is depicted in Fig. 8.1 flow chart that depicts the flow of patients' recruitment at the three centers, Medical University of Graz, Austrian patients with hepatic encephalopathy alcohol-related liver disease; Zhe Jiang First University Hospital, Hang Zhou, Chinese patients with hepatic encephalopathy because of hepatitis B liver disease; and Cairo University Hospital, Egyptian patients with hepatic encephalopathy because of hepatitis C liver disease. We recruited patients from the gastroenterology wards, outpatient clinics, and Critical Care Units (CCUs) of the three centers as well as Cairo University Hospital hematemesis emergency ward [1]. We excluded patients with psychotropic/hypnotic sedative drugs' chronic abuse or patients with neurologic conditions that would affect their mental status such as stupor, dementia, or stroke [1]. We calculated the hepatic dysfunction parameters Child-Pugh score (Table 8.2) [3] and the Model for End-Stage Liver Disease (MELD) score (Mayo Clinic MELD calculator website, http://www.mayoclinic.org/gi-rst/mayomodel5.html) [4] from our patients' laboratory and clinical medical charts. Our study primary endpoint was whether we can use 10-min artifact-free Bispectral Index mean value to grade the severity of hepatic encephalopathy [1]. Our secondary endpoints were to correlate mean Bispectral Index value to patients' various mental state and liver dysfunction parameters such as Glasgow Coma Scale (GCS) (Table 8.3) [5], venous

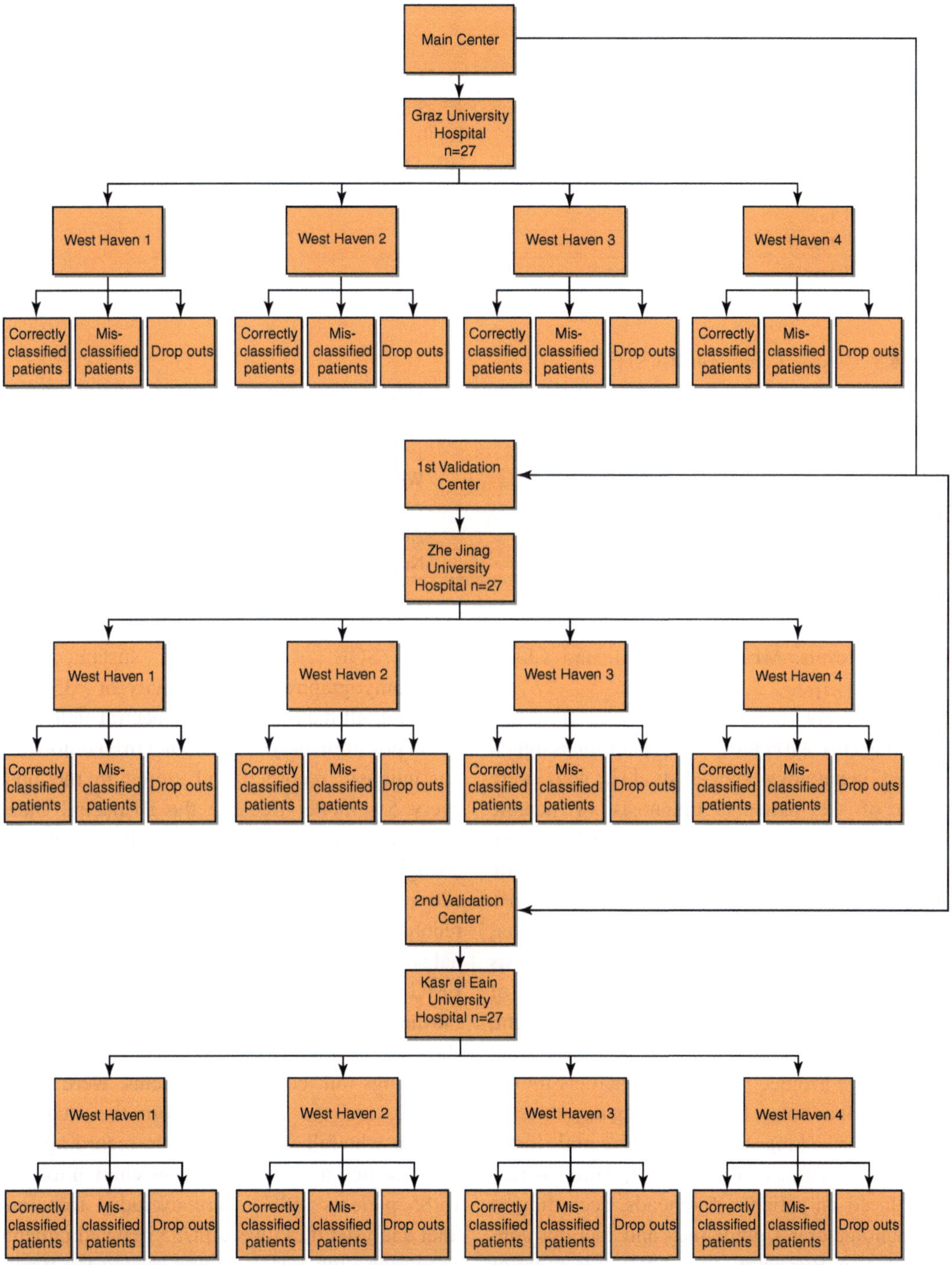

Fig. 8.1 Flow chart depicts our study design and the flow of the study patients' recruitment at the three centers, Medical University of Graz, Austrian patients with hepatic encephalopathy alcohol-related liver disease; Zhe Jiang First University Hospital, Hang Zhou, Chinese patients with hepatic encephalopathy because of hepatitis B liver disease; and Cairo University Hospital, Egyptian patients with hepatic encephalopathy because of hepatitis C liver disease

Table 8.2 Child-Pugh score [3]

Score	1	2	3
Bilirubin (μmol/L)	<35	35–51	>51
Albumin (g/L)	>34	28–34	<28
Prothrombin time (%)	>70	40–70	<40
Ascites	Absent	Moderate	Severe
Encephalopathy	Absent	Grade 1–2	Grade 3–4
Child-Pugh Class A	Total score < 7		
Child-Pugh Class B	Total score 7–9		
Child-Pugh Class C	Total score > 9		

Table 8.3 Glasgow Coma Scale (GCS) [5]

Eye opening	Verbal response	Motor response	Points
		Obeys commands	6
	Oriented	Localizes	5
Spontaneous	Confused	Withdraws	4
To voice	Inappropriate words	Abnormal flexion	3
To pain	Incomprehensible words	Abnormal extension	2
None	None	None	1

ammonia levels, MELD score [4], and Child-Pugh score [3].

Immediately before the Bispectral Index recordings in each center, one of our four assigned blinded hepatology specialists [1], who were aware of the patients' diagnosis of hepatic encephalopathy because of liver cirrhosis but blinded to the grade of their hepatic encephalopathy, assessed the West Haven criteria hepatic encephalopathy grade [2]. Thereafter, one of two blinded assigned anesthesiologists recorded an artifact-free 10 min period of Bispectral Index data in a quiet environment, which lasted up to 15 min of a total recording time [1]. We avoided all verbal or tactile stimulation, as well as noise, hence averting any external stimulus provoking transient Bispectral Index responses [6]. Bispectral Index "Quatro" sensor was placed on patients' foreheads, according to the manufacturer's guidelines and thereby connected to a Bispectral Index BIS-XP monitor (version 3.4, Medtronic, Dublin, Ireland) connected to a laptop computer. All conscious patients were instructed to remain calm and keep their eyes closed but not fall asleep [7], while keeping all their facial muscles completely relaxed [8]. After verifying a high Signal Quality Index (SQI) of >95% and sustained low Electromyography (EMG) activity of <35 decibel (dB), all data from the Bispectral Index and EMG power displayed in decibel (dB) units were continuously recorded and stored in the once every 5 s mode. Because the Bispectral Index spuriously increases with high EMG activity [8], all agitated patients who displayed high EMG activity were excluded from our study, and only recordings without technical artifacts from external stimulation or unexpected noise and with low EMG activity were included in the final analysis. Furthermore, because the Bispectral Index was shown to decline with sleep [7], at the end of the recordings all conscious patients were interviewed, and we excluded from the study all patients who fell asleep during the recordings [1]. As presented in our Fig. 8.1 flow chart, a total of 15/85 patients (18%) were excluded from our final analysis; these were 11/85 agitated patients exhibiting high EMG activity during the whole recording (four Austrian, four Egyptian, and three Chinese patients) and 4/85 patients who fell asleep during the recording (one Egyptian patient and three Chinese patients) [1].

Besides our inter-patient comparisons, we also conducted intra-patient comparisons of all available inpatients, in a longitudinal study, in which we evaluated the evolution of Bispectral Index values within the same patient [1]. Three to four days following the initial assessment, we reassessed all available inpatients, and we recorded a second 10-min artifact-free mean Bispectral Index value whenever there was either an apparent worsening of patients' condition because of complications such as gastrointestinal bleeding/infection or a manifest improvement in the patients' clinical condition with treatment [1]. In the three centers, patients' treatment involved initial management of all precipitating conditions such as electrolyte disturbance, gastrointestinal bleeding, dehydration, or infection. Patients' treatment included prophylactic antibiotics (Medical University of Graz Hospital guidelines were only prophylactic antibiotics in case of esophageal variceal bleeding), restriction of protein intake (Medical University of Graz Hospital guidelines were restriction of protein intake to 1 g kg^{-1} d^{-1}), oral ornithine aspartate (Hepa-Merz®) 6 g 8 h^{-1} for low-grade hepatic encephalopathy, or 20–40 g d^{-1} intravenous for high-grade hepatic encephalopathy, oral or intravenous branched-chain amino acids (Cairo University Hospital and Zhe Jiang First University Hospital guidelines), and oral lactulose (Laevolac®) 20 g 8 h^{-1} (Medical University of Graz Hospital guidelines administered via nasogastric tube and Cairo University Hospital guidelines administered as enema for high-grade hepatic encephalopathy) [1]. Furthermore, selected patients in Medical University of Graz Hospital underwent extracorporeal liver support using Fractionated Plasma Separation and Adsorption (Prometheus®, Fresenius Medical Care, Bad Homburg, Germany) or albumin dialysis using the Molecular Adsorbents Recirculating System (MARS®, Teraklin, Rostock, Germany) [9].

Analyzing Austrian patients' data revealed complete separation between the controls and West Haven grades 1 to 4 in mean Bispectral Index values, with a Bispectral Index cutoff value of 95 for controls versus West Haven grade 1, Bispectral Index cutoff value of 87.8 for grade 1 versus 2, Bispectral Index cutoff value of 70.2 for grade 2 versus 3, and Bispectral Index cutoff value of 54.4 for grade 3 versus 4 [1]. Figure 8.2 depicts patients at the three centers Bispectral Index values, Spectral Edge Frequency (SEF), and Electromyography (EMG) at the four West Haven Criteria Grades. Figure 8.3 depicts West Haven Criteria Grades versus the Bispectral Index cutoff values at the Medical University of Graz, Austria; Zhe Jiang First University Hospital, Hang Zhou, People's Republic of China; and Cairo University Hospital, Egypt. When we used rounded-up cutoff values of 95, 85, 70, and 55, respectively, for the two validation centers, this resulted in the same West Haven grades classification and produced agreement between Bispectral Index classification and West Haven grades with >94% specificity and sensitivity in 40/46 validation patients (87%). Figure 8.4 depicts the Receiver Operating Characteristics (ROC) of the Bispectral Index high predictive power in discriminating between West Haven Grade 1 versus Grades 2–4. Figure 8.5 depicts the ROC of the Bispectral Index high predictive power in discriminating between West Haven Grades 1–2 versus Grades 3–4. Figure 8.6 depicts the ROC of the Bispectral Index high predictive power in discriminating between West Haven Grades 1–3 versus Grade 4. Two Egyptian patients had their West Haven grades 1 and 2 misclassified as West Haven grades 2 and 3. Three Chinese patients had their West Haven grade 2 misclassified as grade 1 in two patients and as grade 3 in one patient. One Chinese patient' West Haven grade 3 was misclassified as grade 4 [1].

There was a significant difference in patients' mean ± SD Bispectral Index value between Bispectral Index 90.2 ± 2.5 West Haven grade 1, Bispectral Index 78.4 ± 6.6 West Haven grade 2, Bispectral Index 63.2 ± 4.8 West Haven grade 3, and Bispectral Index 45.4 ± 5.0 West Haven grade 4 [1]. We have clearly demonstrated that Bispectral Index values could be a beneficial method for assessing and grading the extent of the Central Nervous System (CNS) involvement in patients with chronic hepatic disease [1]. Weighted Kappa statistical analysis showed that the highest agreement between

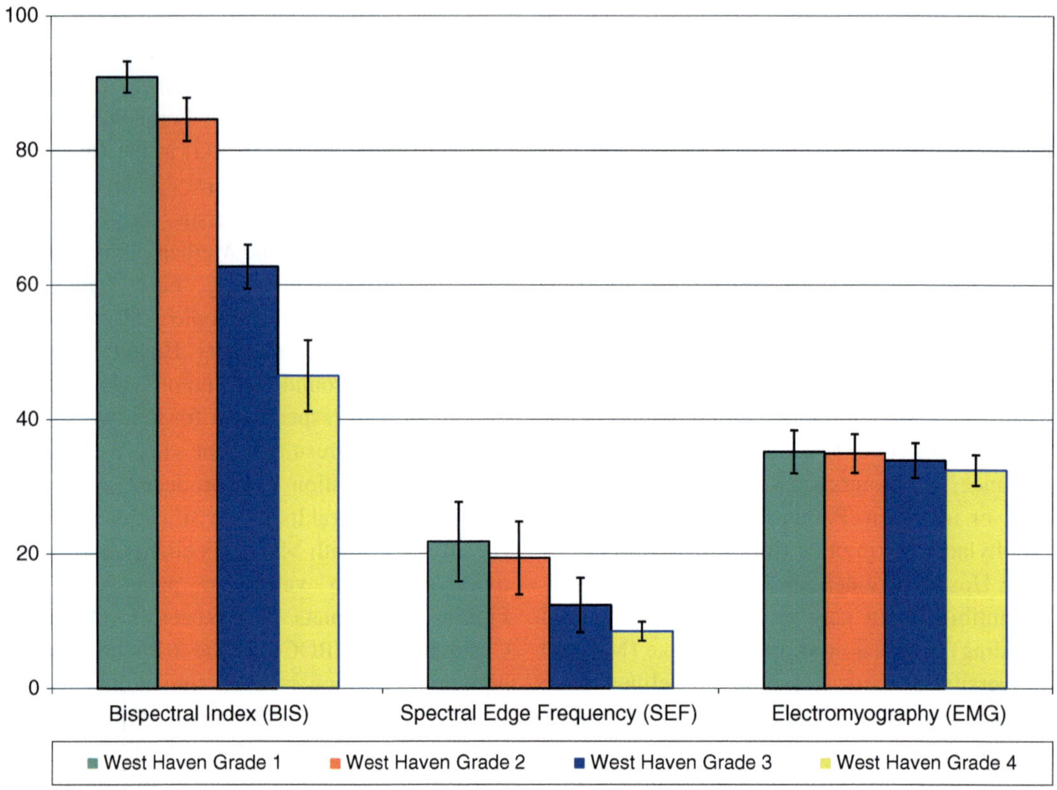

Fig. 8.2 Bispectral Index (BIS) values, Spectral Edge Frequency (SEF), and Electromyography (EMG) at the 4 West Haven Criteria Grades

West Haven grades and Bispectral Index values was in Austrian patients, followed by Egyptian patients [1].

Our multicenter study revealed significant differences between West Haven grades in the mean Bispectral Index values, Child-Pugh score, MELD score, Glasgow Coma Scale, and venous ammonia levels, with no significant center effect [1]. Our study results showed good correlation $r^2 = 0.85$ between West Haven grades and Glasgow Coma Scale, moderate correlation $r^2 = 0.52$ between West Haven grades and Child-Pugh Score, and weak correlation between West Haven grades and MELD score $r^2 = 0.17$ and venous ammonia levels $r^2 = 0.23$ [1]. Mean Bispectral Index values well correlated with West Haven grades $r^2 = 0.90$ and Glasgow Coma Scale $r^2 = 0.78$, moderately correlated $r^2 = 0.47$ with Child-Pugh score, and weakly correlated with MELD score $r^2 = 0.19$ and venous ammonia levels $r^2 = 0.20$ [1].

In our longitudinal study, we reassessed 18 patients' West Haven grades as we recorded a second 10-min artifact-free mean Bispectral Index value (eight Egyptian patients, seven Chinese patients, and three Austrian patients).

Figure 8.7 Three-dimensional (3D) diagram of West Haven grades versus the Bispectral Index values in 18 patients followed up in the longitudinal study at the Medical University of Graz, Austria; Zhe Jiang First University Hospital, Hang Zhou, People's Republic of China; and Cairo University Hospital, Egypt, showing that the good agreement of Bispectral Index values with West Haven grades is retained during the evolution of hepatic encephalopathy. Time interval between the first and second evaluation of hepatic encephalopathy was ≥36 hours. Sagittal planes = Bispectral Index West Haven grades cut-off values, asterisks = first measurements, symbols = second measurements patients. Four longitudinal study patients' West Haven grades

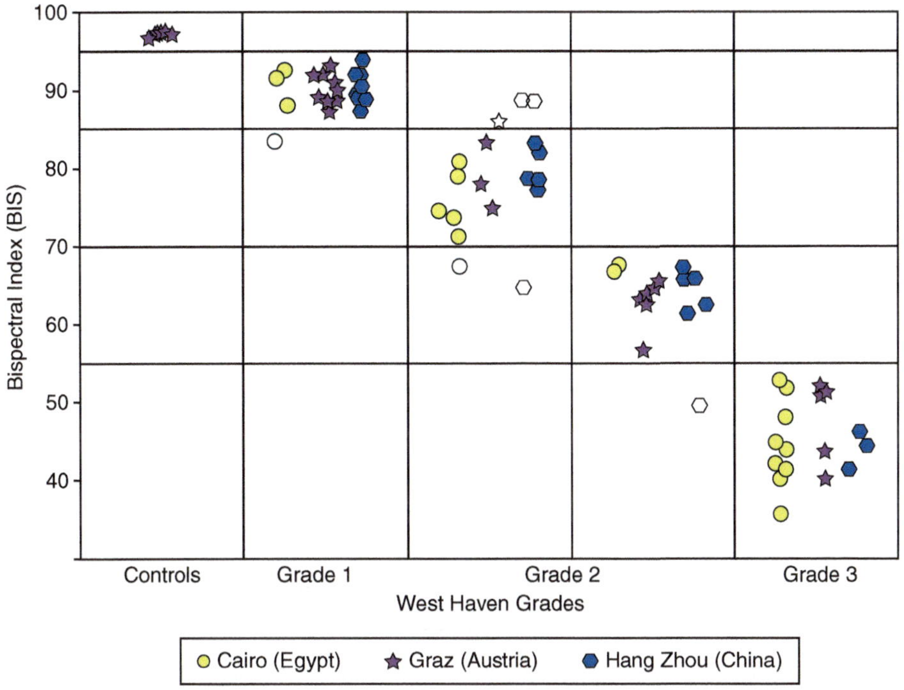

Fig. 8.3 Scatter plot of West Haven Criteria Grades versus the Bispectral Index (BIS) cutoff values at the Medical University of Graz, Austria; Zhe Jiang First University Hospital, Hang Zhou, People's Republic of China; and Cairo University Hospital, Egypt. Horizontal lines = Bispectral Index cutoff values, colored symbols = correctly classified patients, open symbols = misclassified patients. (Reproduced with permission # 5933471378414 from Dahaba et al. [1])

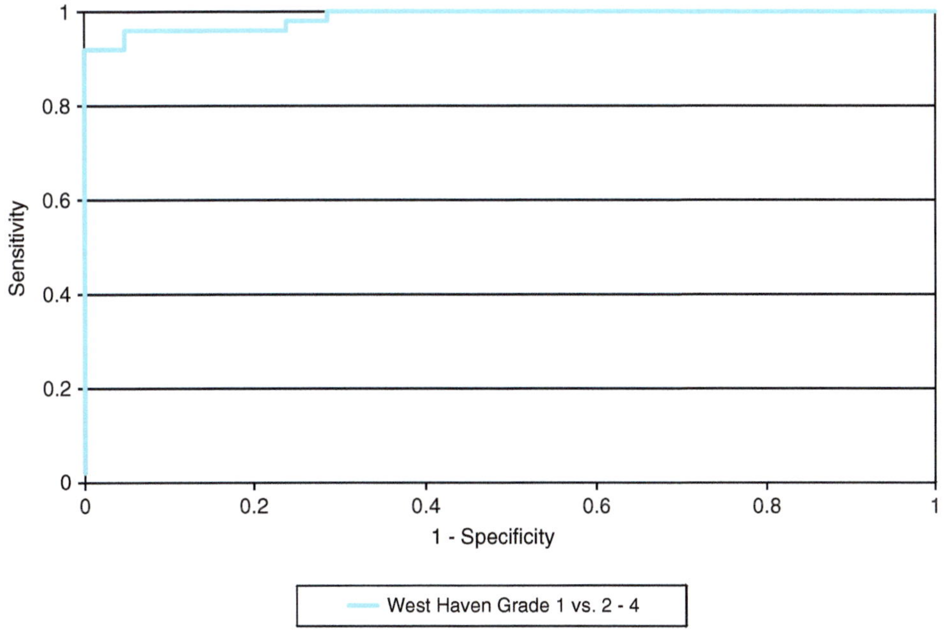

Fig. 8.4 Receiver Operating Characteristics (ROC) curve depict the Bispectral Index (BIS) high predictive power in discriminating between West Haven Grade 1 versus Grades 2–4

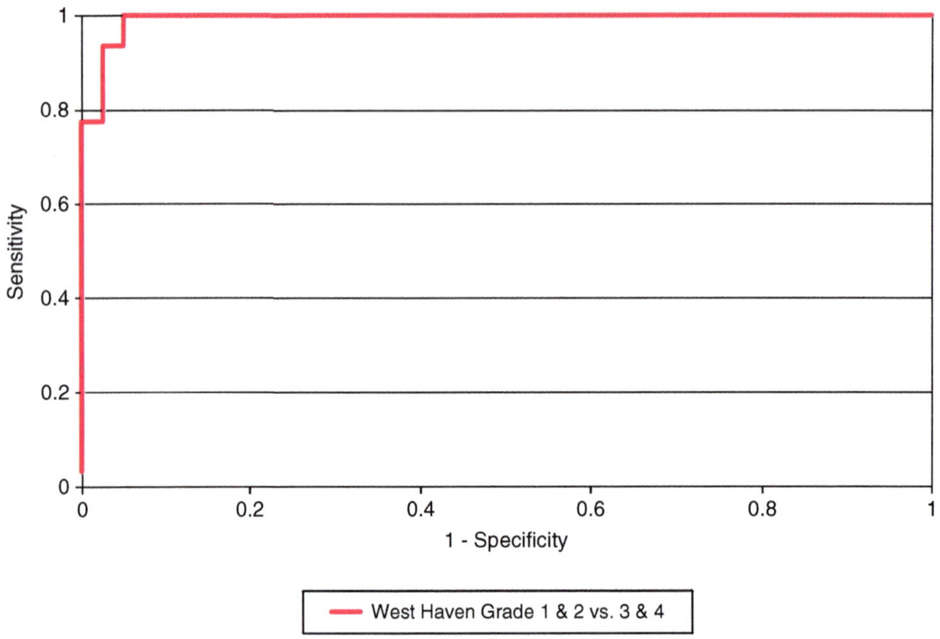

Fig. 8.5 Receiver Operating Characteristics (ROC) curve depict the Bispectral Index (BIS) high predictive power in discriminating between West Haven Grades 1–2 versus Grades 3–4

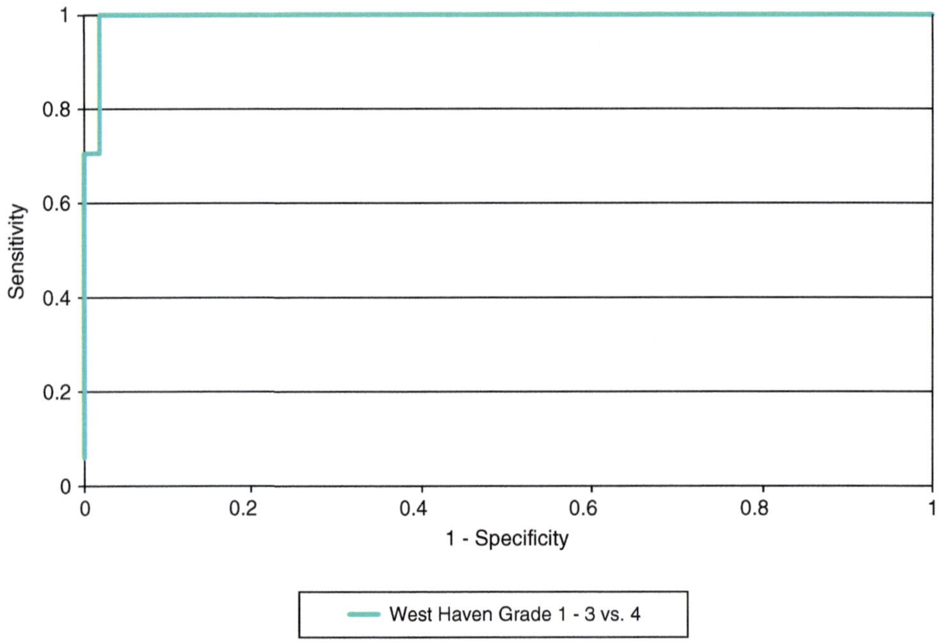

Fig. 8.6 Receiver Operating Characteristics (ROC) curve depict the Bispectral Index (BIS) high predictive power in discriminating between West Haven Grades 1–3 versus Grade 4

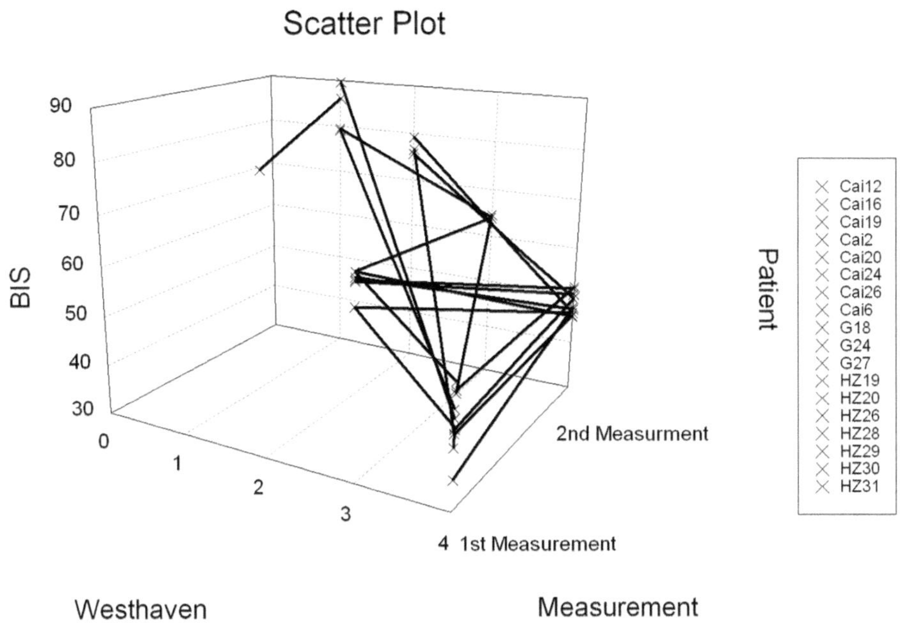

Fig. 8.7 Three-dimensional (3D) diagram of West Haven grades versus the Bispectral Index (BIS) values in 18 patients followed up in the longitudinal study at Medical University of Graz, Austria; Zhe Jiang First University Hospital, Hang Zhou, People's Republic of China; and Cairo University Hospital, Egypt. Time interval between first and second evaluation of hepatic encephalopathy was ≥36 h. Sagittal planes = Bispectral Index West Haven grades cutoff values, asterisks = first measurements and second measurements patients

deteriorated from grade 3 to grade 4 (Fig. 8.8). Five longitudinal study patients remained stable at grade 4 [1]. Four longitudinal study patients showed one West Haven grade improvement, namely, one patient from West Haven grade 2 to 1, one patient from West Haven grade 3 to 2, and two patients from West Haven grade 4 to 3. Five longitudinal study patients showed more than one West Haven grade improvement, namely, one patient improved from West Haven grade 3 to 1, two patients improved from West Haven grade 4 to 2, and two patients improved from West Haven grade 4 to 1 (Fig. 8.9) [1]. Just 2/18 West Haven grade 1 longitudinal study patients, one Egyptian patient and one Chinese patient, were misclassified by the second measurement of the Bispectral Index as West Haven grade 2. Thus, our study results showed an agreement between West Haven grades and the Bispectral Index classification in 16/18 (89%) of our followed up longitudinal study patients. Hence, our longitudinal study confirmed the high discriminative power of the Bispectral Index in both high-grade and low-grade follow-up patients with hepatic encephalopathy [1].

8.1.1 Hepatic Encephalopathy Spectral Electroencephalography (EEG) Neuropsychiatric Analysis

The remarkable ability of the Bispectral Index to "sharply separate" different grades of hepatic encephalopathy West Haven criteria that we showed in our study was previously demonstrated by early pioneer researchers documenting the Electroencephalography (EEG) discrimination of the hepatic encephalopathy mental status.

Several abnormalities in the EEG of hepatic encephalopathy patients have been described

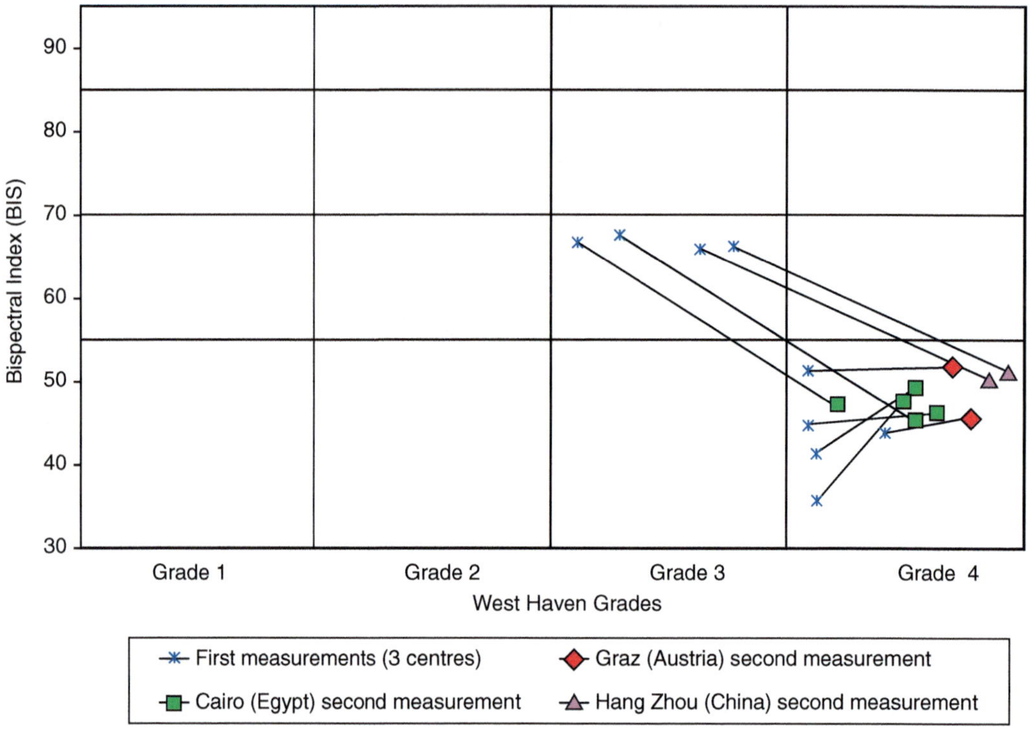

Fig. 8.8 Longitudinal study patients whose West Haven grades deteriorated from Grade 3 to Grade 4 or remained at Grade 4 with no improvement, asterisks = first measurements, symbols = second measurements patients

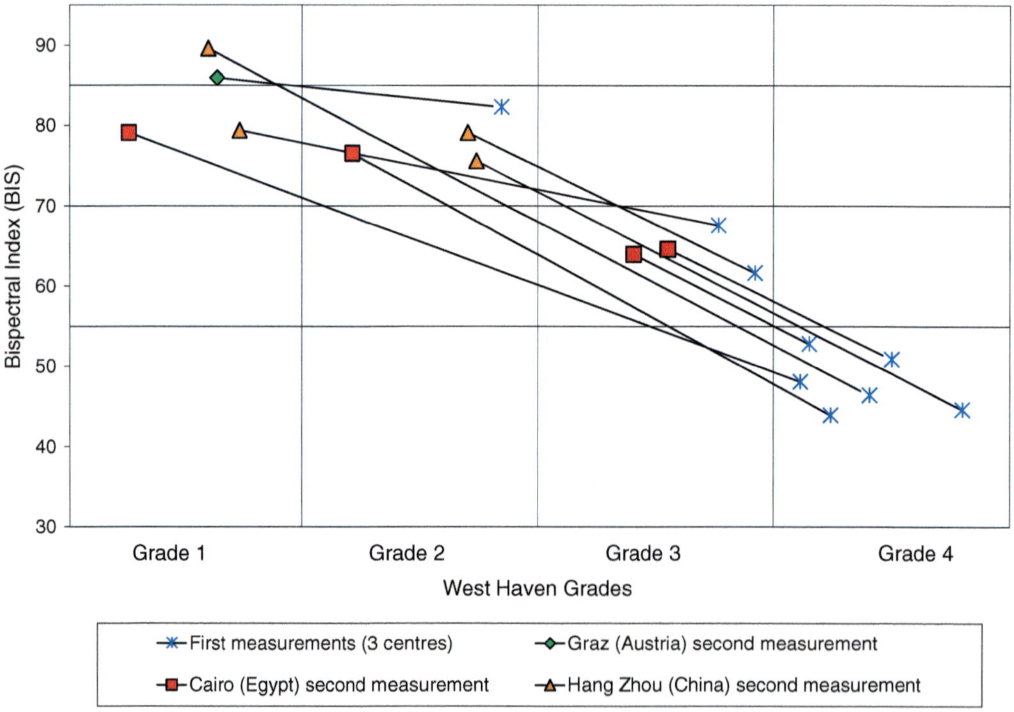

Fig. 8.9 Longitudinal study patients showing one, two, or three West Haven grades improvements, asterisks = first measurements, symbols = second measurements patients

[10]. The severity of liver cirrhosis was found to typically involve an early progressive background EEG tracing slowing due to slowing of the dominant posterior waking α-wave activity, general decline in EEG amplitude, gradual progressive replacement imposition of slower θ-wave and then δ-wave frequencies, and disorganization/de-synchronization of the dominant activity with no changes in β-wave activity [11]. Sometimes this stage of hepatic encephalopathy precoma may show well preserved α-wave rhythm, although somewhat slower in frequency, superimposed on bursts of frontal or more posterior bursts of bilateral paroxysms of θ-wave and/or δ-wave activities [12], or a general amplitude decline [10]. Hepatic coma is characterized by generalized large amplitude δ-wave activity that is maximal over the anterior regions [12]. Further deterioration of hepatic encephalopathy is characterized by a progressive decline in the δ-wave amplitude, periods of flattening, eventually ending in electrocerebral silence [12].

Although EEG recording analysis is an objective approach to the evaluation of hepatic encephalopathy, the simple analysis of the conventional EEG tracings certainly carries a considerable number of variabilities and lack of precision [10]. As expected, these techniques were soon followed by the arrival of more elaborate methods, such as "EEG spectral analysis" [10]. However, the studies undertaken are challenging to compare because of significant notable inter-study differences in the methods used to acquire and analyze the EEG recordings and differences in the patients and controls populations [10]. The studies utilizing spectral analysis varied in terms of choice of derivations, epoching, filters, bandwidth, and equipment [10]. The reported EEG diagnostic sensitivity varies from 83 to 100% for overt hepatic encephalopathy, depending on the technique used and the patients studied [10]. The specificity and sensitivity of the techniques are critically dependent on the performed analysis and the quality of the recordings [10].

Foley and colleagues [13] first reported in the 1950s that abnormalities of δ-wave activity, a feature of the visual EEG, correlated well with patients' hepatic encephalopathy and hepatic coma mental status, as they [13] described the nonspecific slow high-voltage basic EEG rhythm of the δ-wave discharges in patients with hepatic encephalopathy [13]. Since the first description of the EEG changes in hepatic encephalopathy [13], there have been many reports of the usefulness of this technique in monitoring the course of patients with severe liver failure hepatic encephalopathy [14]. Bickford and Butt [15] in 1955 also pointed out that there was an electrical δ-wave activity with hepatic encephalopathy; as the 5–6 Hz θ-wave phase abnormality correlated well with the severity of the mental status varying from normal alertness to deep hepatic encephalopathy coma [15].

Parsons-Smith and colleagues [16] in 1957 later clearly described and graded the conventional visual qualitative electroencephalography (qEEG) abnormality along with the neurological and mental features of hepatic encephalopathy in five separate "stages" (Table 8.4) [16]. They reported that a good correlation exists between the severity of the EEG changes and the clinical condition of patients with hepatic encephalopathy liver disease [16]. The high-voltage slow δ-discharge originally reported by Foley and colleagues [13] was only seen in patients whose consciousness was profoundly disturbed and represents their advanced Grade-E [16]. The earlier 5–6 Hz θ-wave phase abnormality reported by Bickford and Butt [15] corresponded to their Grade-C and Grade-D. However, Parsons-Smith's classification is merely qualitative, meaning that inter- and possibly intra-observer errors may limit its applicability [16].

In the early 1960s, Laidlaw and Read [17] quantified the dominant EEG frequencies in cirrhotic patients and controls using an "electronic EEG wave analyzer" [17]. Quantitative EEG analysis has proven useful in hepatic encephalopathy longitudinal studies [18]. According to Laidlow and Read [19], the EEG "disturbances" provided very valuable information because they are prognostically more reliable than the clinical hepatic encephalopathy evaluation or the ammoniemia [19]. This pioneer research [19] clearly showed the need for precise quantification of hepatic encephalopathy EEG signals [19].

Table 8.4 Parsons-Smith neuropsychiatric qualitative electroencephalography (qEEG) classification [16]

Grade A	No neuropsychiatric abnormality detected
Generalized suppression of α rhythm and its frequent replacement by faster potentials in all leads. The tracings in this grade were generally flat and featureless	
Grade B	Trivial lack of awareness, euphoria, or apathy in the absence of unequivocal neurological abnormality and impaired performance of simple psychological tests, including the construction of a five-pointed star and the ability to add or subtract serial sevens
The α rhythm was very unstable and disturbed by random waves at 5–7 Hz over both hemispheres with an underlying fast activity	
Grade C	Obvious personality changes with definite neurological abnormalities, of which a flapping tremor. The gross facade of personality is preserved
An α rhythm is still present but disturbed over both hemispheres by medium-voltage 5–6 Hz waves that occur in runs and are not paroxysmal	
Grade D	Advanced confusion and disorientation
The 5–6 Hz rhythms seen in Grade C are now constant in all areas and replaced all other cortical activity electroencephalography	
Grade E	Stuporous but responding to stimuli
The 5–6 Hz rhythms are replaced by frontally preponderate bilateral synchronous 2 Hz rhythms, which spread backward over the hemispheres. At times the 6 Hz rhythms might reappear, with the occurrence of diencephalic discharges.	

Demedts and colleagues [20] in 1973 performed simultaneous assessments of neuropsychic changes and EEG patterns of 52 patients with hepatic encephalopathy [20]. They graded EEG changes in five stages (Table 8.5) [20]. The EEG abnormalities consisted of a nonspecific symmetrical slowing of the EEG pattern [20]. They found a very good correlation between patients with chronic or acute liver disease EEG changes and stages of consciousness of the degree of the clinical hepatic encephalopathy [20].

Trewby and colleagues [21] in 1978 performed fronto-temporal bipolar EEG recordings with electrodes applied to the scalp frontal temporal regions at the level of the orbit upper margin and approximately 1 inch anterior to the ear [21]. They reported that at the beginning there was slowing of the α-wave rhythm with hepatic encephalopathy onset. With increasing drowsiness, this was replaced by lower frequency θ-wave activity [21]. As the hepatic encephalopathy coma deepens, high amplitude δ-wave becomes more obvious [21]. Further hepatic encephalopathy coma deepening was associated with a decrease in waves' amplitude and periods of EEG flattening, eventually becoming isoelectric [21].

After that, major efforts were undertaken to optimize EEG spectral analysis and to assess its role in improving the diagnostic and prognostic value of the hepatic encephalopathy EEG changes. Van der Rijt and colleagues [22] in 1984 reported the first trial of an objective classification of EEG changes in hepatic encephalopathy based on EEG parameters obtained by automatic quantitative EEG spectral analysis [22]. They assessed the clinical grades of cirrhotic patients' hepatic encephalopathy using the complex 10/20 Fast Fourier Transform (FFT) power spectrum. The automated EEG spectral analysis uses a combined EEG Mean Dominant Frequency (MDF) for 100 s epochs and the relative power spectral analysis of the θ-wave and δ-wave bands, which provides an automated estimation of the dominant EEG frequencies [22]. They investigated the discriminatory value of the MDF and the so-called power spectrum in controls and patients with different hepatic encephalopathy grades [22]. The parameters calculated in the range of 1–25.6 Hz to minimize the effect of slow eye movements were the MDF, the relative

Table 8.5 Demedts neuropsychiatric electroencephalography (EEG) classification [20]

Stage 0	Normal cerebral function
Normal electroencephalography (EEG) pattern with predominant parieto-occipital α rhythm and minimal transient θ rhythm	
Stage 1	Only minor abnormalities of flapping tremor, hypererflexia, euphoric indifference, and decreased intelligence and memory
Slight slowing or instability of the α rhythm, intermittent and moderate θ dysrhythmia	
Stage 2	Disorientation, agitation, or stupor and often hypertonia
Predominant and diffuse θ dysrhythmia with a slow and reduced α rhythm and minimal transient fronto-centro-temporal δ waves	
Stage 3	Complete coma with or without defense reaction on external stimuli and pupil reflexes; often a positive Babinski sign and decerebration rigidity, especially in patients with acute hepatitis
Predominant diffuse dysrhythmia	
Stage 4	Finally could end in "Coma Dépassé," characterized by abolition of vegetative functions
Flattened EEG tracing with a periodic slow and dyrhythmia	

Table 8.6 Van der Rijt automated quantitative electroencephalography (EEG) classification [22]

	MDF	θ (%)	δ (%)
Grade 0	≥6.4	<35	
Grade 1	≥6.4	≥35	
Grade 2	<6.4		<70
Grade 3–4	<6.4		≥70

δ-wave (1–3.5 Hz), θ-wave (4–8 Hz), α-wave (8.5–12.5), and β-wave (13–25.6 Hz) activities [22]. However, one spectrum was sometimes discarded in case of pronounced muscle activity namely the β frequency band [22]. The MDF in the controls [22] and patients with Grade-0 and Grade-1 hepatic encephalopathy was in the normal ≥6.4 Hz range and clearly differed from patients with Grade-2, Grade-3, and Grade-4 hepatic encephalopathy who had a low <6.4 Hz MDF. Patients with Grade-1 hepatic encephalopathy could be distinguished quite easily from the control group by means of their increased θ-wave activity [22]. The relative powers of θ-wave activity in Grade-1, Grade-2, Grade-3, and Grade-4 patients were clearly ≥35% higher than those of the control group of <35%. Grade-3 and Grade-4 hepatic encephalopathy could be identified by a high power of ≥70% δ-wave activity, whereas in Grade-0 controls and patients with Grade-1 and Grade-2 hepatic encephalopathy the δ-wave activity was almost always <70% (Table 8.6) [22]. Visual examination of the power spectra revealed that hepatic encephalopathy Grade-2 patients could be identified by a biphasic power spectrum peak in the θ-wave and another peak in the δ-wave range [22]. This technique increases the reliability of EEG assessment as it is less inter- and intra-operator dependent and allows the relative contribution of the different rhythms to the overall background frequency to be estimated [22]. However, these hepatic encephalopathy EEG classifications and clinical scores were still characterized by considerable inter- and intra-observer errors [22]. They concluded that automated EEG analysis based on the parameters MDF and the relative powers of the θ-wave and δ-wave bands is very suitable for objective classification of individual patients' hepatic encephalopathy [22].

Sagales and colleagues [23] in 1990 using topographic analysis of the EEG spectral parameter Brain Electrical Activity Mapping (BEAM) to study cerebral electrical activity of 44 patients with chronic liver disease hepatic encephalopathy, demonstrated general spatial distribution EEG slowing [23]. Eight healthy subjects were used as a control group [23]. All subjects were studied in an awake with eyes closed condition using the 10/20 FFT EEG international system. Patients were classified in five groups (Table 8.7) [23]. The spectra were divided into α_1 (8–9.75 Hz), α_2 (10–11.72 Hz), β (12–32 Hz), θ (4–7.75 Hz), and δ (0.5–3.75 Hz). They calculated MDF, the total Power Spectral Density (PSD), and the PSD percentage of different EEG wave bands [23]. An increase in δ-PSD and θ-PSD with a marked decrease in α-PSD, β-PSD, and MDF allowed clear discrimination of Group

Table 8.7 Sagales's neuropsychiatric Brain Electrical Activity Mapping (BEAM) classification [23]

Grade 0	No clinical signs
The highest Power Spectral Density (PSD) is in the α band, particularly in the occipital areas. Much lower PSD percentages in the β, θ, and δ bands. A decrease of the α-PSD band is more evident in the occipital areas. Overall topographic distribution of cerebral activity is not noticeably altered. The MDF decreased, mainly in the occipital regions. A marked increase in the power percentage of the β band, in the anterior area	
Grade 1	Euphoria or apathy, lowered perception. Short attention span. Inverted sleep patterns. Tremor
A >25% increase in the δ-PSD percentage in the occipital areas with a maximum 65% increase in the frontal regions. In the θ band there is central 35% increase and occipital 45% increase. A diffuse increase in δ-PSD, mainly in the parieto-occipital areas and tendency to an increased θ power in the precentral regions. A decrease in α power is more evident in the centro-parietal areas. The MDF decreased, although not statistically significant	
Grade 2	Accentuated grade I signs. Obvious personality changes. Flapping tremor
An increase in the δ-PSD and θ-PSD bands and a decrease in the α-PSD, β-PSD bands and MDF. Some triphasic waves might appear	
Grade 3	Bizarre behavior. Obvious confusion. Disorientation. Somnolence. Stupor
Grade 4	Coma
A significant increase in δ-PSD percentage in all cerebral areas with low α-PSD and β-PSD percentages over all areas. A decrease in the α-PSD, β-PSD bands and MDF, together with an increase in δ-PSD. Triphasic waves appear	

1 and Group 2 patients from control subjects and discrimination of Group 1 from Group 2 patients [23]. Brain mapping in Group 3 and Group 4 patients showed an increase in δ-PSD with a decrease in α-PSD, β-PSD, and MDF that was associated with subjects of these two groups' clinical condition deterioration [23].

Similarly, Kullmann and colleagues [24] in 2001 using BEAM in patients with low-grade hepatic encephalopathy reported a precocious increase in β-wave activity accompanied by a significant shift of the β-wave activity to the cerebral frontocentral area [24]. These are the levels were the Bispectral Index *BetaRatio* subparameter descriptor is the most influential [25]. We believe that the high discriminative power of Bispectral Index monitoring between West Haven grades arises from the ability of the Bispectral Index algorithm to accurately detect and display the graduated increase in the β-wave power [26].

Amodio and colleagues [27] in 1999 used a computerized EEG "Biparietal spectral analysis" to classify, distinguish, and discriminate between hepatic encephalopathy grades' clinical psychometric findings according to the degree of EEG changes [27]. Based on visual assessment of posterior basic rhythm, they graded the EEG background frequency for each patient in a "semi-quantitative" 5-grade classification (Table 8.8) [27]. Spectral analysis was claimed to improve the assessment of hepatic encephalopathy by providing reliable parameters correlated to patients' mental state and by decreasing subjective inter-operator variability [27]. Qualitative visual EEG was blindly evaluated by two electroencephalographers, whereas spectral EEG analysis was performed using three different techniques, namely, spectral EEG Temporo-Occipital MDF analysis, spectral biparietal analysis, and spectral EEG analysis of the Peak Power Frequency (PPF), the frequency which displays the highest absolute power of the relative powers of the rhythms β (13.5–25.5 Hz), α (8.5–13), θ (4–8 Hz), and δ (1–3.5 Hz) [27]. Compared to controls, the cirrhotic subjects with psychometric changes had lower MDF, PPF, α-wave and β-wave relative power, and higher θ-wave relative power [27]. Moreover, compared to cirrhotic subjects without psychometric alterations, the cirrhotic subjects with psychometric changes had lower MDF, PPF, α-wave relative power, and higher θ-wave relative power [27]. The variations

Table 8.8 Amodio's biparietal spectral semi-quantitative electroencephalography (EEG) classification [27]

Grade 0	No signs of encephalopathy
Normal frequency of 8.5–12 Hz, no significant EEG alterations. Well-structured EEG with stable and symmetrical posterior basic rhythm >8 Hz and < 13 Hz dominant in the posterior regions, with medium amplitude of 30–50 μV and is reactive to eye opening. No slow activities or irritative signs	
Grade 1	Normal-limit EEG
Posterior basic rhythm of 7–8 Hz. Unstable or suppressed α rhythm frequently replaced by high prevalence of diffuse β rhythm	
Grade 2	Mild signs of encephalopathy
Mild alterations of 5–6 Hz posterior basic rhythm. Low frequency 8 Hz α rhythm interrupted by random θ range waves over both hemispheres	
Grade 3	Distinctive features of encephalopathy
Background θ range activity, diffuse 3–4 Hz posterior basic rhythm coarse alterations over both hemispheres, random appearance of δ range high waves	
Grade 4	Signs of severe encephalopathy
Posterior basic 1–2 Hz rhythm. Severe disorganization of EEG activity with no normal elements, diffuse asynchronous θ and δ waves over both hemispheres	

of θ-wave and δ-wave relative power appeared to reflect more readily the variations of psychometric performance [27]. The 90% reliability and accuracy of the classification based on biparietal spectral analysis linked to clinical and psychometric findings was higher than that based on the 75% reliability and accuracy of the conventional qualitative visual EEG analysis [27]. In fact, a visual EEG reading was influenced by the subjective electroencephalographer's choice of the most important EEG signal features and their interpretation [27]. Moreover, the agreement between qualitative EEG assessments would be even lower if the electroencephalographer came from different units. Compared with visual EEG, spectral EEG analysis may improve hepatic encephalopathy assessment by decreasing, if not eliminating, the weight of operators' subjectivity by providing quantitative parameters [27].

The latest attempt to use EEG for assessment of hepatic encephalopathy was by Ciancio and colleagues [28], in 2002 when they evaluated the sensitivity of spectral Electroencephalography (sEEG) analysis, compared with visual electroencephalography (vEEG), for quantitating cir rhotic patients with hepatic encephalopathy brain dysfunction. They used computerized sEEG individual channels relative powers of β-wave (12.5–32 Hz), α-wave (8.5–12 Hz), θ-wave (4.5–8 Hz), and δ-wave (0–4 Hz) rhythms, PPF and MDF mean power of each frequency weighted according to frequency rate [28]. Whereas, visual Electroencephalography (vEEG) was scored in four grades with grades ≥2 considered abnormal [28]. The sEEG patterns of healthy controls were distinctly different from sEEG patterns of patients with overt hepatic encephalopathy [28]. Mean sEEG values were significantly lower in patients with chronic or recurrent hepatic encephalopathy than those without a history of hepatic encephalopathy [28]. There was an increase in θ-wave relative power and a decline in α-wave relative power in cirrhotic patients (31%) compared with controls (9.5%), which seems to be a reliable indicator of hepatic encephalopathy [28]. The most prominent difference between healthy controls and cirrhotic patients was that the mean MDF was significantly less in cirrhotic patients with advanced liver disease (8.8 Hz) than in controls (9.7 Hz) [28]. There was a linear correlation between MDF values and Child-Pugh score clinical stages of disease severity that was able to differentiate healthy controls from cirrhotic patients [28]. In their study [28], there was poor correlation between sEEG and psychometric tests [28]. These observations suggest that psychometric and electrophysiological tests may assess different components of hepatic encephalopathy. The discrepancy between psychometric test and sEEG results highlights the concept of hepatic encephalopathy's multidimensional nature [28]. Their study results revealed a good correlation between vEEG and sEEG in the recognition and classification of overt hepatic encephalopathy [28]. The sEEG provided reliable quantitative information to evaluate the degree of hepatic encephalopathy and appears to be more sensitive and superior to vEEG in identifying subclinical stages of hepatic encephalopathy because it showed clinically undetected brain dysfunction in 31% of cirrhotic patients versus 23% by vEEG [28]. They concluded that sEEG analysis might further improve the hepatic encephalopathy recognition by providing quantitative parameters of brain dysfunction and decreasing inter-operator variability [28].

We have noticed that the Bispectral Index discriminative power was slightly higher for patients with hepatic encephalopathy West Haven high grades 3 and 4 than patients with hepatic encephalopathy West Haven low grades 1 and 2. West Haven high grades 3 and 4 generally correspond to Parsons-Smith's EEG classification grades B, C, D, and E [16] and would also correspond to Amodio's EEG classification grades 2, 3, and 4 [27], as they exhibit first "random" then "diffuse" activity in the 5–6 Hz waves range [16], corresponding to the θ-wave and δ-wave activities [25]. This is the level of the Bispectral Index subparameter *SynchFastSlow* descriptor predominance, when the Bispectral Index *Burst Suppression* descriptor would detect deeper EEG suppression [25]. The high discriminative power of the Bispectral Index monitoring at these levels results from the Bispectral Index algorithm ability to accurately detect and display the gradual decrease in the cortical generators of the α-wave and the β-wave activities with a shift toward control by the septal /hippocampal and thalamocortical generators of δ-wave and θ-wave activities [26].

In our multicenter study, hepatic encephalopathy etiology in the Medical University of Graz University Hospital was mostly alcohol-related, whereas hepatic encephalopathy in Cairo University Hospital was mostly due to chronic hepatitis C, and in Hang Zhou Zhe Jiang First University Hospital was mostly chronic hepatitis B related. Because our study results exhibited no significant center effect, this clearly shows that Bispectral Index monitoring possesses high discriminative power for hepatic encephalopathy grading unrelated to the hepatic encephalopathy etiology [1].

Our study [1] clearly shows that taking into consideration the Bispectral Index monitoring confounding factor limitations by ensuring that patients do not fall asleep or exhibit excessive muscular activity, Bispectral Index monitoring signifies an objective and simple neurophysiological tool for monitoring and grading the extent of CNS involvement in patients with chronic liver disease [1].

8.2 Bispectral Index as an Early Intraoperative Indicator of Initial Poor Graft Function After Orthotopic or Split-Graft Liver Transplant

With the worldwide continuous improvement of operative techniques and rising success rates, liver transplantation has become an effective treatment for end-stage liver disease [29]. However, cold or rewarming ischemia is inevitably associated with allograft damage during liver procurement and implantation [30]. Initial Poor Graft Function (IPGF) represents one of the significant complications that could directly impact the survival of the liver allograft. The most serious type of IPGF is primary non-function (PNF) [31], a life-threatening complication occurring in 5.8% of cases [32] and needs urgent hepatic re-transplant [29]. Thus, reliable graft monitoring that enables continuous evaluation of graft function is crucial after liver transplant. Early detection of all IPGFs and specially PNFs could help determine therapy measures including change in medical protocols or performing further diagnostic measures that could prevent PNF or IPGF [29]. Hence, there is a growing need for a simple early indicator by a monitor that is readily used in the operating room (OR) setting that could draw attention to an ensuing developing serious complication [29].

Propofol (2,6-diisopropylphenol) is a rapid acting intravenous anesthetic agent that is worldwide used for induction and maintenance of general anesthesia [33]. Some propofol extrahepatic metabolism has been detected during the liver transplant anhepatic phase [34]. However, hepatic metabolism remains the main route of propofol elimination [33]; hence, patients experiencing liver dysfunction will require less propofol doses, compared with control subjects, when anesthesia is titrated to maintain the same Bispectral Index level [35].

In a pioneer study [29], my research team Dahaba and colleagues at Zhe Jiang First University Hospital, Hang Zhou, People's Republic of China, assessed the utility of using cross clamp-reperfusion gradient of the Bispectral Index value, during constant propofol Target Controlled Infusion (TCI), as an early intraoperative indicator of living-donor or cadaveric-donor graft dysfunction [29]. My research team Dahaba and colleagues correlated the Bispectral Index gradient (Nadir cross clamp–Zenith after reperfusion) with various laboratory parameters in patients with or without IPGF. We also assessed the differences in Bispectral Index values between living-donor and cadaveric-donor transplants as an indicator of the type of histotype injury [29].

We conducted a prospective consecutive pilot study over a period of 18 months in 29 patients who underwent, procured locally without venovenous bypass, Living-Donor Liver Transplantation (LDLT) and 24 patients who underwent Orthotopic Liver Transplantation (OLT) with a Body Mass Index (BMI) >18 <26 kg m^{-2}, suffering from end-stage liver disease as a result of alcohol- or virus-related liver cirrhosis, hepatocarcinoma, or autoimmune hepatitis [29]. Bispectral Index BIS-XP "Quatro" sensor was placed on patients' forehead according to the manufacturer's guidelines and connected to a Bispectral Index BIS-XP monitor (version 3.4) [29]. Bispectral Index recordings were started after verifying electrodes impedance <5 kΩ and Signal Quality Index (SQI) >95%. Data were continuously recorded in the once every 5 s mode with a smoothing window set at 30 s [29].

Oral midazolam 7.5 mg was administered 1 h preoperatively. Remifentanil 0.15–0.3 μg kg^{-1} min^{-1} infusion was started for induction. After preoxygenation and entering patients' anthropometric data, using Diprifusor™ TCI pump (AstraZeneca, Cheshire, UK) [29], incorporating a Marsh [36] 3-compartment pharmacokinetic model, propofol TCI was set to reach a plasma concentration (C_p) of 1.5 μg ml^{-1} and then increased by 0.5 μg ml^{-1} until loss of eyelash reflex. The lungs were mechanically ventilated with 40% oxygen in air, adjusted to maintain end-tidal carbon dioxide at 30–40 mm Hg [29]. Patients were warmed using convection forced hot air blanket to maintain core temperature >35 °C. A stable anesthesia of around Bispectral Index 40 was maintained using propofol ±2 μg ml^{-1} TCI rate adjustments [29]. After a stable Bispectral Index value of around 40 was maintained for 15 min, propofol TCI was kept constant and recorded for the whole transplant surgical procedure with no further TCI rate adjustments whatsoever. Warm Ischemia Time (WIT) was the time from the start of implantation until restoration of arterial and portal blood circulation [29], whereas Cold Ischemia Time (CIT) was the time from donor portal vein flushing until the graft was removed from University of Wisconsin solution [29]. Model for End-Stage Liver Disease (MELD) score (Mayo Clinic MELD calculator website, http://www.mayoclinic.org/gi-rst/mayomodel5.html) [4] and Child-Pugh score [3] were determined from patients' laboratory and clinical charts' data [29].

According to Nanashima's classification [37], IPGF is defined as patients with alanine aminotransferase (ALT) and/or aspartate aminotransferase (AST) levels >1500 IU L^{-1} within 72 h following liver transplant [29]. Whereas PNF is confirmed when ALT and/or AST are continuously elevated in 7 postoperative days, presence of severe coagulopathy, cessation of bile production, or multiple organ failure with the loss of allograft [30]. Otherwise, patients are categorized as non-IPGF patients [29]. Postoperative daily bile flow was recorded. Hepatic allograft biopsies at the end of hepatic cold ischemia were graded, according to the degree of fatty infiltration, as absent, mild (<30%), moderate (30–60%), or severe macrovesicular steatosis (>60%) [38].

After cross-clamping, using paired t test, mean ± SD Bispectral Index values in the LDLT and OLT groups significantly declined to a nadir of 28.6 ± 7.1 and 24.9 ± 6.3, respectively. Afterward, following reperfusion, mean Bispectral Index values significantly increased only in both groups' patients who were non-IPGF [29]. On the contrary, mean ± SD Bispectral Index values of 33.4 ± 4.3 in seven IPGF patients (three LDLT and four OLT patients) did not increase with reperfusion. Figure 8.10 depicts mean ± SEM (standard error of the mean) Bispectral Index values during the paleohepatic, anhepatic, and neohepatic phases in the OLT group. Figure 8.11 depicts mean ± SEM Bispectral Index values during the paleohepatic, anhepatic, and neohepatic phases in the LDLT group. There was no significant difference over time in Bispectral Index values between the two groups. Pearson's correlation statistical analysis revealed that the Bispectral Index "cross clamp nadir–reperfusion zenith gradient" correlated well with postoperative $r = 0.62$ AST, $r = 0.66$ ALT, $r = 0.53$ International Normalized Ratio (INR) but poorly correlated $r = 0.24$ with bilirubin [29].

In our study patients, allograft biopsies revealed <30% mild to absent fatty infiltration [29]. During the first postoperative 72 h, three LDLT patients and two OLT patients exhibited IPGF manifesting as AST and/or ALT levels >1500 IU L^{-1} and their bile production was <100 ml day^{-1}, whereas two OLT patients suffered from PNF with loss of the allograft [29]. Over time there were no statistically significant differences in Bispectral Index, MELD, and Child-Pugh scores between the OLT and LDLT groups. Conversely AST and ALT levels differed widely between the type of transplant as there was a $P < 0.001$ significant difference over time between the OLT and LDLT groups [29]. During

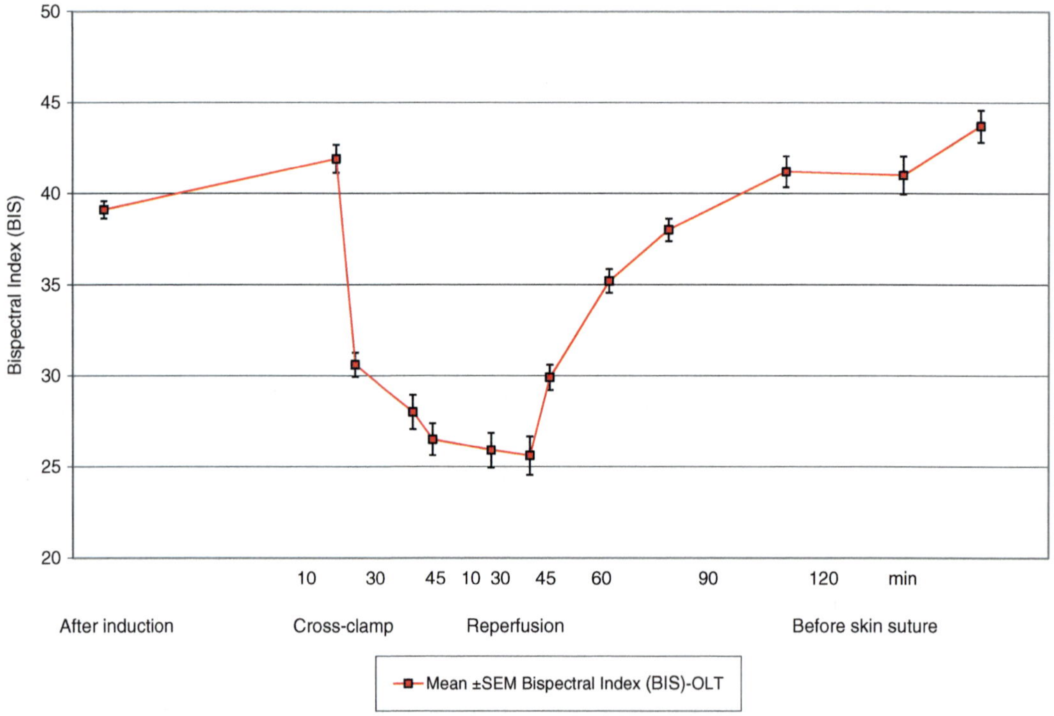

Fig. 8.10 Mean ± SEM Bispectral Index (BIS) during the paleohepatic, anhepatic, and neohepatic phases in the Orthotopic Liver Transplant (OLT) group. Bispectral Index values significantly declined with cross-clamping

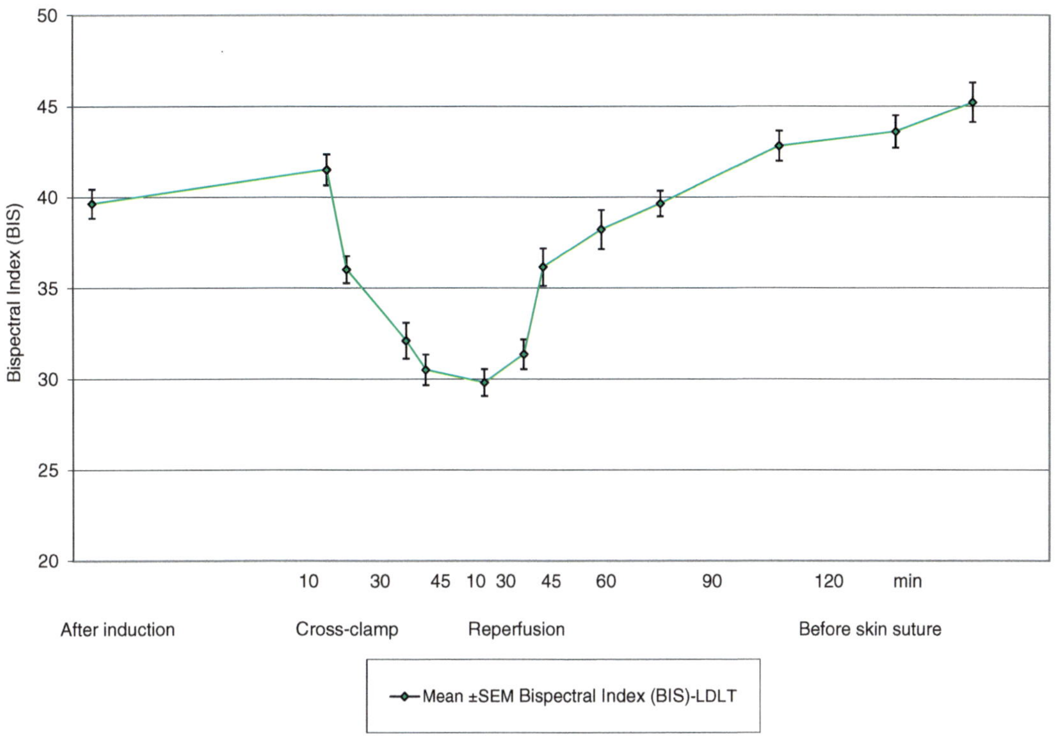

Fig. 8.11 Mean ± SEM Bispectral Index (BIS) during the paleohepatic, anhepatic, and neohepatic phases in the Living-Donor Liver Transplant (LDLT) group. Bispectral Index values significantly declined with cross-clamping

the paleohepatic phase, we maintained our patients' core body temperature well within normal range. However, core body temperature slightly decreased to <35 °C in the anhepatic and neohepatic phases [29], which probably did not significantly influence propofol pharmacokinetics, as Leslie and colleagues [39] demonstrated that propofol TCI infusion rates require no rate adjustments during mild hypothermia [39].

We titrated propofol TCI to maintain Bispectral Index value of around 40 [29], after that propofol TCI was unchanged until the end of all surgical procedures. During the anhepatic phase, the effective hepatic blood flow cessation resulted in termination of all propofol hepatic clearance [33], resulting in an increase in its plasma concentration and consequently a significant decline in Bispectral Index value [29]. Thereafter, during the neohepatic phase, hepatic reperfusion will result in a significant reduction in propofol plasma concentration, restoring Bispectral Index values back to paleohepatic levels in the patients who were non-IPGF, but not in the seven patients with IPGF [29]. The Bispectral Index gradient (Nadir cross clamp–Zenith after reperfusion) correlated well with AST and ALT; that clearly indicated that, in the non-IPGF patients, the cadaveric- and living-donor grafts regained metabolic function [29].

Figure 8.12 demonstrates an OLT hepatic encephalopathy grade 4 patient who exhibited an early functioning transplanted liver graft; Figs. 8.13 and 8.14 demonstrate two OLT hepatic encephalopathy grade 3 patients who exhibited a non-functioning transplanted liver graft, whereas Fig. 8.15 demonstrates an OLT hepatic encephalopathy grade 2 patient who exhibited a delayed functioning transplanted liver graft [29].

At present, IPGF evaluation is based upon static laboratory tests such as serum bilirubin, AST, ALT, and/or graft biopsies at the end of the cold ischemia phase. These tests could only provide indirect measures of a possible graft dys-

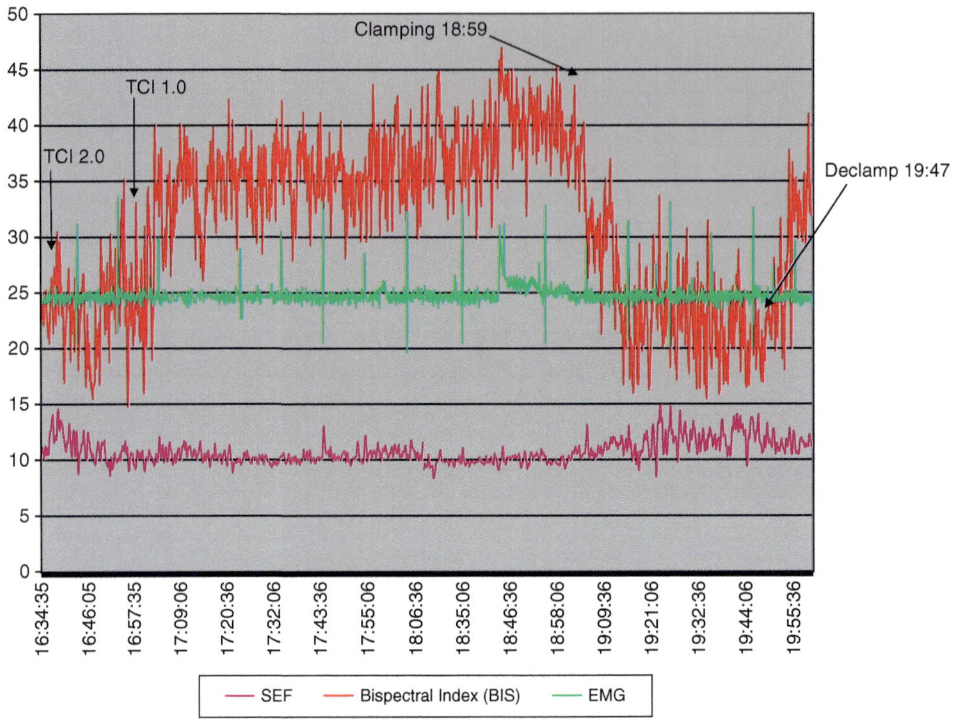

Fig. 8.12 Demonstrates an Orthotopic Liver Transplant (OLT) hepatic encephalopathy grade 4 patient who exhibited an early functioning transplanted liver graft

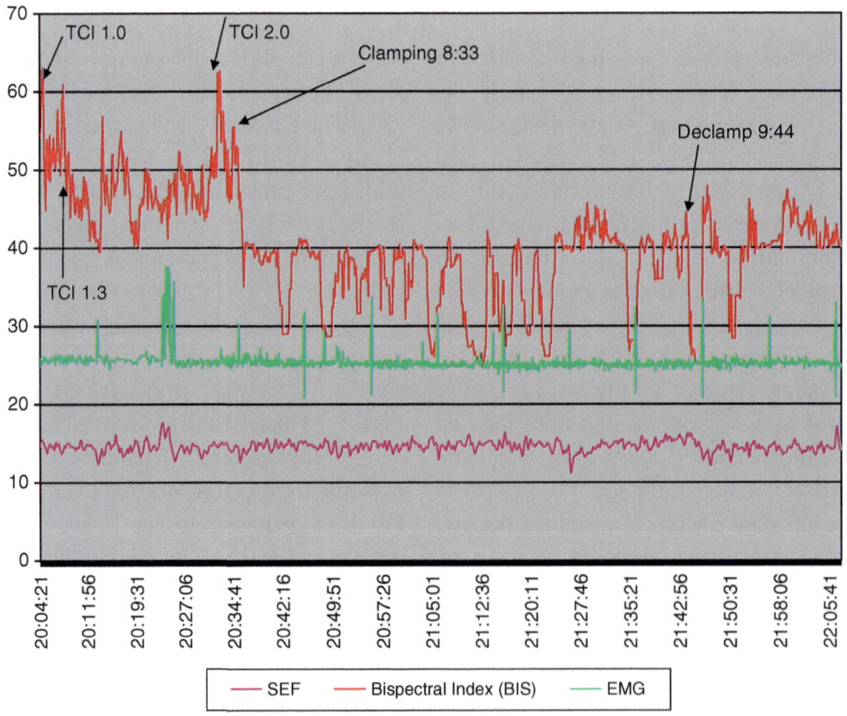

Fig. 8.13 Demonstrates an Orthotopic Liver Transplant (OLT) hepatic encephalopathy grade 3 patients who exhibited a nonfunctioning transplanted liver graft

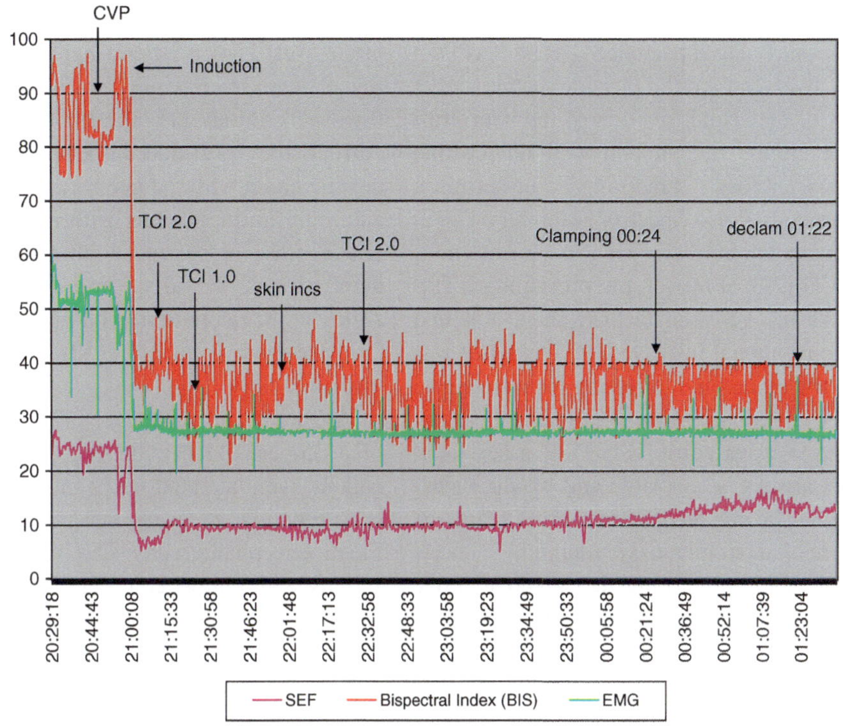

Fig. 8.14 Demonstrates another Orthotopic Liver Transplant (OLT) hepatic encephalopathy grade 3 patients who exhibited a nonfunctioning transplanted liver graft

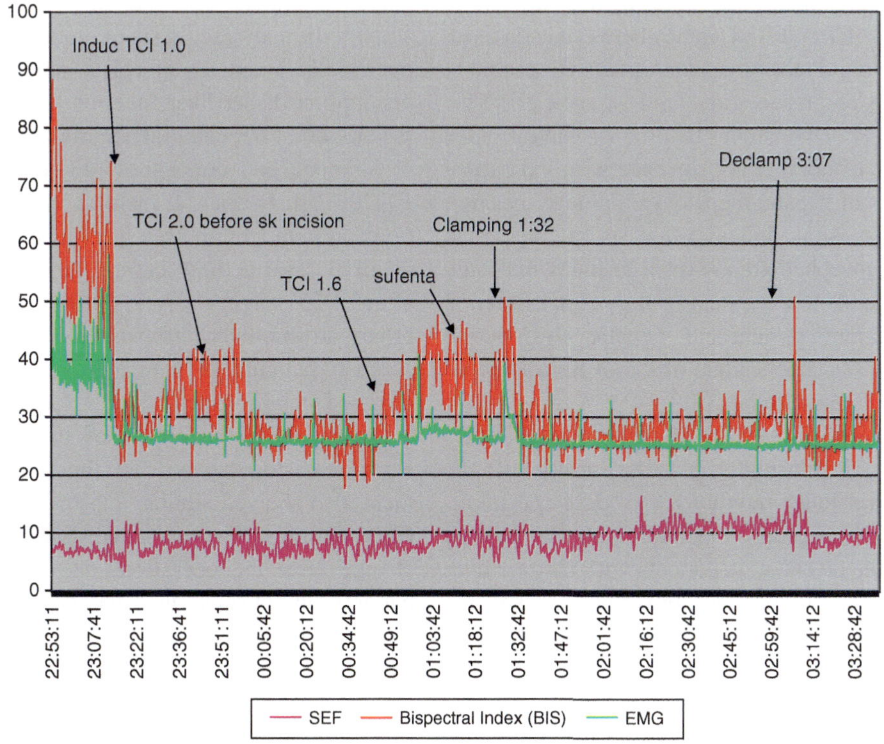

Fig. 8.15 Demonstrates an Orthotopic Liver Transplant (OLT) hepatic encephalopathy grade 2 patient who exhibited a delayed functioning transplanted liver graft

function [29]. Whereas Indocyanine Green (ICG)-Plasma Disappearance Rate (PDR) dynamic liver test could provide better direct measures of the actual function of the liver graft as ICG excretion rate in bile well reflects the hepatic excretory function. Furthermore, Transcutaneous Pulse Spectrophotometry (TPS) is an attractive noninvasive measure for elimination of ICG [29]. However, the precision of TPS was shown to overestimate the values obtained from a conventional invasive ICG clearance test [40]. On the other hand, inserting invasive aortic fiber-optic thermistor via the femoral artery sheath for invasive aortic ICG-PDR assessment might not always be available and would necessitate further involvement of monitoring support [41]. Hence, a simple easy method, such as Bispectral Index monitoring that could identify poor graft function at the early intraoperative period would allow opportune intensive clinical management, which would bring about further improvements in liver transplant outcomes [29].

In our study [29], we found no significant difference between Bispectral Index values during cadaveric- and living-donor transplantations. Conversely, there was a significant difference both in ALT and AST values between cadaveric- and living-donor transplant patients, probably because of preservation fluid injury [29]. This obviously demonstrates that the Bispectral Index value does not directly correlate with indicators/markers of hepatic /hepatocyte damage, because the Bispectral Index value is a processed Electroencephalography (pEEG) index that only corresponds to the extent of the mental function involvement of chronic hepatic dysfunction patients [1], independent of the of hepatic/hepatocyte disease etiology [29].

My research team Dahaba and colleagues previously demonstrated [1] the utility of the Bispectral Index monitor for grading and monitoring the degree of Central Nervous System (CNS) involvement in patients with chronic liver disease. The Bispectral Index could classify the degree of hepatic encephalopathy throughout its whole spectrum and accurately discriminate between low-grade West Haven criteria hepatic encephalopathy patients manifesting as diffuse β-wave and suppressed α-wave activity and high-grade West Haven criteria hepatic encephalopathy patients manifesting as diffuse δ-wave and θ-wave activities [1]. We demonstrated in our current study [29] that the Bispectral Index values increased with the neohepatic phase reperfusion only in the patients with non-IPGF but did not increase in patients with IPGFs. This also demonstrates the Bispectral Index prediction ability of the return of mental function with restoration of the liver functions after successful liver transplant [29].

Aggarwal and colleagues [42], in a case report of a 25-year-old hepatic encephalopathy coma patient, demonstrated early intraoperative cerebral function recovery as preoperatively her Electroencephalography (EEG) displayed δ-wave activity with almost complete lack of EEG high frequency activities [42]. Following her OLT, after 15 min of hepatic reperfusion, 4–5 Hz θ-wave activity started appearing on her EEG tracings. Higher frequency activity of 6–7 Hz θ-wave activity became persistent for 30 min of hepatic reperfusion [42]. Patient regained consciousness 8 h postoperatively as her EEG displayed a gradual increase in 8–9 Hz α-wave activity. In that case, her EEG monitoring intraoperatively acted as an early indicator of the resumption of her liver function that resulted in restoration of her cerebral activity [42].

Shimoda and colleagues [43] examined the 10–20 Fast Fourier Transform (FFT) international system EEG power spectra analysis based on EEG basal activity before the operation and after LDLT surgery as a result of citrullinemia, biliary atresia, liver cirrhosis, and liver cell cancer in 10 patients, six men and four women, who regained consciousness over several days [43]. Following LDLT surgery, 8 of 10 patients who showed postoperative liver function improvement also showed similar improvement in their EEG findings. Two patients showed no EEG change after the operation; one of these two patients did not show improvement in liver function after the operation, while the other patient developed cerebral infarction. They concluded that the EEG is useful for the evaluation of brain function after liver transplant operation [43].

Reeves and colleagues [44] in another study examined the EEG changes before and after OLT surgery in 14 patients [44], who regained consciousness over several days. Following OLT surgery, compared with pre-transplant, post-transplant EEG tracings exhibited an initial early EEG improvement that appeared as an increase in α-wave and β-wave power, a decline in θ-wave power, a rise in the occipital α-θ waves power ratio over anterior, central, and posterior cerebral cortical regions, all come as a confirmation of the reversal of the EEG bioelectric brain dysfunction as normal hepatic function was restored [44].

8.3 The "Plateau" Phenomenon of Stepwise Extracorporeal Albumen-Dialysis Bispectral Index Hepatic Encephalopathy Recovery

My research team Dahaba and colleagues have previously demonstrated that that the Bispectral Index monitor, an EEG derived parameter, could accurately quantify the degree of hepatic encephalopathy Central Nervous System (CNS) involvement and cerebral function recovery, as we validated the discriminative power of Bispectral Index values of 95, 85, 70, and 55 cutoffs [1], for precise discrimination between West Haven from 1 to 4 grades [2]. Molecular Adsorbent Recirculating System™ (MARS™, Gambro AB, Stockholm, Sweden) liver assisted extracorporeal devices offer effective liver detoxifications and efficient removal of both protein-bound and water-soluble toxic metabolites in patients suffering from end-stage liver dysfunction [45].

I report the "Plateau" very peculiar phenomenon manifesting as clear distinct plateaus of "stepwise albumen-dialysis Bispectral Index hepatic encephalopathy recoveries" that my research team Dahaba and colleagues documented in six end-stage liver disease patients who underwent 6–8 h sessions of MARS™ albumin dialysis liver assist extracorporeal-detoxification at the Medical University of Graz Hospital, Graz, Austria [46]. Mean ± SD or median (range) patients demographics and laboratory findings were 62 ± 5 years, 175 ± 4 cm, 93.6 ± 6 kg, Body Mass Index (BMI) 30.9 ± 5.1 kg·m^{-2}, ammonia 135 ± 52.9 μMol/L, Glasgow Coma Scale (GCS) 4 (3–5), Child-Pugh score 12 (11–13), and Model for End-Stage Liver Disease (MELD) score 32 (12–35) [46]. We noticed an unusual, peculiar pattern of Bispectral Index evolution, in conjunction with West Haven grades hepatic encephalopathy expedited recoveries because of their liver assist extracorporeal MARS™ [46]. Contrary to the common belief that hepatic encephalopathy Electroencephalography (EEG) recovery is a slow "gradual" temporal process, we discovered that hepatic encephalopathy clinical recovery manifests as a rather very peculiar series of a Bispectral Index phenomenon of distinct "Plateau" recovery pattern of abrupt Bispectral Index stepwise transitions that precisely coincided with their West Haven grades recoveries as the patients clinically recovered [2], which was paralleled by their biochemical improvements [46].

To assess patients' mental status neurophysiological clinical recovery, we clinically graded West Haven criteria [2] mental state once every 2 hours or when there was an evident change in patients' mental state [46], using psychometric performance standardized intellectual functions' behavior and consciousness test batteries [2]. We continuously digitally recorded Bispectral Index data as we imported Bispectral Index raw EEG ASCII signal into Bitmap format to qualify the EEG dominant waves, which were correlated to continuously graded West Haven criteria [46].

Our continuously recorded Bispectral Index data revealed most unusual "Plateau" neurophysiological phenomenon of distinctive stepwise plateaus of albumen dialysis Bispectral Index hepatic encephalopathy recoveries perfectly corresponding to West Haven grades' return of cerebral activity [46]. Paired *t* test revealed a series of plateaus with abrupt transitions of mean ± SD Bispectral Index (range) 48.7 ± 3.7 (37.6–59.8) Plateau I corresponding to West Haven grade 4; significantly increasing to 63.5 ± 3.5 (53.0–74.0) Plateau II correspond-

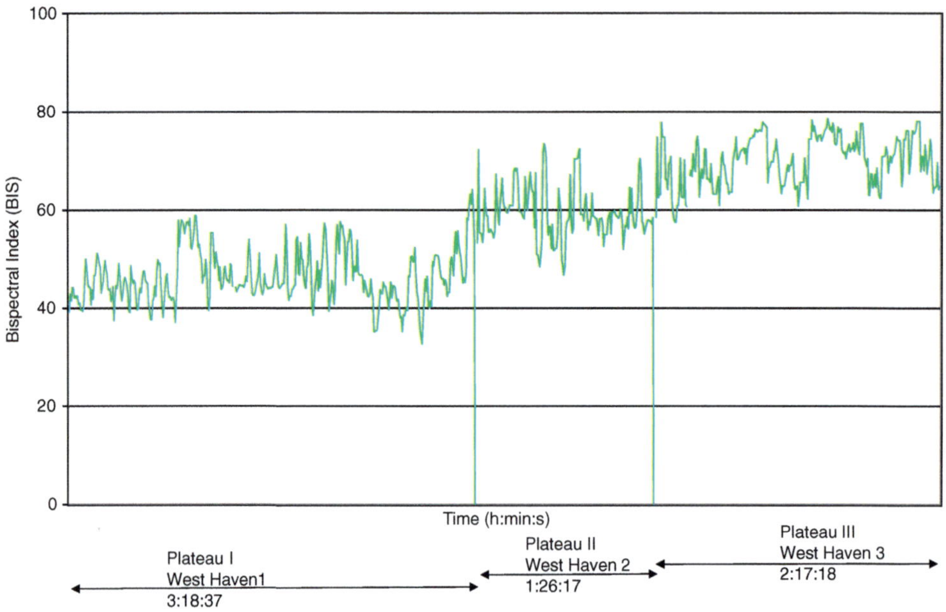

Fig. 8.16 Patient 1 Bispectral Index (BIS) recording during albumen-dialysis. The graph reveals an unexpected pattern of three plateaus, rather than the expected gradual recovery. The step from the first plateau to the second plateau is more distinct than the step from the second plateau to the third plateau. (Reproduced with permission # 5931221246264 from Dahaba et al. [46])

ing to West Haven grade 3; that further significantly increased to 74.3 ± 3.8 (62.9–81.7) Plateau III corresponding to West Haven grade 2; to finally significantly recover to 92.0 ± 2.6 (86.3–97.7) corresponding to West Haven grade 1 [46]. The duration (hr:min) of West Haven grade 4 was 2:10 ± 0:29; duration of West Haven grade 3 was 2:24 ± 0:22; duration of West Haven grade 2 was 2:28 ± 0:22; duration of West Haven grade 1 was 0:47 ± 0:08; rather than a slow gradual "temporal" Bispectral Index recovery [46]. We present raw Bispectral Index trend graphs of patient 1 (Fig. 8.16), patient 2 (Fig. 8.17), patient 3 (Fig. 8.18), patient 4 (Fig. 8.19), patient 5 (Fig. 8.20), and patient 6 (Fig. 8.21) who all displayed the phenomenon of Bispectral Index "plateaus" in conjunction with the 4 West Haven grades stepwise albumen-dialysis hepatic encephalopathy recovery. Spearman's rank determination (r^2) revealed good correlation $r^2 = 0.90$ between the Bispectral Index Plateaus and West Haven clinical grades [46].

At Bispectral Index 38–60 Plateau I, West Haven grade 4 coma manifested as deep EEG suppression. At Bispectral Index 53–74 Plateau II West Haven grade 3 pre-coma manifested as diffuse θ-wave and δ-wave activities. At Bispectral Index 63–82 Plateau III, West Haven grade 2 patients woke up with α-wave activity abruptly replacing diffuse fast β-wave activity. The high discriminative power of Bispectral Index monitoring at various West Haven grades arises from the Bispectral Index algorithm ability to accurately detect and display sharp shifts from control of hippocampal/septal and thalamocortical generators of θ-wave and δ-wave activities toward sharp abrupt increases in cortical generators of α-wave and β-wave activities [26, 47]. In Plateau-I, Bispectral Index "*Burst Suppression Ratio*" subparameter descriptor continuously detects deep EEG suppression of West Haven grade 4 coma state [25]. In Plateau-II pre-coma of West Haven grade 3, "*SynchFastSlow*" Bispectral Index subparameter constantly detects diffuse θ-wave and δ-wave activities [27]. In

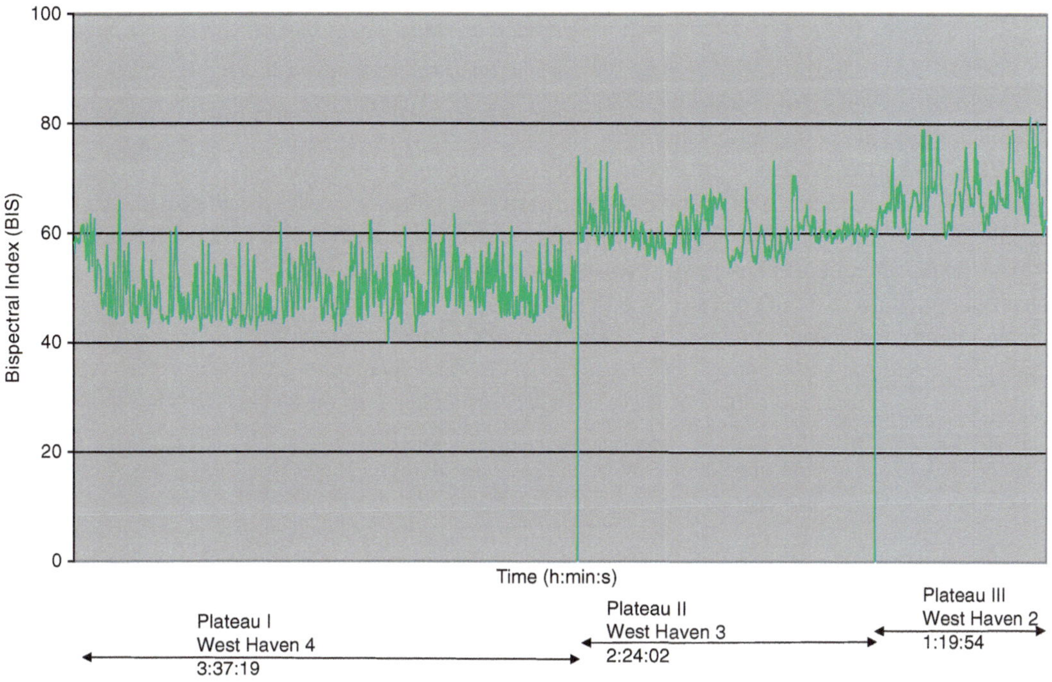

Fig. 8.17 Patient 2 Bispectral Index (BIS) recording during albumen-dialysis. The graph reveals an unexpected pattern of three plateaus, rather than the expected gradual recovery. The step from the first plateau to the second plateau is more distinct than the step from the second plateau to the third plateau. (Reproduced with permission # 5931221246264 from Dahaba et al. [46])

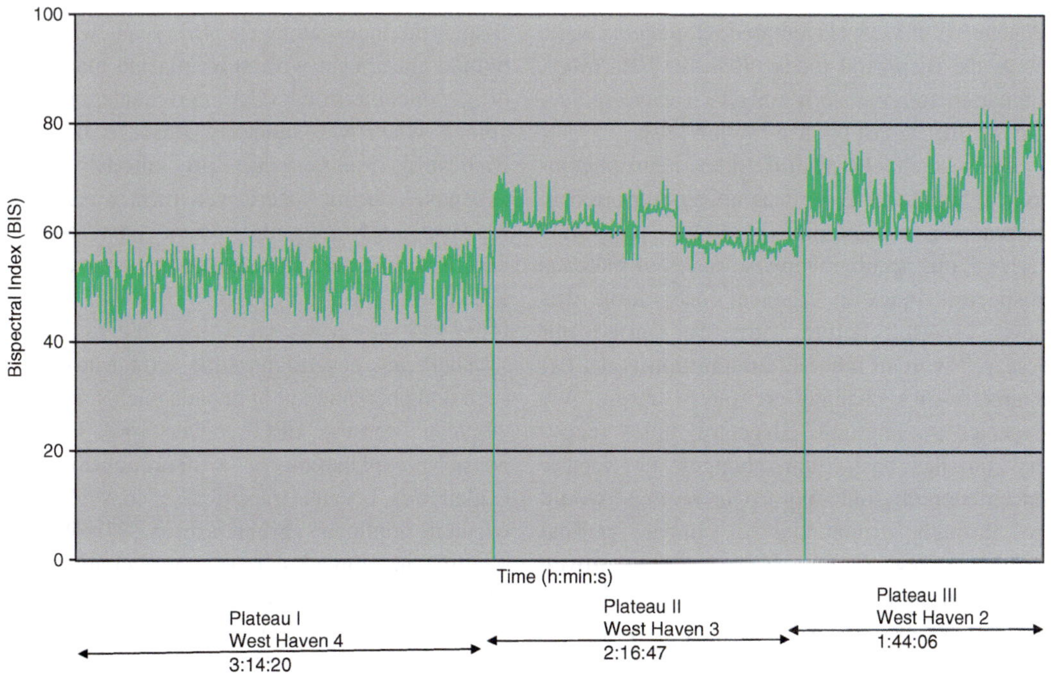

Fig. 8.18 Patient 3 Bispectral Index (BIS) recording during albumen-dialysis. The graph reveals an unexpected pattern of three plateaus, rather than the expected gradual recovery. The step from the first plateau to the second plateau is more distinct than the step from the second plateau to the third plateau. (Reproduced with permission # 5931221246264 from Dahaba et al. [46])

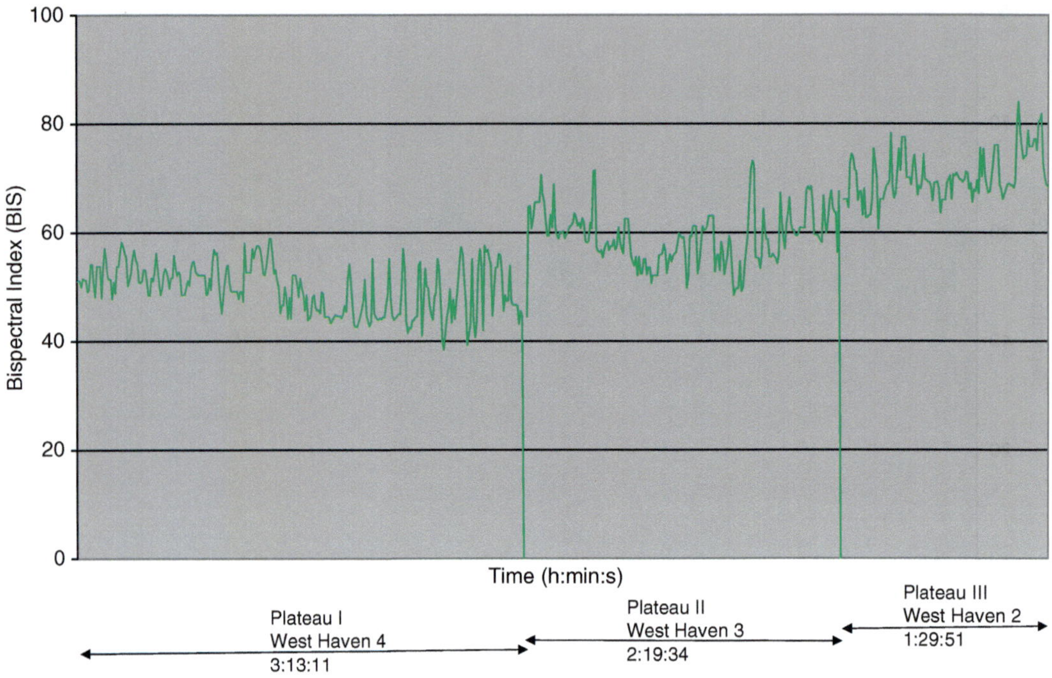

Fig. 8.19 Patient 4 Bispectral Index (BIS) recording during albumen-dialysis. The graph reveals an unexpected pattern of three plateaus, rather than the expected gradual recovery. The step from the first plateau to the second plateau is more distinct than the step from the second plateau to the third plateau. (Reproduced with permission # 5931221440446 from Dahaba et al. [46])

Plateau-III of West Haven grade 2, patients wake up as the Bispectral Index "*Relative BetaRatio*" subparameter constantly reflects α-wave replacing the diffuse fast β-wave rhythm [27].

Because the Bispectral Index neurophysiological substrate so far was always reported to increase in a continuous gradual "temporal" fashion, our peculiar stepwise series of plateaus Bispectral Index emergence observation was never reported before [46]. We started our research with an inherent assumption that EEG characteristics changes are rarely abrupt. We expected a continuous Bispectral Index recovery on the basis that changes in hepatic encephalopathy-inducing toxins removal would go through exponential or linear gradual declines. However, our observation has proven that assumption not to be true [46]. When anesthetic depth changes, we seldom see a real continuous Bispectral Index change. More often, despite anesthetic Target Controlled Infusion (TCI) drug concentrations changing in a continuous fashion, we usually see nonlinear Bispectral Index changes [46]. Even when we induce anesthesia with slow plasma anesthetic drug concentrations (C_p) increments, a point comes when we suddenly observe that the Bispectral Index value precipitously falls. Likewise, during emergence from anesthesia, the Bispectral Index often stays constantly low for some time and then suddenly increases usually around the time of return of consciousness [46].

There are several possible explanations for such odd phenomenon of hepatic encephalopathy "Plateau" stepwise EEG recovery [46]. The first possible explanation is a neurophysiological explanation, because transitions in consciousness occur in nonlinear abrupt fashion [47]. We can see this in clinical practice, as patients could become slightly sedated when we administer low doses of sedatives, but when consciousness is lost, consciousness is "abruptly" lost. My research team Dahaba and colleagues have previously demonstrated something similar during natural physiologic sleep, in sleep-deprived anes-

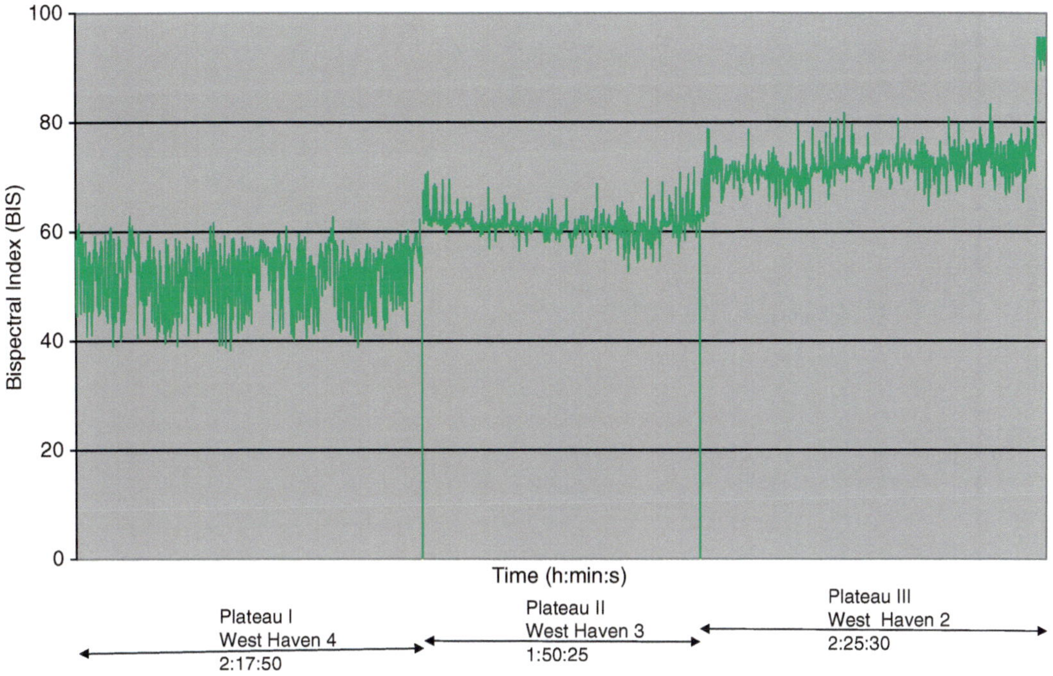

Fig. 8.20 Patient 5 Bispectral Index (BIS) recording during albumen-dialysis. The graph reveals an unexpected pattern of three plateaus, rather than the expected gradual recovery. The step from the first plateau to the second plateau is more distinct than the step from the second plateau to the third plateau. (Reproduced with permission # 5931221440446 from Dahaba et al. [46])

thesia resident anesthesiologists following their nightshift [7]. Their transitions between different sleep stages were abrupt and were associated with abrupt EEG changes particularly transitions in and out of Rapid Eye Movement (REM) sleep [47].

The second possible explanation is that our observed "Plateau" phenomenon [46] is a result of the Bispectral Index calculation algorithms. Despite EEG changes occur gradually, the Bispectral Index is insensitive to these gradual EEG changes, until a certain point when abrupt changes in the weighing of the Bispectral Index descriptors suddenly result in abrupt changes in the Bispectral Index value [47]. An early step in the Bispectral Index recognition algorithm, before calculation of the descriptor of a certain epoch, is a simple pattern recognition process that determines the weighing of each epoch descriptor [25]. The proprietary nature of the Bispectral Index algorithm allows the different Bispectral Index descriptor subparameters to sequentially dominate as the EEG changes its characters [25]. Obviously, the fine details of the weighing of the three Bispectral Index descriptors, *Burst Suppression Ratio*, *SynchFastSlow*, and *Relative BetaRatio*, pattern-recognition process "fuzzy boundaries" has never been fully disclosed, so we cannot conclude that a pattern looks "somewhat" like deep anesthesia, or a pattern looks "somewhat" like a deeper surgical plane of anesthesia [25]. If this is the case, then gradual changes in EEG frequency and phase content will not result in gradual changes in Bispectral Index descriptors weighing; rather there will be thresholds when abrupt changes in weighing occur that will sharply result in sudden change in Bispectral Index value. These two explanations are not mutually exclusive [47].

Furthermore, the third possible explanation of our observed "Plateau" phenomenon [46] is the "quanta removal" of deranged amino acids that alters the neurotransmitter levels in the CNS [45]. Noticeably, all three components of the MARS™ circuit, namely, hemodiafiltration, charcoal, and anion exchange resin hemodiad-

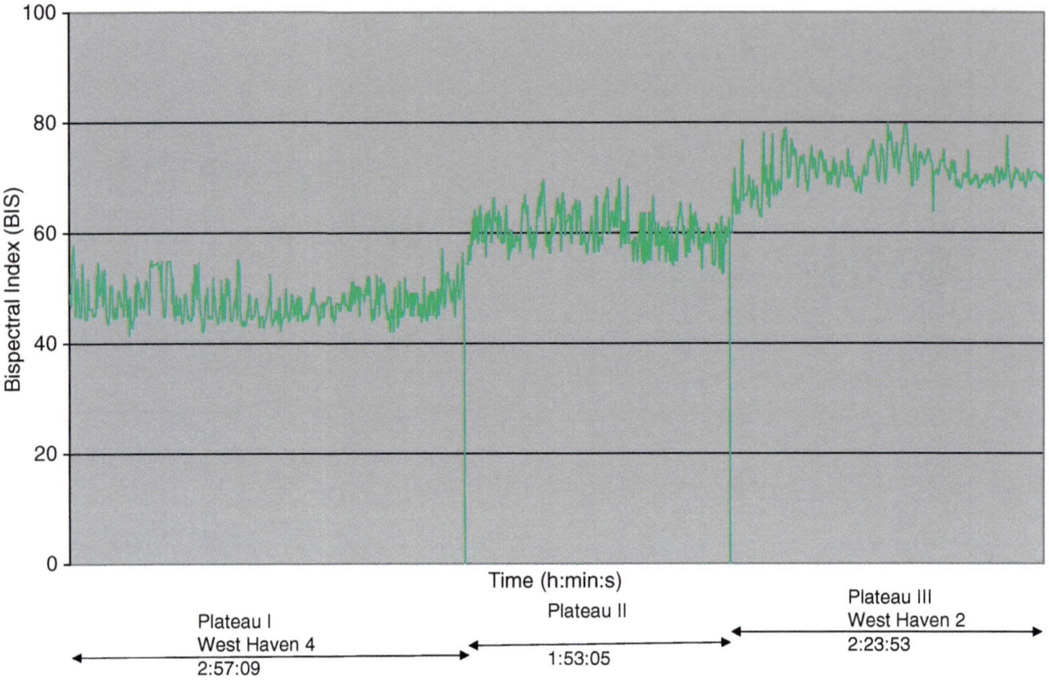

Fig. 8.21 Patient 6 Bispectral Index (BIS) recording during albumen-dialysis. The graph reveals an unexpected pattern of three plateaus, rather than the expected gradual recovery. The step from the first plateau to the second plateau is more distinct than the step from the second plateau to the third plateau. (Reproduced with permission # 5931221440446 from Dahaba et al. [46])

sorption, exhibit a "preferential" removal of Aromatic Amino Acids (AAAs) versus Branched Chain Amino Acids (BCAAs) [48]. This would tend to continuously increase during MARS™ treatment session, resulting in a sharp shift in our study Fischer's Ratio (plasma levels of Σ3 subnormal BCAA; valine, isoleucine, and leucine / Σ2 elevated AAAs; tyrosine and phenylalanine) [48] from 1.44 to 1.75 quantum just during the "first 3 hr" of MARS™ treatment. This was almost halted as our study Fischer's Ratio just slightly shifted to 1.79 in the "following 3 hr" quantum [45].

The fourth possible explanation [47] would be the highly elevated brain concentrations of the precursor of monoamine serotonin neurotransmitter "Tryptophan" free fraction, which plays a strong influential role in the pathogenesis of hepatic coma, and was shown in our study to be extensively removed during the "first 3 hr" quantum, via its "selective" adsorption to the MARS™ charcoal column [45]. Whereas "quantumly" effective removal of the "Glutamine" accumulating in the brain perivascular astrocytes during the "first 3 hr" would result in reversal of astrocyte swelling cerebral edema [45].

Because arousal systems in the brain are numerous and redundant, as are their neurochemical and anatomic substrates, this brilliantly offers a fifth possible explanation, namely, the "consecutive" involvement of different arousal systems, as they "sequentially" recover, not all at once, but at different times [47]. The wake-promoting influence was thought to arise mainly from monoaminergic neurons (serotonergic raphe nuclei, noradrenergic locus coeruleus, and histaminergic tuberomammillary nucleus) as well as cholinergic neurons (pedunculopontine and latero-dorsal tegmental nuclei) in the upper brainstem arousal systems. Recent evidence indicates that these neural arousal systems just play a modulatory role, and the "backbone" of the wake-sleep regulatory system depends upon numerous circuits that all use the "fast" neurotransmitters, Glutamate and γ-amino butyric acid (GABA) [49]. In contrast to these "standard models," in the last few years, a new paradigm of wake–sleep circuitry has emerged that is based on our understanding of these circuits [47]. These circuits include the basal forebrain arousal system that projects directly to the cerebral cortex parabrachial pedunculopontine, and supramammillary arousal systems, that project extensively to the basal forebrain as well as the cerebral cortex [47], whereas the "lateral hypothalamic arousal system" contains the important "peptidergic neuronal system" orexin (hypocretin) neurons that are responsible for driving "rapid" wakefulness as they fire rapidly when activated [49].

Finally, currently there are two entirely different liver assist extracorporeal detoxification devices; one is the MARS™ using albumen-dialysis, which we used in our report, while the other is Prometheus®, from a different manufacturer (Fresenius Kabi, France) that uses "fractionated plasma separation and adsorption" using a totally different principle [9]. We do not know whether the "Plateau" phenomenon will similarly manifest when "Prometheus®" is used [47].

We believe that we have resolved the fundamental problem of the absence of effective EEG monitoring for the strenuously long process of hepatic encephalopathy EEG recovery [47]. With Bispectral Index monitoring and MARS™, we discover that contrary to the common believe that hepatic encephalopathy EEG recovery is a slow "gradual" process, with the expedited MARS™ recovery and Bispectral Index EEG monitoring, we clearly demonstrated that it is not, rather a stepwise series of plateaus with abrupt transitions EEG recovery [47].

References

1. Dahaba AA, Worm HC, Zhu SM, Bao FP, Salah A, Zakaria S, et al. Sensitivity and specificity of bispectral index for classification of overt hepatic encephalopathy: a multicentre, observer blinded, validation study. Gut. 2008;57:77–83.
2. Conn HO, Leevy CM, Vlahcevic ZR, Rodgers JB, Maddrey WC, Seeff L, et al. Comparison of lactulose and neomycin in the treatment of chronic portal-systemic encephalopathy. A double blind controlled trial. Gastroenterology. 1977;72(4 Pt 1):573–83.
3. Pugh RN, Murray-Lyon IM, Dawson JL, Pietroni MC, Williams R. Transection of the oesophagus for bleeding oesophageal varices. Br J Surg. 1973;60:646–9.
4. Kamath PS, Wiesner RH, Malinchoc M, Kremers W, Therneau TM, Kosberg CL, et al. A model to predict survival in patients with end-stage liver disease. Hepatology. 2001;33:464–70.
5. Teasdale G, Jennett B. Assessment of coma and impaired consciousness. A practical scale. Lancet. 1974;2:81–4.
6. Kim DW, Kil HY, White PF. The effect of noise on the bispectral index during propofol sedation. Anesth Analg. 2001;93:1170–3.
7. Dahaba AA, Xue JX, Xu GX, Liu QH, Metzler H. Bilateral Bispectral Index (BIS)-Vista as a measure of physiologic sleep in sleep-deprived anesthesiologists. Minerva Anestesiol. 2011;77:388–93.
8. Bruhn J, Bouillon TW, Shafer SL. Electromyographic activity falsely elevates the bispectral index. Anesthesiology. 2000;92:1485–7.
9. Krisper P, Haditsch B, Stauber R, Jung A, Stadlbauer V, Trauner M, et al. In vivo quantification of liver dialysis: comparison of albumin dialysis and fractionated plasma separation. J Hepatol. 2005;43:451–7.
10. Montagnese S, Amodio P, Morgan MY. Methods for diagnosing hepatic encephalopathy in patients with cirrhosis: a multidimensional approach. Metab Brain Dis. 2004;19(3/4):281–312.
11. Amodio P, Del Piccolo F, Pettenò E, Mapelli D, Angeli P, Iemmolo R, et al. Prevalence and prognostic value of quantified electroencephalogram (EEG) alterations in cirrhotic patients. J Hepatol. 2001;35:37–45.
12. Markand ON. Electroencephalography in diffuse encephalopathies. J Clin Neurophysiol. 1984;1:357–407.
13. Foley JM, Watson CW, Adams RD. Significance of the electroencephalographic changes in hepatic coma. Trans Am Neurol Assoc. 1950;51:161–5.
14. Sherlock S, Summerskill WH, White LP, Phear EA. Portal-systemic encephalopathy; neurological complications of liver disease. Lancet. 1954;267:454–7.
15. Bickford RG, Butt HR. Hepatic coma: the electroencephalographic pattern. J Clin Invest. 1955;34:790–9.
16. Parsons-Smith BG, Summerskill WH, Dawson AM, Sherlock S. The electroencephalograph in liver disease. Lancet. 1957;273:867–71.
17. Laidlaw J, Read AE. The electroencephalographic diagnosis of manifest and latent 'delirium' with particular reference to that complicating hepatic cirrhosis. J Neurol Neurosurg Psychiatry. 1961;24:58–70.
18. Hawkes CH, Macpherson AI, Pryor H, Townsend HR. The value of EEG frequency analysis in hepatic encephalopathy. J R Coll Surg Edinb. 1970;15:151–7.
19. Laidlaw J, Read AE. The E.E.G. in hepatic encephalopathy. Clin Sci. 1963;24:109–20.
20. Demedts M, Pillen E, De Groote J, Van de Woestijne KP. Hepatic encephalopathy: comparative study of EEG abnormalities, neuropsychic disturbances and blood ammonia. Acta Neurol Belg. 1973;73:281–8.

21. Trewby PN, Casemore C, Williams R. Continuous bipolar recording of the EEG in patients with fulminant hepatic failure. Electroencephalogr Clin Neurophysiol. 1978;45:107–10.
22. Van der Rijt CC, Schalm SW, De Groot GH, De Vlieger M. Objective measurement of hepatic encephalopathy by means of automated EEG analysis. Electroencephalogr Clin Neurophysiol. 1984;57:423–6.
23. Sagalés T, Gimeno V, de la Calzada MD, Casellas F, Macià MD, Soriano MV. Brain mapping analysis in patients with hepatic encephalopathy. Brain Topogr. 1990;2:221–8.
24. Kullmann F, Hollerbach S, Lock G, Holstege A, Dierks T, Schölmerich J. Brain electrical activity mapping of EEG for the diagnosis of (sub)clinical hepatic encephalopathy in chronic liver disease. Eur J Gastroenterol Hepatol. 2001;13:513–22.
25. Rampil IJ. A primer for EEG signal processing in anesthesia. Anesthesiology. 1998;89:980–1002.
26. Gugino LD, Chabot RJ, Prichep LS, John ER, Formanek V, Aglio LS. Quantitative EEG changes associated with loss and return of consciousness in healthy adult volunteers anaesthetized with propofol or sevoflurane. Br J Anaesth. 2001;87:421–8.
27. Amodio P, Marchetti P, Del Piccolo F, de Tourtchaninoff M, Varghese P, Zuliani C, et al. Spectral versus visual EEG analysis in mild hepatic encephalopathy. Clin Neurophysiol. 1999;110:1334–44.
28. Ciancio A. Spectral electroencephalogram analysis in hepatic encephalopathy and liver transplantation. Liver Transpl. 2002;8:630–5.
29. Dahaba AA, Feng ZY, Zhu SM, Bornemann H, Rehak PH, Metzler H. The utility of using bispectral index monitoring as an early intraoperative indicator of initial poor graft function after orthotopic or split-graft liver transplantation. Gut. 2009;58:605–6.
30. Brokelman W, Stel AL, Ploeg RJ. Risk factors for primary dysfunction after liver transplantation in the University of Wisconsin solution era. Transplant Proc. 1999;31:2087–90.
31. Chui AK, Shi LW, Rao AR, Anasuya A, Hagl C, Pillay P, et al. Primary graft dysfunction after liver transplantation. Transplant Proc. 2000;32:2219–20.
32. Johnson SR, Alexopoulos S, Curry M, Hanto DW. Primary nonfunction (PNF) in the MELD Era: an SRTR database analysis. Am J Transplant. 2007;7:1003–9.
33. Gepts E, Camu F, Cockshott ID, Douglas EJ. Disposition of propofol administered as constant rate intravenous infusions in humans. Anesth Analg. 1987;66:1256–63.
34. Veroli P, O'Kelly B, Bertrand F, Trouvin JH, Farinotti R, Ecoffey C. Extrahepatic metabolism of propofol in man during the anhepatic phase of orthotopic liver transplantation. Br J Anaesth. 1992;68:183–6.
35. Murata F, Iwade M, Hidano G, Tsunoda C, Nagata O, Ozaki M. Recovery from propofol anesthesia is delayed in hepatectomy patients due to altered pharmacodynamics. Masui. 2006;55:150–7.
36. Marsh B, White M, Morton N, Kenny GN. Pharmacokinetic model driven infusion of propofol in children. Br J Anaesth. 1991;67:41–8.
37. Nanashima A, Pillay P, Verran DJ, Painter D, Nakasuji M, Crawford M, et al. Analysis of initial poor graft function after orthotopic liver transplantation: experience of an australian single liver transplantation center. Transplant Proc. 2002;34:1231–5.
38. Chui AK, Haghighi K, Painter D, Jayasundera M, Hall G, Rao AR, et al. Donor fatty (steatotic) liver allografts in orthotopic liver transplantation. Transplant Proc. 1998;30:3286–7.
39. Leslie K, Bjorksten AR, Ugoni A, Mitchell P. Mild core hypothermia and anesthetic requirement for loss of responsiveness during propofol anesthesia for craniotomy. Anesth Analg. 2002;94:1298–303.
40. Bruegger L, Studer P, Schmid SW, Pestel G, Reichen J, Seiler C, et al. Indocyanine green plasma disappearance rate during the anhepatic phase of orthotopic liver transplantation. J Gastrointest Surg. 2008;12:67–72.
41. Faybik P, Krenn C-G, Baker A, Lahner D, Berlakovich G, Steltzer H, et al. Comparison of invasive and noninvasive measurement of plasma disappearance rate of indocyanine green in patients undergoing liver transplantation: a prospective investigator-blinded study. Liver Transpl. 2004;10:1060–4.
42. Aggarwal S, Bloom M, Kang Y, Kramer D, Martin M, De Wolf AM. EEG and AVDO2 monitoring: adjuncts in the management of acute hepatic encephalopathy during liver transplantation. Transplant Proc. 1991;23:1992–3.
43. Shimoda J, Tunoda Y, Sekiguti K, Fukumura Y, Kubota F, Murakami M. Studies on electroencephalographic changes after liver transplantation. Rinsho Byori. 2003;51:798–804.
44. Reeves RR, Struve FA, Burke RS. Quantitative EEG analysis before and after liver transplantation. Clin EEG Neurosci. 2006;37:34–40.
45. Loock J, Mitzner SR, Peters E, Schmidt R, Stange J. Amino acid dysbalance in liver failure is favourably influenced by recirculating albumin dialysis (MARS). Liver. 2002;22(Suppl 2):35–9.
46. Dahaba AA. The "Plateau" phenomenon of stepwise extracorporeal albumin dialysis bispectral index hepatic encephalopathy recovery. A&A Practice. 2015;5:219–22.
47. Dahaba AA. Possible explanations for the odd phenomenon of liver-assist albumen-dialysis hepatic encephalopathy stepwise "Plateau" electroencephalography recovery. Minerva Anestesiol. 2018;84:262–4.
48. Fischer JE, Rosen HM, Ebeid AM, James JH, Keane JM, Soeters PB. The effect of normalization of plasma amino acids on hepatic encephalopathy in man. Surgery. 1976;80:77–91.
49. Saper CB, Fuller PM. Wake-sleep circuitry: an overview. Curr Opin Neurobiol. 2017;44:186–92.

9 Natural Sleep and Sedation Monitoring in Critically Ill Patients

9.1 Sedation Monitoring of Critically Ill Patients

Critical Care Unit (CCU) patients are exposed to the risk of the so-called "Critical Care Unit Triad" of agitation, pain, and delirium [1]. Critically ill patients are often sedated for a comfortable stay, to comply with mechanical ventilation or to therapeutically reduce intracranial metabolism and hypertension [2]. However, critically ill patients are often over- or under-sedated that might result in adverse complications and poor outcomes [3]. Depth of sedation is assessed via numerous indices like Sedation Agitation Scale (SAS) [4], Richmond Agitation-Sedation Scale (RASS) [5], Observer's Assessment of Alertness/Sedation (OAA/S) Scale [6], or Ramsay Score [7]. These scores have serious shortcomings and limitations like inconsistency, intermittent assessments, needing regular patient stimulation, and depending on patients' behavior and response. Thus, several processed Electroencephalography-Depth Of Anesthesia (pEEG-DOA) monitors have been suggested as possible objective alternatives to provide a real-time, continuous bedside clinical sedation assessment tool [8].

Using pEEG-DOA for monitoring CCU patients is based on several neurophysiologic bases including association with cerebral metabolism; tracking different brain effects of individualized sedatives/hypnotics dosing schemes; detecting neuronal dysfunction; detecting nociception-antinociception imbalance; as well as detecting/evaluating epileptic activity [8–12]. Simplified pEEG-DOA dimensionless indices of hypnosis/consciousness are widely used in guiding sedation/hypnosis in the Operating Rooms (OR) and Critical Care Units (CCU) [13, 14], thus providing valuable insights in high-risk critically ill patients [15], despite some limitations when, at certain occasions, the pEEG-DOA indices do not correctly indicate the depth of consciousness nor correlate with hypnotic agents' concentrations [16, 17]. Even though pEEG-DOA monitors were not adequately validated in the CCU settings, still pEEG-DOA monitors' dimensionless number is better than administering potent hypnotics/sedatives unguided, particularly when there will never be sufficient technical and human resources available for continuously monitoring and interpreting raw Electroencephalography (EEG) 10–20 Fast Fourier Transform (FFT) international system in the CCU [8, 18].

Almost all pEEG-DOA monitors typically display one or more raw real-time EEG channel along with their pEEG-DOA indices parameters, with a dedicated EEG feature displaying raw EEG waves' frequencies, speed, and amplitude, thus offering additional reliable neurophysiological information to compensate for the inadequacy of relying on a single dimensionless pEEG-DOA index [8].

Evidence is growing that using pEEG-DOA monitoring of CCU patients could avoid under-

A. A. Dahaba, *Understanding the Numbers in Depth of Anesthesia Monitors*,
https://doi.org/10.1007/978-3-031-88260-9_9

or over-sedation and could possibly improve clinical outcomes, but much more research is still needed to validate pEEG-DOA monitoring in the CCU settings [8, 15]. The great heterogeneity of different types and levels of sedation, as well as the diverse physio-pathological changes taking place in critical care patients' populations, makes the seek for a "silver bullet" magic-number to guide the CCU patients' sedation even more complex than in anesthesia [16, 19].

However, to further expand this pEEG-DOA monitoring inconsistent technology, with such patients' interpatient variability, to other areas like critical care medicine where there is even less data available, is interesting but still very nascent [20]. When pEEG-DOA monitoring is used in the CCU setting, this would be confronted by extremely compounded/confounded conditions; as the pEEG-DOA monitors would be influenced, in that case, by numerous simultaneously (inter)acting conditions such as polypharmacy of cardiovascular medications, polypharmacy of sedatives/hypnotics medications, blood glucose, blood volume, body temperature, blood pressure, and response to tactile/noxious stimulation like endotracheal intubation, tracheal suction or invasive medical procedures, noisy environment, mechanical ventilation, and disrupted circadian rhythm physiological sleep (or lack of) all "lumped together" [21]. What are pEEG-DOA monitors supposed to indicate in such "unfriendly environment" when all these factors are simultaneously "competing" for the same pEEG-DOA index? Furthermore, what are CCU clinicians supposed to do with such a heavily confounded pEEG-DOA index, just "ride" the numbers [20]? As the golden rule in medicine goes, "any medical parameter is only as good as the doctor who interprets the data". It may certainly be true that the interpretation of these pEEG-DOA indices generated in the CCU is confounded by the heavy intrusion of a great many physiological variables and that this application of all pEEG-DOA in the CCU setting has not been very systematically studied. However, which way we slice it, pEEG-DOA monitors are still being used in the CCU to assess depth of sedation/hypnosis, and I truly believe that is exactly what the manufacturers assert that these devices are intended for [20]. Having said that, I strongly argue against going as far as using pEEG-DOA monitors in an CCU setting; there are too much heterogeneity and many diseases status in the CCU patients other than sedatives/hypnotics, which would strongly influence pEEG-DOA indices' algorithm calculations, in effect rendering the pEEG-DOA indices not only useless most of the time but can also be misleading [20].

9.1.1 Bispectral Index (BIS) Sedation Monitoring of Critically Ill Patients

Using the processed Electroencephalography-Depth Of Anesthesia (pEEG-DOA) monitoring to assess patients under anesthesia in the Operating Room (OR) theater for relatively short periods of time is quite different than monitoring critically ill patients who might be receiving continuous sedation and analgesia in the Critical Care Unit (CCU) for prolonged periods of time [16]. The potential for artifactual contaminations and pitfalls when using pEEG-DOA monitors in the CCU needs to be acknowledged. As I previously mentioned, there have been published reports of artifacts interferences in the signal because of the use of pacemakers or external rewarming blanket devices resulting in erroneous increases in the Bispectral Index [22]. Whereas, infusions of catecholamines that can cross the Blood-Brain Barriers (BBB) may provoke arousal and consequently increase the Bispectral Index [23]. Nevertheless, Bispectral Index monitoring in the CCU seems desirable to avoid awareness during painful invasive procedures, particularly when Neuromuscular Blocking Agents (NMBAs) are used [16].

Critically ill patients are subject to various stressful and painful procedures that would commonly need analgo-sedation, a combination of analgesia and sedation [24]. Nonetheless, analgo-sedation for these interventions might at times require a degree of communication with the patients who are required to respond to several commands while remaining calm and avoiding sudden unexpected movements [24]. Furthermore, critically ill patients condition assessments using clinical scorings that rely on verbal and face expression to evaluate the depth of sedation are often challenging. Another fundamental difficulty lies in the fact that tactile or verbal stimulation for medical grading could be disruptive to both patients and CCU personnel [24]. Using Bispectral Index monitoring to accurately guide administrations of hypnotics/sedatives could assist in resolving these issues. All rapid onset, short-acting agents with predictable onset and offset of action like propofol would enable precise infusion rate adjustments as well as speedy recovery. Nonetheless, respiratory depression, hypotension, and oversedation remain always possible occurrences because of their rather tight therapeutic ranges. Thus, Bispectral Index monitoring could offer a valuable tool to avoid unwarranted oversedation with all its accompanying complications [24].

My research team Dahaba and colleagues [24], at Medical University of Graz, Austria, compared the safety and efficacy of Bispectral Index-guided remifentanil/propofol sedation regime to clinically guided remifentanil/propofol sedation regime for providing optimum sedation [24], defined as Sedation Agitation Scale (SAS) [4] Grade-4, in patients undergoing potentially painful interventional procedures [24]. Our hypothesis was that a Bispectral Index-guided regime has higher efficacy than a SAS-guided regime for providing optimum sedation. Our secondary endpoints were the nominated Bispectral Index range of 80–85 percentage time attained, hemodynamic stability, medication requirements, patients'/operators' satisfactions, and to validate the nominated Bispectral Index range of 80–85 for procedural sedation [24].

Assuming a 15% deviation from previous study of remifentanil 78.3 ± 6.2% mean percentage time of SAS 4 optimum sedation [25], our power calculation resulted in an *a*-priori sample size for two-sided student *t* test (α =0.05) of 27 patients group size with >80% power [24]. In our study, we recruited 54 American Society of Anesthesiologists (ASA) classification physical status I–IV, >40 years of age subjects undergoing potentially painful interventional procedures [24]. We excluded subjects with clinical conditions that might impede assessment of their sedation levels like dementia, stroke, stupor, or subjects deviating with >20% from ideal body weight [24]. According to a computer-generated randomization list, subjects were randomly allocated to the Bispectral Index-guided or the SAS-guided groups [24].

A Bispectral Index "Quatro" sensor was placed on the patients' forehead and attached to a Bispectral Index BIS-XP (version 3.4) monitor [24]. We gave instructions to our patients, so as not to fall asleep, just to remain calm and relaxed. After we verified Signal Quality Index (SQI) of >95% and electrodes' impedance of <5 kΩ, we started our recordings. We continuously collected and stored the Bispectral Index and Electromyography (EMG) values every 5 s, as we set the Bispectral Index smoothing rate at 30 s [24]. All anesthesiologists participating in our study were well experienced and educated in how to assess the level of sedation using the SAS scores for at least 2 years in critically ill patients' SAS score assessment prior to their involvement in our study [24]. Procedures were performed in quiet surroundings to avoid noise from causing spurious Bispectral Index responses. A sizable mobile partition wall separated the patients from the Bispectral Index monitor [24]. For patients randomized to the SAS Group, propofol and remifentanil infusions were positioned on the patients' side of the separation partition wall and the handling anesthesiologists, who were blinded to the Bispectral Index values, adjusted their infusion rates according to their own mean SAS score evaluations over the preceding full 5 min duration [24]. Patients randomized to the Bispectral Index group had their propofol and remifentanil infusion pumps placed on the Bispectral Index side of the separation partition wall, and the handling anesthesiologists adjusted their infusion rates according to their mean Bispectral Index value over the preceding full 5 min duration [24]. Simultaneously a second anesthesiologist, located on the patients' side of the separation partition wall, who was blinded to

Bispectral Index value and the infusion rates, recorded the patient's mean SAS score for the preceding 5 min duration. Patients had a face-mask with 50% oxygen in air insufflation.

To assess patients' pain intensity, we used Pain Intensity scale (PI) of 1 = no pain, 2 = mild pain, 3 = moderate pain, 4 = severe pain, 5 = very severe pain, 6 = worst possible pain [24]. In all patients, remifentanil and propofol infusions were started. The patients randomized allocation recruitment, and dosing algorithms are outlined in the flow chart (Fig. 9.1) [24]. In patients randomized to the SAS group, when the SAS score was 4 and patients were pain-free, then the running infusion rates were upheld [24]. When the SAS score was >4, then infusion rates of propofol and remifentanil were both increased. When patients were in pain, then just the remifentanil infusion was increased [24]. If SAS was <4, both propofol and remifentanil infusion rates are reduced [24]. In patients randomized to the Bispectral Index group, when the Bispectral Index value was in the range of 85–80, then the running infusion rates were upheld. When the

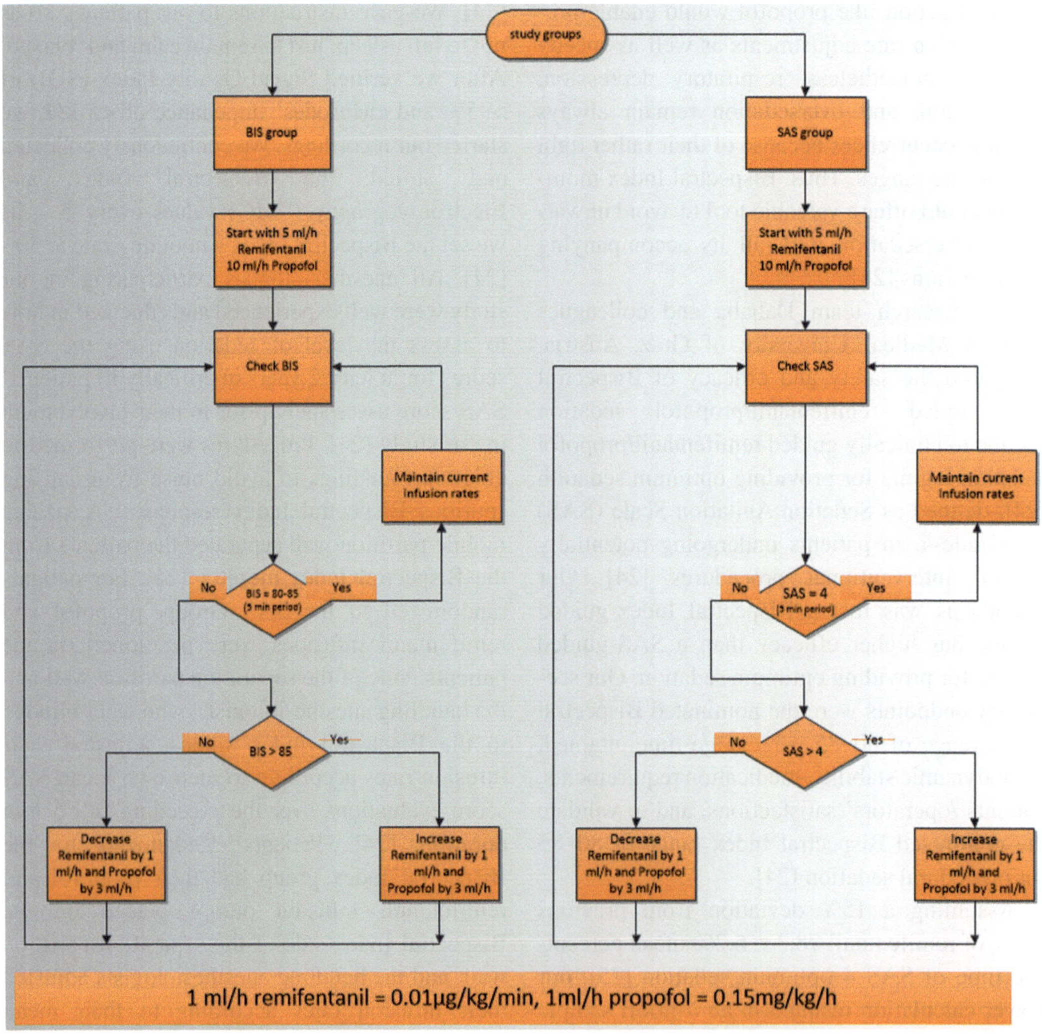

Fig. 9.1 The flow chart outlines patients' randomized allocation recruitment and dosing algorithms. (Reproduced with permission # 5934730525995 from Dahaba et al. Bispectral-index-guided versus clinically guided remifentanil/propofol analgesia/sedation for interventional radiological procedures: an observer-blinded randomized study. Anesth Analg. 2006;103:378–84)

Bispectral Index value was >85, we increased both propofol and remifentanil infusion rates. When patients were in pain, then just the remifentanil infusion rate was increased [24]. When the Bispectral Index was <80, or for a "Rescue Protocol," when subjects were in apnea, they are immediately ventilated via a facemask, and both propofol and remifentanil infusion rates are reduced [24]. Recovery time was time from termination of remifentanil and propofol infusions till patients recalled their Date of Birth (DOB) and completely regained their consciousness. Following end of procedures, Post-Anesthesia Care Unit (PACU) nurses asked the patients and asked the assisting nurses, blinded to the group assignments, to grade their satisfaction degree as excellent = 3, good = 2, acceptable = 1, or Unacceptable = 0 [24].

One-way Analysis Of Variance (ANOVA) revealed that the mean ± SD optimal sedation (SAS 4) percentage time of 76.6 ± 14.7% in the Bispectral Index group was significantly higher than optimal sedation (SAS 4) percentage time of 63.8 ± 16.4% in the SAS group [24]. Furthermore, the nominated Bispectral Index 85–80 percentage time of 75.4 ± 10.9% in the Bispectral Index group was significantly longer than nominated Bispectral Index 85–80 percentage time of 58.5 ± 15.8% in the SAS group [24]. The weighted mean infusion rates of remifentanil and propofol in the Bispectral Index group (0.066 ± 0.027 μg kg^{-1} min^{-1}, 1.59 ± 0.44 mg kg^{-1} h^{-1}) were significantly lower than the weighted mean infusion rates of remifentanil and propofol in the SAS group (0.091 ± 0.036 μg kg^{-1} min^{-1}, 1.92 ± 0.43 mg kg^{-1} h^{-1}) respectively [24], and subjects required fewer infusion rate adjustments in the Bispectral Index group than in the SAS group (Fig. 9.2) [24]. The time of Bispectral Index of <80 in the Bispectral Index group was

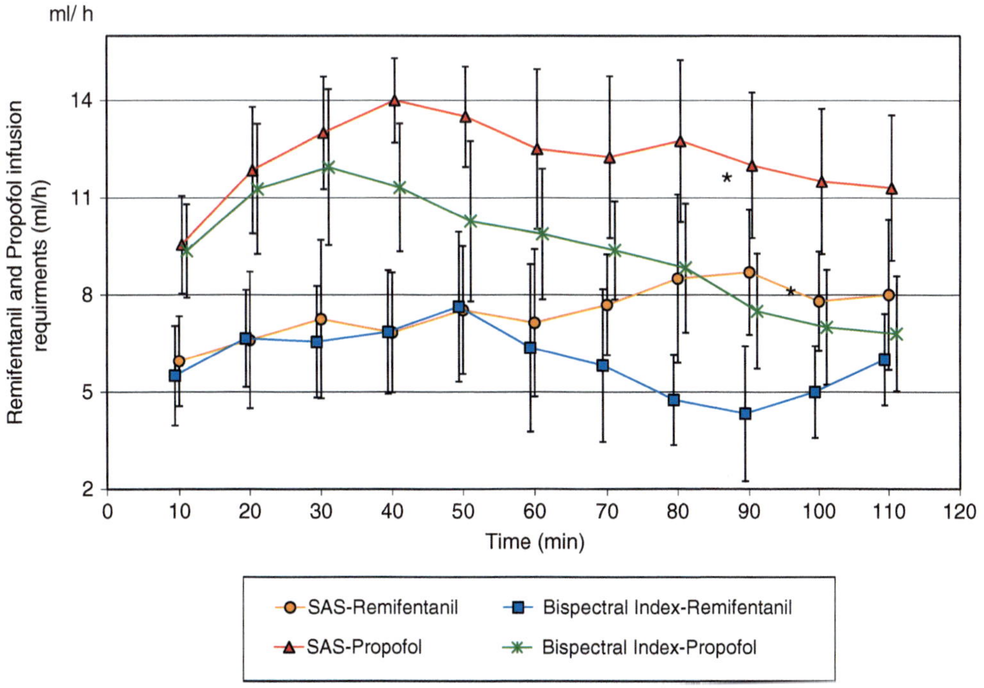

Fig. 9.2 Mean ± SD remifentanil and propofol infusion requirements (ml h^{-1}) in the Bispectral Index (BIS) and Sedation Agitation Scale (SAS) groups. *Significant difference between the two groups. (Reproduced with permission # 5934730525995 from Dahaba et al. Bispectral-index-guided versus clinically guided remifentanil/propofol analgesia/sedation for interventional radiological procedures: an observer-blinded randomized study. Anesth Analg. 2006;103:378–84)

significantly shorter than that in the SAS group, which indicates that subjects in the SAS group were deeper sedated [24]. Pearson's linear determination coefficient (r^2) revealed that Bispectral Index values displayed a clear temporal correlation to the SAS scores ($r^2 = 0.72$) in our overall study population (Fig. 9.3) [24]. Mean ± SD Bispectral Index and EMG values at different SAS grades displayed a clear temporal correlation to the different SAS scores (Fig. 9.4) [24]. Data were normally distributed. Subjects in the SAS group felt no clinically significant pain in 96.4 ± 1.8% of the time compared with 96.6 ± 1.6% of the time in the Bispectral Index group [24]. The "Bispectral Index Group rescue protocol" was never employed. Mann-Whitney *U*-test revealed that median operators' satisfaction in both groups was graded as good (2), whereas patients' satisfaction was graded as excellent (3) in both groups [24].

Our study hypothesis was accepted, as the objective Bispectral Index-guided regimen exhibited higher efficacy for the provision of optimum sedation with less infusion rate adjustments than a subjective clinically guided regimen [24]. This could be accredited to the precise objective nature of the Bispectral Index monitoring that provided real-time data to achieve and maintain a steady sedation level. However, the optimal sedation time difference between the two groups of 13% resembling approximately 5 min, albeit being statistically significant, was still smaller than the 15% difference that we presumed for our power analysis and might be considered clinically irrelevant [24]. Furthermore, our finding of the recovery time 2 min difference between the two groups would present no practical value or advantage of Bispectral Index monitoring over a clinically guided regime [24].

Our study [24] showed a clear temporal correlation between SAS scores and Bispectral Index values. Contrary to propofol that has a powerful effect on EEG; μ-opioids result in far less electrophysiological cerebral cortex changes

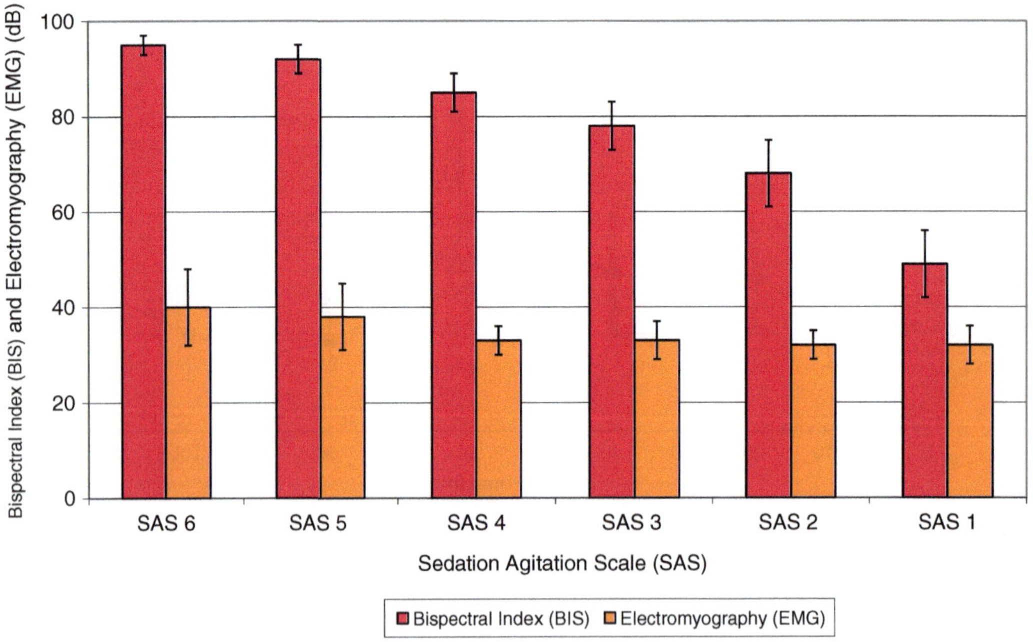

Fig. 9.3 Bispectral Index (BIS) and Electromyography (EMG) values clear temporal correlation to the Sedation Agitation Scale (SAS) in overall study population. (Reproduced with permission # 5934730525995 from Dahaba et al. Bispectral-index-guided versus clinically guided remifentanil/propofol analgesia/sedation for interventional radiological procedures: an observer-blinded randomized study. Anesth Analg. 2006;103:378–84)

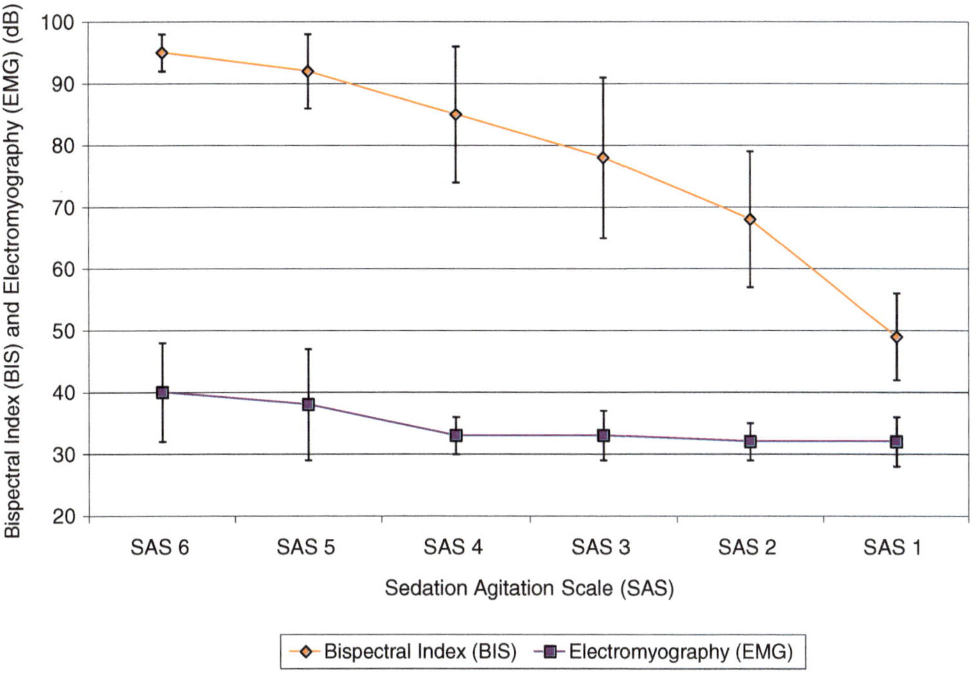

Fig. 9.4 Mean ± SD Bispectral Index (BIS) and Electromyography (EMG) temporal correlation to different Sedation Agitation Scale (SAS) grades

[26]. Nevertheless, remifentanil and μ-opioids could still greatly affect Bispectral Index monitoring in a different fashion, namely, via obtunding noxious stimuli responses. Remifentanil was reported to obtund, in dose-related manner, when propofol infusion was kept constant, rises in Bispectral Index values following endotracheal intubations or laryngoscopy. Thus, Bispectral Index values under constant sedative regime do not just reflect the sedative component of the hypnotic agent but would also indicate the extent of opioids inhibition of the noxious stimulation inducing a pain-provoked cortical arousal reaction [27]. Hence, it is fair to conclude that the Bispectral Index values evidently correlate to both the combined hypnotic and the analgesic elements of our propofol/remifentanil sedation regimen [24].

Bispectral Index values were reported to erroneously rise when there is high EMG activity [28, 29], due to the fact that EMG activity is an artifactual signal occurring in the bispectrum "range of interest" frequency, as $EMG_{30–47\ Hz}$ frequency wave band mimics the *Relative BetaRatio* one of the Bispectral Index descriptor subparameter components and could be mistaken by the Bispectral Index algorithm as EEG activities [30]. In our study [24], throughout the periods where patients were at SAS 5 and SAS 6 light sedation, the mean EMG values were exceeding what could be construed as the EMG cutoff of 35 decibel (dB) [31]; this clearly indicates that high EMG activities might have contributed to the elevated Bispectral Index values when patients were lightly SAS 5 and SAS 6 sedated [24].

Interestingly, the original creators of the SAS score [32, 33] validated the Bispectral Index for grading critically ill patients' sedation, where SAS 4 correlated to mean Bispectral Index value of 79 in one study [32] and to Bispectral Index value of 87 in another study [33] by the same authors [32, 33]. Thus, the designated Bispectral Index 85–80 range of our study's Bispectral

Index group SAS 4 patients corresponded to a slightly narrower Bispectral Index 79–87 range for SAS 4 by the original creators of the SAS score in their two preceding publications where they correlated SAS to Bispectral Index ranges [32, 33].

Our study [24] demonstrated that the Bispectral Index-guided regime was more efficacious than the SAS clinically guided regime for providing optimum sedation in subjects undergoing potentially painful procedures [24]. Utilizing the Bispectral Index for adjusting infusion rate requirements would result in fewer propofol and remifentanil doses [24]. Our study results demonstrated that our nominated Bispectral Index range of 85–80 provided an adequate and efficient sedation level for both patients and personnel [24]. Hence, Bispectral Index monitoring could be a beneficial tool for objectively and precisely providing consistent sedation that could assist in our routine evaluation, as clinical judgment remains vital in ensuring patients safety [24].

Sadhasivam and colleagues [34] demonstrated that the Bispectral Index BIS A-2000 (version 4.0) was more sensitive than just relying on vital signs monitoring in 96 ASA physical status I–II children aged 1–12 year pediatric patients undergoing sedation using OAA/S [6], for procedures like Magnetic Resonance Imaging (MRI) [34]. There was good correlation between Bispectral Index and OAA/S [34].

Kress and colleagues [35] reported large variability in levels of sedation in 128 mechanically ventilated CCU patients, where their levels of arousal range from almost non-arousable at rest to highly arousable during potentially noxious procedures, physiotherapy, medical, or surgical invasive interventions [35]. Oversedation in the CCU seems to be associated with higher rates of ventilator-associated pneumonia and longer CCU stay [16, 35].

Consales and colleagues [36] evaluated sedation, obtained with midazolam or propofol and monitored with Ramsay Score [7], and Bispectral Index BIS A-1000 in 40 patients admitted to CCU for postoperative monitoring after major abdominal or vascular surgery. They reported that using the Bispectral Index in surgical CCU patients who usually receive continuous infusions of sedatives and NMBAs can have both economic and patients' benefits as patients reported less recall of painful frightening events during CCU sedation in a Bispectral Index-guided group [36], in addition to associated drug-cost saving [36]. Whereas using Bispectral Index as an objective guide for titrating dosage of sedatives/hypnotics can decrease medical complications of oversedation, such as decline in cardiac contractility or hypotension, with an added benefit of drug-cost savings and decreased time to extubation [16, 36]. Despite the authors, when using Pearson test, observed a good correlation between Ramsay Score and Bispectral Index values, at the deeper levels of sedation, a wide range of Bispectral Index values corresponded to Ramsay Score 6, indicating various levels of Central Nervous System (CNS) depression that are not identified by clinical evaluation [36].

Since the 2014 publication of Accidental Awareness during General Anaesthesia (AAGA) fifth National Audit Project (NAP5) [37], which found patients receiving NMBAs with Total Intravenous Anesthesia (TIVA) were at higher risk of AAGA [37], many medical institutions are now mandating some sort of pEEG-DOA monitoring for such patients in the OR suite [37]. This logic could be easily extended to patients receiving NMBAs in the CCU setting. This would be especially important during patients' transfer or when painful invasive procedures such as tracheostomy, reintubation, rigid bronchoscopy, or patient prone positioning are performed [16, 38, 39].

Having said that, still Bispectral Index values can greatly vary in critically ill sedated patients and have been shown to poorly correlate with SAS [40–42]. Possible explanations of this wide variation in Bispectral Index values in sedated critically ill patients and how they greatly differ from OR patients undergoing potentially painful surgical procedures under general anesthesia [16] could be the fact that unlike surgical patients, CCU patients only occasionally undergo painful invasive procedures and thus require far less sedation, albeit such intrusive interventions can still lead to sudden arousals and subsequent fluctuations in the Bispectral Index values. Furthermore, the effect of neuromuscular block could be another contributing factor as Vivien and colleagues [43] demonstrated that Bispectral Index BIS A-2000 (version 2.1) values declined in a group of 45 lightly sedated CCU patients who received NMBAs, before Bispectral Index values subsequently increased with the recovery of the lightly sedated CCU patients' neuromuscular block [43]. Here CCU physicians are confronted with the dilemma of what to consider the "true" level of Bispectral Index values that should guide the CCU hypnotics/sedatives administrations. The Bispectral Index value before or after a procedure's noxious stimulation? The Bispectral Index value before or after NMBA administration? It seemed that CCU clinicians who will determine the degree of their critically ill patients' sedation solely based on Bispectral Index monitoring may well expose their CCU patients to inadvertent oversedation [16].

Wang and colleagues [21], in a prospective multicenter validation study, demonstrated that the Bispectral Index BIS-XP A-2000 (version 3.31) reliably detected deep sedation in 90 critically ill mechanically ventilated patients [21]. They compared the Bispectral Index monitored for 24 hours with the Richmond Agitation-Sedation Scale (RASS) [5], evaluated every 4 hours [21]. Deep sedation was defined as ≤ − 3 RASS score. Although deep sedation was only prescribed in six patients (6.7%), 76 patients (84.4%) had at least one episode of deep sedation [21]. Threshold for detecting deep sedation of Bispectral Index BIS 50 Area Under the Receiver Operating Characteristic (ROC) curve was 0.771 (95% Confidence Interval, 0.714–0.828) [21]. The Sensitivity and Specificity of Bispectral Index were 94.0% and 66.5%, respectively [21].

9.1.2 SedLine Sedation Monitoring of Critically Ill Patients

The SedLine® processed Electroencephalography-Depth Of Anesthesia (pEEG-DOA) monitor (Masimo, Irvine, CA, USA) incorporating a raw Electroencephalography (EEG) channel in the Root® monitor that calculates the Patient Sate Index (PSI) [44], was studied in critically ill patients in addition to patients under general anesthesia [44–46]. Idei and colleagues [47] studied the updated PSI algorithm, in an innovative use of SedLine® Root® monitor in critically ill patients [47]. They studied the relationship between SedLine® PSI and the Richmond Agitation-Sedation Scale (RASS), evaluated by Critical Care Unit (CCU) personnel blinded to the PSI value, in 50 sedated and mechanically ventilated critically ill patients [47]. Their results indicated good correlation between the RASS score and the PSI value, where PSI 54 was a cutoff value separating light sedation of ≥ − 2 RASS score from deep sedation of ≤ − 3 RASS score, clearly demonstrating that PSI is an accurate, objective, constant indicator that could help in titrating hypnotics/sedatives for critically ill patients [47]. They concluded that SedLine® PSI could provide appropriate levels of sedation as the PSI detected and accurately distinguish light sedation from deep sedation [47]. Furthermore, the authors found an additional unique advantage in the SedLine® PSI critical care patients' sedation monitoring in a clear strong correlation between RASS clinical sedation and PSI even in critically ill patients who are frequently affected by high Electromyography (EMG) activity [47].

Table 9.1 depicts sedation monitoring in critically ill patients.

Table 9.1 Sedation monitoring of critically ill patients

Reference	Monitor	*n*	Agent /Sedation Score	Index change	Explanation
Dahaba et al. [24]	BIS-XP (3.4)	27 × 2	Propofol/remifentanil BIS vs SAS	Good correlation	BIS detects sedation grades
Sadhasivam et al. [34]	BIS A-2000 (4.0)	48 × 2	BIS vs OAA/S	Good correlation	BIS detects sedation grades
Consales et al. [36]	A-1000	20 × 2	Propofol/midazolam BIS vs Ramsay score	Good correlation	BIS wide variation grade 6
Vivien et al. [43]	A-2000 (2.1)	45	Sedation +NMBAs	BIS ↓ with NMBAs	BIS ↓ with NMBA
Wang et al. [21]	BIS-XP A-2000 (3.31)	90	BIS vs RASS	Good correlation	BIS detect sedation grades
Idei et al. [47]	SEDLine PSI	50	PSI vs RASS	Good correlation	PSI detect sedation grades

n number of patients per group; *BIS* Bispectral Index; *SAS* Sedation Agitation Scale; *OAA/S* Observer's Assessment of Alertness/Sedation Scale; *NMBAs* = Neuromuscular blocking agents; Ramsay Score; *RASS* Richmond Agitation-Sedation Scale; *PSI* Patient Sate Index

9.2 Sleep Monitoring of Critically Ill Patients

Several studies demonstrated that Critical Care Unit (CCU) patients suffer from fragmented sleep [46], as well as altered sleep manner manifesting as extended light sleep, frequent arousals with fewer Electroencephalography (EEG) slow wave, and Rapid Eye Movement (REM) sleep [48–50]. Sleep disruptions could result in cognitive as well as physiological disturbances [46]. Adverse outcomes in patients who suffer from sleep disruptions include impaired immune function resulting in increased infectious diseases [51–53]; negative nitrogen balance and negative protein catabolism [54, 55]; cardiovascular changes such as hypertension and increased sympathetic activity [56, 57]; decreased cognitive function like memory deficiency, slow reaction time [58, 59]; and decline in quality of life parameters [60]. Sleep disturbances have also been proposed as one of the major factors contributing to hospital patients' delirium [61], as 20–80% of CCU patients were reported to suffer from delirium [62]. Even though sleep disruption in adult critically ill patients results in acute deterioration in cognitive functions, including delirium, still it is underdiagnosed in critically ill patients [46]. Up till now reducing CCU delirium via improving sleep quantity and quality has not been thoroughly studied [46].

9.2.1 Polysomnography (PSG) and Processed Electroencephalography (pEEG) Sleep Monitoring

Sleep stages can be assessed using Electroencephalography (EEG)-based Polysomnography (PSG) [63]. Standard PSG implies patients' multiple electrodes head montage mapping system that collects a combination of muscle tone, eye movement, and EEG [14]. However, acquiring these EEG data needs PSG knowhow experience, and because the PSG electrodes require constant maintenance and monitoring that makes them not suitable for use over prolonged periods of time for patients who would require complex hospital care, allowing just a limited monitoring period until EEG electrodes must be replaced. Furthermore, the EEG electrodes' replacement and removal could result in skin irritations. Whereas, Critical Care Unit (CCU) patients populations who are regularly undergoing numerous tests, procedures, and intervention over their CCU stay would make it particularly impossible to regularly maintain PSG monitoring over extended periods of time [46].

Visual PSG scoring of 30-sec epochs sleep data is not always enough to show in sufficient details the dynamic structure of sleep that is also prone to considerable inter- and intra-variability [64]. Furthermore, PSG acquisition and interpretation is costly, requires rigorous training, requires considerable experience, very time-consuming, tedious, labor intensive, hard to perform over extended time periods for CCU patients, require several days of uninterrupted recordings, and could hinder care of critical care patients [46]. Hence, automated sleep staging could indeed overcome all these limitations. Because normal physiological sleep Non-Rapid Eye Movement (NREM) and anesthesia have been reported to possess numerous neurophysiological EEG similarities, processed Electroencephalography-Depth Of Anesthesia (pEEG-DOA) monitors have started to penetrate sleep research [64].

All portable pEEG-DOA, except the Narcotrend-Compact M monitor [46], are primarily designed and validated to monitor the Depth of Anesthesia (DOA) or sedation and not natural physiological sleep [65]. Currently another class of pEEG-DOA monitors that involve the use of processed Electroencephalography (pEEG) such as Bispectral Index BIS™ [66], SEDLine® [44], Entropy™ [67], or Narcotrend®-Compact M [68] are used for sleep monitoring [46]. While they are simple and easy to use, pEEG-DOA monitors are designed and validated for improving general anesthesia efficiency and safety. Because of their portability and small size, these pEEG-DOA monitors have considerable advantages over the traditional PSG equipment; that makes them beneficial for collecting data in the CCU challenging environment. Because these pEEG-DOA monitors require minimum setup and experience, they result in little interference with the clinical care of CCU patients. The pEEG-DOA indices are dimensionless indices generated by proprietary algorithms from EEG waveforms. Albeit these pEEG-DOA indices do not provide any specific neurophysiological information and their numbers do not necessarily correspond to the depth of sedation, hence pEEG-DOA indices might be less helpful for titrating sedative/hypnotic agents' administrations [14]; still these small portable pEEG-DOA monitors allow some kind of quick access to the EEG waveform [46].

9.2.2 Bispectral Index (BIS) Sleep Monitoring

One possible useful application for Bispectral Index monitoring of physiologic sleep could be in critical care patients who almost uniformly suffer from sleep deprivation and sleep disruption as they lose their circadian rhythm in a Critical Care Unit (CCU) harsh frightening noisy environment that abolishes their circadian rhythm [61], with the numerous nurse-patient interactions [17, 61]. The Italian Society for Anaesthesia, Analgesia, and Intensive Care (ISAAIC) proposed using objective tools such as Bispectral Index monitoring [17], once validated for physiological sleep monitoring in CCU patients, which could significantly improve patients' morbidity and outcomes via increasing their sleeping time [69].

The Bispectral Index BIS-Vista brain monitoring system, a processed Electroencephalography-Depth Of Anesthesia (pEEG-DOA) parameter derived from an empirical database using a complex proprietary algorithm that combines three subparameters into a single metric, provides two channels or four Electroencephalography (EEG) channels monitoring [46, 70]. The Bilateral Bispectral Index BIS-Vista monitor offers new modifications over earlier Bispectral Index models, such as multiple-channel real-time EEG waveform display, and version 1.4 new algorithm, claimed by the manufacturer to be better suited to improved filtering of the Electromyography (EMG) artifacts [71]. Because EMG artifacts are probably a constant component during stages of Rapid Eye Movement (REM) sleep, we hypothesized that the new 1.4 version Bilateral Bispectral Index algorithm might be better suited to depict natural physiologic sleep [71].

My research team Dahaba and colleagues [71], in Xuan Wu Hospital, Capital Medical University, Beijing, People's Republic of China, used Bilateral Bispectral Index BIS-Vista (version 1.4) to measure natural physiologic sleep in sleep-deprived anesthesiologists [71]. Young anesthesiologists who work extended hours suffer from acute and chronic sleep deprivation [72, 73]. It was shown that anesthesia residents on post-call exhibit a level of daytime sleepiness like that in patients with narcolepsy or sleep apnea [72, 73]. We hypothesized [71] that Bilateral Bispectral Index BIS-Vista (version 1.4) could quantify sleep depth and would exhibit a temporal decline corresponding to natural physiologic sleep stages in healthy sleep-deprived anesthesiologists on post-call anesthesia nightshift [71]. We also assessed Bispectral Index interhemispheric differences when using the Bilateral Bispectral Index BIS-Vista monitor [71].

Based upon a previous study [74], in which awake Bispectral Index values of 92 ± 3 declined to 81 ± 9 with light sleep (S1, S2) and to 83 ± 6 with REM sleep [74], our *a*-priori power analysis one-sample paired *t* test (α =0.05) showed that a group size of eight volunteers for light sleep (S1, S2) and nine volunteers for REM sleep would be required to reveal a statistically significant difference with >90% power [71]. We then increased our sample size to 10 volunteers to allow for possible dropouts [71].

We chose post-call anesthesia nightshift sleep-deprived young anesthesiologists as they were easier to enroll in the study and would readily fall asleep, rather than trying to find healthy volunteers in the general population who might have difficulty in falling asleep [71]. We enrolled five male and five female healthy young (29 ± 8 years) anesthesia residents, who had 3–5 years anesthesia experience and no history of sleep disorders, head injuries, neurological disease, or head operations [71]. Our recruited anesthesiologists did not sleep during their 24 h anesthesia shift, and those who had a chance to sleep on their anesthesia shift were excluded from the study [71]. All subjects were instructed to refrain from caffeine-containing beverages 8 h prior to the recordings. Bispectral Index BIS-Vista bilateral sensors were mounted on the forehead according to the manufacturer's guidelines [71]. Bilateral Bispectral Index values were continuously recorded till the study anesthesiologists naturally woke up. Following a 10 min baseline recording with our study anesthesiologists awake with eyes open, they were instructed to remain relaxed with their eyes closed in a quiet room [71]. Because noise was shown to transiently provoke a Bispectral Index response, there was no tactile or verbal communications with our study anesthesiologists [75], who were instructed to keep their facial muscles completely relaxed and avoid making any facial expressions as EMG activity was shown to spuriously elevate Bispectral Index values [30].

The raw EEG signals were band-pass filtered to 0.2–30 Hz and processed in real time using version 1.4 of the Bispectral Index BIS-Vista algorithm. In addition, the Bispectral Index BIS-Vista calculates the EMG power in the 70–110 Hz frequency band displayed in decibel (dB) [71]. Bilateral Bispectral Index recordings were started after verifying a Signal Quality Index (SQI) >95% and electrodes impedance <5 kΩ. The EEG variables were collected and stored with the smoothing window set at 15 s [71]. Recorded data were analyzed in the off-line mode. Using the conventional Rechtschaffen and Kales criteria [76], recorded real-time EEG signals from all subjects were analyzed by the same trained sleep technician [71]. We did not use the American Academy of Sleep Medicine (AASM) classification [77], which would have required a minimum of three simultaneous EEG-derived recordings from the frontal, central, and occipital regions [77]. Taking into consideration the 15 s latency of the Bispectral Index smoothing window, each recording was divided in epochs of 30 s and classified as one of the following stages: awake, S1, S2, S3, S4, or REM sleep [76]. For intra-patient reproducibility, we used the criteria proposed by Niedhart and colleagues [78] that the two hemispheres are considered similar when the Bispectral Index values on the left and right hemispheres readings that differed by >10 Bispectral Index units were <6% of the recording time [71].

At around 8 A.M., volunteer anesthesiologists slept for 6.1 ± 2.2 h till they spontaneously woke up [71]. All anesthesiologists reported that they deeply slept, as they were very tired from their previous all-night shift. Sleep onset time was 11 ± 5 min [71]. They readily descended to sleep cycles of S1 (6 ± 5 min), S2 (33 ± 14 min), S3 (4 ± 3 min), S4 (14 ± 9 min), and REM sleep (14 ± 12 min) [71]. Using two-way Analysis Of Variance (ANOVA) to compare volunteers' data recorded on the left and right hemispheres over time, mean ± SD awake, S1, S2, S3, S4, REM sleep, and after waking up Bispectral Index values significantly declined on the left (96 ± 2, 82 ± 11, 73 ± 10, 53 ± 13, 43 ± 11, 66 ± 9) and right (97 ± 2, 78 ± 11, 69 ± 9, 50 ± 10, 39 ± 13,

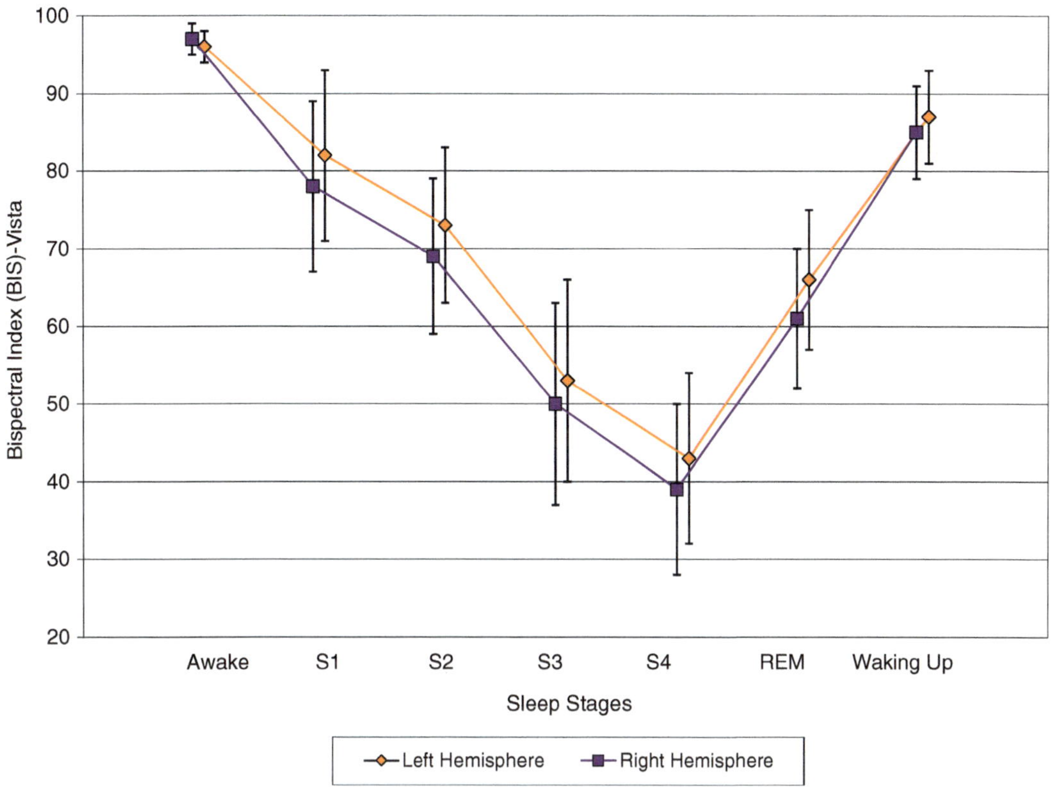

Fig. 9.5 Mean ± SD Bispectral Index (BIS) sleep stages at S1, S2, S3, and S4 and Rapid Eye Movement (REM) sleep cycles on the left and right hemispheres. (Reproduced with Minerva Anestesiologica permission from Dahaba et al. Bilateral Bispectral Index (BIS)-Vista as a measure of physiologic sleep in sleep-deprived anesthesiologists. Minerva Anestesiol. 2011;77:388–93)

61 ± 14) hemispheres, respectively, as depicted in (Fig. 9.5) [71]. The hypothesis of our study was accepted, as we demonstrated that, albeit the Bispectral Index was not designed nor validated for monitoring natural physiologic sleep, Bispectral Index monitor depicted a temporal decline to Bispectral Index values of around 40 that commensurate with natural physiologic sleep stages when sleep-deprived anesthesiologists fell asleep [71]. But that does not seem to be always the case; despite Bispectral Index depicting a temporal decline with sleep, Bispectral Index cutoff values at awake, S1, S2, S3, S4, and REM natural physiologic sleep stages greatly overlapped with wide interindividual differences [71]. There were no significant differences between Bispectral Index values on the left and right hemispheres over time [71]. Bispectral Index-Vista and EMG of volunteer anesthesiologist #4 depicted in (Fig. 9.6) is a typical volunteer anesthesiologist who represents sleep cycles of all nine volunteers recorded on the left and right hemispheres [71].

In our study, despite the fact that in nine volunteers the two cerebral hemispheres gave comparable Bispectral Index and EMG recordings [71], however, in one subject (volunteer #3), his right and left Bispectral Index and EMG values were mostly discordant throughout the whole sleep cycle recordings, in spite of high SQI and low electrodes impedance as depicted in (Fig. 9.7) [71]. When recorded using two Bispectral Index-Vista monitors, our volunteer #3 recordings would show total discrepancy between Bispectral Index value of 55 with slow δ- and θ-waves on one side and concurrent Bispectral Index value of 95 with fast α and β-waves on the other side as depicted in (Fig. 9.8) [71], or our volunteer #3

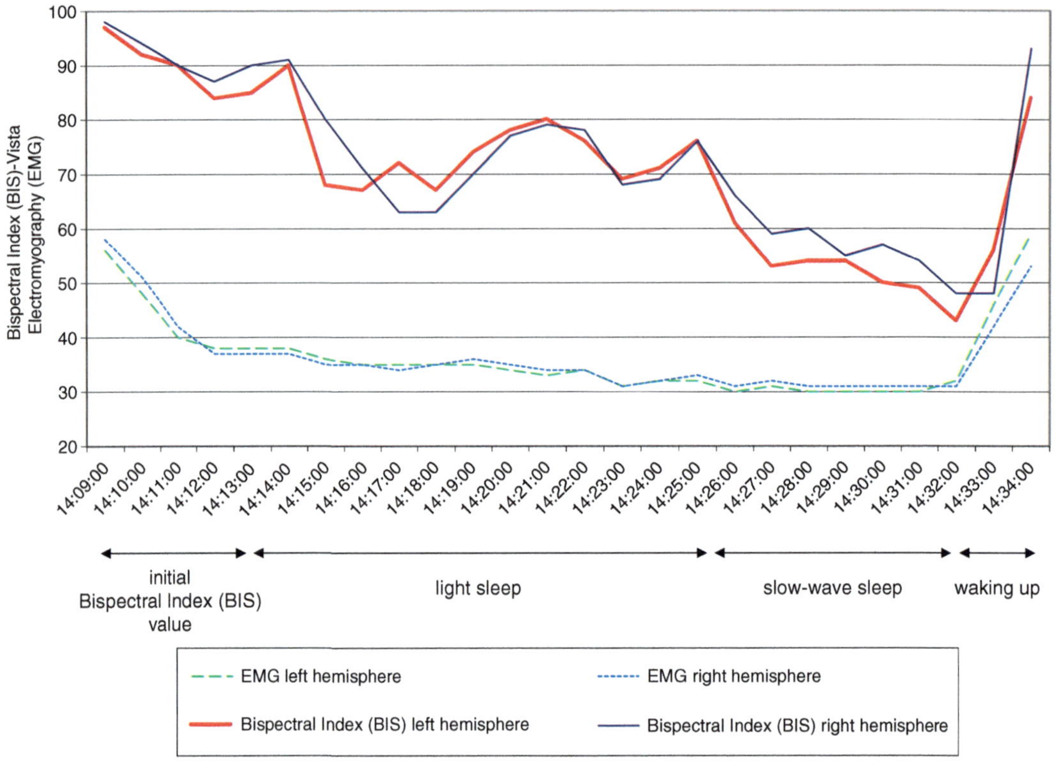

Fig. 9.6 Bispectral Index (BIS) and Electromyography (EMG) sleep stages of volunteer #4 is a typical volunteer anesthesiologist who represents sleep cycles of all nine volunteers recorded on the left and right hemispheres. (Reproduced with Minerva Anestesiologica permission from Dahaba et al. Bilateral Bispectral Index (BIS)-Vista as a measure of physiologic sleep in sleep-deprived anesthesiologists. Minerva Anestesiol. 2011;77:388–93)

recordings would show total discrepancy between Bispectral Index value of 65 with slow δ- and θ-waves on one side and concurrent Bispectral Index value of 97 with fast α and β-waves on the other side as depicted in (Fig. 9.9) [71]. Our volunteer #3 subsequently underwent neurological testing that revealed he has absolutely no abnormalities [71].

Previous studies using earlier Bispectral Index models to depict natural physiologic sleep produced widely conflicting results [74, 79]. The way Bispectral Index values declined in our sleep-deprived anesthesiologists was like subjects previously reported by Sleigh and colleagues [74], using Bispectral Index A-1000 (predates version 3.0) in five volunteers. They reported that Bispectral Index monitoring was a good indicator of sleep depth [74], as Bispectral Index values gradually decreased with natural physiological sleep stages. Light sleep corresponded to Bispectral Index values of 90–75, slow-wave sleep corresponded to Bispectral Index values of 70–20, and REM sleep corresponded to Bispectral Index values of 92–75 [74]. They concluded that Bispectral Index values well correlated with natural physiological sleep stages [74].

Conversely, our results were in sharp contrast to Nieuwenhuijs and colleagues [79] study results using Polysomnography (PSG) and Bispectral Index A-2000 (version 3.3) in 10 volunteers. They reported 67 ± 20.2 Bispectral Index values at arousal, 50 ± 16.5 Bispectral Index values at light sleep, 42 ± 11.2 Bispectral Index values at slow wave sleep, and 48 ± 7.1 Bispectral Index values at REM sleep, respectively [79]. They reported a delay in the Bispectral Index decline with widely distributed overlapping Bispectral Index values over different sleep stages. The authors concluded that Bispectral Index monitoring did not reliably indicate sleep stages [79].

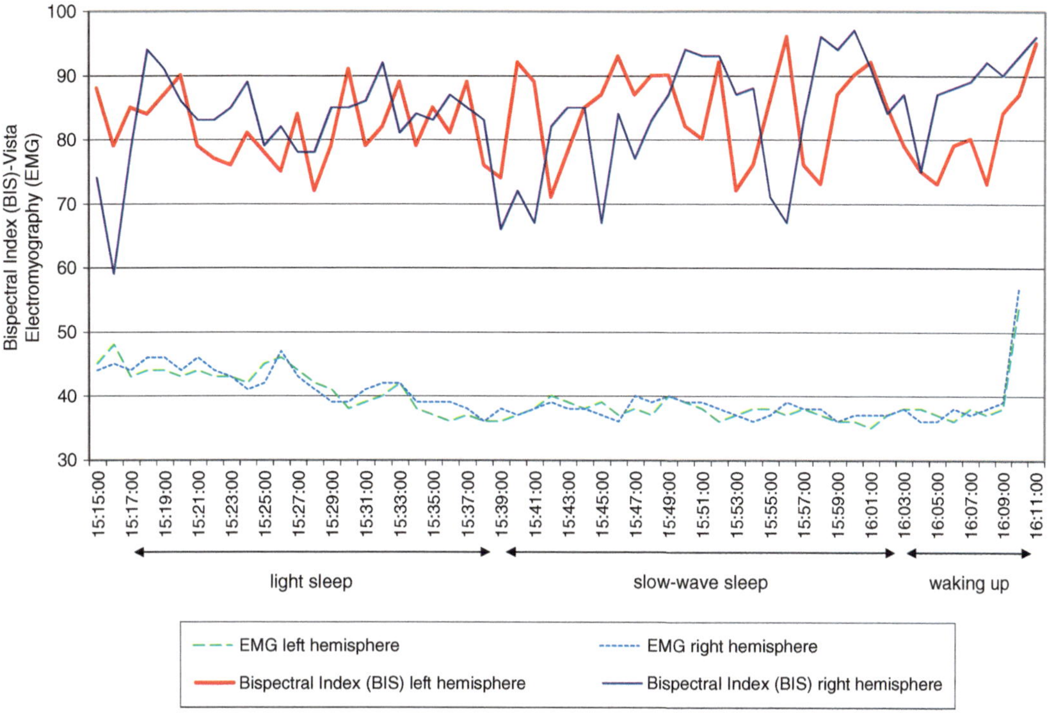

Fig. 9.7 Bispectral Index (BIS) and Electromyography (EMG) sleep cycles of volunteer #3 recorded on the left and right hemispheres appear mostly discordant (Reproduced with Minerva Anestesiologica permission from Dahaba et al. Bilateral Bispectral Index (BIS)-Vista as a measure of physiologic sleep in sleep-deprived anesthesiologists. Minerva Anestesiol. 2011;77:388–93)

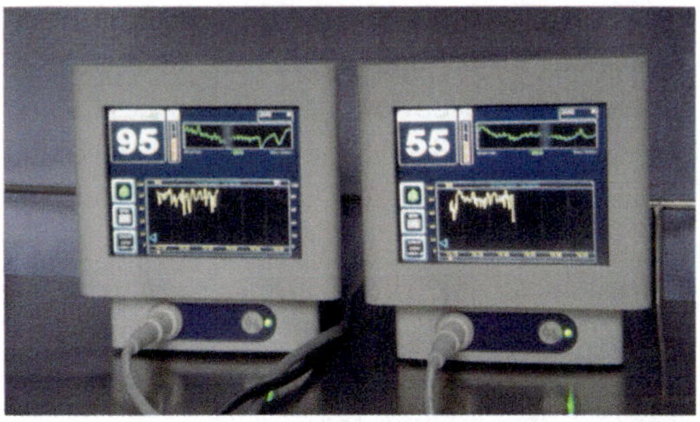

Fig. 9.8 Using two Bispectral Index-Vista monitors depicted volunteer #3 recordings of total discrepancy between Bispectral Index value of 55 with slow δ- and θ-waves on one side and concurrent Bispectral Index value of 95 with fast α and β-waves on the other side. (©2025 Medtronic. All rights reserved. Used with the permission of Medtronic)

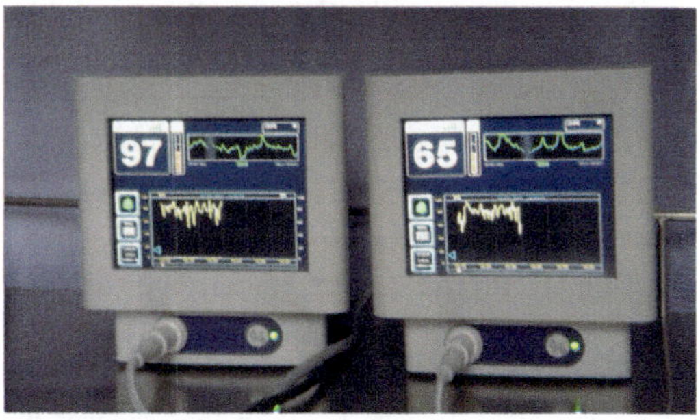

Fig. 9.9 Using 2 Bispectral Index-Vista monitors depicted volunteer #3 recordings of total discrepancy between Bispectral Index value of 65 with slow δ- and θ-waves on one side and concurrent Bispectral Index value of 97 with fast α and β-waves on the other side

Nicholson and colleagues [80] used Bispectral Index and submental electromyogram overnight recording to monitor sleep patterns of 27 minimally sedated critical care patients in their recovery phase [80]. No patient showed a completely normal sleep pattern, but about half showed 40–90 min cycle periods of slow-wave sleep that approached a normal sleep pattern [80]. They concluded that traditional classifications of EEG sleep staging are deficient when used to describe sleep in critical care unit patients [80].

Even though Ira Rampil [70] published the general principles of the Bispectral Index, still to date the complete details of the proprietary Bispectral Index algorithms core technology are only incompletely known outside of the manufacturer [70]. Although each release of a new Bispectral Index algorithm is probably substantially based on a sizeable component of previous algorithms, still potentially varied computational Bispectral Index algorithms used in earlier studies assessing natural physiologic sleep [74, 79] might be considerably different than the Bispectral Index BIS-Vista (version 1.4) algorithm. Hence one should not assume that historical Bispectral Index numbers recording physiologic sleep [74, 79] could be treated as invariant constants that carry identical meanings.

One would not necessarily expect significant differences in interhemispheric EEGs of healthy subjects. Thus, it is not surprising that two cerebral hemispheres of nine volunteer anesthesiologists in our study gave comparable recordings of mostly similar bilateral Bispectral Index BIS-Vista values [71]. However, one very interesting observation was with the mostly dissimilar Bispectral Index values and EEGs patterns of volunteer #3 whose interhemispheric differences were substantially greater than the intra-patient variability [78]. It may well be that bilateral EEGs could differ for a variety of reasons such as anatomical differences or underlying disease. However, volunteer #3 was presumably healthy with no history of head injuries, brain operations, or neurological disease [71]. In that regard, we cannot totally rule out unusual Bispectral Index recording due to abnormal EEG patterns, as there is a well-known precedent of a fully conscious volunteer with a genetically determined low-voltage EEG variant, who exhibited, on three different occasions, an unreasonably low Bispectral Index value of 40 while fully awake [81]. However, this is a very unlikely explanation of volunteer #3 as EEG variants would presumably occur bilaterally and will not be confined to a single hemisphere [71].

One possible explanation would be that the large interhemispheric differences, observed in volunteer #3 [71], could be simply attributed to EMG artifacts resulting from REM eyeball movement that were not recognized as artifact by the Bispectral Index BIS-Vista algorithm [28]. EMG activities of eyeball movements are strong artifactual signals which appear in the frequency "Range of Interest" of the bispectrum [30] and would result in spurious Bispectral Index values [28]. This raised the question of whether version 1.4 of the Bispectral Index algorithm is vulnerable to the same EMG artifacts. This seems to be the case, as we learned from the Bispectral Index monitor evolution that different Bispectral Index software versions detect EMG to greater or lesser degrees [82].

Interestingly, such unique finding of mostly concordant uni-hemispheric sleep was observed when bilateral Bispectral Index sensors were mounted over dolphins' heads, who have human sized brains, when dolphins were sleeping out of the water, e.g., dolphins' left side Bispectral Index value of 60 with right side Bispectral Index value of 90 [83]. A plausible explanation would be a real natural physiologic interhemispheric difference as the mechanisms processing EEG patterns derived from two separate sources might not always be similar and would thus yield different values. John and colleagues [84] demonstrated that at deep loss of consciousness levels there is a clear lack of coherence between homologous regions of our two hemispheres, with significant uncoupling in their interhemispheric relationships [84]. This may result in interhemispheric differences in signal processing speed and sampling in the Bispectral Index values, a near real-time monitor [84].

9.2.3 SEDLine Sleep Monitoring

The SEDLine monitor, formerly Patient State Index (PSI, Baxter, Deerfield, Illinois, US) [46], displays four channels derived from symmetrical bilateral array. Using Bifrontal electrodes, the SEDLine assesses four channels of raw Electroencephalography (EEG), while separately displaying Electromyography (EMG) artifacts such as patients' movements and Burst Suppression Ratio (BSR). The SEDLine monitor uses frontal electrode montage that closely resembles frontal electrode placements recommended by the American Academy of Sleep Medicine. Thus, the SEDLine monitor could possibly generate enough EEG recordings to identify different stages of sleep, something that would be very valuable in a challenging setting such as the Critical Care Unit (CCU) for conducting the "gold standard" Polysomnography (PSG). The SEDLine monitor has the capacity to directly record four channels of continuously monitored raw EEG in the CCU settings [46].

Vacas and colleagues [46], in a prospective pilot cohort study in CCU patients scheduled for sleep laboratory, compared the precision of continuously using the SEDLine portable monitor utility and feasibility to monitor the wake-sleep cycles sleep measurement of sleep stages concurrently monitored via standard PSG monitor [46]. They first performed a baseline comparison between SEDLine and PSG recordings in three sleep laboratory patients. In a separate patient cohort of 23 CCU patients, they continuously monitored raw EEG and the SEDLine for sleep disruptions. After recording the raw EEG, board-certified specialists evaluated sleep stages and arousal, whereas trained research nurses evaluated delirium using the "Confusion Assessment Method (CAM)" specially devised for CCU patients [46].

The SEDLine monitor could differentiate between different sleep stages, between sleep stages' transitions, as well as capturing arousals, compared with PSG that was performed in a sleep laboratory. The wake stage agreement percentage was 67%, and agreement percentage for the Non-Rapid Eye Movement (NREM) stages was 77% (N_1 was 29%, N_2 was 88%, N_3 was 6%), and agreement percentage for the stage of Rapid Eye Movement (REM) was 89% [46]. Weighted Kappa overall agreement was 0.61 (95% Confidence Interval 0.58–0.64). For the 23 enrolled patients, mean recording time was 19.1 h. Patients exhibited several signs indicative of poor-quality sleep, when sleep was distributed throughout the day with less time spent in REM (1.38 ± 2.74% of total sleep time) or stage N3 (2.17 ± 5.53% of total sleep time) coupled with a high arousal index (34.63 ± 19.04 arousals /h) [46]. The occurrence of CCU delirium was not significantly different between patients with and without sleep disruption [46]. Their results suggest the utility of SEDLine portable EEG monitor to measure different sleep stages, transitions, and arousals. However, compared with PSG, the accuracy in measuring different sleep stages by the SEDLine varies. Their results indicate that critically ill patients suffer from fragmented sleep [46].

9.2.4 Narcotrend Sleep Monitoring

The Narcotrend-Compact M developed by Schultz and colleagues [85], a clinically and scientifically validated depth of anesthesia monitor [46, 65, 85], automatically interprets and classifies the Electroencephalography (EEG) during anesthesia. The Narcotrend classification algorithms are based on visually classified EEG epochs [85]. The Narcotrend classification scheme, used for these visual assessments, has its origin in sleep analysis, which was further adapted for EEG during anesthesia [85]. From the awake state to very deep anesthesia, Narcotrend has 15 distinguished stages [A, B (0-2), C (0-2), D (0-2), E (0-2), F (0-1)] [85]. The transformation of these stages into a numerical scale from 100 to 0 is a further Narcotrend monitor refinement for a differentiated presentation of EEG effects [85]. The Narcotrend uses an automatic classification multivariate discriminant function that further incorporates age-related changes of the EEG [85]. The device Narcotrend contains artifacts' identification functions. The Narcotrend typically records EEG from a frontal channel using standard ECG electrodes, whereas anesthesiologists can choose other electrode positions and types [85].

9.2.5 NeuroSENSE Sleep Monitoring

Agrawal and colleagues [64] investigated the ability of WAV_{CNS} index of the NeuroSENSE

Table 9.2 Sleep monitoring of critically ill patients

Reference	Monitor	n	Sleep stages	Index change	Explanation
Dahaba et al. [71]	Bilateral BIS (1.4)	10	S1, S2, S3, S4, REM sleep stages	BIS ↓ with sleep stages	BIS detects sleep stages
Sleigh et al. [74]	BIS A-1000 (pre 3.0)	5	Light, slow, REM sleep	BIS ↓ with sleep stages	BIS detects sleep stages
Nieuwenhuijs et al. [79]	PSG, BIS A-2000 (3.3)	10	Light, slow, REM sleep	BIS delay and wide range	BIS delay wildly distributed
Nicholson et al. [80]	BIS, submental EMG	27	No normal sleep cycles in CCU	No sleep classification	No sleep classification
Vacas et al. [46]	PSG, SEDLine	23	Sleep stages, transitions, arousals	SedLine could differentiate	SedLine detects sleep stages
Schultz et al. [85]	Narcotrend	Cohort	Based on sleep analysis	Uses EEG epochs	15 distinguished sleep stages
Agrawal et al. [64]	PSG, NeuroSENSE	24	Wake, NREM sleep	Strong temporal correlation	No overlap sleep stages

n number of patients per group; *BIS* Bispectral Index; *REM* Rapid Eye Movement; *NREM* Non-Rapid Eye Movement; *CCU* Critical Care Unit; *EEG* Electroencephalography; SEDLine, Narcotrend, NeuroSENSE, *PSI* Patient Sate Index; *PSG* Polysomnography

processed Electroencephalography-Depth Of Anesthesia (pEEG-DOA) monitor, to detect Non-Rapid Eye Movement (NREM) sleep stages and wake state obtained from 24 adolescents [64]. Retrospective analysis for full overnight Polysomnography (PSG) sleep data, scored by a registered PSG technologist for different sleep stages, was performed on a single frontal channel using the WAV_{CNS} algorithm. Using non-parametric descriptive statistics, the authors demonstrated $P = 0.9458$ strong correlation between the NeuroSENSE WAV_{CNS} index and NREM sleep stages, with the NeuroSENSE WAV_{CNS} index values decreasing with increasing sleep stages [64]. Moreover, there was no significant overlap between different NREM sleep stages as classified by the NeuroSENSE WAV_{CNS} index, which was able to significantly differentiate between all pairs of Awake and different NREM stages [64]. The authors demonstrated that changes in the depth of natural physiologic NREM sleep are sensitively reflected by changes in the NeuroSENSE WAV_{CNS} index. Hence, WAV_{CNS} index may serve as an automatic real-time indicator of depth of natural physiologic sleep with high temporal resolution and can possibly be of great use for automated sleep staging in routine /postoperative somnographic studies [64].

Table 9.2 depicts sleep monitoring in critically ill patients.

9.3 Bispectral Index (BIS) Monitoring of Narcolepsy-Cataplexy

Gélineau syndrome or narcolepsy is an unknown etiology extremely incapacitating chronic sleep condition of overwhelming deep sleep attacks that is usually accompanied by abrupt loss of muscular tone cataplexy [86]. Because narcolepsy syndrome is a rare condition, anesthesia administration recommendations and/or monitoring of narcolepsy patients undergoing surgical procedures are not yet determined. The postoperative complications that could occur in a narcolepsy patient following inhalation anesthesia have been reported due to increased sensitivity to general anesthesia as well as interactions of anesthetic agents with drugs used for narcolepsy therapy [87]. On the other hand, regional anesthesia, an attractive alternative to general anesthesia, equally entails substantial risk to narcolepsy patients as Pasteau and colleagues reported two cataplectic/narcoleptic spells in a patient undergoing surgery under spinal anesthesia [88]. The Bispectral Index, a processed Electroencephalography-Depth Of Anesthesia (pEEG-DOA) monitor, displayed good correlations to numerous conditions shown to influence the Electroencephalography (EEG) [89]. We chose the Bispectral Index for monitoring a patient with a long medical history of narcolepsy-cataplexy who underwent surgical procedure under regional anesthesia, since a pEEG-DOA derived Bispectral Index might probably convey possible narcolepsy episodes EEG changes [90]. Furthermore, Bispectral Index monitor comprises Electromyography (EMG) activity in decibel (dB) monitoring that could provide anesthesiologists another advantage of possibly predicting an ensuing cataplectic spell that would entail associated hazards of complete loss of muscular tone [90]. My research team Dahaba and colleagues evaluated the utility of using Bispectral Index monitoring of potential narcolepsy-cataplexy episodes and whether a distinct Bispectral Index profile might provide anesthesiologists with an early alarm of an imminent narcolepsy-cataplexy spell [90].

During a narcolepsy-cataplexy episode, my research team Dahaba and colleagues [90], at Xuan Wu Hospital, Capital Medical University, Beijing, People's Republic of China, bilaterally recorded the Bispectral Index of both hemispheres of a 57-year-old man with a prolonged medical history of daytime recurring 2–3 hours of deep sleep attacks accompanied by an abrupt total loss of muscular tone, who underwent a lower limb surgical procedure under regional anesthesia femoral nerve block [90]. Bispectral Index data, real-time raw EEG, and EMG in db units were recorded [90]. Our patient did not receive any Central Nervous System (CNS) stimulants or amphetamine-like agents for his narcolepsy-cataplexy condition that might alter the Bispectral Index monitoring [90]. He did not receive any premedication before surgery. We used paired *t*-test to compare the change in parameters at different sleep stages. We assessed patients' alertness/attention via a simple stimulation response test of just calling the patient by name [90]. Patient's alertness status was categorized as "High" when patient could swiftly respond; "Low" when patient was sleepy, dazed, and response was staccato speech or slow; and "Non-Responsive" when patient was in stupor [90]. The patient's regional anesthesia was ropivacaine 0.5% 30 ml femoral nerve block identified via 0.1 ms nerve stimulator with 0.4 mA stimulations [90]. After confirming a successful nerve block, surgical procedures were started without any sedative agents administrations during surgery [90].

Twenty min from the beginning of his Bispectral Index recordings, the patient experienced a narcolepsy-cataplexy incident as he experienced three stages: first phase prodroma of "Intermittent Low Alertness" as the patient sporadically drifted from low alertness to high muscular activity [90]. Mean ± SD Bispectral Index significantly fluctuated from 55.5 ± 5.8 to 93.8 ± 2.3 that was accompanied by significantly fluctuating EMG activity from 32.9 ± 2.0 to 50.1 ± 3.8 dB [90]. This was followed by "Continuous Low Alertness" second phase in which the patient was calm with Bispectral Index of around 75 (right Bispectral Index 72.7 ± 5.9, left Bispectral Index 76.3 ± 6.6) with no muscular activity (right EMG 31.9 ± 1.1 dB, left EMG 31.6 ± 1.3) [90]. Raw EEG tracings revealed a decline in the fast α-wave and β-wave with an increase in slow θ-wave and δ-wave [90]. This ended with a third phase "Non-Responsiveness" of a full-blown narcolepsy-cataplexy episode as Bispectral Index significantly declined to about 45 at both sides (right Bispectral Index 44.5 ± 3.3, left Bispectral Index 44.1 ± 2.5) with full cataplexy loss of muscular tone (right EMG 32.0 ± 0.6 dB, left EMG 31.1 ± 0.9) [90]. Ultimately the Bispectral Index increased with the patient's recovery from his narcolepsy-cataplexy incident with the end of his surgical procedure. Bispectral Index values obtained from the right and left hemispheres displayed similar trend patterns (Fig. 9.10) [90]. The patient had an uneventful postoperative recovery, following which he experienced no change in the duration or frequency of his narcolepsy-cataplexy episodes [90].

To the best of my knowledge, this was the first case report of Bispectral Index monitoring of an active 2–3-hour narcolepsy-cataplexy episode in a patient undergoing a surgical procedure under regional anesthesia [90]. During the preliminary narcolepsy prodroma of "Intermittent Low Alertness" phase, the Bispectral Index exhibited a very characteristic Bispectral Index profile of rapid alternating low and high Bispectral Index values accompanied by alternating low and high

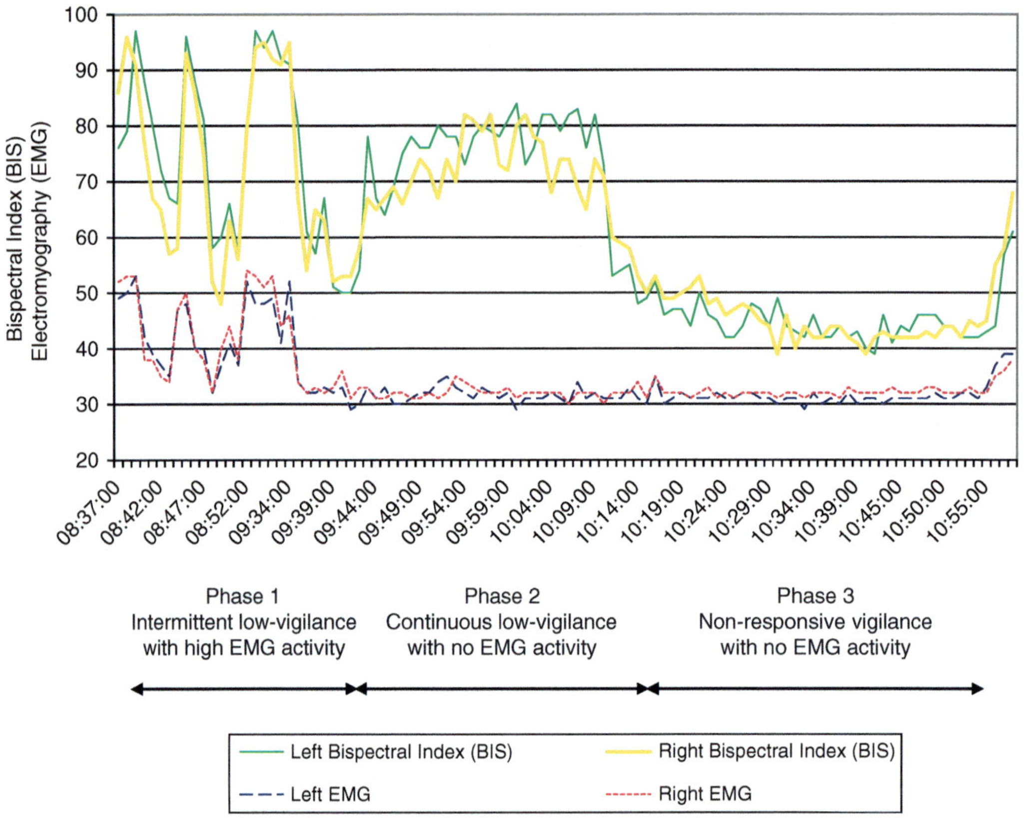

Fig. 9.10 Narcolepsy Bispectral Index (BIS) and Electromyography (EMG) recordings of 2–3 hours three narcolepsy-cataplexy episodes of intermittent low-alertness, continuous low-alertness, and complete non-responsive. (Reproduced with permission # 5934700154301 from Dahaba et al. Bispectral index monitoring of a narcolepsy-cataplexy episode during regional anesthesia. Anesth Analg. 2009;108:613–5)

EMG values, which could provide anesthesiologists with an early alarm of an imminent possibly dangerous narcolepsy-cataplexy spell with complete loss of muscle tone under regional anesthesia [90]. After that there was a relatively stable phase of Bispectral Index value of about 75 during the narcolepsy "Continuous Low Alertness" phase, which ended with the possibly dangerous phase of a full-blown narcolepsy episode with abrupt complete loss of muscle tone during which Bispectral Index value declined to about 45 for around 30 min [90].

At the first narcolepsy phase [90], the preliminary Bispectral Index decrease depicted a narcolepsy prodroma of low alertness status disrupted by spuriously elevated Bispectral Index values because of the high EMG activity [30]. The patient's high EMG activity at the narcolepsy prodroma clearly implies that our narcolepsy patient was trying to make numerous attempts to counter his imminent narcolepsy low alertness state [90]. Electromyography activities are artifactual signals occurring in the frequency "range of interest" of the bispectrum, as the EMG frequency wave could mimic the *Relative BetaRatio*, the Bispectral Index component descriptor, and the recurring EMG activities increases in the first narcolepsy prodroma phase would be interpreted by the Bispectral Index algorithm as EEG activities and allocated a high Bispectral Index value [30].

This was followed by the second narcolepsy phase of diurnal low alertness status accompanied by a higher sleep urgency [90]. Saletu and colleagues [91] previously described this diurnal state in an EEG brain mapping study, manifesting in patients suffering from narcolepsy as rise in θ-wave power with slowing of the total power spectrum, β-wave and α-wave [91]. This characteristic EEG pattern was identified by the Bispectral Index algorithm and presented as a Bispectral Index value of about 75.

The drastic Bispectral Index decline in a full-blown narcolepsy episode as our patient went into stupor [90] was reported in a previous trial by Mukai and colleagues [92], of eight patients suffering from narcolepsy sleep disorder, compared with eight matched healthy volunteers [92]. Narcoleptic patients' significant EEG depression manifested as decline in central arousal activities, decline in β-wave EEG power density, and slow waves predominance [92], manifesting in our patient's Bispectral Index drastic decline to about 45 from as high as 75 [90]. In the second narcolepsy phase, our patient appeared relaxed as he displayed no high muscular activity, but he was not atonic [90], before the third phase of a full-blown cataleptic total loss of muscle power [90]. We have very clearly demonstrated that our patient with narcolepsy exhibited a totally different Bispectral Index profile than that of natural physiologic sleep, whereas the distinct narcolepsy first phase prodroma of fast Bispectral Index /EMG fluctuations could provide an early alarm of an imminent narcolepsy-cataplexy spell with all its related dangers [90].

Narcolepsy is an under-diagnosed neurologic disorder because the narcolepsy syndrome incidence is estimated to be as low as 0.025%. Once manifest, the disorder turns chronic, with no remissions [93]. Narcolepsy syndrome appears to have a genetic origin and was linked to human leukocyte DQwl and DR2 antigens [94]. Narcolepsy is a severely incapacitating debilitating condition that presently has no cure, as current management is focused on the symptoms' control via amphetamine-like CNS-stimulants for treatment of excessive day sleepiness and the use of antidepressants for treatment of cataplexy [86]. Hence prolonged anesthesia recovery with extended postoperative apnea, cataplectic spells, and an increase in the duration or frequency of narcoleptic episodes following general anesthesia are all considerable concerns [88]. On the other hand, regional anesthesia appears to be an attractive alternative that could avoid interactions between general anesthesia agents and patients' narcolepsy medications or disorder. Unfortunately, our patient still suffered a narcolepsy-cataplexy episode while undergoing surgical procedure under regional anesthesia [90].

In our study [90], we demonstrated that patients with narcolepsy can still run the risk of losing consciousness with atonia when undergoing medical procedures under regional anesthesia, and such an undesirable complication should not be underestimated. Bispectral Index monitoring is an easy method that could provide an early alarm of an impending narcolepsy episode, with all its accompanying dangers in patients with a known medical history of narcolepsy-cataplexy.

9.4 Effect of Flumazenil on Bispectral Index (BIS) Monitoring

Flumazenil, an imidazobenzodiazepine, can swiftly reverse via competitive inhibition [95], the hypnotic/sedative effect of benzodiazepines on γ-aminobutyric acid (GABA) receptors [96]. Flumazenil demonstrated a useful analeptic stimulant effect in patients suffering from hepatic encephalopathy, as a meta-analysis study [97], including numerous randomized double-blind placebo-controlled trials, reported recovery of hepatic encephalopathy Electroencephalography (EEG) signs following flumazenil administrations in patients who did not receive sedative/hypnotic agents [97]. Endogenous benzodiazepines ligands, "Endozepines" [98], were detected in cerebrospinal fluid (CSF) [99], urine [100], and breast milk [101] of women who did not receive benzodiazepines. It has been suggested that there might be a beneficial analeptic stimulant effect in subjects who did not receive benzodiazepines, because flumazenil displaces those endogenous benzodiazepine ligands [95].

The Bispectral Index a processed Electroencephalography-Depth Of Anesthesia (pEEG-DOA) monitor, widely used for hypnotics/anesthetics effect monitoring [95], was reported to correlate to numerous conditions that could affect the EEG [89]. However, there is lack of information about the effect of Central Nervous System (CNS) stimulants on Bispectral Index monitoring when they are administered under general anesthesia. Because flumazenil is a readily available medicament that exerts an enhancing recovery stimulant effect, my research group Dahaba and colleagues [95], at Peking University First Hospital, Beijing, People's Republic of China, examined the stimulant effect of flumazenil 0.5 mg administration to unpremedicated healthy patients, during Total Intravenous Anesthesia (TIVA) deep surgical remifentanil/propofol anesthesia maintained around Bispectral Index value of 40. We hypothesized that Flumazenil administration might increase the Bispectral Index value to a lighter plane of anesthesia and might expedite anesthesia recovery [95].

We randomly allocated 60 healthy unpremedicated patients to the flumazenil or control groups [95]. A Bispectral Index "Quatro" sensor was placed on the patients' forehead and connected to a Bispectral Index BIS-XP monitor (version 3.4). After entering patients' anthropometric data, propofol Target Controlled Infusion (TCI) using a Diprifusor™ infusion pump (AstraZeneca, Cheshire, UK), incorporating a three-compartment Marsh pharmacokinetic model [102], was set to gradually reach an estimated plasma concentration (C_p) of 3 μg ml^{-1} over 2-min period as patients were spontaneously breathing via a facemask. Time to Loss of Consciousness (LOC) was defined as time till patients lost their eyelash reflex and did not respond to verbal commands [95]. Propofol TCI rate and Bispectral Index values at LOC were recorded [95]. Remifentanil 0.1–0.3 μg kg^{-1} min^{-1} continuous infusion was started, and rocuronium 600 μg kg^{-1} was administered for tracheal intubation followed by rocuronium 200 μg kg^{-1} top-up doses to maintain first twitch (T_1%, T_1:T_0) of Train-Of-Four (TOF) at 25% [95]. Intraoperatively patients did not receive any benzodiazepines or sedative agents during the operation [95]. At deep surgical anesthesia, a stable Bispectral Index value of 40 [103] was maintained with propofol TCI ±0.2 μg ml^{-1} rate adjustments [95]. Thirty min before the expected end of surgery, a blinded research anesthesiologist administered the randomly allocated study drugs flumazenil 0.5 mg or placebo [95]. Following anesthesia termination, a dedicated blinded research anesthesiologist asked, in a standard fashion, an oral command every 20 s to assess recovery times parameters, namely, time to return of spontaneous respiration, time to opening of eyes, time to squeeze anesthesiologist's hand on oral command, extubation time, and time to Date Of Birth (DOB recollection [95].

There were no significant differences between the two groups in patients' demographics, propofol TCI rate at LOC, time to LOC or Bispectral Index values at LOC [95]. Bispectral Index baseline values in the flumazenil group (39 ±4) significantly increased 15 min following flumazenil administration (53 ± 5), whereas Bispectral Index baseline values in the control group (40 ± 3) did not significantly increase 15 min following placebo administration (42 ± 5) (Fig. 9.11) [95]. Two-way Analysis Of Variance (ANOVA, group x time) showed a significant difference between the two groups in Bispectral Index values over time [95]. Dunnett's two-sided multiple-comparison post hoc test showed significant differences between the two groups starting 6 min from flumazenil administration [95]. Similarly following discontinuation of anesthesia, there were significant differences in Bispectral Index values over time (Fig. 9.12) [95]. Mean recovery time parameters comprising the time to spontaneous breathing, time to eye opening on verbal command, time to hand squeezing on oral command, extubation time, and time to DOB recollection were all significantly shorter in the flumazenil group (4.6 ± 1.8, 5.4 ± 2.8, 6.4 ± 3.1, 8.7 ± 2.5, 9.8 ± 3.2 min) compared with the control group (5.7 ± 2.1, 7.2 ± 3.8, 10.4 ± 2.9, 12.8 ± 3, 13.1 ± 3.1 min), respectively [95].

Our main finding [95] that flumazenil's administration in unpremedicated healthy subjects resulted in a significant increase in Bispectral Index value could be ascribed to flumazenil's inhibition of GABA-benzodiazepine receptor

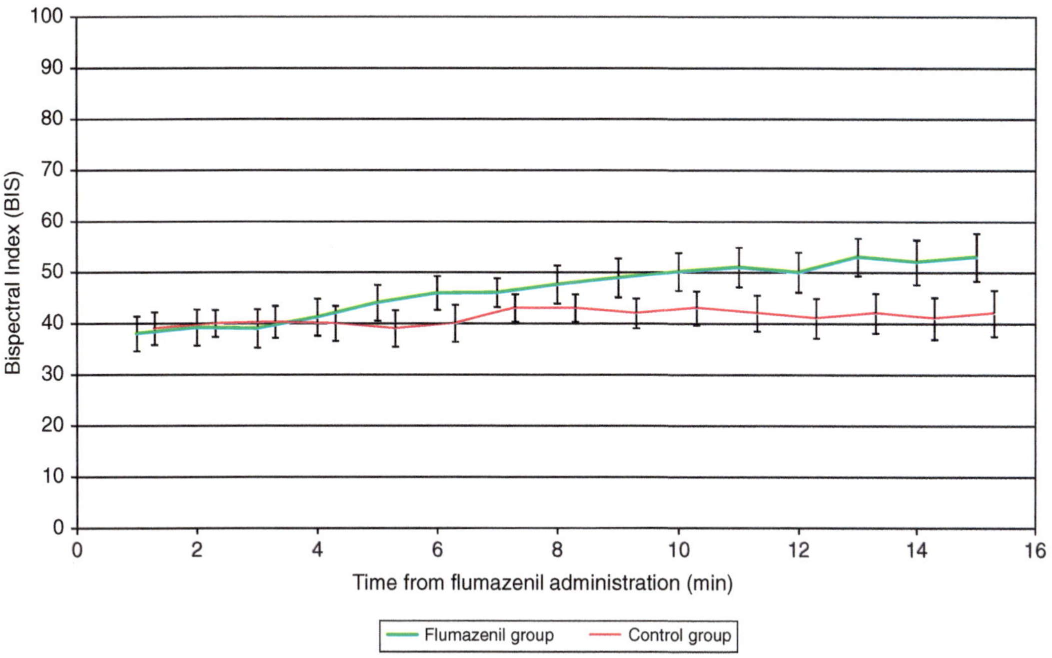

Fig. 9.11 Mean ± SD Bispectral Index (BIS) after flumazenil or placebo administration. Two-way Analysis Of Variance (ANOVA, group x time) followed by Dunnett's post hoc test revealed significant differences between the two groups starting 6 min from flumazenil administration. (Reproduced with permission # 5934730140803 from Dahaba et al. Effect of flumazenil on bispectral index monitoring in unpremedicated patients. Anesthesiology. 2009;110:1036–40)

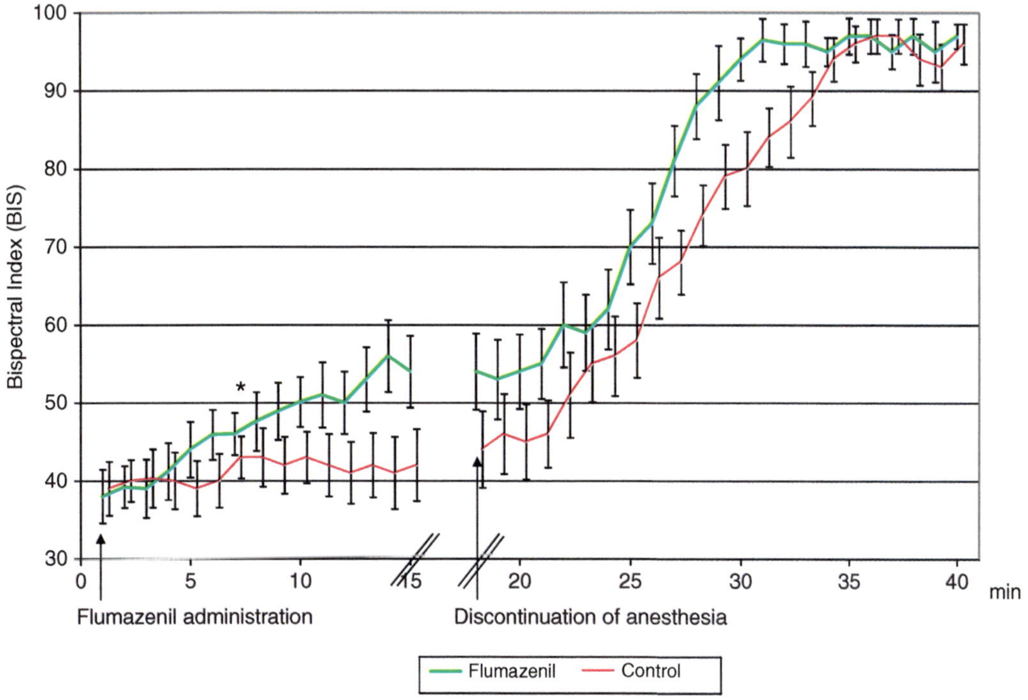

Fig. 9.12 Mean ± SD Bispectral Index (BIS) following discontinuation of anesthesia. There were significant differences in Bispectral Index values between the two groups over time

complex of the endogenous benzodiazepine-like ligands, "Endozepines" in subjects who did not receive any benzodiazepines, or sedative agents during surgery [95]. Discovery of high affinity "benzodiazepine receptor binding sites" in cerebral cellular tissues of humans [104] marked the rise of the comprehensive search for human putative endogenous receptor ligands [105]. Naturally occurring "Endozepines" that did not originate from environmental contaminations with pharmaceutical benzodiazepines or benzodiazepines derivatives [105] were detected in different tissues of mammals' origins [95, 106]. Endogenous benzodiazepines [105], in physiological concentrations of patients who did not receive benzodiazepines, were found in human plasma [107], serum [108], ultrafiltrates [100], CSF [99], urine [100], or healthy newly delivered women's breast milk that did not take any benzodiazepines [101, 105, 109]. Those "Endozepines" [105] included an array of 1,4-benzodiazepines like N-desmethyldiazepam, diazepam, lorazepam, or oxazepam commercially available benzodiazepines [95, 105]. Amazingly, in the early 1990s, genuine benzodiazepines were discovered in brains of humans that were paraffin preserved from the early 1940s, well preceding the period of the wide industrial synthesis of benzodiazepine [95, 105, 110].

In 1987 [111], vegetal cells synthesis of naturally occurring benzodiazepines "Endozepines" was first proposed [95, 105]. This was later confirmed in preserved sterile cultures showing that during germination, potatoes and wheat can create benzodiazepine molecules that are indeed pharmacologically active [95, 105, 112, 113]. Moreover, confirming the theory of benzodiazepines' "*de novo* biosynthetic pathway formation" [105], numerous benzodiazepines not currently present for medical use like isodiazepam and deschlorodiazepam were detected in plants [95, 105, 112]. Moreover, natural benzodiazepines [105], like temazepam and delorazepam that are present in animal tissue, soil, and plants [105], are all almost impossible to differentiate from the industrial origin benzodiazepines regarding their chemical structures and pharmacological activity [95, 105]. To date, Mass Spectrometry analyses [105, 112, 113] confirmed the presence of minimum nine different pharmacologically active natural 1,4-benzodiazepines [105], in numerous nutritive plant food products or plants that are frequently consumed by humans [105], like mushrooms, cherries, wheat, potato tuber, rice, soy beans, maize, lentils, and grapes [105, 112, 114, 115], all this convincingly suggesting that those Endozepine substances [105] are at present fully integrated in our human food chains' consumption [95, 105]. Thus, flumazenil reversing these "naturally occurring" Endozepines both exogenous from food or endogenous benzodiazepine-like agents might be another probable mechanism [95, 105].

In our study [95], we used Bispectral Index monitoring as a surrogate parameter that indicated "light anesthesia" following flumazenil administration. Similarly, prior trials used the "surrogate" Bispectral Index monitoring to accurately quantify the CNS-stimulant agents' analeptic effects. These studies showed significant increases in Bispectral Index values following doxapram or aminophylline administrations under steady-state volatile anesthetic agents [95, 116, 117].

In unpremedicated healthy subjects, we demonstrated that flumazenil significantly increased the Bispectral Index values and resulted in faster recovery in patients who did not receive benzodiazepines or sedatives. I believe that our study of flumazenil analeptic effect could offer valuable medical knowledge regarding the "enhanced awakening" analeptic effect of a readily available medication that could be utilized, on a case-by-case basis, to reverse exogenous or endogenous "Endozepines", thus possibly playing a significant role in critical care patients whose recovery is unexpectedly prolonged [95].

References

1. Arroyo-Novoa CM, Figueroa-Ramos MI, Puntillo KA. Occurrence and practices for pain, agitation, and delirium in intensive care unit patients. P R Health Sci J. 2019;38:156–62.
2. Reade MC, Finfer S. Sedation and delirium in the intensive care unit. N Engl J Med. 2014;370:444–54.

3. Fuller BM, Pappal RD, Mohr NM, Roberts BW, Faine B, Yeary J, et al. Awareness with paralysis among critically ill emergency department patients: a prospective cohort study. Crit Care Med. 2022;50:1449–60.
4. Riker RR, Picard JT, Fraser GL. Prospective evaluation of the sedation-agitation scale for adult critically ill patients. Crit Care Med. 1999;27:1325–9.
5. Sessler CN, Gosnell MS, Grap MJ, Brophy GM, O'Neal PV, Keane KA, et al. The Richmond agitation-sedation scale: validity and reliability in adult intensive care unit patients. Am J Respir Crit Care Med. 2002;166:1338–44.
6. Chernik DA, Gillings D, Laine H, Hendler J, Silver JM, Davidson AB, et al. Validity and reliability of the Observer's assessment of alertness/sedation scale: study with intravenous midazolam. J Clin Psychopharmacol. 1990;10:244–51.
7. Ramsay MA, Savege TM, Simpson BR, Goodwin R. Controlled sedation with alphaxalone-alphadolone. Br Med J. 1974;2:656–9.
8. Lobo FA, Robba C, Lamperti M, Romagnoli S, Rasulo FA. Processed EEG monitoring in critical care: a black swan or a shining star? J Clin Monit Comput. 2023;37:3–5.
9. Purdon PL, Sampson A, Pavone KJ, Brown EN. Clinical electroencephalography for anesthesiologists: part I: background and basic signatures. Anesthesiology. 2015;123:937–60.
10. Montupil J, Defresne A, Bonhomme V. The raw and processed electroencephalogram as a monitoring and diagnostic tool. J Cardiothorac Vasc Anesth. 2019;33(Suppl 1):S3–10.
11. García PS, Kreuzer M, Hight D, Sleigh JW. Effects of noxious stimulation on the electroencephalogram during general anaesthesia: a narrative review and approach to analgesic titration. Br J Anaesth. 2021;126:445–57.
12. Lobo FA, Shander A. Modern anesthetic noninvasive monitoring: a deep look into perioperative care. J Cardiothorac Vasc Anesth. 2019;33(Suppl 1):S1–2.
13. Fahy BG, Chau DF. The technology of processed electroencephalogram monitoring devices for assessment of depth of anesthesia. Anesth Analg. 2018;126:111–7.
14. Escallier KE, Nadelson MR, Zhou D, Avidan MS. Monitoring the brain: processed electroencephalogram and peri-operative outcomes. Anaesthesia. 2014;69:899–910.
15. Rasulo FA, Hopkins P, Lobo FA, Pandin P, Matta B, Carozzi C, et al. Processed electroencephalogram-based monitoring to guide sedation in critically ill adult patients: recommendations from an international expert panel-based consensus. Neurocrit Care. 2023;38:296–311.
16. Hajat Z, Ahmad N, Andrzejowski J. The role and limitations of EEG-based depth of anaesthesia monitoring in theatres and intensive care. Anaesthesia. 2017;72(Suppl 1):38–47.
17. Lobo FA, Schraag S. Limitations of anaesthesia depth monitoring. Curr Opin Anaesthesiol. 2011;24:657–64.
18. Young GB, Mantia J. Continuous EEG monitoring in the intensive care unit. Handb Clin Neurol. 2017;140:107–16.
19. Introna M, Gemma M, Carozzi C. Improving the benefit of processed EEG monitors: it's not about the car but the driver. J Clin Monit Comput. 2023;37:723–5.
20. Dahaba AA. Thinking outside the box. Off-label use of Bispectral index within context and limitations for conditions other than depth of anesthesia. Minerva Anestesiol. 2019;85:189–93.
21. Wang Z-H, Chen H, Yang Y-L, Shi Z-H, Guo Q-H, Li Y-W, et al. Bispectral index can reliably detect deep sedation in mechanically ventilated patients: a prospective multicenter validation study. Anesth Analg. 2017;125:176–83.
22. Gallagher JD. Pacer-induced artifact in the bispectral index during cardiac surgery. Anesthesiology. 1999;90:636.
23. Andrzejowski J, Sleigh JW, Johnson IA, Sikiotis L. The effect of intravenous epinephrine on the bispectral index and sedation. Anaesthesia. 2000;55:761–3.
24. Dahaba AA, Lischnig U, Kronthaler R, Bornemann H, Georgiev V, Rehak PH, et al. Bispectral-index-guided versus clinically guided remifentanil/propofol analgesia/sedation for interventional radiological procedures: an observer-blinded randomized study. Anesth Analg. 2006;103:378–84. table of contents
25. Dahaba AA, Grabner T, Rehak PH, List WF, Metzler H. Remifentanil versus morphine analgesia and sedation for mechanically ventilated critically ill patients: a randomized double blind study. Anesthesiology. 2004;101:640–6.
26. Lysakowski C, Dumont L, Pellegrini M, Clergue F, Tassonyi E. Effects of fentanyl, alfentanil, remifentanil and sufentanil on loss of consciousness and bispectral index during propofol induction of anaesthesia. Br J Anaesth. 2001;86:523–7.
27. Guignard B, Menigaux C, Dupont X, Fletcher D, Chauvin M. The effect of remifentanil on the bispectral index change and hemodynamic responses after orotracheal intubation. Anesth Analg. 2000;90:161–7.
28. Bruhn J, Bouillon TW, Shafer SL. Electromyographic activity falsely elevates the bispectral index. Anesthesiology. 2000;92:1485–7.
29. Baldesi O, Bruder N, Velly L, Gouin F. Spurious bispectral index values due to electromyographic activity. Eur J Anaesthesiol. 2004;21:324–5.
30. Sleigh JW, Steyn-Ross DA, Steyn-Ross ML, Williams ML, Smith P. Comparison of changes in electroencephalographic measures during induction of general anaesthesia: influence of the gamma frequency band and electromyogram signal. Br J Anaesth. 2001;86:50–8.

31. Mathews DM, Kumaran KR, Neuman GG. Bispectral index-derived facial electromyography-guided fentanyl titration in the opiate-exposed patient. Anesth Analg. 2003;96:1062–4.
32. Simmons LE, Riker RR, Prato BS, Fraser GL. Assessing sedation during intensive care unit mechanical ventilation with the Bispectral index and the sedation-agitation scale. Crit Care Med. 1999;27:1499–504.
33. Riker RR, Fraser GL, Simmons LE, Wilkins ML. Validating the sedation-agitation scale with the Bispectral index and visual analog scale in adult ICU patients after cardiac surgery. Intensive Care Med. 2001;27:853–8.
34. Sadhasivam S, Ganesh A, Robison A, Kaye R, Watcha MF. Validation of the bispectral index monitor for measuring the depth of sedation in children. Anesth Analg. 2006;102:383–8.
35. Kress JP, Pohlman AS, O'Connor MF, Hall JB. Daily interruption of sedative infusions in critically ill patients undergoing mechanical ventilation. N Engl J Med. 2000;342:1471–7.
36. Consales G, Chelazzi C, Rinaldi S, De Gaudio AR. Bispectral index compared to Ramsay score for sedation monitoring in intensive care units. Minerva Anestesiol. 2006;72:329–36.
37. Pandit JJ, Andrade J, Bogod DG, Hitchman JM, Jonker WR, Lucas N, et al. The 5th National Audit Project (NAP5) on accidental awareness during general anaesthesia: summary of main findings and risk factors. Anaesthesia. 2014;69:1089–101.
38. Bould MD, Mahtani DG, Davies R, Roughton M, Hunter DN, Kelleher A. Bispectral index values during elective rigid bronchoscopy: a prospective observational pilot study. Anaesthesia. 2007;62:438–45.
39. Prottengeier J, Moritz A, Heinrich S, Gall C, Schmidt J. Sedation assessment in a mobile intensive care unit: a prospective pilot-study on the relation of clinical sedation scales and the bispectral index. Crit Care. 2014;18:615.
40. De Deyne C, Struys M, Decruyenaere J, Creupelandt J, Hoste E, Colardyn F. Use of continuous bispectral EEG monitoring to assess depth of sedation in ICU patients. Intensive Care Med. 1998;24:1294–8.
41. Gélinas C, Tousignant-Laflamme Y, Tanguay A, Bourgault P. Exploring the validity of the bispectral index, the critical-care pain observation tool and vital signs for the detection of pain in sedated and mechanically ventilated critically ill adults: a pilot study. Intensive Crit Care Nurs. 2011;27:46–52.
42. Nasraway SASA, Wu EC, Kelleher RM, Yasuda CM, Donnelly AM. How reliable is the Bispectral index in critically ill patients? A prospective, comparative, single-blinded observer study. Crit Care Med. 2002;30:1483–7.
43. Vivien B, Di Maria S, Ouattara A, Langeron O, Coriat P, Riou B. Overestimation of Bispectral index in sedated intensive care unit patients revealed by administration of muscle relaxant. Anesthesiology. 2003;99:9–17.
44. Drover D, Ortega HR. Patient state index. Best Pract Res Clin Anaesthesiol. 2006;20:121–8.
45. Schneider G, Heglmeier S, Schneider J, Tempel G, Kochs EF. Patient state index (PSI) measures depth of sedation in intensive care patients. Intensive Care Med. 2004;30:213–6.
46. Vacas S, McInrue E, Gropper MA, Maze M, Zak R, Lim E, et al. The feasibility and utility of continuous sleep monitoring in critically ill patients using a portable electroencephalography monitor. Anesth Analg. 2016;123:206–12.
47. Idei M, Seino Y, Sato N, Yoshida T, Saishu Y, Fukui K, et al. Validation of the patient state index for monitoring sedation state in critically ill patients: a prospective observational study. J Clin Monit Comput. 2023;37:147–54.
48. Aurell J, Elmqvist D. Sleep in the surgical intensive care unit: continuous polygraphic recording of sleep in nine patients receiving postoperative care. Br Med J (Clin Res Ed). 1985;290:1029–32.
49. Friese RS, Diaz-Arrastia R, McBride D, Frankel H, Gentilello LM. Quantity and quality of sleep in the surgical intensive care unit: are our patients sleeping? J Trauma. 2007;63:1210–4.
50. Beecroft JM, Ward M, Younes M, Crombach S, Smith O, Hanly PJ. Sleep monitoring in the intensive care unit: comparison of nurse assessment, actigraphy and polysomnography. Intensive Care Med. 2008;34:2076–83.
51. Irwin M, McClintick J, Costlow C, Fortner M, White J, Gillin JC. Partial night sleep deprivation reduces natural killer and cellular immune responses in humans. FASEB J. 1996;10:643–53.
52. Oztürk L, Pelin Z, Karadeniz D, Kaynak H, Cakar L, Gözükirmizi E. Effects of 48 hours sleep deprivation on human immune profile. Sleep Res Online. 1999;2:107–11.
53. Lange T, Perras B, Fehm HL, Born J. Sleep enhances the human antibody response to hepatitis a vaccination. Psychosom Med. 2003;65:831–5.
54. Scrimshaw NS, Habicht JP, Pellet P, Piché ML, Cholakos B. Effects of sleep deprivation and reversal of diurnal activity on protein metabolism of young men. Am J Clin Nutr. 1966;19:313–9.
55. Kant GJ, Genser SG, Thorne DR, Pfalser JL, Mougey EH. Effects of 72 hour sleep deprivation on urinary cortisol and indices of metabolism. Sleep. 1984;7:142–6.
56. Kato M, Phillips BG, Sigurdsson G, Narkiewicz K, Pesek CA, Somers VK. Effects of sleep deprivation on neural circulatory control. Hypertension. 2000;35:1173–5.
57. Zhong X, Hilton HJ, Gates GJ, Jelic S, Stern Y, Bartels MN, et al. Increased sympathetic and decreased parasympathetic cardiovascular modulation in normal humans with acute sleep deprivation. J Appl Physiol. 1985;2005(98):2024–32.
58. Misra S, Malow BA. Evaluation of sleep disturbances in older adults. Clin Geriatr Med. 2008;24:15–26, v.

59. Walker MP. Cognitive consequences of sleep and sleep loss. Sleep Med. 2008;9(Suppl 1):S29–34.
60. Eddleston JM, White P, Guthrie E. Survival, morbidity, and quality of life after discharge from intensive care. Crit Care Med. 2000;28:2293–9.
61. Mistraletti G, Carloni E, Cigada M, Zambrelli E, Taverna M, Sabbatini G, et al. Sleep and delirium in the intensive care unit. Minerva Anestesiol. 2008;74:329–33.
62. McNicoll L, Pisani MA, Zhang Y, Ely EW, Siegel MD, Inouye SK. Delirium in the intensive care unit: occurrence and clinical course in older patients. J Am Geriatr Soc. 2003;51:591–8.
63. Hollingworth P, Harold D, Sims R, Gerrish A, Lambert J-C, Carrasquillo MM, et al. Common variants at ABCA7, MS4A6A/MS4A4E, EPHA1, CD33 and CD2AP are associated with Alzheimer's disease. Nat Genet. 2011;43:429–35.
64. Agrawal G, Modarres M, Zikov T, Bibian S. NREM sleep staging using WAV(CNS) index. J Clin Monit Comput. 2011;25:137–42.
65. Ellerkmann RK, Soehle M, Kreuer S. Brain monitoring revisited: what is it all about? Best Pract Res Clin Anaesthesiol. 2013;27:225–33.
66. Punjasawadwong Y, Phongchiewboon A, Bunchungmongkol N. Bispectral index for improving anaesthetic delivery and postoperative recovery. Cochrane Database Syst Rev. 2014;2014:CD003843.
67. Viertiö-Oja H, Maja V, Särkelä M, Talja P, Tenkanen N, Tolvanen-Laakso H, et al. Description of the entropy algorithm as applied in the Datex-Ohmeda S/5 entropy module. Acta Anaesthesiol Scand. 2004;48:154–61.
68. Kreuer S, Wilhelm W. The Narcotrend monitor. Best Pract Res Clin Anaesthesiol. 2006;20:111–9.
69. Mattia C, Savoia G, Paoletti F, Piazza O, Albanese D, Amantea B, et al. SIAARTI recommendations for analgo-sedation in intensive care unit. Minerva Anestesiol. 2006;72:769–805.
70. Rampil IJ. A primer for EEG signal processing in anesthesia. Anesthesiology. 1998;89:980–1002.
71. Dahaba AA, Xue JX, Xu GX, Liu QH, Metzler H. Bilateral Bispectral index (BIS)-vista as a measure of physiologic sleep in sleep-deprived anesthesiologists. Minerva Anestesiol. 2011;77:388–93.
72. Howard SK, Gaba DM, Rosekind MR, Zarcone VP. The risks and implications of excessive daytime sleepiness in resident physicians. Acad Med. 2002;77:1019–25.
73. Howard SK, Gaba DM, Smith BE, Weinger MB, Herndon C, Keshavacharya S, et al. Simulation study of rested versus sleep-deprived anesthesiologists. Anesthesiology. 2003;98:1345–55; discussion 5A.
74. Sleigh JW, Andrzejowski J, Steyn-Ross A, Steyn-Ross M. The bispectral index: a measure of depth of sleep? Anesth Analg. 1999;88:659–61.
75. Kim DW, Kil HY, White PF. The effect of noise on the bispectral index during propofol sedation. Anesth Analg. 2001;93:1170–3.
76. Rechtschaffen Allan KA. A manual of standardized terminology, techniques and scoring system for sleep stages of human subjects. Washington DC: Public Health Service, US Government Printing Office; 1968.
77. Schulz H. Phasic or transient? Comment on the terminology of the AASM manual for the scoring of sleep and associated events. J Clin Sleep Med. 2007;3:752.
78. Niedhart DJ, Kaiser HA, Jacobsohn E, Hantler CB, Evers AS, Avidan MS. Intrapatient reproducibility of the BISxp monitor. Anesthesiology. 2006;104:242–8.
79. Nieuwenhuijs D, Coleman EL, Douglas NJ, Drummond GB, Dahan A. Bispectral index values and spectral edge frequency at different stages of physiologic sleep. Anesth Analg. 2002;94:125–9. table of contents
80. Nicholson T, Patel J, Sleigh JW. Sleep patterns in intensive care unit patients: a study using the bispectral index. Crit Care Resusc. 2001;3:86–91.
81. Schnider TW, Luginbühl M, Petersen-Felix S, Mathis J. Unreasonably low bispectral index values in a volunteer with genetically determined low-voltage electroencephalographic signal. Anesthesiology. 1998;89:1607–8.
82. Dahaba AA, Mattweber M, Fuchs A, Zenz W, Rehak PH, List WF, et al. The effect of different stages of neuromuscular block on the Bispectral index and the Bispectral index-XP under remifentanil/Propofol anesthesia. Anesth Analg. 2004;99:781–7.
83. Howard RS, Finneran JJ, Ridgway SH. Bispectral index monitoring of unihemispheric effects in dolphins. Anesth Analg. 2006;103:626–32.
84. John ER, Prichep LS, Kox W, Valdés-Sosa P, Bosch-Bayard J, Aubert E, et al. Invariant reversible QEEG effects of anesthetics. Conscious Cogn. 2001;10:165–83.
85. Schultz B, Kreuer S, Wilhelm W, Grouven U, Schultz A. The Narcotrend monitor. Development and interpretation algorithms. Anaesthesist. 2003;52:1143–8.
86. Keam S, Walker MC. Therapies for narcolepsy with or without cataplexy: evidence-based review. Curr Opin Neurol. 2007;20:699–703.
87. Gómez-Garrido M, Torcal E, Nevado E, Ibarra M, Cuesta J. Inhalation anesthesia in Gelineau syndrome (cataplexy-catalepsy). Rev Esp Anestesiol Reanim. 2002;49:338–9.
88. Pasteau F, Graftieaux JP, Gomis P, Scavarda D, Léon A. Problems posed by spinal anesthesia in a patient with Gélineau disease. Ann Fr Anesth Reanim. 1996;15:669–72.
89. Dahaba AA. Different conditions that could result in the bispectral index indicating an incorrect hypnotic state. Anesth Analg. 2005;101:765–73.
90. Dahaba AA, Xu GX, Liu QH, Xue JX, Metzler H. Bispectral index monitoring of a narcolepsy-cataplexy episode during regional anesthesia. Anesth Analg. 2009;108:613–5.

91. Saletu MT, Anderer P, Saletu-Zyhlarz GM, Mandl M, Arnold O, Nosiska D, et al. EEG-mapping differences between narcolepsy patients and controls and subsequent double-blind, placebo-controlled studies with modafinil. Eur Arch Psychiatry Clin Neurosci. 2005;255:20–32.
92. Mukai J, Uchida S, Miyazaki S, Nishihara K, Honda Y. Spectral analysis of all-night human sleep EEG in narcoleptic patients and normal subjects. J Sleep Res. 2003;12:63–71.
93. Mesa A, Diaz AP, Frosth M. Narcolepsy and anesthesia. Anesthesiology. 2000;92:1194–6.
94. Mignot E, Lin L, Rogers W, Honda Y, Qiu X, Lin X, et al. Complex HLA-DR and -DQ interactions confer risk of narcolepsy-cataplexy in three ethnic groups. Am J Hum Genet. 2001;68:686–99.
95. Dahaba AA, Bornemann H, Rehak PH, Wang G, Wu XM, Metzler H. Effect of flumazenil on bispectral index monitoring in unpremedicated patients. Anesthesiology. 2009;110:1036–40.
96. Amrein R, Hetzel W, Hartmann D, Lorscheid T. Clinical pharmacology of flumazenil. Eur J Anaesthesiol Suppl. 1988;2:65–80.
97. Goulenok C, Bernard B, Cadranel JF, Thabut D, Di Martino V, Opolon P, et al. Flumazenil vs. placebo in hepatic encephalopathy in patients with cirrhosis: a meta-analysis. Aliment Pharmacol Ther. 2002;16:361–72.
98. Rothstein JD, Garland W, Puia G, Guidotti A, Weber RJ, Costa E. Purification and characterization of naturally occurring benzodiazepine receptor ligands in rat and human brain. J Neurochem. 1992;58:2102–15.
99. Deckert J, Kuhn W, Przuntek H. Endogenous benzodiazepine ligands in human cerebrospinal fluid. Peptides (NY). 1984;5:641–4.
100. Beaumont K, Cheung AK, Geller ML, Fanestil DD. Inhibitors of peripheral-type benzodiazepine receptors present in human urine and plasma ultrafiltrates. Life Sci. 1983;33:1375–84.
101. Dencker SJ, Johansson G, Milsom I. Quantification of naturally occurring benzodiazepine-like substances in human breast milk. Psychopharmacology. 1992;107:69–72.
102. Marsh B, White M, Morton N, Kenny GN. Pharmacokinetic model driven infusion of propofol in children. Br J Anaesth. 1991;67:41–8.
103. Høymork SC, Raeder J, Grimsmo B, Steen PA. Bispectral index, predicted and measured drug levels of target-controlled infusions of remifentanil and propofol during laparoscopic cholecystectomy and emergence. Acta Anaesthesiol Scand. 2000;44:1138–44.
104. Möhler H, Okada T. Benzodiazepine receptor: demonstration in the central nervous system. Science. 1977;198:849–51.
105. Sand P, Kavvadias D, Feineis D, Riederer P, Schreier P, Kleinschnitz M, et al. Naturally occurring benzodiazepines: current status of research and clinical implications. Eur Arch Psychiatry Clin Neurosci. 2000;250:194–202.
106. Medina JH, Peña C, Piva M, Paladini AC, De Robertis E. Presence of benzodiazepine-like molecules in mammalian brain and milk. Biochem Biophys Res Commun. 1988;152:534–9.
107. Wildmann J, Niemann J, Matthaei H. Endogenous benzodiazepine receptor agonist in human and mammalian plasma. J Neural Transm. 1986;66:151–60.
108. Duthel JM, Constant H, Vallon JJ, Rochet T, Miachon S. Quantitation by gas chromatography with selected-ion monitoring mass spectrometry of "natural" diazepam, N-desmethyldiazepam and oxazepam in normal human serum. J Chromatogr. 1992;579:85–91.
109. Peña C, Medina JH, Piva M, Diaz LE, Danilowicz C, Paladini AC. Naturally occurring benzodiazepines in human milk. Biochem Biophys Res Commun. 1991;175:1042–50.
110. Klotz U. "Natural" benzodiazepines in man. Lancet. 1990;335:922.
111. Wildmann J, Möhler H, Vetter W, Ranalder U, Schmidt K, Maurer R. Diazepam and N-desmethyldiazepam are found in rat brain and adrenal and may be of plant origin. J Neural Transm. 1987;70:383–98.
112. Wildmann J. Increase of natural benzodiazepines in wheat and potato during germination. Biochem Biophys Res Commun. 1988;157:1436–43.
113. Kavvadias D, Abou-Mandour AA, Czygan FC, Beckmann H, Sand P, Riederer P, et al. Identification of benzodiazepines in Artemisia dracunculus and Solanum tuberosum rationalizing their endogenous formation in plant tissue. Biochem Biophys Res Commun. 2000;269:290–5.
114. Unseld E, Krishna DR, Fischer C, Klotz U. Detection of desmethyldiazepam and diazepam in brain of different species and plants. Biochem Pharmacol. 1989;38:2473–8.
115. Klotz U. Occurrence of "natural" benzodiazepines. Life Sci. 1991;48:209–15.
116. Wu C-C, Mok MS, Chen J-Y, Wu G-J, Wen Y-R, Lin C-S. Doxapram shortens recovery following sevoflurane anesthesia. Can J Anaesth. 2006;53:456–60.
117. Wu C-C, Lin C-S, Wu G-J, Lin Y-H, Lee Y-W, Chen J-Y, et al. Doxapram and aminophylline on bispectral index under sevoflurane anaesthesia: a comparative study. Eur J Anaesthesiol. 2006;23:937–41.

10 Artificial Intelligence-Machine Learning (AI-ML), Concerto 3-Channels Closed-Loop Automated System

In the future, Artificial Intelligence-Machine Learning (AI-ML) will not take your job, but for sure your job will be taken by somebody who can use Artificial Intelligence (AI). Nowadays we are living in the era of newly developed technologies in anesthesiology [1]. We titrate anesthetic agents' administration to the exact patients' physiological requirements and the undergoing surgical needs using a range of modern sophisticated anesthetic equipment [1]. Several processed Electroencephalography-Depth Of Anesthesia (pEEG-DOA) monitors have been suggested as possible objective alternatives to provide real-time Depth of Anesthesia (DOA) monitoring clinical assessment tools [1]. In modern anesthesia, patients' safety, which takes the highest priority, is coupled with innovative advanced technologies [1]. These advanced technologies laid the groundwork for innovative practices such as Target Controlled Infusion (TCI) systems and 3-Channel Closed-Loop administration of hypnotic anesthetic agents, Nociception Anti-Nociception Balance (NANB) analgesic agents, and Neuromuscular Blocking Agents (NMBAs) [1].

In our anesthesia practice, we usually administer fast acting drugs. Basically, we are interacting with three different domains, namely, consciousness (hypnosis), pain (NANB), and muscle relaxation (NMBAs), with all three exhibiting predictable interactions [1]. Our pharmacological and physiological knowledge of our anesthetic agents together with our acquired experiences are the solid foundations of our ability of delivering state-of-the-art care for our patients [1]. Over the decades, innovative new devices are currently used by anesthesiologists in their clinical practice as well as in research, aiming at assisting anesthesiologists in performing a better anesthesia practice and better handle the growing sophisticated environment [1].

Nowadays Artificial Intelligence (AI) can take the burden of doing the dull repetitive labor off the shoulders of anesthesiologists and let them get along with their other pressing Operating Room (OR) tasks. In daily anesthesia practice, several monitors are currently used for monitoring anesthesia and neuromuscular block. These include Bispectral Index (BIS) or Entropy for monitoring effects of anesthetics such as propofol or sevoflurane, whereas Neuromuscular Transmission Module (M-NMT) or TOF-Watch are used for monitoring the effects of NMBAs such as cisatracurium or rocuronium. Several NMBAs Closed-Loop infusion systems appeared in the literature [2, 3]. One of the NMBAs Closed-Loop systems by Illman and colleagues [2], used in 70 subjects, basically consisted of a conventional neuromuscular monitoring device the Electromyography (EMG) Relaxograph (Biometer, Denmark) with a NMBA infusion pump for maintaining rocuronium neuromuscular block at T_1% first twitch of the Train of Four (TOF) percent T_1: control response (T_1: T_0) at T_1

A. A. Dahaba, *Understanding the Numbers in Depth of Anesthesia Monitors*,
https://doi.org/10.1007/978-3-031-88260-9_10

10% set-point. The other study by Schumacher and colleagues [3] in 15 subjects basically also consisted of a conventional neuromuscular monitoring device the Neuromuscular Transmission Module (M-NMT, General Electric GE Healthcare, Chicago, Illinois, United States) [4], with a NMBA infusion pump for maintaining rocuronium neuromuscular block at T_1% 10% set-point [2, 3], both connected to central controllers [2, 3]. Albeit possibly being efficient and reliable, such experimental NMBAs Closed-Loop systems could experience numerous impediments as they could occupy a large space in the OR, take time to set up, and operate in an often-crowded OR. Thus, there is a need for efficacious, less laborious, safe, reliable, easy to set and operate Closed-Loop infusion delivery devices.

10.1 Total Intravenous Anesthesia (TIVA)-Target Controlled Infusion (TCI)

Because Total Intravenous Anesthesia (TIVA) is perceived to have the advantage of offering a rapid turnover anesthesia and recovery, it is becoming very popular in Europe, as nowadays almost 50% of general anesthesia are performed under TIVA [5]. Advantages TIVA offers over volatile anesthesia include drastically reducing environmental pollution, rapid patients' cognitive recovery, early patients discharge, and reduced cost [6–9]. However, the disadvantages of TIVA are the infusion syringes' requiring repeated re-fillings as well as infusion rate adjustments that require great efforts when anesthesiologists could be confronted with a critical situation. Furthermore, when TIVA technique is used for long surgical procedures, then up to three different infusion pumps would require to be separately set and continuously managed, i.e., a hypnotic anesthetic like propofol, an Nociception Anti-Nociception Balance (NANB) analgesic like remifentanil, and a Neuromuscular Blocking Agent (NMBA) like rocuronium or cis-atracurium all at the same time, which would add to the mental and emotional anesthesiologists' strain. Furthermore, mental pressure for Depth of Anesthesia (DOA) monitoring would be required from anesthesiologists using TIVA, when for inhalational volatile anesthesia, End-Tidal monitoring could suffice [1].

Target Controlled Infusion (TCI) is a form of TIVA available worldwide except in the United States. This method of TCI aims at reducing the workload of anesthesiologists [10, 11]. The TCI system delivers the anesthetic agents via pharmacokinetic infusion algorithms that maintain given rate concentrations, which could be an estimated plasma concentration (C_p) or a predicted effect-site body compartment (C_e); in our case that would be the brain [1]. Instead of the anesthesiologist programming determined infusion rates, the anesthesiologist could choose either a certain plasma or a brain target concentration [1]. The TCI system uses certain pharmacokinetic models to determine the infusion rates of the estimated plasma concentration or predicted brain effect-site concentration, as well as patient's anthropometric data like sex, age, or weight [1]. The TCI infusion pumps' microprocessors then estimate the infusion rates and all the associated involved parameters such as wake-up times at any given future moment [1]. The end user, usually the anesthesiologist, would set the anesthetic/hypnotic, NANB analgesic agent, NMBA desired level targets, and the TCI systems then aim at establishing and maintaining the set variables as precise as possible to be continuously maintained over any given periods of time. Obviously for measuring drugs' plasma concentrations, blood samples cannot be continuously taken; thus, TCI pharmacokinetic models are derived as of population studies either from volunteers or patients [1]. Such infusion pharmacokinetic models have evolved over the years from a simple basic two-compartment model [12], to far complicated models that are used by the microprocessor derived TCI infusion pump [13]. Those more advanced three-compartments pharmacokinetic models account for numerous parameters such as bolus dose administration, anesthetic agents' elimination, and transfer inside different body compartments [1]. The TCI concept is now widely applied to other hypnotic/anesthetic and

NANB analgesic agents such as remifentanil, ketamine, sufentanil, and dexmedetomidine [1].

However, all these pharmacokinetic models were originally derived back in the 1980s, and since that time the populations' demographic profiles have vastly transformed because of the inhabitants' growing overweight, advanced age, increasing prescription drugs intake, or illicit drug abuse [1]. Thus, new pharmacokinetic models are required that would take into consideration all these demographic, pharmacological, and physiological changes [1]. Moreover, all the original pharmacokinetic models were mostly derived from regions of predominantly White Caucasian populations [1]. Hence new revision updates are due for the pharmacokinetic models to include the current widely diverse populations [1]. Remarkably, Eleveld and colleagues's study [14] is the first available update for propofol pharmacokinetic models that simply reanalyzed a combination of most of the earlier pharmacokinetic models like Marsh and colleagues [15], Schnider and colleagues [16], as well as other pharmacokinetic models that were all based on a small sample size group of 20 and 24 volunteers, respectively [15, 16], to a total of 1033 subjects [14]. The model analyzes a wide range of covariates, namely, age, sex, weight, height, postmenstrual age, and the simultaneous administration or not of anesthetic agents [14]. Independent research statisticians compared Eleveld and colleagues' pharmacokinetic model [14], with Marsh and colleagues [15] and Schnider and colleagues [16] for bias, via the Median Absolute Performance Error (MDAPE) and the Median Performance Error (MDPE) [1]. In Eleveld and colleagues' model [14], the MDAPE was 22% (range 4%–50%); in Marsh and colleagues' model [15], the MDAPE was 25% (range 6%–58%); and in Schnider and colleagues' model [16], the MDAPE was 26% (range 2%–54%). The MDPEs were − 18%, 9%, and − 20%, respectively [14–16]. This simply means that Eleveld and colleagues' [14] and Schnider and colleagues' [16] pharmacokinetic models are overestimating, whereas Marsh and colleagues' [15] model is underestimating the real measured plasma concentrations [1]. This comes as no surprise because propofol metabolism is vastly influenced by a wide range of individual factors, all of which were not accounted for in all three pharmacokinetic models [1]. Moreover, when the three pharmacokinetic models were tested in different Geographic-Ethnic population such as Japan, their accuracies were not reproducible [1].

Hara and colleagues [17] compared 11 propofol TCI pharmacokinetic models for children [17]. They demonstrated that Short and colleagues' [18] pharmacokinetic model had the best performance over long-term steady-state propofol administration [18]. Limitations to their study were the fact that they induced anesthesia with either sevoflurane or propofol, thus introducing bias in their study design [1, 18]. Furthermore, the authors did not use any processed Electroencephalography-Depth Of Anesthesia (pEEG-DOA) monitoring in their study [17, 18], while their 21 recruited children sample size was very low [1, 17, 18].

On the other hand, pharmacokinetic models for remifentanil TCI pumps are currently widely used [1, 19]. Kim and colleagues [20] acknowledged absence of any overweight subjects in the remifentanil original Minto and colleagues' [19] TCI pharmacokinetic model [19]. Thus, they estimated Kim-Obara-Egan's [20] remifentanil TCI pharmacokinetic model that included much more versatile patients' populations [20]. Even though, in the simulations, the three remifentanil TCI pharmacokinetic models, Eleveld and colleagues [21], Minto and colleagues [19], and Kim-Obara-Egan [20], all lead to incomparable remifentanil administrations [19–21].

A major TCI development is using TCI for dexmedetomidine or propofol sedation outside the Operating Room (OR) in a non-hospital setting by non-anesthesiologist care givers personnel [1, 22]. In a trial of 101 subjects, TCI remifentanil and propofol were administered to induce "moderate" sedation by two TCI-trained Certified Registered Nurses (CRN) with one of the two nurses exclusively assigned the responsibility of watching the dental clinic patients [22]. None of the subjects had Serious Adverse Events (SAE) like desaturation to less than 94% or complete loss of responsiveness [1, 22].

Thus, a single pharmacokinetic model that is fitting for all Geographic-Racial patients' populations throughout the world still does not exist [1], as bias still exists among various pharmacokinetic models' Geographic-Racial patients' populations. Here the key to success is combining TCI with DOA monitoring in an "individualized" patient orientated anesthesia [1]. As I have previously mentioned; true variations in "potency" of individual responses to anesthetic hypnotic agents is *stricto* sensu a pharmacodynamic parameter [1]. Even when a patient's model could produce a degree of estimated plasma concentration (C_p) precision with minimal "bias," still when compared with the actual measured plasma concentrations (C_m), this does not guarantee that a satisfactory DOA will be attained [1]. Hence DOA monitoring is always necessary, keeping in mind that artifacts could still falsify, alter, or interfere with DOA monitoring [1]. Thus, when artifacts' disturbances are kept to a minimum while NANB analgesic is kept optimal, then pEEG-DOA monitoring seems to be the most appropriate method to guide TCI titration to the desired effect. In the future, a complex pharmacokinetic model incorporating a wide range of multivariate parameters might be able to cover every possible variable among various Geographic-Racial populations [1].

10.2 Artificial Intelligence-Machine Learning (AI-ML)

Artificial Intelligence-Machine Learning (AI-ML) applications are becoming increasingly prevalent in healthcare [23]. This in turn is challenging clinicians, physicians, hospital administrators, as well as health policymakers to better understand such emerging technologies and to develop suitable frameworks for clinical AI-ML applications that are both safe and sustainable [23]. Within the discipline of anesthesiology, there are challenges with huge opportunities for AI-ML applications that will support patients' care arising from the ever more available massive amounts of evolving perioperative real-time electronic healthcare information data [23]. Such colossal amounts of healthcare information are obtained from a wide-ranging variety of sources that are all being rapidly and continuously first collected and then instantaneously interpreted to be rapidly acted upon by anesthesiologists in modern daily anesthesia practice [23].

To address such challenges and opportunities, recent applications relevant to anesthesiology, including Depth of Anesthesia (DOA) monitoring as conceptual examples exploring strengths and limitations of AI-ML within healthcare, characterize this evolving landscape [23]. Reviewing such applications, I will introduce basic AI-ML concepts and methodologies, as well as practical considerations and ethical concerns for initiating and maintaining safe clinical utilization of AI-ML-based algorithms for anesthesia patient care [23]. Because of the high-volume available data, the fast nature, and wide variation of such clinical information being ever available in the Operating Room (OR) and the Critical Care Unit (CCU), those settings are becoming the perfect environment for applications of Artificial Intelligence (AI) and Machine Learning (ML) algorithms that will eventually lead to the greatest usage of all this huge amount of electronically collected medical data [23–25]. Currently, significant progress is being made towards the implementation of AI-ML applications in the clinical anesthesia practice [23]. The most prominent AI-ML applications example, which is relevant to anesthesiology, would be in the field of DOA monitoring [23]. Through reviewing such modern concepts, I will introduce the most fundamental AI-ML concepts, methodologies, safety and sustainability considerations for clinical anesthesiology AI-ML applications [23].

10.2.1 Guidelines for Developing and Reporting Machine Learning Models

Even with its ever-increasing recognition, it is still hard to get a "globally recognized definition" of what Machine Learning (ML) exactly is. Several ML techniques originated more than 40 years ago in *Breiman LF* 1984 book

Classification and Regression Trees [26]. However, ML began having such a wide influence just about 10 years ago. The two review articles by Jordan and Mitchell [27] and Ghahramani [28] clarify the ML outlines. An increasing number of scientists are resorting to large datasets in search of novel biomedical discoveries prospects. Thus, ML models are becoming ever more widely utilized as the solid foundation for analyzing large datasets [29]. Nevertheless, because of the characteristically complex nature of ML approaches, they are often vulnerable to being misused. Because of the wide variability in formulating the ML models, their outcomes are frequently inadequately described, thus hampering the accurate evaluation of the ML models' validity or reliably understanding the ML models' outputs [29]. To obtain solid guidelines on how to use ML predictive models in a medical setting, thus ensuring the proper application of ML models and to adequately report the ML true findings that are clearly distinguishable from just random coincidences, a panel of ML experts from various disciplinaries, traditional statisticians and clinicians, were all questioned, using an "iterative procedure" according to the *Delphi scheme* [29]. The procedure generated a complete set of guidelines that enabled precise implementation of ML models as well as reliable reporting of the models' characteristics and outcomes [29]. The guidelines consist of first a list of reporting items to be included in a research article and second a set of practical sequential steps for developing ML predictive models [29]. Such guidelines would enhance more adoptions of large datasets' analysis using ML techniques in the biomedical clinical research communities [29]. All ML predictive models including the renowned "Random Forest Tree Algorithms" [30] and "Support Vector Machines" [31], among others, differ from the conventional statistical methods like Logistic Regressions (LR) or Analysis of Variance (ANOVA) in their ability of "making precise predictions on hidden data" [29]. This lets ML process huge numbers of variables at the same time, which is common among most large data tasks [29]. Along with the flexibility of the evolving ML methods comes inconsistency and doubt in using ML models [29]. Owing to Machine Learning's underlying algorithmic and mathematical complexities, ML is frequently deemed the "Black Magic" that involves a very delicate balance between many conflicting considerations [29].

10.2.2 Basic Methodologies for Artificial Intelligence-Machine Learning (AI-ML) Applications

According to the "Guidelines for Developing and Reporting Machine Learning Models" [29], the key steps of creating and training Artificial Intelligence-Machine Learning (AI-ML) models should be detailed. Featured learning curves during model training and validation should clearly demonstrate the performance metrics [29]. The featured learning curves should allow the detection of variance-bias in the form of overfitting or underfitting, by plotting the model training and validation performances as function of the training set size [29]. In any AI-ML model, a good variance-bias balance must be attained, i.e., the models are not overfitting nor underfitting the training data [29]. To better understand which features have higher influence, the famous "Random Forest Tree Algorithm" model [30] should be trained to create features-ranking based on their relative importance [29, 30]. The ranking of these features is expressed in "Gini importance score" (mean decrease in impurity) [32]. This score tells us "How crucial each factor is" in making accurate predictions [29, 32]. The higher the "Gini importance score" [32], the more influential the factor is. As such, AI-ML algorithms could relieve the end users' burden and could potentially increase monitoring compliance [29].

Basically, AI-ML would entail mathematical algorithms that would allow a computer to learn from experiences for performing certain tasks without being specifically programmed to perform such complex tasks. There are two main Machine Learning approaches; one of them is the "Supervised Machine Learning," while the other is the "Unsupervised Machine Learning." The main difference between these two approaches is in how the data, which the computer algorithm is exposed to during the learning process, is structured [23, 33, 34].

Supervised Machine Learning

That would take a "Goal Oriented" learning approach, where the Supervised Machine Learning algorithms are exposed to multiple datasets with clearly labeled outputs. Machine Learning would then "train" from these datasets and hence learn how to predict a correct "Output Label" based exclusively on these provided features [33]. Supervised Machine Learning algorithms do this by first initially randomly assigning an outcome and, after identifying all associations between outputs and inputs, would gradually improve Machine Learning (ML) predictive ability with more additional data being further introduced [33]. Supervised Machine Learning is very well suited to research of clinical outcome but needs a vast large library of "labeled data" from which the Supervised Machine Learning algorithms could be first modeled and then tested [23].

Unsupervised Machine Learning

Contrary to Supervised Machine Learning, Unsupervised Machine Learning takes a different approach of first taking an "Exploratory Learning" approach. Unsupervised Machine Learning algorithms are exposed from the beginning to datasets with no "Explicit Output Labels". Hence, Unsupervised Machine Learning algorithms are allowed to find patterns or structures that the algorithm will use for classifying such data [33]. The "Unsupervised" nature of such an approach would then refer to absence of any direct training or annotations by humans, thus making the unsupervised training less resource intensive, because it is not restricted to any beforehand specified data. In anesthesiology, Unsupervised Machine Learning is far less commonly utilized but could be highly beneficial to bundle datasets into clusters as part of the data preprocessing and simplifying before the Supervised Machine Learning algorithms are then utilized [23].

Deep Learning

The well-known "Deep Learning" describes one of the three Machine Learning categories which was originally inspired by the human nervous system's neural network and could be employed to both Supervised Machine Learning and Unsupervised Machine Learning abovementioned approaches [34]. This Deep Learning approach introduces an Artificial Neural Network, where all the input data are "transformed" via sequentially interconnected layers of neurons to finally predict an output. An Artificial Neural Network is deemed "Deep" if it includes, between its input and output layers, at least one "hidden" intermediate layer. Deep Learning algorithms apply different weights for different neurons' interconnections that would influence the data flow through each final output "layer". These "Intermediate Layers" are essential in simplifying the complicated relationships between inputs and outputs, because each layer functions, on one hand, via amplifying the relevant features while, at the same time, suppressing irrelevant features [23, 35].

10.2.3 Artificial Intelligence-Machine Learning (AI-ML) in Depth of Anesthesia Monitoring

Precisely quantifying Depth of Anesthesia (DOA) remains a challenging task simply because the complete details of what happens when patients are under anesthesia are still not fully understood [36–38]. Nevertheless, the concept of DOA monitoring remains very useful in helping anesthesiologists titrate their anesthetic agents' administration [23, 39–41], whereas processed Electroencephalography-Depth Of Anesthesia (pEEG-DOA) monitors' algorithms that estimate in real-time DOA are still vulnerable to various Electroencephalography (EEG) conditions that are different than those of normal healthy volunteers' EEG from which they were originally derived. Different factors like age, medical conditions such as hypoglycemia, hypothermia, hypovolemia, as well as specific anesthetic agents such as Nitrous Oxide (N_2O) or ketamine are all known for altering the pEEG-DOA monitors displayed numbers, hence preventing precise titration of anesthetic medications and possibly not even preventing adverse outcomes such as Accidental Awareness under General Anesthesia (AAGA) [42–45]. The pEEG-DOA monitors are underperforming, especially in the cardiac surgery patients' population, because many of these algorithms were originally derived from non-cardiac surgical subjects [46], in addition to all the recognized cardiac surgical confounding factors during and after Cardiopulmonary Bypass (CPB), like cardiac pacemakers, hypothermia, and hypoperfusion, all having profound effects on the pEEG-DOA monitors. Here Machine Learning holds great potential for improving pEEG-DOA monitoring in such unfavorable conditions [23, 47, 48].

The first contribution of Machine Learning (ML) innovation is in the field of raw EEG signals' processing [49]. Here Deep Machine Learning models were successfully employed in EEG processing for extracting signals from an EEG source with low signal-to-noise ratio, thus vastly enhancing the EEG processing performances [50]. The second contribution of ML innovations is in the "interpretive algorithms" conventionally used in the pEEG-DOA monitors EEG processing [51]. A commonly used approach is using Supervised Machine Learning models, such as *Convolutional Neural Networks (CNN)* [35, 51], "Forest Domain Decision Tree", or "Random Forest Tree Algorithm" [30], and then training these models using few selected key EEG-derived inputs like *Relative BetaRatio* descriptor, *SynchFastSlow* descriptor, 95% Spectral Edge Frequency (SEF_{95}), 4-band spectral power, or Entropy [23, 52–56]. This method grants adequate training of Supervised Machine Learning models on small datasets but would be restrained by the bounds of these selected inputs' performances. Here a sensible "Objective Selection" for identifying those inputs with relative importance, using Unsupervised Machine Learning models for estimating pEEG-DOA levels, proved to be very valuable [57]. Finally, as more and more ever-increasing access to bigger datasets is growing, complex Machine Learning models that could directly process the whole digital raw EEG waveforms, thus bypassing the necessity for Supervised Machine Learning models' explicit EEG-features-extractions, are being currently under development and are showing great potential [23, 58].

One serious limitation for developing Machine Learning for pEEG-DoA monitors is that, up till now, none of them currently incorporate intraoperative vital signs data that could be an outstanding source of pEEG-DoA data. Ranta and colleagues [59] have demonstrated that Machine Learning Models that could predict AAGA using a Neural Network that is trained on just using patients' general anesthesia records' source data could be constructed [60]. The authors believe that in the future, the techniques that most probably will perform the best are those that will incorporate all three data sources, namely, the EEG dataset, intraoperative vital signs like hemodynamic/Electrocardiographic (ECG) data, to be combined with information from the patients' Electronic Health Records (EHR) [23, 60].

10.2.4 Anesthetic Endpoints to Develop Artificial Intelligence-Machine Learning (AI-ML) Models

The search for the neuroscientific basis of anesthetic loss and return of consciousness has been the core of research for decades [36]. Great advances toward discovering the underlying mechanisms of loss and return of consciousness have been accomplished using multichannel EEG, Magnetoencephalography, and functional Magnetic Resonance Imaging (fMRI). By careful dosing of various volatile and intravenous anesthetic agents to the level of clinically judged "Behavioral Unresponsiveness," common specific changes in functional effective connectivity across large-scale brain networks have been discovered and further interpreted in the context of how the synthesis of neural information is altered under anesthesia [36]. Most investigations to date tend to conclude that there is a "common neural correlate of anesthetic-induced unresponsiveness" that consistently depressed functional connections of the "Lateral Frontoparietal Networks," which are thought to be crucial for "consciousness of the environment" around us [36]. This is achieved via anesthetic agents' disruption of "large-scale information integration" resulting in unconsciousness. The critical neural events that account for the transition between Responsive and Unresponsive states should be assessed at the same anesthetic doses just below and above the loss or return of consciousness [36]. Finally, finding a behavior-independent, subjective, clinically judged, covert cognition measures in unresponsive subjects with all their delineating causal factors and correlated events will eventually enable us to understand the neuronal basis of human consciousness and unconsciousness [36].

A big limitation for developing the Artificial Intelligence-Machine Learning (AI-ML) models for Depth of Anesthesia (DOA) algorithms is the absence, to date, of a Gold-Standard reference comparator for DOA. Many current research AI-ML models were criticized for using the Bispectral Index values for training, validation, and testing their AI-ML models [58]. To overcome such enormous limitation, a small number of investigators have trained their AI-ML models on clinicians-annotated anesthetic status [53, 57]. As more data of actual high-quality annotation anesthetic information becomes available, there is a great potential to combine various processed Electroencephalography (pEEG) data with clinical information sources to use such AI-ML methods for producing a widely approved universally accepted composite measure of DOA [23, 61, 62].

Thus, AI-ML could significantly advance our understanding of many complex scenarios that can then be used to solve many challenges facing us in the DOA monitoring fields [23]. However, a word of caution here, as in the absence of rigorous testing of multicenter "Clinical Trials" data, we must exercise caution to avoid bias or even possible unintentional harmful outcomes when implementing such AI-ML models using not properly vetted datasets [23]. Such AI-ML models to be applied requires cooperation between practicing front anesthesiologists and oversight committees for a constant rigorous watchful eye to maintain the highest quality data acquisition and performance for training of AI-ML algorithms as new medical interventions are continuously evolving [23].

10.3 Closing the Loop

A classic Closed-Loop system [1] would typically consist of three main components: an output of a patient's specific effect that is detected and accurately quantified by reliable monitoring of feedback from sensitive sensors; the input is the continuous variable that is supposed to produce a very specific desired effect on the patient in the form of an anesthetic/hypnotic agent, analgesic Nociception Anti-Nociception Balance (NANB) and Neuromuscular Blocking Agent (NMBA), that are continuously and accurately titrated to produce the precise desired effect; and a computer processor actuator control that would continuously manipulate the input infusion pump according to the constantly changing sensors' input feedback [1]. The Closed-Loop system then aims at establishing the set variables as precise as possible, to be continuously maintained over any given period of time [1].

The "Closed-Loop" is a relatively new concept. The first attempt at closing-the-loop was as recent as 1989 when Schwilden and colleagues [63] published what could be considered the "first" working Closed-Loop for propofol administration in 11 subjects, guided by the Electroencephalography (EEG) Median Frequency. Based on an "adaptive feedback system," the Closed-Loop control system was as follows: when Median EEG Frequency was maintained between 2 and 3 Hz, then propofol was continuously administered following a Bolus-Elimination-Transfer (BET) infusion scheme. But when the Median EEG Frequency was outside the range of 2–3 Hz, then the differences between the "measured" and the "predicted" values were used to adjust propofol infusion rates either by increasing or decreasing propofol infusion rates. Schwilden and colleagues [63] thus laid the foundation for formulating the fundamental principles of the Closed-Loop anesthesia, stating that "by means of the recognized Pharmacokinetic/ Pharmacodynamic models, we can formulate preprogrammed-schematics for drugs administrations based upon an estimation of patient's individualized drugs' requirements. Then the Closed-Loop controller identifies the individualized adjustments on the proposed administration schemes". Schwilden and colleagues [63] also emphasized, as a safety measure, the need to always be able to instantaneously manually override the running Closed-Loop system.

Liu and colleagues [64], 2-channel Closed-Loop system using a Proportional-Integral-Derivative (PID) algorithm compared 83 Closed-Loop patients with 84 open-loop patients, for remifentanil and propofol continuous infusions, titrated via Bispectral Index monitoring continuous output. The controller utilizes three output signals from the Bispectral Index A-2000 XP (version 3.11), namely, the measured Bispectral Index values, Electromyography (EMG) activities, and the Burst Suppression Ratio (BSR) percentage to maintain a Bispectral Index set-point between 40 and 60. Their algorithms use a Rule-Base (RB) system to choose whether remifentanil's infusion rate should be changed, or propofol and remifentanil infusion rates should both be changed [1, 64].

To avoid the "*Cliff-Edge*" phenomenon [1], namely, when anesthesiologists encounter a situation such as when at Bispectral Index score of 60 the patient is "awake" and with Bispectral Index score of 59 the patient is "asleep," then fuzzy logic should be used like when Hemmerling and colleagues [1] employed the "Bayesian Principle," first reported by De Smet and colleagues [65], to be used in Closed-Loop control systems. This system's algorithm grades patient's induction "sensitivity reaction" and accordingly modifies the system's hemodynamic target values [66]. In contrast to Liu and colleagues [64], using this system during induction allows the system to take into consideration patients' reaction to an induction given dose and accordingly adjusts the model algorithms for each individual patient. Additionally, the system's algorithm lets anesthesiologists specify the noxious surgical stimulations with a push of a button, hence allowing the system to use such a valuable input information when the anesthesiologists designates an anticipated noxious surgical stimulation, the system's algorithm then delivers an opioid bolus dose without need for changing the running infusion rates [1].

A two-channel Closed-Loop system described by Zikov and colleagues [67] used a frontal single-channel EEG signal for analyzing the Stationary Wavelet Transform (SWT) [67]. The wavelet EEG coefficients are pooled into a statistical representation of the two well-defined states, namely, the awake state with normal EEG activity and the isoelectric state with maximal cortical depression [67]. The resulting index, referred to as the Wavelet-based Anesthetic Value for Central Nervous System (WAV_{CNS}) monitoring, quantifies the Depth of Anesthesia (DOA) between these two extremes [67].

West and colleagues Closed-Loop [68], used output of the NeuroSENSE wavelet-based (WAV_{CNS}) value for propofol as well as remifentanil from a single WAV_{CNS} sensor [68]. Propofol and remifentanil infusions were continuously administered in 127 patients via a Dual-Channel Closed-Loop "slow actuator" that maintains the designated fundamental WAV_{CNS} targets ±10% of the set-point [68]. Anesthesia induction and maintenance were titrated via the "icontrol-RP system" [68]. The addition of remifentanil's infusion channel to make a Dual-channel Closed-Loop system greatly enhanced the system's precision. That was attributed to the fact that simultaneously directing the two remifentanil and propofol infusions' inputs would result in attaining both consciousness and NANB analgesia being managed concurrently [1, 68].

The *McSleepy* system [69] is a three-channel Closed-Loop system for propofol, remifentanil, and rocuronium [66, 69]. Hemmerling and colleagues compared 93 Closed-Loop patients with 93 open-loop patients using Bispectral Index Vista set-point value of 45 for hypnosis and the Analgoscore as well as vital signs for NANB analgesia set-point [69]. The *McSleepy* utilizes all three Closed-Loop components of control, namely, hypnosis, NANB analgesia, and neuromuscular block [69–71]. It uses the so-called "Expert Rule-Based" system that focuses mainly on artifact's rejection in the hypnotic component of anesthesia [69]. Numerous predetermined pushbuttons allow anesthesiologists to inform the system of surgical events at the key time points [69]. The system stops when the last button is pushed, allowing wakeup when administering NMBAs was stopped 20 minutes before end of surgery [69]. These systems could also be utilized for performing anesthesia at distant locations [72].

Table 10.1 depicts Closed-Loop 1-Channel, 2-Channel, and 3-Channel Closed-Loop systems.

10.3.1 Description of the Concerto

We developed the "Concerto" an integrated automated 3-Channel Closed-Loop system using continuous Target Controlled Infusion (TCI) estimated plasma concentration (C_p) or predicted effect site concentration (C_e) rate constants' delivery technique. *Concerto* 3-Channel Closed-Loop Dashboard and Screen (Fig. 10.1) depicts how the delivery system could simultaneously administer a constant steady infusion of the intravenous anesthetic propofol, the Nociception Anti-Nociception Balance (NANB) analgesic remifentanil, and Neuromuscular Blocking Agent (NMBA) rocuronium or cisatracurium. The "Concerto" is smaller in size than any currently available multiple TCI pump devices like Orchester (Fresenius kabi, Bad Homburg vor der Höhe, Germany). My *Concerto* 3-Channel Closed-Loop algorithms, whose details appear on *Concerto* dashboard (Fig. 10.2), could interface any anesthetic monitor such as the Bispectral Index or Entropy with our designated Closed-Loop Channel A propofol intravenous anesthetic infusion pump. Channel B is designated for Closed-Loop remifentanil delivery algorithms. Simultaneously, our Concerto algorithms could interface any conventional neuromuscular monitor, such as Neuromuscular Transmission Module (M-NMT, General Electric GE Healthcare, Chicago, Illinois, United States), TOF-Watch, or its built in Compressomyograph (CMG) device [73], with our designated Closed-Loop Channel C for cisatracurium or rocuronium NMBA infusion pumps to deliver the set rates of NMBAs continuous infusion. Our *Concerto* could take over delivering, titrating, and maintaining the

Table 10.1 Closing the loop

Reference	Year	Monitor	*n*	Agent	Closed-Loop	Set-Point
Schwilden et al. [63]	1989	EEG	11	Propofol	1-channel	Range 2–3 Hz
Illman et al. [2]	2008	Relaxograph EMG	35 × 2	Rocuronium TCI	1-channel	T_1 10%
Schumacher et al. [3]	2006	NMT-module EMG	15	Mivacurium	1-channel	T_1 10%
Liu et al. [64]	2011	BIS XP A-2000 (3.11), EMG, BSR	83/84	Propofol/ remifentanil	2-channel	BIS 40–60
West et al. [68]	2018	NeuroSENSE (WAV_{CNS})	127	Propofol/ remifentanil	2-channel	WAV_{CNS} ± 10%
Hemmerling et al. [69]	2013	McSleepy, BIS vista	93/93	Propofol/ remifentanil/ Rocuronium	3-channel	BIS 45, Analgoscore
Dahaba et al. [73]	2011	Concerto, BIS vista	24 × 4	Propofol/ remifentanil/ NMBAs	3-channel	BIS 45, vital signs, T_1 7%

n number of patients per group; *BIS* Bispectral Index; *EEG* Electroencephalography; *EMG* Electromyography; NeuroSENSE, WAV_{CNS} Wavelet-based Anesthetic Value for Central Nervous System; *NMBAs* neuromuscular blocking agents; *T1%* First Twitch of the Train of Four (TOF)

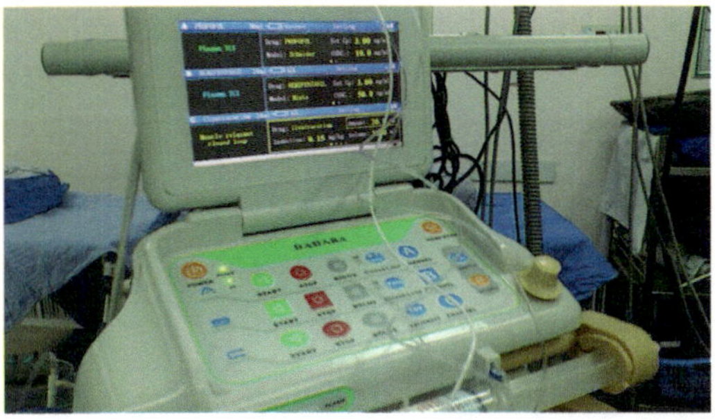

Fig. 10.1 Concerto 3-Channel Closed-Loop dashboard and screen

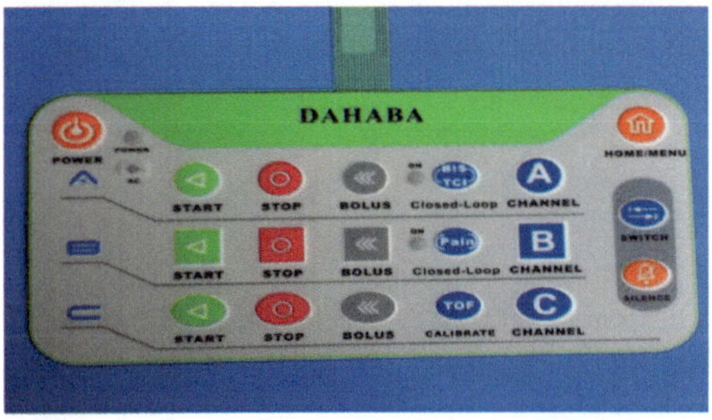

Fig. 10.2 Concerto 3-Channel Closed-Loop dashboard details

hypnosis anesthesia, NANB analgesia, and neuromuscular block. Our *Concerto* algorithms maintain a steady C_p or C_e rate constants of the propofol/remifentanil/NMBA in an efficacious, safe, reliable Closed-Loop fully automated infusion delivery system that is highly useful in an often-crowded, busy Operating Room (OR) (Figs. 10.3, 10.4, 10.5, and 10.6).

The two most used Neuromuscular Blocking Agents (NMBAs) in today's anesthesia practice, the intermediate-acting rocuronium and long-acting cisatracurium, present an additional challenge to a Closed-Loop infusion device as they are harder to precisely titrate, adjust, and maintain than a short-acting NMBA such as mivacurium [3]. My Compressomyograph (CMG) that I developed [73] quantifies the neuromuscular function through measuring the signal generated from the thumb and digits squeezing pressure on a handheld balloon in response to ulnar nerve stimulation at the wrist [73]. This creates a positive voltage of 200 ms, followed by a negative voltage, thus generating a voltage potential which is directly proportional to the displacements of the thumb and digits force [73]. The Area Under the positive voltage wave Curve (AUC) over time is calculated and quantified [73]. The signal is then amplified, filtered, displayed, and recorded [73].

10.3.2 Safety and Efficacy of the Concerto

My research group evaluated the safety and efficacy of the *Concerto* that is, to the best of our knowledge, the first integrated automated 3-Channel Closed-Loop drug infusion delivery system in patients undergoing surgery under general anesthesia. We compared *Concerto* 3-Channel Closed-Loop drug infusion delivery system with an open-loop for the provision of optimal propofol anesthesia, optimal remifentanil Nociception Anti-Nociception Balance (NANB) analgesia, and optimal rocuronium or cisatracurium neuromuscular block. We compared the weighed percentage time with optimal anesthesia defined as Bispectral Index set-point value of 45 maintained at Bispectral Index value between 35 and 55 in the Closed-Loop and open-loop groups. Using the Compressomyograph (CMG) [73], we also compared the weighed percentage time with optimal neuromuscular block defined as First Twitch (T_1) of Train-of-Four (TOF) set-point value of 7% (93% block) maintained between T_1 1–15% in the Closed-Loop and open-loop groups. The secondary endpoints included the weighted mean propofol/remifentanil/Neuromuscular Blocking Agent (NMBA) infusion rate requirements, anesthetics, NANB analgesics and NMBA costs, hemodynamic stability, anesthesiologists', anesthesia nurses', and surgeons' satisfactions and to validate the nominated neuromuscular block range of T_1 1–15% for surgical relaxation.

To estimate the asymptotic steady-state rates of infusion (I_{ss}), we used nonlinear curve fitting for the cumulative dose curve of cisatracurium/rocuronium during 90-min study period [74]. The possible differences between the Closed-Loop and open-loop groups in infusion requirements of

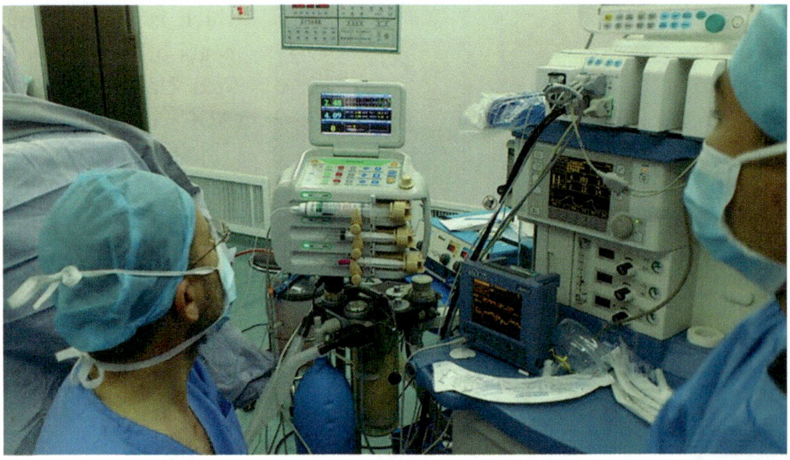

Fig. 10.3 Me after Concerto (kitty) taking over the hypnosis, analgesia, and neuromuscular block. ()

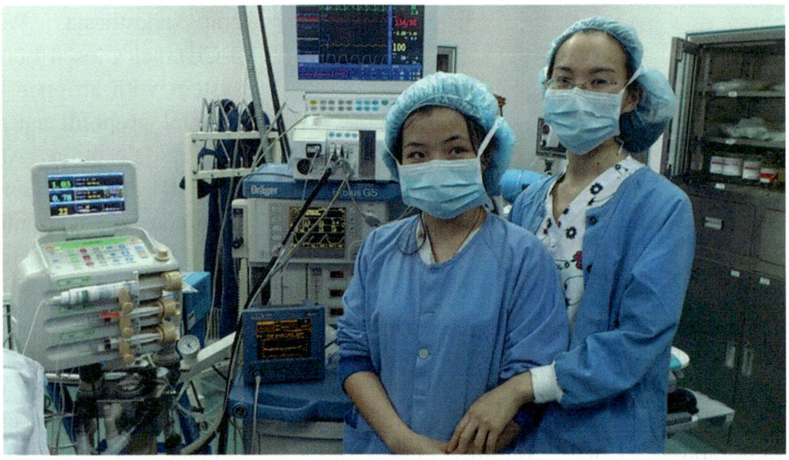

Fig. 10.4 My three assistants Concerto (Kitty) on the left, Yy in the middle, and JJ on the right

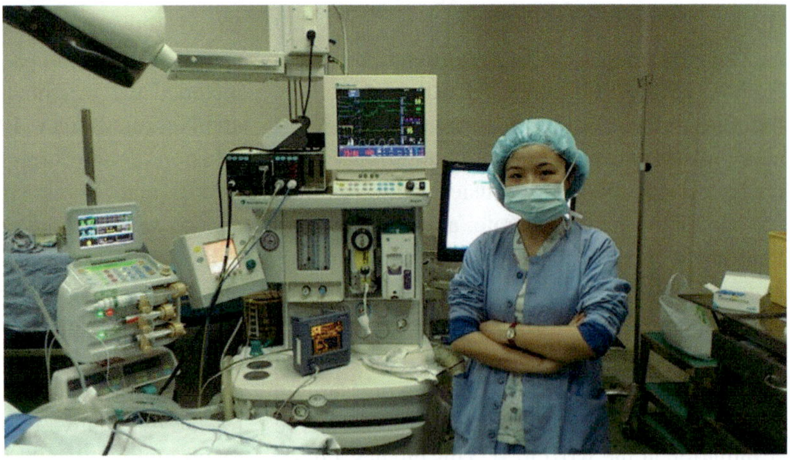

Fig. 10.5 My two assistants Concerto (Kitty) and Yy

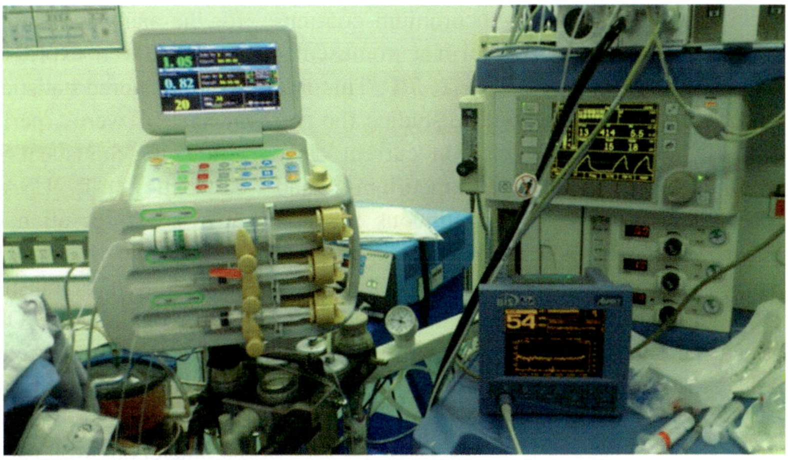

Fig. 10.6 Concerto (Kitty) completely taking over hypnosis, analgesia, and neuromuscular block. (©2025 Medtronic. All rights reserved. Used with the permission of Medtronic)

cisatracurium or rocuronium were quantitated by comparing the asymptotic steady-state infusion rates for T_1 7% (93%) block between the two groups [74]:

Cumulative dose of cisatracurium/rocuronium = $D\ (1 - e^{-kT}) + I_{ss}t$ where D is the amount of cisatracurium/rocuronium in its apparent Volume of distribution (V_d), k is the relative rate of distribution of cisatracurium/rocuronium, I_{ss} is the asymptotic steady-state infusion rate of cisatracurium/rocuronium, and t is the duration of administration of cisatracurium/rocuronium [74]. The asymptotic steady-state infusion rates were given as actual values and per kilogram Ideal Body Weight (I_{ss}/IBW) [74]. Subjects were allowed to recover spontaneously from neuromuscular block.

Our study was registered at European Community Clinical Trials Database EudraCT (http://eudract.emea.eu), trial registration number: **2011-002583-24**. A prospective consecutive study was conducted in accordance with the 2005 Stockholm revision of the guidelines of the "Good Clinical Research Practice (GCRP) in Pharmacodynamic Studies of Neuromuscular Blocking Agents III: The 2023 Geneva revision" [75] and "Standards for Reporting of Diagnostic Accuracy" (STARD) criteria [76]. A prospective clinical consecutive study was conducted in conformity with the guidelines of the "Consolidated Standards of Reporting Trials (CONSORT)-statement" [77]. After Xi Jing Hospital, Xi'an, People's Republic of China ethics committee approval (**20110707-5** at 7 July 2011 meeting chaired by Prof Ming Quan Li) and Sir Run Run Shaw Hospital (SRRSH), Hang Zhou, People's Republic of China ethics committee approval (20111227-10 at 27 December 2011 meeting chaired by Prof Guoqiang Pan), all patients who agreed to participate in the study gave written informed consent. Exclusion criteria were Body Mass Index (BMI) <18.5 or > 24.9 kg m^{-2} [75], patients suffering from small joint arthritis, history of alcohol or drug abuse, hepatic, renal, or neuromuscular diseases, medical conditions that could affect their level of consciousness such as stroke, or dementia, and patients on treatment with medications thought to interfere with neuromuscular transmission.

Ninety-six consecutive patients aged 18–59 years, American Society of Anesthesiologists (ASA) classification physical status I–II, undergoing elective surgical procedures of expected duration >3 h were included in the study. Anesthesiologists who participated in the study were adequately informed and experienced in using Compressomyograph (CMG), Neuromuscular Transmission Module (M-NMT), or TOF-Watch for neuromuscular block monitoring as well as Bispectral Index or Entropy DOA monitoring for at least 10 years prior to their involvement in the study. Patients, stratified for gender with equal number of males and females, were randomly allocated to one of four groups: the Closed-Loop-cisatracurium, open-loop-cisatracurium, Closed-Loop-rocuronium or open-loop-rocuronium continuous infusions. The desired level of neuromuscular block (i.e., the set-point) was T_1 7%. This is a 2 × 2 *Latin square design*. We studied the two NMBAs that are most used in current daily anesthesia practice, cisatracurium and rocuronium, and the two continuous infusion delivery systems, Closed-Loop and open-loop. Thus, we assessed the type of NMBA (intermediate-acting rocuronium versus long-acting cisatracurium) and mode of continuous infusion delivery systems (Closed-Loop and open-loop).

10.3.3 Concerto Study Statistical Analysis

Because the *Concerto* Closed-Loop algorithm was developed in our center, there is no data available on the Closed-Loop algorithm to enable an *a*-priori sample size to be calculated. Kopman and colleagues [3, 78] determined the slopes of the dose–response relationships of eight Neuromuscular Blocking Agents (NMBAs) using regression analysis [78]. Based on their results of the retrospective analysis, they concluded that a total sample size of not less than 24 subjects should be adequate for determining NMBAs' clinical potency parameters with a reasonable degree of assurance [78]. Thus, we decided that a sample size of 24 patients would be required to reveal a statistically significant difference between the two groups. In addition, based upon infusion rates data obtained in a previous study [79], during major non-cardiac surgery, in which cisatracurium 1.5 μg kg^{-1} min^{-1} or rocuronium 10 μg kg^{-1} min^{-1} [79], assuming a 15 percentage points deviation from the assumed 70 ± 10% mean percentage time of NMBAs optimal relaxation to maintain First Twitch of Train-of-Four (TOF) T_1 between 5–15%, our *a*-priori power analysis (α =0.05) showed that a sample size of 24 patients would be adequate to reveal a statistically significant difference between the two groups with >80% power.

We used Student's *t*-test one way Analysis of Variance (ANOVA) to compare the differences between the groups and Mann-Whitney *U*-test for the analysis of nonparametric data parameters. Data were expressed as mean ± SD. $P < 0.05$ was considered statistically significant. Statistical analyses were performed using Number Crunching Statistical System 2007 (NCSS Inc., Kaysville, UT, USA) and StatXact (Cytel Software Corporation, Cambridge, MA, USA).

In our study, although open-loop Neuromuscular Transmission Module (M-NMT) kinemyography or TOF-Watch acceleromyography are the most used methods for quantifying the neuromuscular block, they are still prone to experimental errors. All the pharmacodynamic parameters as well as maintenance and recovery from neuromuscular block data were further described based on Bland and Altman format [80] that deems any two simultaneously used methods subject to experimental errors and hence consider the two measurements' average as the true value estimation and does not assume any of the two simultaneously used methods as the true standard [80]. The "Bias" is defined as the "mean" of the differences between the two methods [80]. The "Limits of Agreement (LOA)" are defined as the Bias ±1.96 Standard Deviation (SD) where 95% of the difference between the two methods are expected to lie [80].

10.3.4 The Concerto Study

We administered oral midazolam 7.5 mg 1 h. preoperatively for premedication. In the open-loop group after entering patients' anthropometric data, induction and maintenance of anesthesia was with remifentanil 4 ηg mL^{-1} and propofol 4 μg ml^{-1} Target Controlled Infusion (TCI) using Orchester (Fresenius kabi, Bad Homburg vor der Höhe, Germany), incorporating propofol Schnider and colleagues [81] pharmacokinetic model and remifentanil Minto and colleagues [19] pharmacokinetic model. Patients were ventilated via a facemask. In an ascending sequence of electric stimulations starting from 10 mA, supramaximal current was defined as the stimulation intensity 20% higher than the maximal response current [75]. To determine the supramaximal current, we stimulated the ulnar nerve using silver/silver chloride surface electrodes 4 cm apart on the ulnar nerve at the wrist, by Train-Of-Four (TOF) stimuli for 2-s stimulations of 2-Hz square wave pulse width 200 µs at 20-s intervals. To evaluate neuromuscular block, we used the percentage of First Twitch (T_1): Control Response (T_1: T_0) and TOF ratio (T_4: T_1).

We administered rocuronium 0.6 mg kg^{-1} (twice the 95% Effective Dose 2 x ED_{95}) [82] or cisatracurium 0.2 mg kg^{-1} (four times 95% Effective Dose 4 x ED_{95}). Endotracheal intubation was attempted at T_1% maximal suppression. The lungs were ventilated mechanically with 40% oxygen in air. Ventilation was adjusted to maintain 30–40 mm Hg End-Tidal carbon dioxide. Lactated Ringer's solution was infused, at a rate sufficient to replace fluid losses. Patients were warmed using a forced-hot-air-blanket to keep the temperature of both hands constant above 32 °C and the core temperature above 35 °C. The skin temperature at the thenar area of the palm of the hand and the core temperature were monitored by skin and esophageal temperature probes.

The neuromuscular block monitoring arm was comfortably placed on an arm-board. To ensure good contact, we rinsed with alcohol-swab the wrist's skin area above the ulnar nerve for electrodes' placements. To balance the right- or left-hand dominance effect, the neuromuscular monitor was randomly allocated to the right- or left-hand according to a separate computer-generated scheme than the group randomization scheme. The *Concerto* and neuromuscular monitor Neuromuscular Transmission Module (M-NMT) data were collected and stored on a laptop computer via a USB port using the *Concerto* software "*Dahaba 1.0*," or the General Electric S/5 "Collect" (General Electric GE Healthcare, Chicago, Illinois, United States) data collection software programs [83].

For depth of anesthesia monitoring, a Bispectral Index Quatro sensor was placed on patients' forehead. In the open-loop group, the Bispectral Index data were collected and stored on a laptop computer. My software "Dahaba 1.0" can record each patient's required primary and secondary parameters directly from the *Concerto* device or from the laptop in the open-loop group. The software calculates time of the Closed-Loop infusion, percentage time with optimal neuromuscular block defined as T_1 maintained between 1–15%, percentage time with optimal anesthesia defined as Bispectral Index values maintained between 35–55, the weighted mean propofol/remifentanil/ Neuromuscular Blocking Agents (NMBAs) infusion rate requirements and anesthetic agents and NMBAs costs. This enabled us to deduct all the required primary and secondary parameters of each patient easily, rapidly and accurately.

10.3.5 Concerto Dosing Algorithms

The *Concerto* Closed-Loop algorithms can interface Depth of Anesthesia (DOA) and neuromuscular monitors by quantifying the DOA, Nociception Anti-Nociception Balance (NANB), or neuromuscular block using the signal generated from DOA and neuromuscular monitors that is then relayed to Channel A, Channel B, and Channel C Target Controlled Infusion (TCI) pumps that titrate, adjust, and maintain the DOA, NANB analgesia, and neuromuscular block set-points of hypnotic anesthetic, NANB analgesic agents, and Neuromuscular Blocking Agents (NMBAs) continuous infusions. The *Concerto* Closed-Loop delivery systems could take over the anesthesia induction (Fig. 10.7), and the so-called "valley of deep anesthesia" that follows anesthesia induction

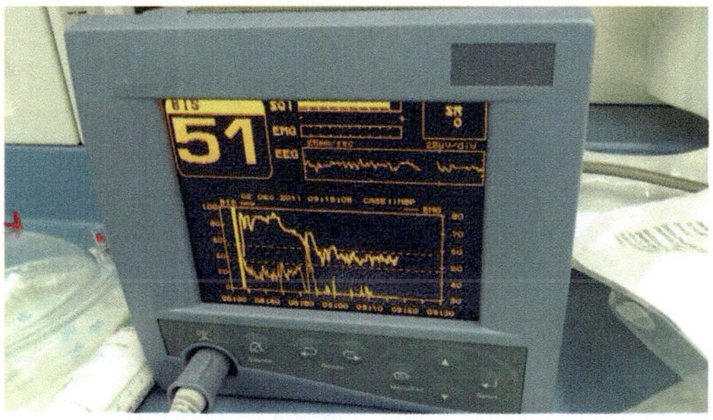

Fig. 10.7 Concerto (Kitty) doing the induction of anesthesia. (©2025 Medtronic. All rights reserved. Used with the permission of Medtronic)

(Fig. 10.8). The *Concerto* Closed-Loop delivery systems could then take over delivering, titrating, and maintaining the hypnosis anesthesia, NANB analgesia, and neuromuscular block, thus giving anesthesiologist two free hands for attending to other tasks like placing a central venous line among other tasks (Figs. 10.9 and 10.10). The *Concerto* Closed-Loop delivery systems could keep our study patients around the Bispectral Index value of 45 set-point in patient #1 (Fig. 10.11), in patient #2 (Fig. 10.12), in patient #3 (Fig. 10.13), in patient #4 (Fig. 10.14), in patient #5 (Fig. 10.15), and in patient #6 in the dark (Fig. 10.16, over extended periods of time in all patients,

In subjects allocated to the open-loop groups, experienced attending anesthesiologists adjusted the propofol infusion rate to maintain Bispectral Index values at 45 via increase/decrease propofol infusion rates when 3-min Bispectral Index consecutive values were outside the range of Bispectral Index values of 35–55, i.e., the Bispectral Index set-point of 45. Similarly, the open-loop groups infusion syringes were prepared with cisatracurium or rocuronium, so that even small changes in infusion rates led to significant changes of volume delivered, and thus a quick alteration in effect [79]. Experienced attending anesthesiologists, in the open-loop groups, adjusted the infusion rates to maintain T_1 First Twitch: Control Response (T_1: T_0) of Train-Of-Four (TOF) at T_1 7% via increase/decrease dose adjustments when three consecutive signals were outside the range of T_1 1–15% of the desired level of neuromuscular block, i.e., the set-point of T_1 7%. After recovery of T_1 to 7% from the endotracheal intubation neuromuscular block in the open-loop groups, an infusion was started at a rate of cisatracurium 2.5 μg kg^{-1} min^{-1} or rocuronium 30 μg kg^{-1} min^{-1} and adjusted to maintain T_1% at 1–15%. These doses were based on data obtained during major non-cardiac surgery [79].

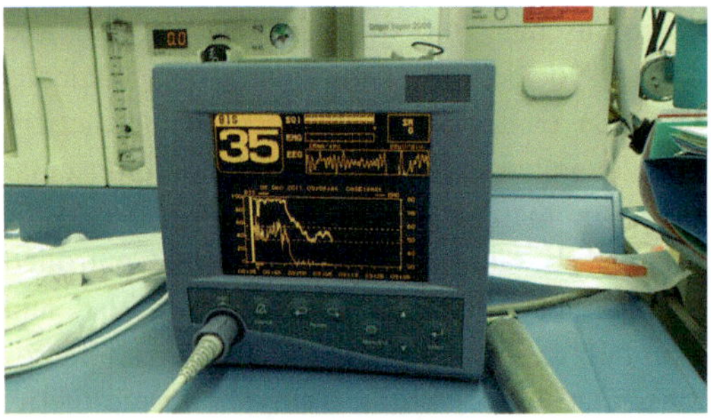

Fig. 10.8 Concerto (Kitty) doing the valley of deep anesthesia induction. (©2025 Medtronic. All rights reserved. Used with the permission of Medtronic)

In subjects allocated to the Closed-Loop groups, a lock out mechanism ensured that the *Concerto* Closed-Loop algorithm did not deliver a dose that is more than 30% of the set propofol, remifentanil, cisatracurium, or rocuronium infu-

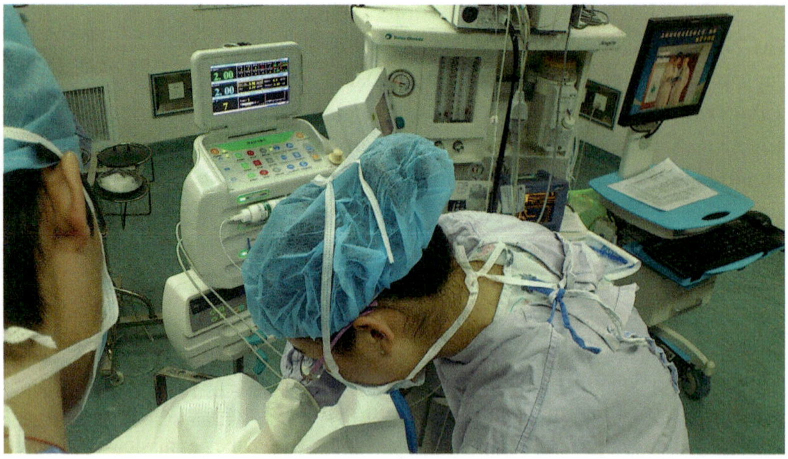

Fig. 10.9 Concerto (Kitty) giving anesthesiologist two free hands to place a central venous line

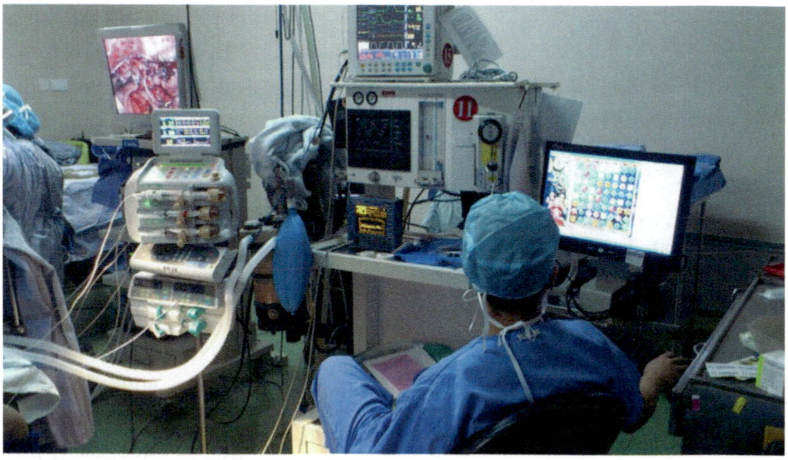

Fig. 10.10 Concerto (Kitty) giving anesthesiologist two free hands to play video games

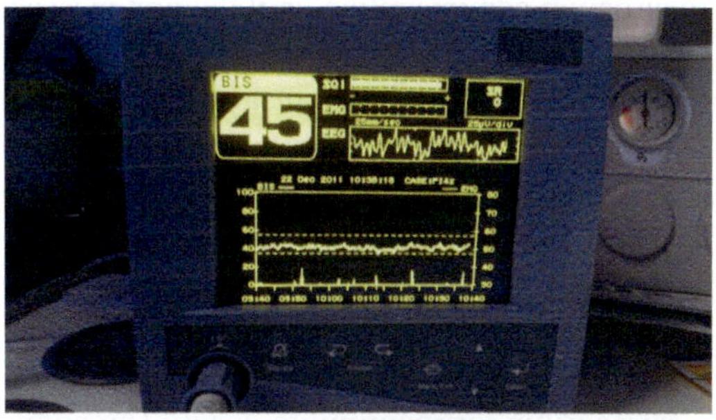

Fig. 10.11 Concerto (Kitty) doing Bulls' eye Bispectral Index 45 set-point in patient #1. (©2025 Medtronic. All rights reserved. Used with the permission of Medtronic)

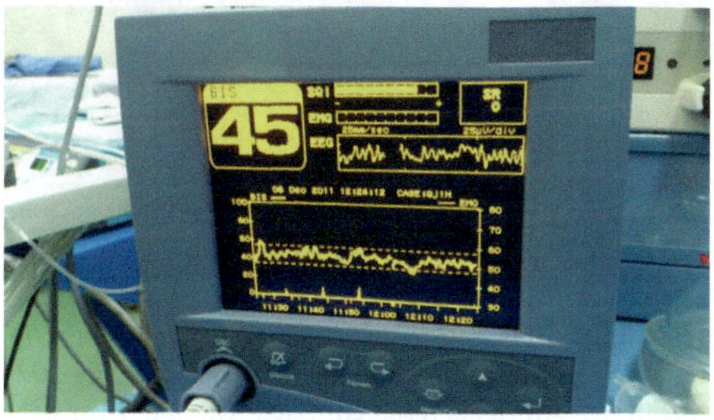

Fig. 10.13 Concerto (Kitty) doing Bulls' eye Bispectral Index 45 set-point in patient #3. (©2025 Medtronic. All rights reserved. Used with the permission of Medtronic)

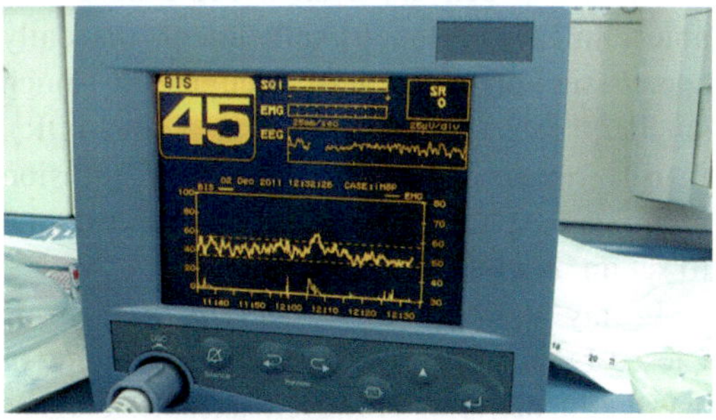

Fig. 10.12 Concerto (Kitty) doing Bulls' eye Bispectral Index 45 set-point in patient #2. (©2025 Medtronic. All rights reserved. Used with the permission of Medtronic)

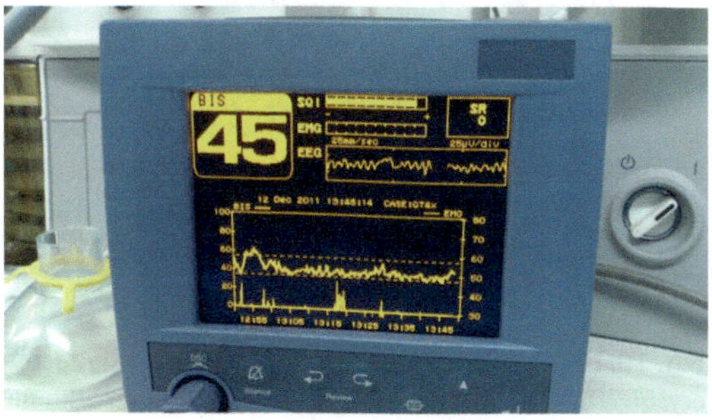

Fig. 10.14 Concerto (Kitty) doing Bulls' eye Bispectral Index 45 set-point in patient #4. (©2025 Medtronic. All rights reserved. Used with the permission of Medtronic)

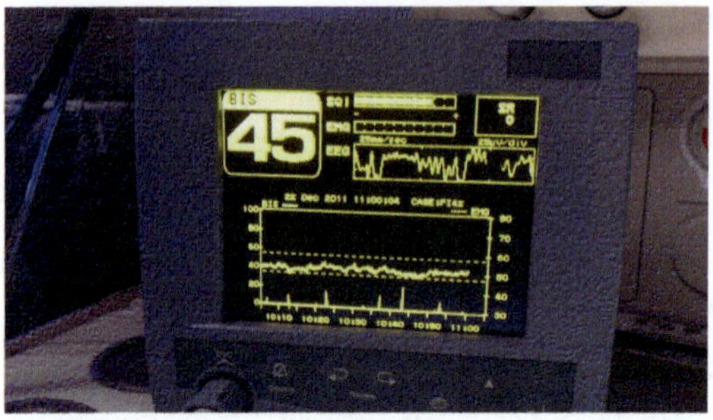

Fig. 10.15 Concerto (Kitty) doing Bulls' eye Bispectral Index 45 set-point in patient #5. (©2025 Medtronic. All rights reserved. Used with the permission of Medtronic)

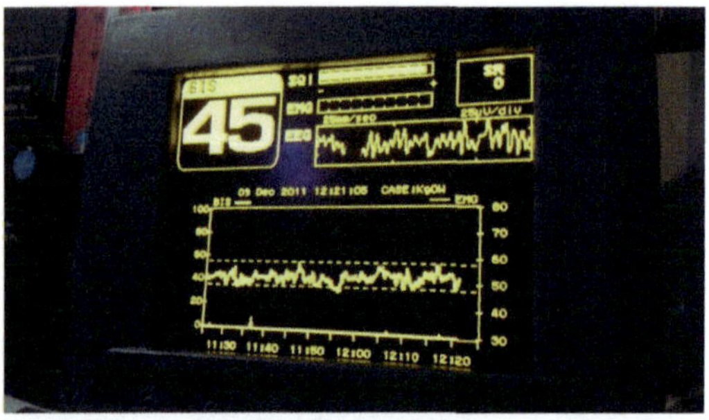

Fig. 10.16 Concerto (Kitty) doing Bulls' eye Bispectral Index 45 set-point in patient #6 in the dark. (©2025 Medtronic. All rights reserved. Used with the permission of Medtronic)

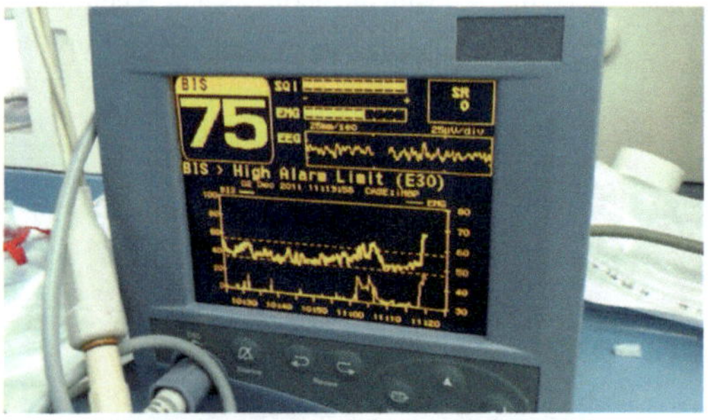

Fig. 10.17 Concerto (Kitty) doing the anesthesia recovery. (©2025 Medtronic. All rights reserved. Used with the permission of Medtronic)

sion rates, whereas for subjects allocated to the Closed-Loop group, a *Concerto* "rescue protocol" of bolus doses of propofol 1 mg kg^{-1}, cisatracurium 0.03 mg kg^{-1}, or rocuronium 0.15 mg kg^{-1} [79] is ready at the *Concerto* "Bolus-button" when patients' gross movements indicate patients were not adequately anesthetized or not adequately relaxed and not properly controlled by the *Concerto* Closed-Loop delivery system.

After the end of the surgical procedures, the *Concerto* Closed-Loop delivery systems could do the recovery from anesthesia (Fig. 10.17). After that Post Anesthesia Care Unit (PACU) nurses asked the anesthesiologists, anesthesia nurses, and surgeons who were blinded to the technique used, to grade their degree of satisfaction as Excellent (3), Good (2), Acceptable (1), or Unacceptable (0). Induction parameters lag time and onset time (time from start of cisatracurium/rocuronium administration until the first measurable and maximal neuromuscular blocks, respectively), and recovery parameters Dur_{25-75} and $Dur_{25-0.9}$ (time from 25% T_1 to 75% T_1 and 0.9 TOF ratio recoveries) were calculated [75].

The *Concerto* Closed-Loop delivery system was more efficacious than an open-loop regimen. The use of *Concerto* Closed-Loop delivery system result in fewer propofol, remifentanil, cisatracurium, or rocuronium doses. The nominated neuromuscular block range of 1–15 T_1% provided a sufficient functional level of neuromuscular block for anesthesiologists, anesthesia nurses,' and surgeons' satisfaction. *Concerto* Closed-Loop delivery system could be a reliable clinical continuous infusion in the daily anesthesia practice that occupies a much smaller space in the Operating Room (OR), than a regular routinely used multi-infusion pump, is easy to use and does not require time to set up or adjust.

In my opinion, my *Concerto* 3-Channels Closed-Loop system-wide use is the future of anesthesia delivery. However, anesthesiologists are not ready to sit in the "backseat" while technology does the driving! What we have achieved so far is to have a very efficient "autopilot" at our disposal, but we are always "present" and ready to take charge at any moment. That is where we are with today's available technology.

References

1. Hemmerling TM, Jeffries SD. Robotic anesthesia: a vision for 2050. Anesth Analg. 2024;138:239–51.
2. Illman H, Antila H, Olkkola KT. Quantitation of the effect of nitrous oxide on rocuronium infusion requirements using closed-loop feedback control. Anesthesiology. 2008;108:388–91.
3. Schumacher PM, Stadler KS, Wirz R, Leibundgut D, Pfister CA, Zbinden AM. Model-based control of neuromuscular block using mivacurium: design and clinical verification. Eur J Anaesthesiol. 2006;23:691–9.
4. Dahaba AA, Rehak PH, List WF. Assessment of accelerography with the TOF-GUARD: a comparison with electromyography. Eur J Anaesthesiol. 1997;14:623–9.
5. Anesi GL, Kerlin MP. The impact of resource limitations on care delivery and outcomes: routine variation, the coronavirus disease 2019 pandemic, and persistent shortage. Curr Opin Crit Care. 2021;27:513–9.
6. McGain F, Muret J, Lawson C, Sherman JD. Environmental sustainability in anaesthesia and critical care. Br J Anaesth. 2020;125:680–92.
7. Miller TE, Gan TJ. Total intravenous anesthesia and anesthetic outcomes. J Cardiothorac Vasc Anesth. 2015;29(Suppl 1):S11–5.
8. Miller D, Lewis SR, Pritchard MW, Schofield-Robinson OJ, Shelton CL, Alderson P, et al. Intravenous versus inhalational maintenance of anaesthesia for postoperative cognitive outcomes in elderly people undergoing non-cardiac surgery. Cochrane Database Syst Rev. 2018;8:CD012317.
9. Goh A-CN, Bagshaw O, Courtman S. A follow-up survey of total intravenous anesthesia usage in children in the U.K. and Ireland. Paediatr Anaesth. 2019;29:180–5.
10. Schnider TW, Minto CF, Struys MMRF, Absalom AR. The safety of target-controlled infusions. Anesth Analg. 2016;122:79–85.
11. Absalom AR, Glen JIB, Zwart GJC, Schnider TW, Struys MMRF. Target-controlled infusion: a mature technology. Anesth Analg. 2016;122:70–8.
12. Krüger-Thiemer E. Continuous intravenous infusion and multicompartment accumulation. Eur J Pharmacol. 1968;4:317–24.
13. Schwilden H. A general method for calculating the dosage scheme in linear pharmacokinetics. Eur J Clin Pharmacol. 1981;20:379–86.
14. Eleveld DJ, Colin P, Absalom AR, Struys MMRF. Pharmacokinetic-pharmacodynamic model for propofol for broad application in anaesthesia and sedation. Br J Anaesth. 2018;120:942–59.
15. Marsh B, White M, Morton N, Kenny GN. Pharmacokinetic model driven infusion of propofol in children. Br J Anaesth. 1991;67:41–8.
16. Schnider TW, Minto CF, Gambus PL, Andresen C, Goodale DB, Shafer SL, et al. The influence of method of administration and covariates on the pharmacokinetics of propofol in adult volunteers. Anesthesiology. 1998;88:1170–82.
17. Hara M, Masui K, Eleveld DJ, Struys MMRF, Uchida O. Predictive performance of eleven pharmacokinetic models for propofol infusion in children for long-duration anaesthesia. Br J Anaesth. 2017;118:415–23.
18. Short TG, CST A, Tan P, Wong J, Tam YH, Oh TE. A prospective evaluation of pharmacokinetic model controlled infusion of propofol in paediatric patients. Br J Anaesth. 1994;72:302–6.
19. Minto CF, Schnider TW, Egan TD, Youngs E, Lemmens HJ, Gambus PL, et al. Influence of age and gender on the pharmacokinetics and pharmacodynamics of remifentanil. I Model Develop Anesthesiol. 1997;86:10–23.
20. Kim TK, Obara S, Egan TD, Minto CF, La Colla L, Drover DR, et al. Disposition of remifentanil in obesity: a new pharmacokinetic model incorporating the influence of body mass. Anesthesiology. 2017;126:1019–32.
21. Eleveld DJ, Colin P, Absalom AR, Struys MMRF. Target-controlled-infusion models for remifentanil dosing consistent with approved recommendations. Br J Anaesth. 2020;125:483–91.
22. Lobb D, MiriMoghaddam M, Macalister D, Chrisp D, Shaw G, Lai H. Safety and efficacy of target controlled infusion administration of propofol and remifentanil for moderate sedation in non-hospital dental practice. J Dent Anesth Pain Med. 2023;23:19–28.
23. Mathis M, Steffner KR, Subramanian H, Gill GP, Girardi NI, Bansal S, et al. Overview and clinical applications of artificial intelligence and machine learning in cardiac anesthesiology. J Cardiothorac Vasc Anesth. 2024;38:1211–20.
24. Lonsdale H, Jalali A, Gálvez JA, Ahumada LM, Simpao AF. Artificial intelligence in anesthesiology: hype, Hope, and hurdles. Anesth Analg. 2020;130:1111–3.
25. Hashimoto DA, Witkowski E, Gao L, Meireles O, Rosman G. Artificial intelligence in anesthesiology: current techniques, clinical applications, and limitations. Anesthesiology. 2020;132:379–94.
26. Breiman Leo FJORASC. Classification and regression trees. 1st ed. New York: Chapman and Hall/CRC; 1984.
27. Jordan MI, Mitchell TM. Machine learning: trends, perspectives, and prospects. Science. 2015;349:255–60.
28. Ghahramani Z. Probabilistic machine learning and artificial intelligence. Nature. 2015;521:452–9.
29. Luo W, Phung D, Tran T, Gupta S, Rana S, Karmakar C, et al. Guidelines for developing and reporting machine learning predictive models in biomedical research: a multidisciplinary view. J Med Internet Res. 2016;18:e323.
30. Kotsiantis SB. Decision trees: a recent overview. Artif Intell Rev. 2013;39:261–83.
31. Cortes C, Vapnik V. Support-vector networks. Mach Learn. 1995;20:273–97.

32. Menze BH, Kelm BM, Masuch R, Himmelreich U, Bachert P, Petrich W, et al. A comparison of random forest and its Gini importance with standard chemometric methods for the feature selection and classification of spectral data. BMC Bioinformatics. 2009;10:213.
33. Deo RC. Machine learning in medicine. Circulation. 2015;132:1920–30.
34. Esteva A, Robicquet A, Ramsundar B, Kuleshov V, DePristo M, Chou K, et al. A guide to deep learning in healthcare. Nat Med. 2019;25:24–9.
35. LeCun Y, Bengio Y, Hinton G. Deep learning. Nature. 2015;521:436–44.
36. Hudetz AG, Mashour GA. Disconnecting consciousness: is there a common anesthetic end point? Anesth Analg. 2016;123:1228–40.
37. Mashour GA. Cognitive unbinding: a neuroscientific paradigm of general anesthesia and related states of unconsciousness. Neurosci Biobehav Rev. 2013;37(10 Pt 2):2751–9.
38. Mashour GA, Orser BA, Avidan MS. Intraoperative awareness: from neurobiology to clinical practice. Anesthesiology. 2011;114:1218–33.
39. Gounon LJ, Davies C, Milan Z, Kunst G. Concordance of depth of anesthesia monitors during cardiopulmonary bypass? J Cardiothorac Vasc Anesth. 2020;34:308–10.
40. Jarry S, Halley I, Calderone A, Momeni M, Deschamps A, Richebé P, et al. Impact of processed electroencephalography in cardiac surgery: a retrospective analysis. J Cardiothorac Vasc Anesth. 2022;36:3517–25.
41. Maksimow A, Särkelä M, Långsjö JW, Salmi E, Kaisti KK, Yli-Hankala A, et al. Increase in high frequency EEG activity explains the poor performance of EEG spectral entropy monitor during S-ketamine anesthesia. Clin Neurophysiol. 2006;117:1660–8.
42. Anderson RE, Jakobsson JG. Entropy of EEG during anaesthetic induction: a comparative study with propofol or nitrous oxide as sole agent. Br J Anaesth. 2004;92:167–70.
43. Dahaba AA. Different conditions that could result in the bispectral index indicating an incorrect hypnotic state. Anesth Analg. 2005;101:765–73.
44. Obert DP, Schweizer C, Zinn S, Kratzer S, Hight D, Sleigh J, et al. The influence of age on EEG-based anaesthesia indices. J Clin Anesth. 2021;73:110325.
45. Schmidt GN, Bischoff P, Standl T, Jensen K, Voigt M, Esch SA, J. Narcotrend and Bispectral index monitor are superior to classic electroencephalographic parameters for the assessment of anesthetic states during propofol-remifentanil anesthesia. Anesthesiology. 2003;99:1072–7.
46. Gallagher JD. Pacer-induced artifact in the bispectral index during cardiac surgery. Anesthesiology. 1999;90:636.
47. Isley MR, Edmonds HL, Stecker M, American Society of Neurophysiological Monitoring. Guidelines for intraoperative neuromonitoring using raw (analog or digital waveforms) and quantitative electroencephalography: a position statement by the American Society of Neurophysiological Monitoring. J Clin Monit Comput. 2009;23:369–90.
48. Kaiser HA, Hirschi T, Sleigh C, Reineke D, Hartwich V, Stucki M, et al. Comorbidity-dependent changes in alpha and broadband electroencephalogram power during general anaesthesia for cardiac surgery. Br J Anaesth. 2020;125:456–65.
49. Jeleazcov C, Egner S, Bremer F, Schwilden H. Automated EEG preprocessing during anaesthesia: new aspects using artificial neural networks. Biomed Tech (Berl). 2004;49:125–31.
50. Wang Q, Liu F, Wan G, Chen Y. Inference of brain states under anesthesia with meta learning based deep learning models. IEEE Trans Neural Syst Rehabil Eng. 2022;30:1081–91.
51. Liu Q, Cai J, Fan S-Z, Abbod MF, Shieh J-S, Kung Y, et al. Spectrum analysis of EEG signals using CNN to model Patient's consciousness level based on anesthesiologists' experience. IEEE Access. 2019;7:53731–42.
52. Gu Y, Liang Z, Hagihira S. Use of multiple EEG features and artificial neural network to monitor the depth of anesthesia. Sensors (Basel). 2019:19.
53. Ortolani O, Conti A, Di Filippo A, Adembri C, Moraldi E, Evangelisti A, et al. EEG signal processing in anaesthesia. Use of a neural network technique for monitoring depth of anaesthesia. Br J Anaesth. 2002;88:644–8.
54. Saadeh W, Khan FH, Bin AMA. Design and implementation of a machine learning based EEG processor for accurate estimation of depth of anesthesia. IEEE Trans Biomed Circuits Syst. 2019;13:658–69.
55. Shalbaf A, Saffar M, Sleigh JW, Shalbaf R. Monitoring the depth of anesthesia using a new adaptive Neurofuzzy system. IEEE J Biomed Health Inform. 2018;22:671–7.
56. Shalbaf R, Behnam H, Jelveh MH. Monitoring depth of anesthesia using combination of EEG measure and hemodynamic variables. Cogn Neurodyn. 2015;9:41–51.
57. Sanz-García A, Pérez-Romero M, Pastor J, Sola RG, Vega-Zelaya L, Vega G, et al. Potential EEG biomarkers of sedation doses in intensive care patients unveiled by using a machine learning approach. J Neural Eng. 2019;16:026031.
58. Afshar S, Boostani R, Sanei S. A combinatorial deep learning structure for precise depth of anesthesia estimation from EEG signals. IEEE J Biomed Health Inform. 2021;25:3408–15.
59. Ranta SOV, Hynynen M, Räsänen J. Application of artificial neural networks as an indicator of awareness with recall during general anaesthesia. J Clin Monit Comput. 2002;17:53–60.
60. Roy Chowdhury M, Madanu R, Abbod MF, Fan S-Z, Shieh J-S. Deep learning via ECG and PPG signals for prediction of depth of anesthesia. Biomed Signal Process Control. 2021;68:102663.
61. Pastorino R, De Vito C, Migliara G, Glocker K, Binenbaum I, Ricciardi W, et al. Benefits and chal-

lenges of big data in healthcare: an overview of the European initiatives. Eur J Pub Health. 2019;29 Supplement_3:23–7.
62. Goldberger AL, Amaral LA, Glass L, Hausdorff JM, Ivanov PC, Mark RG, et al. PhysioBank, PhysioToolkit, and PhysioNet: components of a new research resource for complex physiologic signals. Circulation. 2000;101:E215–20.
63. Schwilden H, Stoeckel H, Schüttler J. Closed-loop feedback control of propofol anaesthesia by quantitative EEG analysis in humans. Br J Anaesth. 1989;62:290–6.
64. Liu N, Chazot T, Hamada S, Landais A, Boichut N, Dussaussoy C, et al. Closed-loop coadministration of propofol and remifentanil guided by bispectral index: a randomized multicenter study. Anesth Analg. 2011;112:546–57.
65. De Smet T, Struys MMRF, Neckebroek MM, Van den Hauwe K, Bonte S, Mortier EP. The accuracy and clinical feasibility of a new bayesian-based closed-loop control system for propofol administration using the bispectral index as a controlled variable. Anesth Analg. 2008;107:1200–10.
66. Wehbe M, Arbeid E, Cyr S, Mathieu PA, Taddei R, Morse J, et al. A technical description of a novel pharmacological anesthesia robot. J Clin Monit Comput. 2014;28:27–34.
67. Zikov T, Bibian S, Dumont GA, Huzmezan M, Ries CR. Quantifying cortical activity during general anesthesia using wavelet analysis. IEEE Trans Biomed Eng. 2006;53:617–32.
68. West N, van Heusden K, Görges M, Brodie S, Rollinson A, Petersen CL, et al. Design and evaluation of a closed-loop anesthesia system with robust control and safety system. Anesth Analg. 2018;127:883–94.
69. Hemmerling TM, Arbeid E, Wehbe M, Cyr S, Taddei R, Zaouter C. Evaluation of a novel closed-loop total intravenous anaesthesia drug delivery system: a randomized controlled trial. Br J Anaesth. 2013;110:1031–9.
70. Zaouter C, Hemmerling TM, Mion S, Leroux L, Remy A, Ouattara A. Feasibility of automated Propofol sedation for Transcatheter aortic valve implantation: a pilot study. Anesth Analg. 2017;125:1505–12.
71. Zaouter C, Hemmerling TM, Lanchon R, Valoti E, Remy A, Leuillet S, et al. The feasibility of a completely automated Total IV anesthesia drug delivery system for cardiac surgery. Anesth Analg. 2016;123:885–93.
72. Hemmerling TM, Arbeid E, Wehbe M, Cyr S, Giunta F, Zaouter C. Transcontinental anaesthesia: a pilot study. Br J Anaesth. 2013;110:758–63.
73. Dahaba AA, Bornemann H, Holst B, Wilfinger G, Metzler H. Comparison of a new neuromuscular transmission monitor compressomyograph with mechanomyograph. Br J Anaesth. 2008;100:344–50.
74. Keéri-Szanto M. Drug consumption during thiopentone-nitrous oxide-relaxant anaesthesia: the preparation and interpretation of time/dose curves. Br J Anaesth. 1960;32:415–23.
75. Fuchs-Buder T, Brull SJ, Fagerlund MJ, Renew JR, Cammu G, Murphy GS, et al. Good clinical research practice (GCRP) in pharmacodynamic studies of neuromuscular blocking agents III: the 2023 Geneva revision. Acta Anaesthesiol Scand. 2023;67:994–1017.
76. Bossuyt PM, Reitsma JB, Bruns DE, Gatsonis CA, Glasziou PP, Irwig LM, et al. The STARD statement for reporting studies of diagnostic accuracy: explanation and elaboration. Clin Chem. 2003;49:7–18.
77. Moher D, Schulz KF, Altman DG. The CONSORT statement: revised recommendations for improving the quality of reports of parallel-group randomised trials. Lancet. 2001;357:1191–4.
78. Kopman AF, Lien CA, Naguib M. Neuromuscular dose-response studies: determining sample size. Br J Anaesth. 2011;106:194–8.
79. Cammu G, Coddens J, Hendrickx J, Deloof T. Dose requirements of infusions of cisatracurium or rocuronium during hypothermic cardiopulmonary bypass. Br J Anaesth. 2000;84:587–90.
80. Bland JM, Altman DG. Statistical methods for assessing agreement between two methods of clinical measurement. Lancet. 1986;1:307–10.
81. Schnider TW, Minto CF, Shafer SL, Gambus PL, Andresen C, Goodale DB, et al. The influence of age on propofol pharmacodynamics. Anesthesiology. 1999;90:1502–16.
82. Agoston S. Onset time and evaluation of intubating conditions: rocuronium in perspective. Eur J Anaesthesiol Suppl. 1995;11:31–7.
83. Dahaba AA, von Klobucar F, Rehak PH, List WF. The neuromuscular transmission module versus the relaxometer mechanomyograph for neuromuscular block monitoring. Anesth Analg. 2002;94:591–6. table of contents

Appendix

A. A. Dahaba, *Understanding the Numbers in Depth of Anesthesia Monitors*,
https://doi.org/10.1007/978-3-031-88260-9

Medizinische Universität Graz

Verleihung der Lehrbefugnis an Herrn

Dr. med. univ. Ashraf DAHABA

Die vom Senat der Medizinischen Universität Graz eingesetzte Habilitationskommission zur Behandlung Ihres Ansuchens um Verleihung der Lehrbefugnis hat am

9. Juli 2009

rechtswirksam beschlossen, Ihnen die Lehrbefugnis als Privatdozent (Venia docendi) für das Fach

„Anästhesiologie und Intensivmedizin"

zu erteilen.

Ich darf Ihnen im Namen der Medizinischen Universität Graz dazu herzlich gratulieren.

Univ.-Prof. Dr. Josef Smolle
Rektor

Graz, am 10. Juli 2009

Medizinische Universität Graz
Das Rektorat
Auenbruggerplatz 2/4 , A-8036 Graz

GZ.: 335 ex 2008/09

Graz, den 10.7.2009

Herrn
PD Dr. med. Ashraf Dahaba
Statteggerstraße 181
8046 G r a z

Bescheid:

Die mit der Behandlung Ihres Ansuchens vom 12.11..2008 um Erteilung der Lehrbefugnis (venia docendi) betraute Habilitationskommission hat am 9.7. 2009 über Ihr Ansuchen entschieden. Es ergeht gemäß § 103 Universitätsgesetz 2002 (UG 2002), BGBl. I Nr. 120/2002, i.d.g.F., folgender

Spruch:

Auf Grund des Beschlusses der Habilitationskommission wird Ihnen die Lehrbefugnis für das Habilitationsfach

„Anästhesiologie und Intensivmedizin"

Gemäß § 103 Abs. 11 UG 2002, BGBl. I Nr. 120/2002, i.d.g.F., wird durch die Erteilung der Lehrbefugnis (venia docendi) weder ein Arbeitsverhältnis begründet, noch ein bestehendes Arbeitsverhältnis zur Universität verändert (Privatdozentin oder Privatdozent).

Rechtsmittelbelehrung:

Gegen diesen Bescheid ist gemäß § 103 Abs. 9 UG 2002, BGBl. I Nr. 120/2002, i.d.g.F., kein ordentliches Rechtsmittel zulässig. Gegen diesen Bescheid kann binnen 6 Wochen ab Zustellung eine Beschwerde beim Verfassungsgerichtshof gemäß Art. 144 B-VG und/oder beim Verwaltungsgerichtshof gemäß Art. 131 Abs. 1 Z 1 B-VG eingebracht werden.

Für den Rektor:

Josef Smolle

Univ.-Prof. Dr. med. Josef Smolle

SALUS AEGROTI SUPREMA LEX · MEDIZINISCHE UNIVERSITÄT GRAZ

Zustellverfügung:
1. Herrn PD Dr. Ashraf Dahaba
2. Personalabteilung

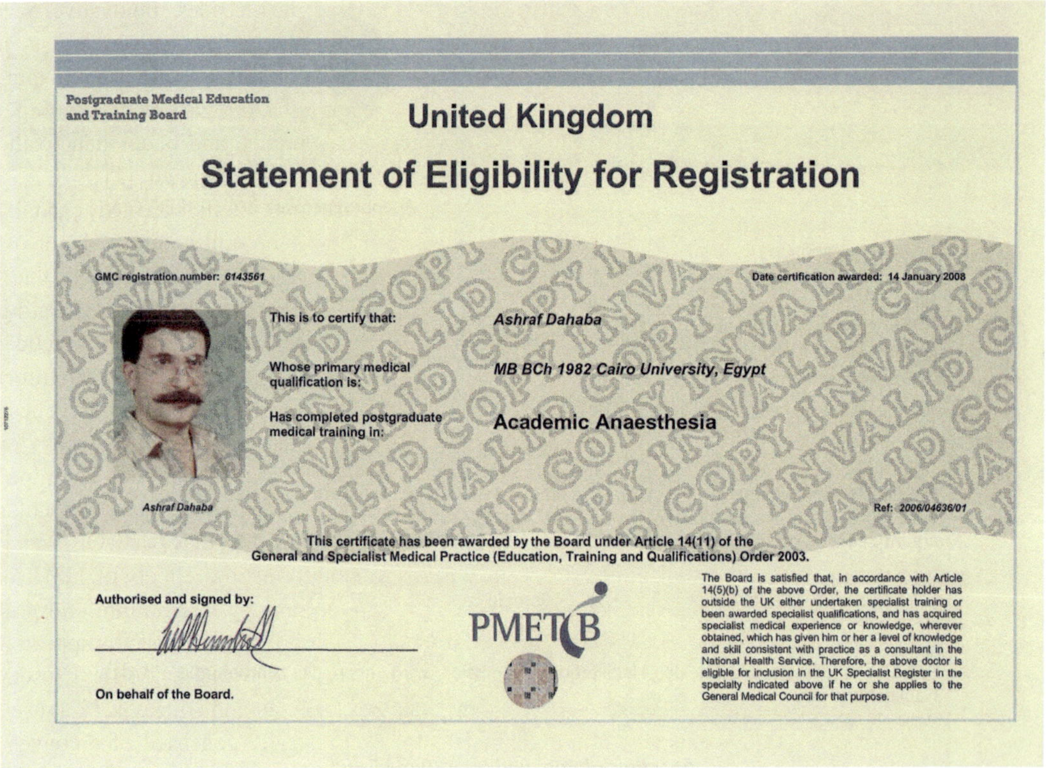

Postgraduate Medical Education and Training Board

United Kingdom

Statement of Eligibility for Registration

GMC registration number: *6143561*

Date certification awarded: 14 January 2008

This is to certify that: *Ashraf Dahaba*

Whose primary medical qualification is: *MB BCh 1982 Cairo University, Egypt*

Has completed postgraduate medical training in: **Academic Anaesthesia**

Ashraf Dahaba

Ref: *2006/04636/01*

This certificate has been awarded by the Board under Article 14(11) of the General and Specialist Medical Practice (Education, Training and Qualifications) Order 2003.

Authorised and signed by:

On behalf of the Board.

PMETB

The Board is satisfied that, in accordance with Article 14(5)(b) of the above Order, the certificate holder has outside the UK either undertaken specialist training or been awarded specialist qualifications, and has acquired specialist medical experience or knowledge, wherever obtained, which has given him or her a level of knowledge and skill consistent with practice as a consultant in the National Health Service. Therefore, the above doctor is eligible for inclusion in the UK Specialist Register in the specialty indicated above if he or she applies to the General Medical Council for that purpose.

General
Medical
Council

Regulating doctors
Ensuring good medical practice

Principal List of the register of medical practitioners

Certificate of Full Registration as a Medical Practitioner with Specialist Registration

Description of certificate:	Fully registered medical practitioner with specialist registration
GMC reference number:	6143561
Date of issue:	14/07/2008
Practitioner's name:	Dahaba, Ashraf (formerly Mohamed Ashraf Dahaba) **Gender:** Man
Date of full registration:	08/05/2006
Date of current entry from:	14/07/2008
Practitioner's address:	Deapartment of Anaesthesiology, Graz Medical University, Auenbruggerplatz 29, A-8036, Austria
Registered qualifications:	MB BCh 1982 Cairo University
Annual fee due date:	11/07/2009

Entry in the Specialist Register:	**Specialty**	**Sub-specialty**	**Date of entry**
	Academic Anaesthesia		14/07/2008

Description of employment

Grade: Any including honorary, substantive or fixed term consultant in the NHS. Confirmation of a doctor's eligibility to work in general practice. Doctors must have confirmation of their eligibility to work in general practice. This confirmation must be: i. a CCT or Article 11 certificate awarded by PMETB from September 2005; or ii. a certificate awarded by the JCPTGP before 30 September 2005. The Certification Unit at the RCGP will be able to confirm these awards if necessary; iii. confirmation by the GMC in the case of an EEA doctor. In addition, some GPs are eligible to work in general practice by virtue of being a principal in NHS general practice on or after 15 February 1981. They should have a certificate of acquired rights issued by PMETB (or the JCPTGP prior to 30 September 2005) or a certificate of exemption issued by their Health Authority at the time.

Chief Executive Finlay Mc Scott

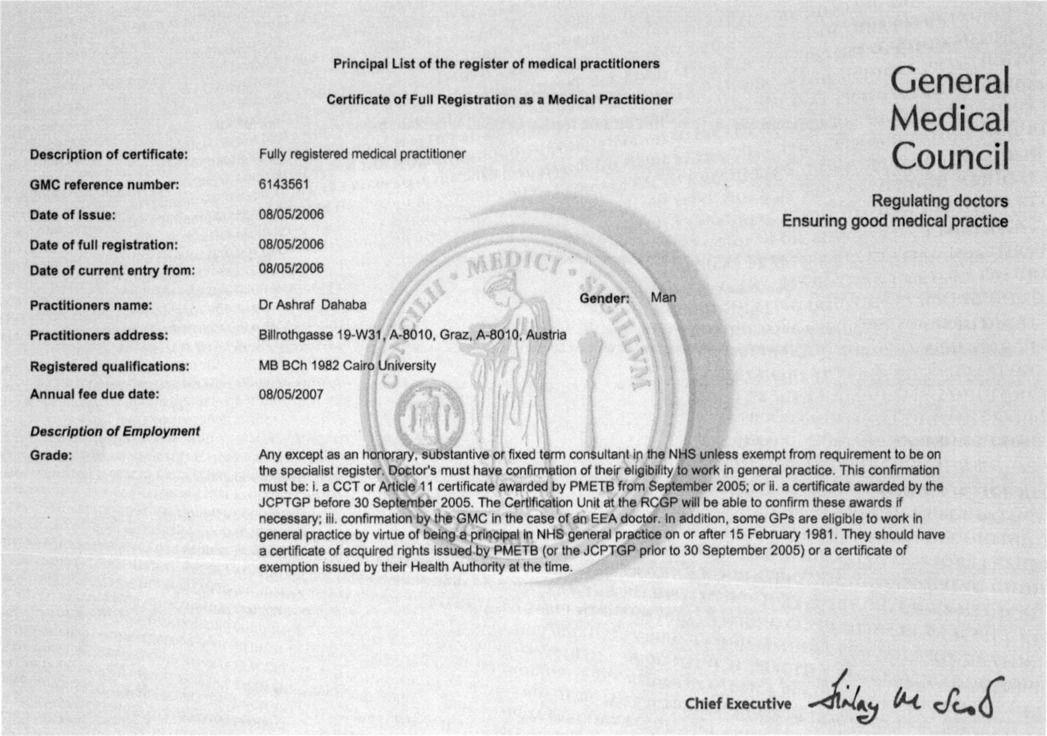

Principal List of the register of medical practitioners

Certificate of Full Registration as a Medical Practitioner

General Medical Council

Regulating doctors
Ensuring good medical practice

Description of certificate:	Fully registered medical practitioner
GMC reference number:	6143561
Date of Issue:	08/05/2006
Date of full registration:	08/05/2006
Date of current entry from:	08/05/2006
Practitioners name:	Dr Ashraf Dahaba **Gender:** Man
Practitioners address:	Billrothgasse 19-W31, A-8010, Graz, A-8010, Austria
Registered qualifications:	MB BCh 1982 Cairo University
Annual fee due date:	08/05/2007

Description of Employment

Grade: Any except as an honorary, substantive or fixed term consultant in the NHS unless exempt from requirement to be on the specialist register. Doctor's must have confirmation of their eligibility to work in general practice. This confirmation must be: i. a CCT or Article 11 certificate awarded by PMETB from September 2005; or ii. a certificate awarded by the JCPTGP before 30 September 2005. The Certification Unit at the RCGP will be able to confirm these awards if necessary; iii. confirmation by the GMC in the case of an EEA doctor. In addition, some GPs are eligible to work in general practice by virtue of being a principal in NHS general practice on or after 15 February 1981. They should have a certificate of acquired rights issued by PMETB (or the JCPTGP prior to 30 September 2005) or a certificate of exemption issued by their Health Authority at the time.

Chief Executive

Faculty Of Medicine
Suez Canal University
WHO Collaborating Center
For Health Manpower Development

Suez Canal University
Office of University Rector

وزارة خارجية جمهورية مصر العربية
مكتب تصديقات الإسماعيلية

Executive Decision
Number (744) dated 10/6/2006

-According to act number 49 for the year 1972 governing the organisation of Universities, its executive regulation and amendments.

-According to the approval of the Suez Canal University Board Meeting dated 30/5/2006 granting the Academic Title of:

Professor of Anesthesiology and Intensive Care Medicine, to

Dr Ashraf Dahaba

Currently Ass. Professor of Anesthesiology and ICM at the Faculty of Medicine

Decision

Article 1 – Starting from 30/5/2006, date of the approval of the Suez Canal University Board Meeting, Dr Ashraf dahaba is accordingly granted the Academic Title of Professor of Anesthesiology and Intensive Care Medicine, Currently Ass. Professor of Anesthesiology and ICM at the Faculty of Medicine.

Article 2 – All concerned parties at the Suez Canal University have to carry out the above-mentioned executive decision.

Rector of Suez Canal University

F.M. Abdel

Prof. Dr. Farouk Abdel Kadder

• Telephone: +20 (64) 329702, 327125 • Fax: +20 (64) 325207 •
• Faculty of Medicine, Suez Canal University, Ismailia 41522, Egypt •

浙江大学
1897
ZHEJIANG UNIVERSITY

聘 書

特聘请 Dr. Ashraf Dahaba

为浙江大学医学院

客座教授

Guest Professor

浙 江 大 学

校 长

聘书编号 2009 - 23 2009 年 07 月 2 日

聘客字第1109号

聘 书

兹聘 阿史瑞夫.达哈巴 教授

为第四军医大学客座教授

Prof. **Ashraf Dahaba**

is invited to be visiting professor of

the Fourth Military Medical University

第四军医大学校长

Dai–Ming Fan
President of FMMU
Xi'an,shaanxi
15 /04 / 2011

Abbreviations

3D	Three-Dimensional
AAA	Ascending Aortic Aneurysm
AAAs	Aromatic Amino Acids
AAGA	Accidental Awareness During General Anaesthesia
AAGBI	Association of Anaesthetists of Great Britain and Ireland
AAI	A-Line Autoregressive Index
AASM	American Academy of Sleep Medicine
ACA	Anterior Cerebral Artery
ACA	Anterior Communicating Artery
AEP	Auditory Evoked Potentials
AEPIndex	Auditory Evoked Potential index
AERRPS	Acute Encephalitis with Refractory, Repetitive, Partial Seizures
AI	Artificial Intelligence
AI-ML	Artificial Intelligence-Machine Learning
AIC	Akaike's Information Criterion
ALS	Amyotrophic Lateral Sclerosis
ALT	Alanine Aminotransferase
AmE	American English
ANH	Acute Normovolemic Hemodilution
ANOVA	Analysis of Variance
AOB	Aortic Occlusion Balloon
APA	American Psychiatric Association
ASA	American Society of Anesthesiologists
ASD	Atrial Septal Defect
AST	Aspartate Aminotransferase
AUC	Area Under Curve
AV	Atrioventricular
BBB	Blood-Brain Barrier
BCAAs	Branched Chain Amino Acids
BEAM	Brain Electrical Activity Mapping
BET	Bolus Elimination Transfer
BFM	Brain Function Monitors
BIS	Bispectral Index
BIS_{50}	Bispectral Index Value of 50
BIS-BSR	Bispectral Index-Burst Suppression Ratio
BISi	Bispectral Index Identification
BIS_{LOBR}	Bispectral Index value at Loss-of-Behavioral-Responses
BIS_{ROBR}	Bispectral Index value at Return-of-Behavioral-Responses
BMI	Body Mass Index
BMJ	British Medical Journal
BP	Blood Pressure
bpm	Beats per Minute
BrE	British English
BS	Burst Suppression
BSR	Burst Suppression Ratio
CABG	Coronary Artery Bypass Graft
CAM	Confusion Assessment Method
CBF	Cerebral Blood Flow
CBP	Cardiopulmonary Bypass
CCU	Critical Care Unit
C_e	Predicted Effect-Site Concentration
CEA	Carotid Endarterectomy
CH12	Channel 12

A. A. Dahaba, *Understanding the Numbers in Depth of Anesthesia Monitors*,
https://doi.org/10.1007/978-3-031-88260-9

CI	Confidence Interval
CIPA	Congenital Insensitivity to Pain with Anhidrosis
CIT	Cold Ischemia Time
Cl	Clearance
C_m	Measured Plasma Concentrations
CMG	Compressomyograph
$CMRO_2$	Cerebral Metabolic Rate of Oxygen
CNS	Central Nervous System
CO	Cardiac Output
CO_2	Carbon Dioxide
CONSORT	Consolidated Standards of Reporting Trials
C_p	Estimated Plasma Concentration
CPB	Cardiopulmonary Bypass
CPC	Cerebral Performance Category
CPMR	Cerebral Palsy Mentally Retarded
CPP	Cerebral Perfusion Pressure
CPR	Cardiopulmonary Resuscitation
CRN	Certified Registered Nurses
CRNAs	Certified Registered Nurse Anesthetists
CSF	Cerebrospinal Fluid
CSI	Cerebral State Index
CSM	Cerebral State Monitor
CT	Computerized Tomography
CVC	Central Venous Catheter
CVI	Composite Variability Index
dB	Decibel
DHCA	Deep Hypothermic Circulatory Arrest
DOA	Depth of Anesthesia
DOB	Date of Birth
DSA	Density Spectral Arrays
E	Effect
E_0	Baseline BIS Effect
EC_{50}	Effective Concentration$_{50}$
EC_{50LMA}	Effective Concentration for 50% of the Attempts to Secure Laryngeal Mask Insertion
ECG	Electrocardiography
ECoG	Electrocorticography
ECT	Electroconvulsive Therapy
ED	Emergency Department
EEG	Electroencephalography
EEG-BS	Electroencephalography-Burst Suppression
EF	Ejection Fraction
EHR	Electronic Health Records
E_{max}	Maximum BIS decline
EMG	Electromyography
EMG-ETT	Electromyograph Endotracheal Tube
Entropy-R	Response Entropy
Entropy-S	State Entropy
ERAS	Enhanced Recovery After Anesthesia
ERD	Event-Related Desynchronization
ESWT	High-Energy Extracorporeal Shockwave Lithotripsy Therapy
ETAG	End-Tidal Anesthetic Gas
$EtCO_2$	End-Tidal Carbon Dioxide
ET_{sev}	End-Tidal Sevoflurane Concentrations
FDA	Food and Drug Administration
fEMG	Frontal Electromyography
FFT	Fast Fourier Transform
FIO_2	Fractional Inspired Oxygen
fMRI	Functional Magnetic Resonance Imaging
GABA	γ-Amino-Butyric-Acid
GCRP	Good Clinical Research Practice
GCS	Glasgow Coma Scale
GOF	Goodness of Fit
GOS	Glasgow Outcome Scale
GPS	Global Positioning System
Hb	Hemoglobin
HCPB	Hypothermic Cardiopulmonary Bypass
HES	Hydroxyethyl Starch
HR	Heart Rate
IABP	Intra-Aortic Balloon Pump
IC_{50}	Inhibitory Propofol Concentration at 50% BIS Decline
ICA	Internal Carotid Artery
ICG	Indocyanine Green
ICP	Intracranial Pressure
IFT	Isolated Forearm Technique

IJV	Internal Jugular Vein
IMMM	Inhibitory Monophasic Mathematical Model
INR	International Normalized Ratio
IoC	Index of Consciousness
IPGF	Initial Poor Graft Function
IPP	Individual Pharmacokinetic Parameters
IQ	Intelligence Quotient
ISAAIC	Italian Society for Anaesthesia, Analgesia, and Intensive Care
I_{ss}	Asymptotic Steady-State Rates of Infusion
ITT	Intention to Treat
IWP	Ice Water Pain test
K_{10}	Rate Constant
K_{12}, K_{21}, K_{13}, K_{31}	Rate Constants
$\boldsymbol{K_{e0}}$	First-Order Rate Constant
k_{e0}	First-Order Rate Constant
LBM	Lean Body Mass
LDLT	Living-Donor Liver Transplantation
LIS	Locked-in-State
LOA	Limits of Agreement
LOBR	Loss-of-Behavioral-Responses
LOC	Loss of Consciousness
LR	Likelihood Ratio
LR	Logistic Regressions
LVAD	Left Ventricular Assist Device
LMAC	Low-Minimum Alveolar Concentration
MAC	Minimum Alveolar Concentration
$MAC_{1.3}$	1.3 MAC
$MAC_{1.3}$	1.3 Minimum Alveolar Concentration
MACs	Equi-minimum Alveolar Concentrations
MAP	Mean Arterial Pressure
MCA	Marginal Coronary Artery
MCA	Middle Cerebral Artery
MCBFV	Middle Cerebral Artery Blood Flow Velocity
MCS	Minimally Conscious Subject
MDAPE	Median Absolute Performance Error
MDF	Mean Dominant Frequency
MDPE	Median Performance Error
MELD	Model for End-Stage Liver Disease
MH	Malignant Hyperthermia
MLAEP	Middle Latency Auditory Evoked Potential
MMG	Mechanomyograph
MMSE	Mini-mental State Examination
M-NMT	Neuromuscular Transmission Module
MRI	Magnetic Resonance Imaging
N_2O	Nitrous Oxide
NANB	Nociception Anti-Nociception Balance
NAP5	fifth National Audit Project
NCT	Narcotrend
NDE	Near-Death Experience
NEJM	New England Journal of Medicine
NHS	National Health Service
NICE	National Institute for Health and Care Excellence
NICU	Neurointensive Care Unit
NIRS	Near-Infrared Spectroscopy
NLR	Non-linear Regression Analysis
NMBA	Neuromuscular Blocking Agent
NMDA	N-Methyl-D-Aspartate
NONMEM	Non-linear Mixed Effects Modeling
NPV	Negative Predictive Value
NREM	Non-rapid Eye Movement
O_2	Oxygen
OAA/S	Observer's Assessment of Alertness/Sedation Scale
OLT	Orthotopic Liver Transplantation
OPCAB	Off-pump Coronary Artery Bypass Graft
OR	Operating Room
PAC	Pulmonary Artery Catheter
P_aCO_2	Partial Pressure of Carbon Dioxide
PACU	Post-Anesthesia Care Unit
PCA	Patient Controlled Analgesia
PCA	Posterior Cerebral Artery
PDR	Plasma Disappearance Rate

pEEG	Processed Electroencephalography
pEEG-DOA	Processed Electroencephalography-Depth Of Anesthesia
PEG	Percutaneous Endoscopic Gastrostomy
PI	Gosling Pulsatility Index
PI	Pain Intensity Scale
PID	Proportional-Integral Derivative
PK/PD	Pharmacokinetic/Pharmacodynamic
PNF	Primary Non-function
POCD	Postoperative Cognitive Dysfunctions
PopPK-PD	Population Pharmacokinetic-Pharmacodynamic
PP	Per Protocol
PPF	Peak Power Frequency
PPV	Positive Predictive Value
PSD	Power Spectral Density
PSG	Polysomnography
PSI	Patient State Index
PTSD	Posttraumatic Stress Disorder
qEEG	Qualitative Electroencephalography
QEEG	Quantitative Electroencephalography
RAS	Reticular Activating System
RASS	Richmond Agitation-Sedation Scale
RCT	Randomized Controlled Trials
RE	Response Entropy
REM	Rapid Eye Movement
RLN	Recurrent Laryngeal Nerve
ROBR	Return-of-Behavioral-Responses
ROC	Receiver Operating Characteristic
ROC	Return of Consciousness
ROC-AUC	Receiver Operating Characteristic Area Under the Curve
RR	Respiratory Rate
RSE	Root Square Error
SAE	Serious Adverse Events
SAS	Sedation Agitation Scale
sBIS	Standard Deviation of the Bispectral Index
SD	Standard Deviation
SE	State Entropy
sEEG	Spectral Electroencephalography
SEF	Spectral Edge Frequency
SEF_{95}	95% Spectral Edge Frequency
sEMG	Standard Deviation of Electromyography
SMA	Supplementary Motor Area
SO_2	Oxygen Saturation
SOC	Standard of Care
SpO_2	Oxygen Saturation
SQI	Signal Quality Index
SR	Smoothing Rate
SrO_2	Cerebral Hemoglobin Oxygen Saturation
STARD	Standards for Reporting of Diagnostic Accuracy
$StcO_2$	Transcranial Oxygen Saturation
SWT	Stationary Wavelet Transform
T_1	First Twitch
T_4:T_1	TOF Ratio
TBI	Traumatic Brain Injury
TCD	Transcranial Doppler
TCI	Target Controlled Infusion
TDE	Time Delay Estimation
TEE	Transesophageal Echocardiography
TEN	Total Enteral Nourishment
TENS	Transcutaneous Electrical Nerve Stimulation
TIVA	Total Intravenous Anesthesia
T_{LOBR}	Time to Loss-of-Behavioral-Responses
TOF	Train-of-Four
TPPV	Tracheostomy Positive-Pressure Ventilation
TPS	Transcutaneous Pulse Spectrophotometry
T_{ROBR}	Time to Return-of-Behavioral-Responses
TVR	Tricuspid Valve Replacement

USAF	United States Air Force
V_1	Central Compartment
V_2, V_3	Peripheral Compartments
VAS	Visual Analogue Scale
V_d	Volume of distribution
Vd_{ss}	Volume of distribution at Steady State
vEEG	Visual Electroencephalography
VSD	Ventricular Septal Defect
WAV_{CNS}	Wavelet-Based Anesthetic Value for Central Nervous System
WFNS SAH	World Federation of Neurosurgical Societies Subarachnoid Hemorrhage Score
WIT	Warm Ischemia Time

GPSR Compliance

The European Union's (EU) General Product Safety Regulation (GPSR) is a set of rules that requires consumer products to be safe and our obligations to ensure this.

If you have any concerns about our products, you can contact us on ProductSafety@springernature.com

In case Publisher is established outside the EU, the EU authorized representative is:

Springer Nature Customer Service Center GmbH
Europaplatz 3
69115 Heidelberg, Germany

Batch number: 10732801

Printed by Printforce, the Netherlands